IET TELECOMMUNICATIONS SERIES 100

Antenna Booster Technology for Wireless Communications

Other volumes in this series:

Volume 9	**Phase Noise in Signal Sources** W.P. Robins
Volume 12	**Spread Spectrum in Communications** R. Skaug and J.F. Hjelmstad
Volume 13	**Advanced Signal Processing** D.J. Creasey (Editor)
Volume 19	**Telecommunications Traffic, Tariffs and Costs** R.E. Farr
Volume 20	**An Introduction to Satellite Communications** D.I. Dalgleish
Volume 26	**Common-Channel Signalling** R.J. Manterfield
Volume 28	**Very Small Aperture Terminals (VSATs)** J.L. Everett (Editor)
Volume 29	**ATM: The Broadband Telecommunications Solution** L.G. Cuthbert and J.C. Sapanel
Volume 31	**Data Communications and Networks, 3rd Edition** R.L. Brewster (Editor)
Volume 32	**Analogue Optical Fibre Communications** B. Wilson, Z. Ghassemlooy and I.Z. Darwazeh (Editors)
Volume 33	**Modern Personal Radio Systems** R.C.V. Macario (Editor)
Volume 34	**Digital Broadcasting** P. Dambacher
Volume 35	**Principles of Performance Engineering for Telecommunication and Information Systems** M. Ghanbari, C.J. Hughes, M.C. Sinclair and J.P. Eade
Volume 36	**Telecommunication Networks, 2nd Edition** J.E. Flood (Editor)
Volume 37	**Optical Communication Receiver Design** S.B. Alexander
Volume 38	**Satellite Communication Systems, 3rd Edition** B.G. Evans (Editor)
Volume 40	**Spread Spectrum in Mobile Communication** O. Berg, T. Berg, J.F. Hjelmstad, S. Haavik and R. Skaug
Volume 41	**World Telecommunications Economics** J.J. Wheatley
Volume 43	**Telecommunications Signalling** R.J. Manterfield
Volume 44	**Digital signal Filtering, Analysis and Restoration** J. Jan
Volume 45	**Radio Spectrum Management, 2nd Edition** D.J. Withers
Volume 46	**Intelligent Networks: Principles and applications** J.R. Anderson
Volume 47	**Local Access Network Technologies** P. France
Volume 48	**Telecommunications Quality of Service Management** A.P. Oodan (Editor)
Volume 49	**Standard Codecs: Image Compression to Advanced Video Coding** M. Ghanbari
Volume 50	**Telecommunications Regulation** J. Buckley
Volume 51	**Security for Mobility** C. Mitchell (Editor)
Volume 52	**Understanding Telecommunications Networks** A. Valdar
Volume 53	**Video Compression Systems: From first principles to concatenated codecs** A. Bock
Volume 54	**Standard Codecs: Image compression to advanced video coding, 3rd Edition** M. Ghanbari
Volume 59	**Dynamic Ad Hoc Networks** H. Rashvand and H. Chao (Editors)
Volume 60	**Understanding Telecommunications Business** A Valdar and I Morfett
Volume 65	**Advances in Body-Centric Wireless Communication: Applications and state-of-the-art** Q.H. Abbasi, M.U. Rehman, K. Qaraqe and A. Alomainy (Editors)
Volume 67	**Managing the Internet of Things: Architectures, theories and applications** J. Huang and K. Hua (Editors)
Volume 68	**Advanced Relay Technologies in Next Generation Wireless Communications** I. Krikidis and G. Zheng
Volume 69	**5G Wireless Technologies** A. Alexiou (Editor)
Volume 70	**Cloud and Fog Computing in 5G Mobile Networks** E. Markakis, G. Mastorakis, C.X. Mavromoustakis and E. Pallis (Editors)
Volume 71	**Understanding Telecommunications Networks, 2nd Edition** A. Valdar
Volume 72	**Introduction to Digital Wireless Communications** Hong-Chuan Yang
Volume 73	**Network as a Service for Next Generation Internet** Q. Duan and S. Wang (Editors)
Volume 74	**Access, Fronthaul and Backhaul Networks for 5G & Beyond** M.A. Imran, S.A. R. Zaidi and M.Z. Shakir (Editors)
Volume 75	**Digital Television Fundamentals** S. Mozar and K. Glasman (Editors)
Volume 76	**Trusted Communications With Physical Layer Security for 5G and Beyond** T.Q. Duong, X. Zhou and H.V. Poor (Editors)

Volume 77 **Network Design, Modelling and Performance Evaluation** Q. Vien
Volume 78 **Principles and Applications of Free Space Optical Communications** A.K. Majumdar, Z. Ghassemlooy, A.A.B. Raj (Editors)
Volume 79 **Satellite Communications in the 5G Era** S.K. Sharma, S. Chatzinotas and D. Arapoglou
Volume 80 **Transceiver and System Design for Digital Communications, 5th Edition** Scott R. Bullock
Volume 81 **Applications of Machine Learning in Wireless Communications** R. He and Z. Ding (Editors)
Volume 83 **Microstrip and Printed Antenna Design, 3rd Edition** R. Bancroft
Volume 84 **Low Electromagnetic Emission Wireless Network Technologies: 5G and beyond** M.A. Imran, F. Héliot and Y.A. Sambo (Editors)
Volume 86 **Advances in Communications Satellite Systems Proceedings of the 36th International Communications Satellite Systems Conference (ICSSC-2018)** I. Otung, T. Butash and P. Garland (Editors)
Volume 87 **Real Time Convex Optimisation for 5G Networks and Beyond** T.Q. Duong, L. D. Nguyen and H.D. Tuan
Volume 89 **Information and Communication Technologies for Humanitarian Services** M.N. Islam (Editor)
Volume 90 **Communication Technologies for Networked Smart Cities** S.K. Sharma, N. Jayakody, S. Chatzinotas and A. Anpalagan (Editors)
Volume 91 **Green Communications for Energy-Efficient Wireless Systems and Networks** Himal Asanga Suraweera, Jing Yang, Alessio Zappone and John S. Thompson (Editors)
Volume 92 **Flexible and Cognitive Radio Access Technologies for 5G and Beyond** H. Arslan and E. Başar (Editors)
Volume 93 **Antennas and Propagation for 5G and Beyond** Q. Abbasi, S.F. Jilani, A. Alomainy and M.A. Imran (Editors)
Volume 94 **Intelligent Wireless Communications** G. Mastorakis, C.X. Mavromoustakis, J. M. Batalla and E. Pallis (Editors)
Volume 95 **ISDN Applications in Education and Training** R. Mason and P.D. Bacsich
Volume 96 **Edge Caching for Mobile Networks** H. Vincent Poor and Wei Chen (Editors)
Volume 98 **Artificial Intelligence Applied to Satellite-based Remote Sensing Data for Earth Observation** M.P. Del Rosso, A. Sebastianelli and S.L. Ullo (Editors)
Volume 99 **Metrology for 5G and Emerging Wireless Technologies** T.H. Loh (Editor)
Volume 101 **Wireless Mesh Networks for IoT and Smart Cities: Technologies and applications** L. Davoli and Gi. Ferrari (Editors)
Volume 104 **Enabling Technologies for Social Distancing: Fundamentals, concepts and solutions** D.N. Nguyen, D.T. Hoang, T.X. Vu, E. Dutkiewicz, S. Chatzinotas and B. Ottersten
Volume 105 **Non-Geostationary Satellite Communications Systems** E. Lagunas, S. Chatzinotas, K. An and B.F. Beidas
Volume 107 **Physical Layer Security for 6G Networks** T.Q. Duong, J. Zhang, N. Yang, X. Zhou, V. Sharma
Volume 108 **The Role of 6G and Beyond on the Road to Net-Zero Carbon** M. Ali Imran, A. Taha, S. Ansari, M. Usman and Q.H. Abbasi (Editors)

Antenna Booster Technology for Wireless Communications

Jaume Anguera and Aurora Andújar

The Institution of Engineering and Technology

Published by The Institution of Engineering and Technology, London, United Kingdom

The Institution of Engineering and Technology is registered as a Charity in England & Wales (no. 211014) and Scotland (no. SC038698).

First published 2024

The Institution of Engineering and Technology
Futures Place
Kings Way, Stevenage
Herts, SG1 2UA, United Kingdom

www.theiet.org

British Library Cataloguing in Publication Data
A catalogue record for this product is available from the British Library

ISBN 978-1-83953-300-6 (hardback)
ISBN 978-1-83953-301-3 (PDF)

Typeset in India by MPS Limited
Printed in the UK by CPI Group (UK) Ltd

Cover Image: Cover Image from Ignion, edited by Eugenio Fernández / Antenna booster embedded in a wireless device

Contents

About the authors

Dr. Jaume Anguera was born in Vinaròs, Spain, in 1972. He is the co-founder and CTO at Ignion; a Professor at Universitat Ramon Llull, both institutions in Barcelona, Spain.

He received a Technical Engineering degree in Electronic Systems and Engineering, degree in Electronic Engineering, both from the Ramon Llull University (URL), Barcelona, Spain, in 1994 and 1998, respectively, and a Telecommunication Engineering degree (5 years degree) and a Ph.D. degree in Telecommunications, both from the Polytechnic University of Catalonia (UPC), Barcelona, Spain, in 1998 and 2003, respectively. From 1997 to 1999, he joined the Electromagnetic and Photonic Engineering Group of the Signal Theory and Communications Department of the UPC as a researcher in microstrip fractal-shaped antennas. In 1999, he was a researcher at Sistemas Radiantes, Madrid, Spain, where he was involved in designing dual-band, dual-polarized fractal-inspired microstrip patch arrays for mobile communications. In the same year, 1999, he became Assistant Professor at the Department of Electronics and Telecommunications, Universitat Ramon Llull-Barcelona, Associate Professor in 2016 and Professor in 2024, where he teaches Antenna Theory and related courses and belongs to the recognised Research Group Smart Society. Since 2001, he has led research projects in the antenna field for wireless applications in a frame of Industry-University collaboration: Ignion and the Department of Electronics and Telecommunications of Universitat Ramon Llull-Barcelona, Spain. The Spanish Ministry and other institutions have awarded several of his supervised students Best Bachelor's and Master's Thesis.

From 1999 to 2017, he was with Fractus (founder partner) in Barcelona, Spain, where he held the position of R&D manager and developed various cutting-edge antenna technologies. At Fractus, he led projects for antenna products for base station antennas, automotive, and mobile phones.

Between 2003 and 2006, he was also assigned to Fractus in South Korea to head up the research team. One of his main tasks was to provide training, education, and development of the team's core competencies and provide an R&D vision to address the rapidly growing mobile device market. Under his technical leadership, the company had secured major contracts with companies such as Samsung, LG, and Bellwave, to name a few. He published a book about "Korean Experiences" in 2015.

Since 2017, he has been with Ignion in the role of CTO. He leads the company's R&D activity to create new products and patents, envisages new technologies, and provides technology strategy and direction to scale the company's business.

He holds over 180 granted invention patents (the USA, Asia, Europe) in the antenna field, many of which have been licensed and adopted by the wireless industry. Among his most outstanding contributions is that of the inventor of Antenna Booster Technology (commercially known as Virtual Antenna® technology), a technology that fostered the creation of Ignion. The wireless industry has adopted millions of these products globally to facilitate wireless connectivity for IoT (Internet of Things) devices, utilising a miniature component known as an antenna booster.

He is the author of more than 300 journal articles and international and national conference papers (h-index = 55 with more than 9,200 citations based on Google Scholar, in the top 1.4% of the ranking of Spanish researchers). He has given over 80 antenna lectures worldwide (the USA, China, South Korea, India, the UK, France, Germany, Poland, Czech Republic, Denmark, The Netherlands, Malta, Tunisia, Nigeria, Perú, Brazil, Canada, and Spain). He has directed over 170 bachelor's, master's, and Ph.D. theses. He has authored seven books.

Dr. Anguera has participated in over 24 national/international projects and research grants valued at over $14 million, many of which he was the principal researcher.

His research interests include antenna booster technology, multiband and small antennas, diversity antenna systems/MIMO, electromagnetic dosimetry, microstrip antennas, fractals, and genetically optimised antenna systems.

Dr. Anguera was a member of the fractal team that, in 1998, received the European Information Technology Grand Prize for Applied Science and Engineering for the fractal-shaped antenna application to cellular telephony. 2003 Finalist for the Best Doctoral Thesis on UMTS (Fractal and Broadband Techniques on Miniature, Multifrequency, and High-Directivity Microstrip Patch Antennas); prize promoted by "Technology plan of UMTS promotion" given by Telefónica Móviles España). New Faces of Engineering 2004 (promoted by IEEE and IEEE foundation). In the same year, he won the Best Doctoral Thesis (Ph.D.) in "Network and Broadband Services" (XXIV Prize Edition "Ingenieros de Telecomunicación") organised by Colegio Oficial de Ingenieros de Telecomunicación (COIT) and the Company ONO. In 2011, he received the Alè Vinarossenc recognition from Fundació Caixa Vinaròs. In 2014, together with four other Fractus inventors, he received the "2014 Finalist to European Patent Award". As a team in Fractus, we received the Technology Pioneer distinction from the World Economic Forum in 2005 and the Academiae Dilecta award from the Royal Academy of Engineering in 2015. As a team, Ignion has been recognised as the 2023 Innovative Company, awarded by Enginyers BCN in recognition of its disruptive Virtual Antenna® technology.

He is a reviewer for several IEEE journals. He is an associate editor at the *IEEE Open Journal on Antennas and Propagation* and *Electronics Letters*. His biography is listed in Who's Who in the World, Who's Who in Science and Engineering, Who's Who in Emerging Leaders, and IBC (International Biographical Center, Cambridge, England). He is an IEEE Antennas and Propagation Distinguished Lecturer. He is the working group "Software and Modelling" vice-chair at EurAAP.

His detailed information can be found at http://users.salleurl.edu/~jaume.anguera/

Dr. Aurora Andújar was born in Barcelona, Spain, in 1984. She is the co-founder at Ignion (Director of Engineering, 2020, VP of Engineering, 2024 – up to date). She received her Bachelor's and Master's degrees in Telecommunications Engineering in 2005 and 2007, respectively. The Master's degree of Science in Telecommunications Engineering and Management in 2007, and the Ph.D. degree in 2013 from the Polytechnic University of Catalonia (UPC). In 2004–2005, she received a research fellowship from the UPC in the field of electromagnetic compatibility. In 2005, she worked as a Software Test Engineer in Tempos21. In 2006, she worked as a software engineer for Digital Campus at UPCnet. In 2007, she joined Fractus as an R&D + I Engineer, where she was in charge of the development of technological projects focused on the design of miniature and multiband antennas for wireless devices. She was also involved in the maintenance and growth of the company's patent portfolio, including the development, writing, and prosecution of new inventions and patents (2007–2013). In 2013, she moved to the Product & Services Department; in 2015, she became the Product Manager of the division; and in 2018, she was promoted to Senior Product and Business Development Manager. She was in charge of the development and release of new antenna products and led customer-oriented projects. In 2020, she became Director of Engineering, and she is now in charge of leading the company's engineering department, ensuring good engineering methodologies and practices to allow the company to scale fast. Since 2009, she has led research projects in the collaborative university-industry framework. In 2011, she became an Assistant Professor at the Open University of Catalonia (UOC). Later, in 2013, she joined the Universitat Ramon Llull (URL) as an assistant professor.

Dr. Andújar has more than 16 years of experience in the antenna industry. Author of more than 58 patents in the antenna field. She has published more than 125 papers in scientific journals and at international and national conferences. She has directed more than 80 bachelor's and master's theses.

In 2015, Dr. Andújar received the award for the best doctoral thesis of Spain by the Spanish Association of Telecommunications Engineers (COIT-AEIT). As a research team in Fractus, she received the Academiae Dilecta distinction from the Engineering Academy. Later, in 2016, she received the award for the best doctoral thesis in the scope of TIC Engineering by the UPC. In 2021, she became a finalist of the DonaTIC awards in the category of entrepreneur, organised by the department of the vice-presidency and digital and territorial policies of Catalonia, which have the dual aim of recognising the fundamental role of women in the professional, business, and academic world while offering references to the sector. In early 2023, IGNION was the winner of the Innovative Company Award from the College of Graduate Engineers and Industrial Technical Engineers of Barcelona (ENGINYERS BCN) because of its contributions to the redesign of the Internet of Things (IoT) connectivity, thanks to the creation of the next generation of multi-band, multi-purpose, and ready-to-use antenna chips.

Preface

In a world driven by interconnectedness and constant communication, the power of wireless technology has become an indispensable force shaping our lives. From the early days of radio waves to the vast networks of the Internet age, the evolution of antennas has played a crucial role in enabling seamless communication across the globe.

Conventional wisdom has long dictated the design of small-multiband antennas, each tailored meticulously to cater to specific frequency bands. Engineers would tirelessly labour over intricate, complex geometries, striving to achieve the desired coverage and signal quality. However, as technology advances and the demand for wireless communication grows, the limitations of traditional antenna approaches become increasingly apparent.

In this landscape, the concept of Antenna Booster Technology emerges where complex antenna designs are substituted by small and simple elements called antenna boosters. Antenna boosters excite the ground plane of a wireless device, becoming the main antenna contributor and, thus, determining the performance in terms of bandwidth, efficiency, and radiation patterns. With the addition of a matching network, the frequency bands are easily accommodated to any wireless protocol without the need to design a specific antenna for each device or frequency band. Gone are the days of laborious redesigns to accommodate different frequency bands; here, innovation thrives on simplicity, where the only necessary adaptation lies in the artful design of the matching network. The book shows that this is much simpler than designing antennas from scratch. In fact, the book teaches more than antenna design but how wireless devices can be designed by embedding antenna boosters. Thanks to this approach, we unravel the power of antenna boosters with matching network synthesis tools. Witness how these modern wonders enable engineers to streamline the design process and optimise performance with unprecedented ease.

Join us as we embark on this awe-inspiring voyage through the world of Antenna Booster Technology, where innovation knows no bounds and the future of wireless communication lies within our grasp.

How to use the book

The book can be used for many professionals, from undergraduate students with basic RF knowledge to experts in the field, graduate students, Ph.D. scholars, as well as professionals related to the wireless world who need to understand the essential benefits that Antenna Booster Technology may bring to the wireless industry. To facilitate readers' comprehension and assimilation of the material, Table P.1 provides recommendations on which sections of the book should be explored, spanning from a foundational level to that of an expert. This classification is also valuable for teaching purposes, allowing university professors to create courses on Antenna Booster Technology. Suppose the book is used for teaching, such as in a university course. In that case, it is recommended to allocate three hours of lecture per week, in addition to the time students will spend on exercises at the end of each chapter.

The book is accompanied by over a hundred exercises that will consolidate the reading as you go through.

Table P.1 Four levels of learning Antenna Booster Technology

Chapter	Basic: 4 weeks	Intermediate: 8 weeks	Advanced: Half year	Expert: Full year
Chapter 1: Introduction	All	All 1,2	1st semester[1,2](1S)	1S[1,2]
Chapter 2: State of the Art	–	–	–	1S
Chapter 3: Antenna Boosters: Fundamentals	–	–	1S	1S
Chapter 4: Antenna Booster Technology for Single–Band Operation	–	Sections 4.1–4.5, 4.13, 4.14. Exercises 2, 3, 4, 5	1S	1S
Chapter 5: Antenna Booster Technology for Multiband Operation	–	Sections 5.1, 5.6, and 5.9. Exercises 2, 3, 4, 6	1S	1S
Chapter 6: Antenna Booster Technology for Multi-radio Operation	–	–	–	1S
Chapter 7: Reconfigurable Multiband Architectures for Wireless Devices Embedding Antenna Boosters	–	–	–	2S
Chapter 8: Wireless Devices Embedding Antenna Booster Near Conductive Bodies	–	–	–	2S
Chapter 9: Matching Network Synthesis Considering Pad Layout	–	–	1S	2S
Chapter 10: Antenna Booster Technology Versus Other Antenna Technologies	–	–	1S	2S
Chapter 11: From Zero to Hero in Antenna Booster Technology	All	All	1S	2S

[1]Skip if knowledgeable in basic antenna parameters.

[2]It is recommended to allocate 3 h of lecture per week, in addition to the time students will spend on exercises at the end of each chapter.

The learning journey is divided into four different levels, where the basic and intermediate levels focus on *how*, and advanced and expert levels also include the *why* for more profound knowledge. The four levels are:

- Basic level: The introductory course is recommended for non-experts to acquire general knowledge of RF and antennas. It includes an introduction to basic concepts of antennas and a chapter devoted to all you should know about Antenna Booster Technology in a very descriptive manner.
- Intermediate level: a crash course of eight weeks covering the essentials of Antenna Booster Technology for designing wireless devices for single and multiband operation. In this case, Chapter 1 may be skipped from the agenda depending on the knowledge of the fundamentals of antennas. This crash course also includes a set of exercises aimed at reinforcing the fundamental design rules. It is highly recommended for wireless or electronic engineers who must design wireless devices.
- Advanced level: this half-year course builds upon the topics covered in the intermediate level while also delving into the fundamental aspects of Antenna Booster Technology to provide a deeper understanding of the *why* behind the concepts. This level and the next level are intended for researchers and engineers to master Antenna Booster Technology to become leaders in designing wireless devices embedding antenna boosters.
- Expert level: this full-year course encompasses all the book chapters, guiding students to attain an expert level. It is particularly recommended for graduate students who intend to conduct advanced research in this field or take on roles as design experts within high-tech companies to design novel wireless devices incorporating Antenna Booster Technology.

Acknowledgements

The authors express their gratitude to the following professionals, companies, and institutions:

- Ignion.
- The institutions that funded research projects allowing to push antenna booster technology to the next level: Spanish Ministry, CDTI (Centro Desarrollo Tecnológico e Industrial), CIDEM Catalunya, HORIZON 2020 Europe Commission, European Innovation Council EIC Accelerator through the following grants:
    1. Project AVANZA no. TSI-020100-2010-0349
       "Tecnología Antennaless para la futura generación de teléfonos móviles sin antena" (Antennaless Technology for the future generation of mobile phones)
       Granting Institution: Ministerio de Industria, Energía y Turismo
    2. 2009 TEM 0028 (Programa Talent Empresa – Program Industry Talent)
       Granting Institution: Agència de Gestió d'Ajuts Universitaris i de Recerca
    3. Project FINRD08-2-0002-00
       "Recerca de noves antenes i sistemes miniaturitzats per a plataformes wireless mòbils de nova generació" (Research of new antennas and miniaturized systems for the next generation of mobile wireless platforms)
    4. Project TSI-020601-2012-27
       "Tecnologia Antennaless para dispositivos Wireless MIMO" (Antennaless Technology for MIMO wireless devices)
       Granting Institution: Ministerio de Industria, Energía y Turismo
    5. Project TSI-100101-2013-29
       "Tecnología de Antena Virtual Integrando Planos de Masa Inteligente" (Virtual Antenna® Technology Embedding Intelligent Ground Planes)
       Granting Institution: Ministerio de Industria, Energía y Turismo
    6. Project HORIZON 2020 (grant agreement no. 674491)
       "Virtual Antenna® Technology: The Next Generation Antenna for Wireless Devices"
       Granting Institution: HORIZON 2020, European Commission

7. Project TSI-100103-2015-39
   "Smart Virtual Antenna® Technology"
   Granting Institution: Ministerio de Industria, Energía y Turismo
8. Project IDI-20190285
   "Virtual Antenna® Technology para Soluciones Compactas GNSS, Bluetooth y 5G" (Virtual Antenna® Technology for GNSS, Bluetooth and 5G Compact Solutions)
   Granting Institution: CDTI (Centro para el Desarrollo Tecnológico e Industrial) – Ministerio de Ciencia, Innovación y Universidades
9. Project HORIZON 2020 (grant agreement no. 674491)
   "Variant™: Multiband Miniature Antenna for Open Hardware IoT Platforms"
   Granting Institution: HORIZON 2020, European Commission
10. Project TIC-20200069
    "Smart Matching Architectures for IoT Devices embedding Virtual Antenna®" (Arquitecturas de Adaptación Inteligentes para Dispositivos IoT con Virtual Antenna®)
    Granting Institution: CDTI (Centro para el Desarrollo Tecnológico e Industrial) – Ministerio de Ciencia, Innovación y Universidades
11. Project Industrial Ph.D. – 2021 DI 23
    "Nova generació de productes Virtual Antenna® i arquitectures reconfigurables per a sensors IOT" (New generation of Virtual Antenna® products and reconfigurable architectures for IoT sensors)
    Granting Institution: Industrial Doctorate Plan of the Secretariat of Universities and Research of the Department of Business and Knowledge of the Generalitat of Catalonia (Reference 2021 DI 23).
12. Project Industrial Ph.D. – 2021 DI 115
    "Nova generació de productes Virtual Antenna® i arquitectures reconfigurables per a sensors IOT" (New generation of Virtual Antenna® products and reconfigurable architectures for IoT sensors)
    Granting Institution: Industrial Doctorate Plan of the Secretariat of Universities and Research of the Department of Business and Knowledge of the Generalitat of Catalonia (Reference 2021 DI 115)
13. Project EIC Accelerator
    "TRIO++ – The First Miniaturized Non-Resonant Antenna Tailored to Internet of Things Applications"
    Granting Institution: EIC – European Innovation Council
14. Project CPP2022-010034
    "Adaptive Architectures Embedding Virtual Antenna® Technology: Towards a Smart Generation of Low Consumption and Easy-to-Design IoT Devices"
    Granting Institution: Spanish Ministry of Science and Innovation

- To the students who, with passion and commitment developed their engineering, master's, and Ph.D. thesis in the field of Antenna Booster Technology.

- To the following companies for sharing images of their wireless devices embedding antenna booster technology (following an alphabetical order):
    1. Blues Wireless.
    2. Cactus.
    3. Cavli Wireless.
    4. CWD Limited.
    5. Devalirian Engineering.
    6. Floox.
    7. HardwareMe Electronics.
    8. Nordic Semiconductor.
    9. Nuvathings.
    10. Riosh.
    11. Samea.
    12. Sequans.
    13. Sodaq.
    14. The Things Industries.
    15. Velavu.
- Of the above list, a special thanks to Brandon Satrom, Israel Blanco, Aaron Pica, Thomas Fesquet, Ajit Thomas, Shariq Jalaluddin, Rohit Raju, Gopinath Krishnamurthy, Tejas Kothari, Francesc Escudero, Joaquim Porte, Martí Salvador, Juan J. Vicente, Joaquim Tønnesen, Joan Vidal, Siva Sankar, Sébastien Amiot, Jeremy Gosteau, Kimberly Tassin, Ollie Smeenk, Ella Forester, Orkhan Amiraslanov, Dominic Peters.
- Ignion customers worldwide for trusting in antenna booster technology and enabling connectivity to their devices.
- Antti Lautanen, Sherry Hess, and David Vye from Cadence.
- Mikael Svard and Mark Fitter from Qorvo.
- Joao Costa previously at Dassault Systèmes CST.
- Jussi Rahola from Optenni.
- Coauthors in journal papers on the topic of antenna booster technology.
- Fractus.
- La Salle Engineering at Universitat Ramon Llull.
- To IEEE Antennas and Propagation Distinguished Lecturer program. To IEEE Antennas and Propagation Society, Microwave Theory and Techniques Society, IEEE Internet of Things, IEEE AP-S Education Young Professionals and Industry Initiatives.
- To the following IEEE Chapters: IEEE Vizag Bay, IEEE APS Chapter Northeast Brazil Section, IEEE MTTS Student Chapter Mumbai, IEEE Perú Chapter, IEEE ComSoc Chapter Pune Section, IEEE AP/MTT Benelux Chapter, IEEE Antennas and Propagation Society, Student Branch Chapter, IEEE APS Joint Section Chapter-Silchar National Institute of Technology Silchar, India, IEEE Madras Section, IEEE Ottawa Section, IEEE MTT/AP/EMC Chapter, IEEE Hyderabad Section, IEEE Manipal Student Branch.

- To the Indian Institute of Technology (IIT), Kanpur, India, Sathyabama Institute of Science and Technology, Sri Sivasubramaniya Nadar College of Engineering – Chennai, TU Eindhoven, European School of Antennas and Propagation, European Association on Antennas and Propagation, Chennai Tamilnadu India, Dassault Systems, Optenni, University of Malta, Chip Select, Manchester Metropolitan University, Meta, Universitat Pompeu Fabra, Universitat Politècnica de Catalunya, Jeppiaar Institute of Technology, Shri Vishnu Engineering College for Women Bhimavaram, India, School of Electrical, Electronics and Communication Engineering (SEEC) Manipal University Jaipur, India, Université Côte d'Azur, CMR College of Engineering & Technology, Hyderabad, Telangana, India, Manipal Institute of Technology, Salesians de Sarrià, College of Engineering Guindy, Anna University, Chennai, India, National Institute of Technology Mizoram, India, Vellore Institute of Technology, Vellore, India.
- Thank you for reading this book. Do not hesitate to reach out to the authors for comments.
- To our families for their patience and support during these years of hard work.

Chapter 1
Introduction

This chapter serves as an essential and fundamental introduction to understanding the crucial role of antennas in wireless devices. It lays the foundation for comprehending the subsequent chapters focused on Antenna Booster Technology. Written from a basic perspective, the chapter includes concise technical details, making it a simple yet comprehensive introduction suitable for a wide range of professionals in the wireless communications field who require knowledge of embedded antennas.

This chapter is highly recommended for non-antenna experts, from electronics engineers to business executives, marketing professionals, and sales personnel. It provides valuable insights for those seeking a better understanding of antennas' significance in wireless communication. Additionally, antenna experts can easily navigate to refresh essential concepts utilised throughout the book.

The chapter is divided as follows: the impact of small and multiband antennas on society is described (Section 1.1). Radiofrequency (RF) components, including antennas, are introduced to understand the essential elements in transmitting and receiving information in wireless devices (Section 1.2). The role of an antenna in a telecommunication system is explained to analyse its impact on a communication link (Sections 1.3 and 1.4). An introduction to simulation tools, including full electromagnetic solvers and microwave simulators, is given (Section 1.5). Section 1.6 presents measurement equipment such as vector network analysers, reverberation, and anechoic chambers. A description of typically small and multiband antenna techniques and Antenna Booster Technology is addressed in Section 1.7. Section 1.8 explains the wireless device design approach when embedding Antenna Booster Technology compared to a classic antenna route. All book chapters are introduced in Section 1.9. The conclusion part summarises the main findings (Section 1.10). Besides the contents, a list of exercises is proposed at the end of the chapter (Section 1.11), as well as references for further reading (References).

The chapter has the following objectives:

1. Know the RF components that play a role in the communication chain. Besides the antenna, other components are discussed, for instance, matching networks, inductors and capacitors, conductors, substrates, ground planes, transmission lines, tunable components, transceivers, filters, multiplexers, and amplifiers.
2. Basic antenna parameters (reflection coefficient, bandwidth, efficiency, radiation pattern, polarisation, gain) and their impact on wireless communication, such as range.

3. Limitations on bandwidth and efficiency of small antenna systems.
4. To familiarise yourself with simulation and measurement design tools.
5. Understand Antenna Booster Technology compared to other classic antenna technologies to design wireless devices.

1.1 Impact of antennas on our daily life

Since the apparition of 1st generation mobile phones back in the 1980s to connect people with people to the Internet of Things (IoT) to connect things with things, if somebody or something is not connected, it *does not exist*. The ancient philosophical approach *cogito ergo sum* (I think, therefore I am) by René Descartes in 1637 can now be changed to *connected, therefore, it exists*. It is amazing how many humans, animals, plants, and things are wirelessly connected among them. The type of wireless devices has grown from well-known mobile phones to other connected devices such as smartwatches, tablets, trackers, washing machines, fridges, water and gas meters, parking sensors, containers, vehicles, and buildings, to name a few [1,2]. The number of applications becomes humongous: monitoring waste containers for efficient collection; sensors for farming to improve watering in agriculture; industrial applications to monitor temperature, air quality, humidity, and light to minimise risks and be less energy-consuming; tracking goods for efficient transportation and security reasons; medical devices, and much more. All such data must be sent to a cloud, thanks to an antenna. The number of applications relying on wireless is proportional to human imagination. Therefore, the role of an antenna is vital to ensuring efficient connectivity in terms of power-consuming, good-quality, long ranges, keeping battery life long, and enabling small wireless devices to be hidden or installed easily.

Imagine a device in a field where the temperature and humidity need to be controlled. The temperature and humidity sensors of the device capture information, which is sent to the cloud thanks to a wireless connection. Besides, the device may be controlled with a mobile phone using a short-range communication protocol. Even we can imagine that the device includes geo-localisation features to obtain its position in real time. This basic example integrates three radio protocols, each one needing an antenna and the electronics associated with transmitting/receiving data. Overall, this puts a considerable responsibility on the antenna: to be small enough to fit in the reduced space of a wireless device, to be able to transmit at many frequency bands, efficient to ensure battery life, easy to embed, cost-competitive, and robust to ensure long life. All these constraints are addressed thanks to Antenna Booster Technology.

A tangible illustration of how connected devices can be beneficial to drivers is through the use of a smart flashlight [3]. This innovative flashlight serves as a modern alternative to the traditional reflective red triangle, which is typically used to signal other vehicles. The connected flashlight eliminates the need for the driver to step out of the car to set up the triangle, thus reducing the risk of accidents involving pedestrians or other vehicles.

Instead, the flashlight is securely attached to the roof of the car using magnets. When activated, it emits a highly visible intermittent light that can be seen from a distance of up to 1 kilometre, effectively alerting nearby drivers to exercise caution. Additionally, this device is equipped with geolocation and cellular antenna capabilities, leveraging the antenna booster technology discussed in this book.

Through its cellular connection, this smart flashlight communicates with other drivers' vehicle navigation systems, notifying them to drive cautiously due to the presence of a stalled vehicle on the road. This practical example incorporates two antenna boosters: one for receiving satellite signals and another for ensuring seamless cellular connectivity (Figure 1.1).

Since there are more things to be connected than humans, IoT connections have surpassed the number of non-IoT connections (smartphones, laptops/computers). As a reference, 12 billion IoT connections were reported in 2020 [6]. Indeed, the term IoT (Internet of Things) has witnessed a significant surge in popularity since 2010, experiencing consistent growth throughout the entire decade from 2010 to 2020. This growth has garnered substantial attention, even in the face of other prominent topics like quantum computing, virtual reality, and global warming, all while competing closely with the field of artificial intelligence (Figure 1.2).

Overall, the importance of antennas cannot be overstated, as they are essential for enabling wireless communication and facilitating the growth of the connected world. Connections mean antennas; therefore, an antenna is a key player in this multi-billion industry pushing technology to its limits for a better-connected society. Antenna technology will continue to be a critical area of research and

Figure 1.1 Illustration of a wirelessly connected flashlight device that securely attaches to the roof of a vehicle using magnets. The flashlight has a distinctive yellow top button that activates a flashing light, visible from a distance of up to 1 kilometre. It embeds antenna booster technology with two tiny antenna boosters: one for providing cellular connectivity and the other for the reception of GPS signal [4,5].

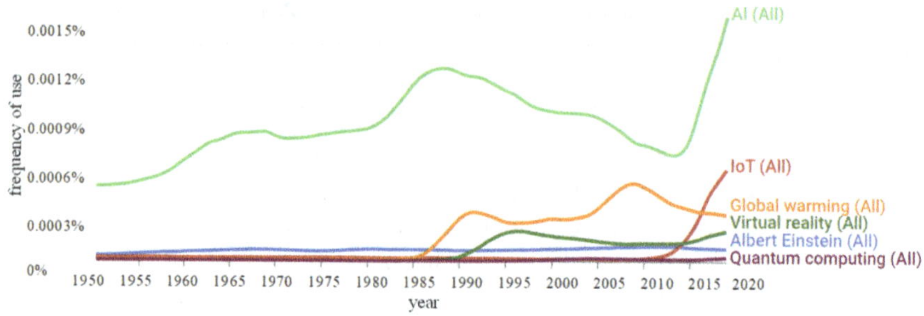

Figure 1.2 Usage of the terms IoT (Internet of Things), AI (Artificial Intelligence), global warming, Albert Einstein, and quantum computing from 1950 to 2020. Data obtained with Google's Ngram Viewer.

development, driving the advancement of future IoT applications and enabling new and innovative use cases.

1.2 RF components

This section outlines the key components crucial to radiofrequency operations within a wireless device, encompassing antennas, matching network, inductors, capacitors, conductors, substrates, ground planes, PCB (Printed Circuit Board), transmission lines, tunable components, transceivers, filters, multiplexers, and amplifiers. The description is basic but enough to progress along with the contents of the following chapters throughout the book.

1.2.1 Antenna

Following the standard definition from IEEE, an antenna is "that part of a transmitting or receiving system that is designed to radiate or to receive electromagnetic waves" [7].

In a bid to further understand the origins of the antenna wireless revolution, an elementary introduction is presented. It starts with J. C. Maxwell, who postulated the existence of electromagnetic waves. The Scottish physicist James Clerk Maxwell (1831–1879) first proposed the equations that are now associated with his name in the Philosophical Transactions of the Royal Society of London in 1865. With his equations, Maxwell summarised the previous work of Oersted, Gauss, Ampère, Faraday, and others [8] (Figure 1.3).

It was followed by H. R. Hertz, who demonstrated the existence of electromagnetic propagation. Heinrich Rudolf Hertz (1857–1894) was a German physicist who demonstrated in 1888 the existence of electromagnetic waves from a dipole charged with metallic spheres. Although he theorised about the propagation equations, Maxwell did not indicate any method of generating them. Furthermore, he died without seeing Hertz's brilliant experimental demonstration, which made Maxwell's theory universally accepted.

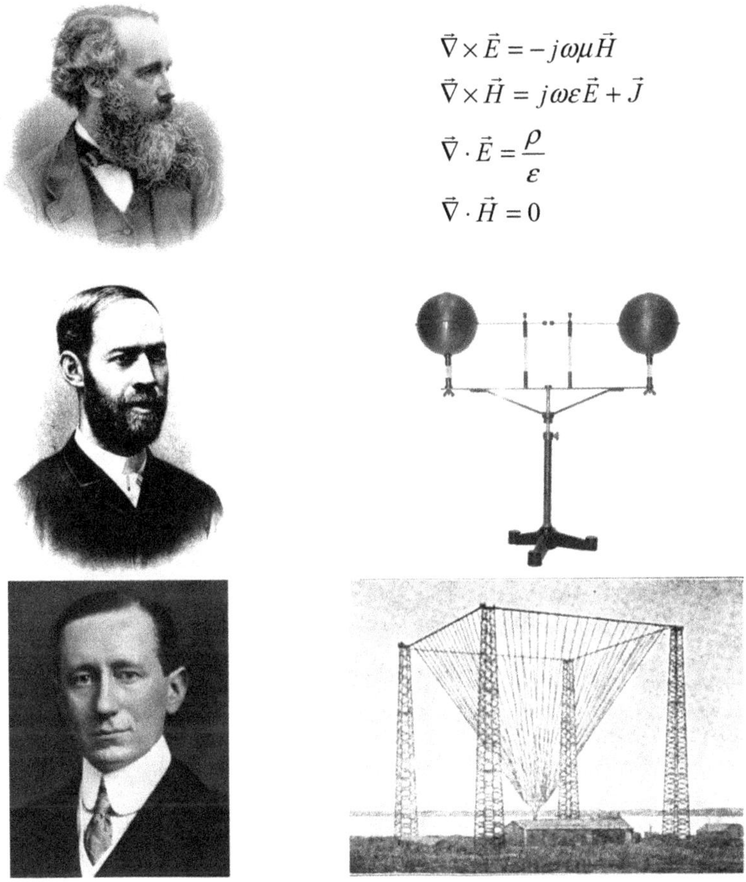

Figure 1.3 (Top) The Scottish physicist James Clerk Maxwell (1831–1879); (middle) Heinrich Rudolf Hertz (1857–1894), German physicist; (c) Guglielmo Marconi (1874–1937), Italian engineer

Continues with G. Marconi as the first transatlantic wireless communication. For further details on the evolution of wireless, the reader should refer to [8], and for a more detailed description, consult [9]. Guglielmo Marconi (1874–1937) was an Italian engineer who won the Nobel Prize in Physics. He carried out the first transatlantic transmission of electromagnetic waves between Wales and Newfoundland, covering a total distance of 3,000 km [9].

In essence, the size of the antenna depends on the operating wavelength λ given by:

$$\lambda = \frac{c}{f} \tag{1.1}$$

where c is the speed of the electromagnetic wave in the medium where the wave is propagating, and f is the frequency of the electromagnetic wave. The speed c in free space is approximately 300,000 km/s.

One of the most common antennas is the half-wavelength dipole, a thin linear conductor with a length of $L = \lambda/2$ (Figure 1.4). The physical size will depend on the frequency. For example, if $f = 600$ MHz, the size results in 50 cm. However, if the frequency increases to $f = 6$ GHz, the antenna size decreases to 5 cm. Thus, an antenna can be physically small as the frequency increases. In the present book, frequency ranges from a few hundred MHz up to 10 GHz because this frequency range allocates most of the communication protocols for cellular communication, geolocalisation, short-range wireless, etc.

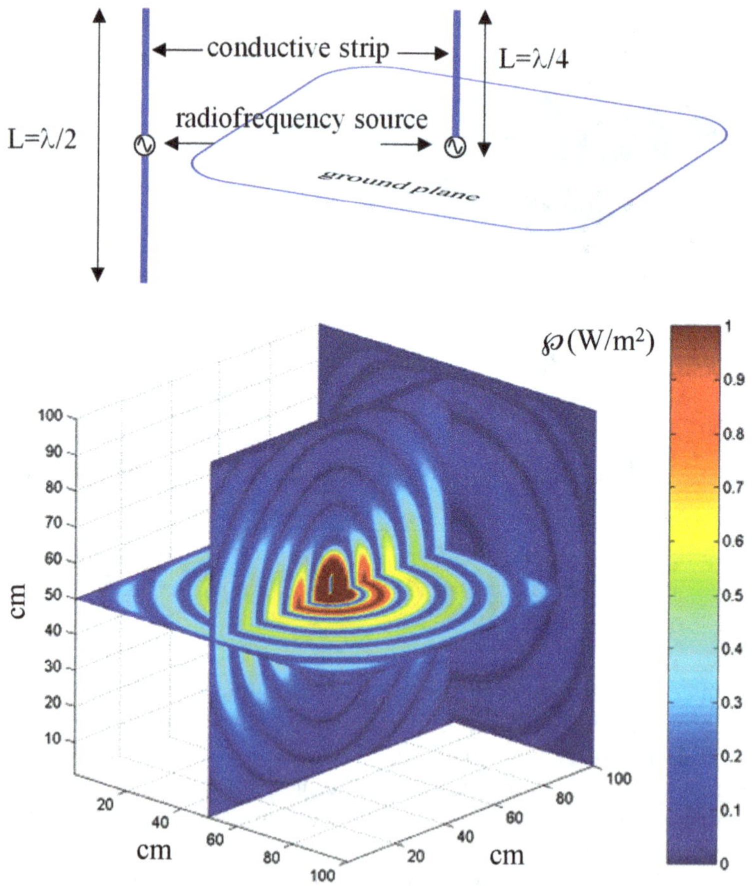

Figure 1.4 (Top left) A half-wavelength dipole; (top right) a quarter-wave monopole which is half the dipole on a conductive plane called ground plane; (bottom) a vertical half-wavelength dipole antenna transmitting power into space. Red colour means more transmitted power, whereas dark blue represents no power.

A $\lambda/2$ dipole transmits and receives electromagnetic signals exhibiting an intensity pattern that remains consistent across the horizontal plane, while showing a minimal intensity along its axis (Figure 1.4). This kind of pattern, called a radiation pattern, receives the name omnidirectional radiation. This is reciprocal: if the antenna transmits the maximum power in the horizontal plane, the antenna receives the maximum power in the same place. If the antenna transmits no power along the axis, it gets no power along the axis, even if a radio frequency signal impinges on the antenna along the axis.

As far as the electromagnetic spectrum is concerned, the frequency range in the radio and microwave regions is most used for wireless communications (Figure 1.5). In particular, the Antenna Booster Technology described in this book shines in the frequency region between 400 MHz and 10.6 GHz. However, the technology can be scaled to another part of the spectrum, as explained in Section 1.4.9.

1.2.2 Matching network

A matching network is a radiofrequency circuit that maximises the transfer of power from one circuit to another, for example, between an antenna and a transceiver and vice versa. It is not always needed, but in many situations, when the antenna is connected to a transceiver or any other component such as an amplifier or a filter, if there is no proper impedance matching (Section 1.4), not all the output power of a transmitter is delivered to an antenna (Figure 1.6). A matching network typically comprises inductors and capacitors or transmission lines and combinations thereof.

1.2.3 Inductor and capacitors

Inductors and capacitors are passive components (do not need power to operate) used in antenna engineering mainly to help the antenna be matched to the transceiver (Figure 1.7). However, it is not always required. For example, an antenna can be designed to be directly connected to a transceiver to radiate or receive power to/from space because the antenna can capture most of the power from the transceiver. However, in other situations, the antenna needs some help from the

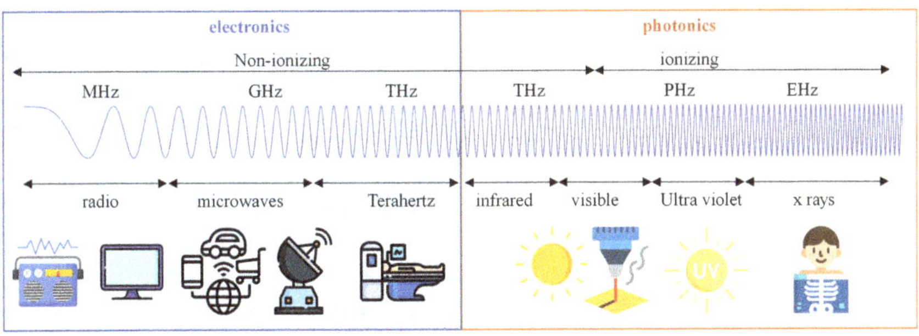

Figure 1.5 Electromagnetic spectrum from radio waves (~10^6 Hz) to x-rays (~10^{20} Hz). MHz (mega Hz, 10^6 Hz), GHz (giga Hz, 10^9 Hz), THz (tera Hz, 10^{12} Hz), PHz (peta Hz, 10^{15} Hz), EHz (exa Hz, 10^{18} Hz)

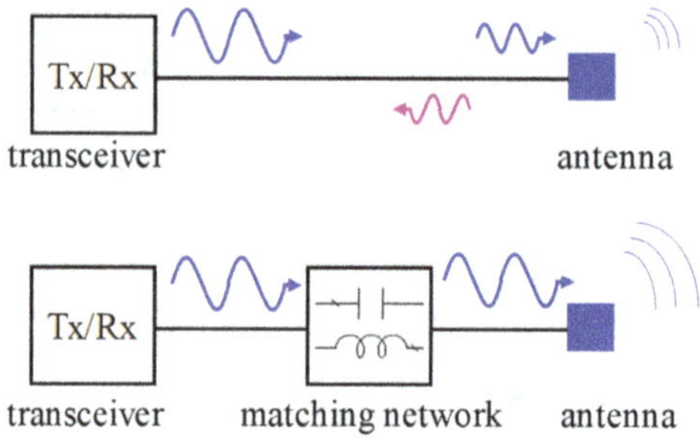

Figure 1.6 *(Top) An antenna not matched to the transceiver does not capture all the output power; (bottom) a matching network matches the antenna to the transceiver to allow capturing the output power from the transceiver, maximising the radiated power into space and vice versa*

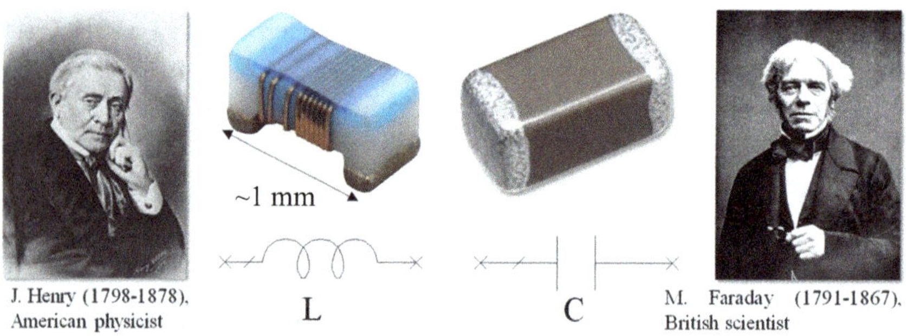

Figure 1.7 *SMD inductor L and capacitor C*

matching network, which includes inductors and capacitors. This helps form the so-called matching network, a circuit between the antenna and typically the transmitting/receiving part (transceiver) to optimise the power transmitted to space and the power received from space (Figure 1.6).

Ideal inductors and capacitors do not dissipate power, but in practice, they introduce some losses that decrease the transmitted/received power. However, those losses can be minimised if the component's quality is high. This quality term is known as quality factor Q:

$$Q = \frac{|X|}{R} \tag{1.2}$$

where X is the reactance associated with an inductor or a capacitor, and R is the resistance associated with losses. For an ideal inductor or capacitor, $R = 0$, so Q becomes infinite and, thus, lossless. However, R is not equal to zero. Consequently, to minimise the losses introduced by the inductor and capacitor, the higher the Q, the lesser the losses. For example, a 10 nH inductor at 900 MHz presents an $R = 0.07\ \Omega$

due to losses in the wiring to create the inductance. With (1.2), $X = j2\pi \cdot f \cdot L = j6.5\ \Omega$, resulting in $Q = 88$. The impact of Q is deeply analysed in Chapter 4.

Inductors and capacitors take the form of tiny components called SMD (Surface Mounted Device), easily assembled into a PCB using a pick-and-place process. The capacitors and inductors' units are Farads and Henry in honour of Michael Faraday and Josep Henry. In the frequency range of hundreds of MHz up to 10 GHz, typical values are of the order of pico farads (1 pF = 10^{-12} F) and nano henry (1 nH = 10^{-9} H).

Inductors and capacitors are reciprocal, meaning both ports can be interchanged, maintaining the same performance.

1.2.4 Conductor

A conductor is a material that allows the flow of charge in any direction. Such materials are characterised by high conductivity σ (Table 1.1). Antennas are typically made out of conductors, dielectrics, and combinations thereof. A conductor is to an electromagnetic wave, what concrete is to a water wave. Depending on the shape of a concrete wall, a water wave can be conducted and directed toward a particular direction.

Due to the high conductivity of copper, it is common in the microwave domain. Copper is coated to prevent it from corrosion.

1.2.5 Dielectric substrate

Following the IEEE definition, a substrate is a "supporting material upon or within which a planar transmission line is fabricated or to which it is attached. A substrate can be composed of one or more non-conducting layers" [10]. As far as antenna performance is concerned, a dielectric substrate is characterised by two parameters: relative permittivity ε_r and loss tangent tanδ (Table 1.2). Relative permittivity, in

Table 1.1 Conductivity of some materials

Material	**σ (S/m) at 20 °C**
Copper	5.9×10^7
Aluminium	3.8×10^7
Brass	2.5×10^7
Bronze	1.0×10^7

Table 1.2 Relative permittivity and loss tangent of some dielectric substrates

Material	ε_r	**tanδ**
FR4	3.9–4.7	0.02
Teflon	2.08	0.0004
Alumina	9.5–10.0	0.0003
Ceramic (A-35)	5.6	0.0041

simple terms, is equal to 1 for free space and increases for more dense materials. On the other hand, to minimise losses, the loss tangent should be close to zero. FR4 is a common substrate for PCB since the cost is competitive. For some applications requiring large transmission lines, such as those for antenna arrays, low-loss substrates ($\tan\delta < 0.001$) are more convenient. However, since a wireless device is centimetres in size, the transmission lines are rarely larger than a centimetre. Thus, losses are admissible. Some substrates are biodegradable, making them attractive for sustainable radiofrequency solutions [11].

Dielectric can be used as a merely supporting material to print a conductor when designing antennas or even to make antennas smaller for a particular set called microstrip patch antennas (Section 1.7).

1.2.6 Ground plane and Printed Circuit Board

IEEE defines a ground plane as "a conducting surface or plate used as a common reference point for circuit returns and electric or signal potentials" [10].

A PCB comprises a substrate material with several layers, including a ground plane, a power plane to feed the electronics, and communication layers with traces to connect components like antennas, integrated circuits, and SMD components (Figure 1.8). Those components are located at the top and bottom layers of the PCB. For example, an antenna is placed on the top or bottom of the PCB and connected to other microwave devices, such as a matching network or a communication circuit.

A ground plane of a PCB is usually a continuous conductive layer placed at the top and bottom of the PCB.

1.2.7 Transmission lines

A transmission line is a radiofrequency structure to guide electromagnetic waves. Transmission lines are then helpful for connecting an antenna to a transceiver and

Figure 1.8 A PCB comprising an antenna booster and electronics components [12]

other components. The transmission line with minimal losses guides the output power of the transceiver. It then reaches the antenna to radiate the power into space as an electromagnetic wave. As in the case of inductors and capacitors, transmission lines must present low losses.

Besides being a guide to waves, transmission lines are also employed as elements of a matching network to help the antenna transmit/receive, similar to inductors and capacitors [13]. However, transmission lines have a size comparable to the wavelength of operation, and thus, for wireless devices, where space is a constraint, matching networks are designed with lumped components at least up to a few GHz (~10 GHz).

Among many kinds of transmission lines, three types are used throughout the book: a micro coaxial, a microstrip, and a coplanar waveguide (Figures 1.9 and 1.10).

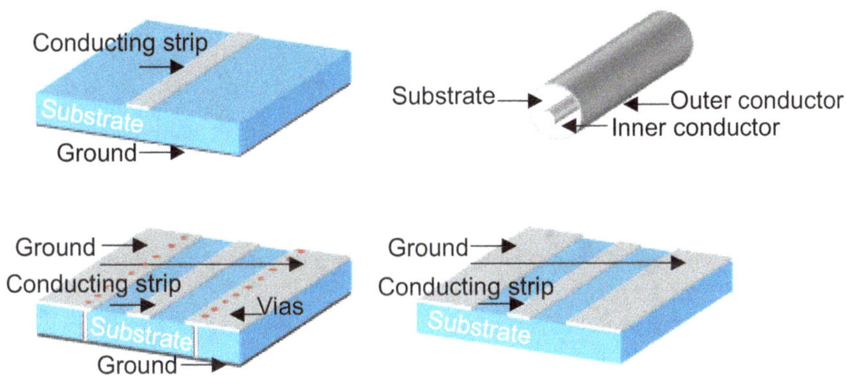

Figure 1.9 Transmission lines: microstrip (top left), coaxial (top right), coplanar with bottom ground (bottom left), and coplanar without bottom ground (right)

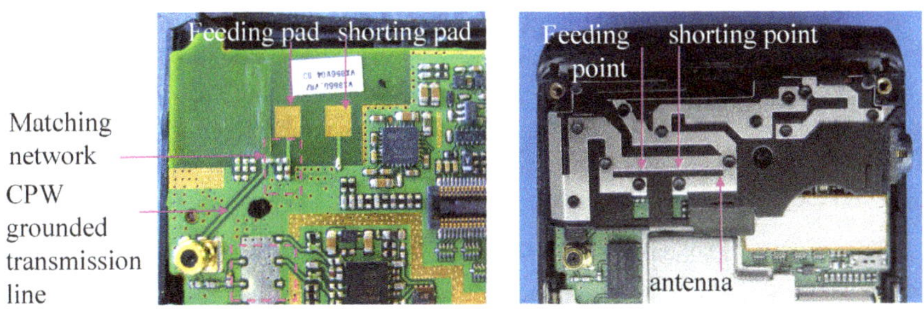

Figure 1.10 Example of a coplanar waveguide (CPW) grounded transmission line connecting an antenna (right) to a multiplexer. The antenna has a π-type matching network (shunt-series-shunt) comprising SMD components.

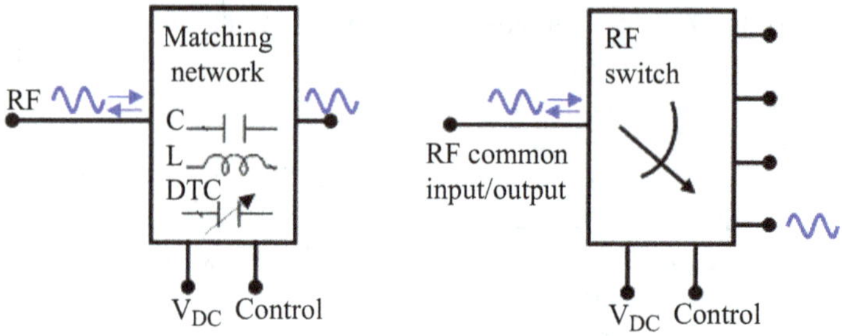

Figure 1.11 *(Left) A matching network comprising capacitors, inductors, and tunable capacitors (DTC: Digitally Tunable Capacitor); (right) an RF switch. Both the DTC and the switch need DC power to operate. The control line allows setting a particular state of the DTC and the RF switch to connect the common input/output port to another port.*

1.2.8 Tunable components

Besides inductors, capacitors, and transmission lines, other components can be included to help match the antenna with a transceiver. Tunable components allow for the provision of a reconfigurable set of inductors and capacitors. For example, a tunable capacitor offers a different set of capacitances. The range of values may be continuous or discrete. Either way, the tunable capacitor needs a control line to set a specific value of a capacitor (Figure 1.11).

Radiofrequency switches are other components that help to provide a set of different values of inductors and capacitors. A switch typically presents a common input/output port and several input/output ports. Control lines are employed to set a particular state of a switch. Both tunable capacitors and switches need to be fed (continuous voltage V_{DC} in Figure 1.11) and, thus, are active components consuming power. Such power may come from the battery of a wireless device, and thus, the power consumption of these devices is an important parameter to consider to ensure long battery life.

1.2.9 Filters

A filter is a passive radiofrequency circuit with two ports blocking a radio signal at a given frequency band to go from one port to the other (Figure 1.12). It is a reciprocal circuit; it filters the same way from ports one to two and vice versa [14].

There are four types of filters:

- Low pass-band filter where the filter allows low frequencies to pass and rejects high frequencies. For Figure 1.12, a low pass-band filter allows f_1 and f_2 to pass and rejects f_3 and f_4 (note that $f_1<f_2<f_3<f_4$).
- High pass-band filter where the filter allows high frequencies to pass and rejects low frequencies. For Figure 1.12, a high pass-band filter allows f_4 to pass and rejects f_1, f_2, and f_3.

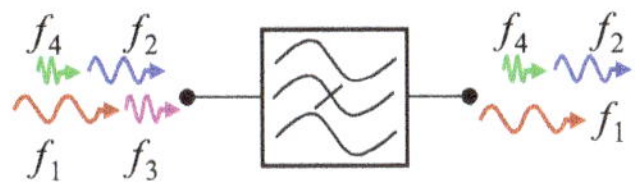

Figure 1.12 A reject filter attenuates some frequency bands, allowing other frequency bands to pass. Reject filters can also reject several frequency bands. Each frequency carries information and, therefore, has a specific bandwidth Δf. Note that behaviour is reciprocal; if f_1, f_2, f_3, and f_4 are incoming signals to the port on the right, f_3 is rejected and, therefore, highly attenuated when leaving to the left port.

- Pass-band filter where the filter allows a frequency band Δf to pass. For Figure 1.12, a pass-band filter at f_2 would allow $f_2 \pm \Delta f$ to pass (Δf represents the bandwidth of the filter, which are frequencies transmitted from one port to the other with a similar magnitude), and f_1, f_3, and f_4 will be rejected.
- Reject filter where the filter rejects a frequency band to pass. For Figure 1.12, the filter is a reject filter at f_3.

1.2.10 Multiplexers

A multiplexer is a passive device with one common input/output port and several input/output ports. For a diplexer, there are two different input/output ports besides the common port; for a triplexer, there are three, and so forth. The purpose of a multiplexer is to combine or split radio signals of different frequencies. For example, let us imagine a radio signal comprising three different frequencies, f_1, f_2, and f_3, where $f_1 < f_2 < f_3$ (Figure 1.13). The signal reaches the common port of the multiplexer, where each radio signal is split through each different port at the right of the multiplexer, as shown in Figure 1.13. The same reasoning applies the other way around; all different radio signals reaching the ports on the right of the multiplexer are combined into the common port.

Note that a multiplexer can be understood using the filter device explained before, cascading several pass-band filters. For example, for a triplexer shown in Figure 1.13, a first pass-band filter allows f_1 to pass from the left to the right port. Two more pass-band filters at f_2 and f_3 allow f_2 and f_3 to pass from left to right, respectively.

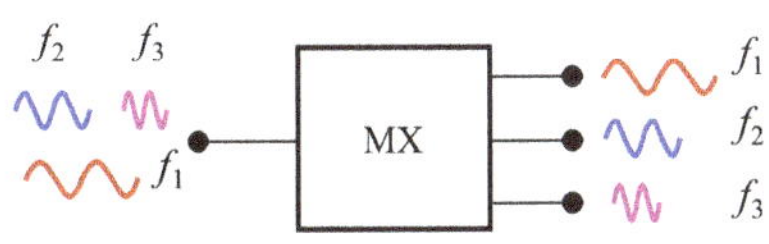

Figure 1.13 A multiplexer splits/combines radio signals of different frequencies

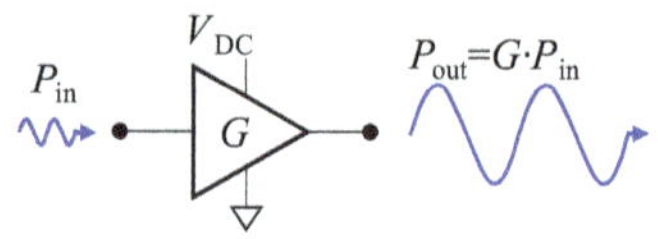

Figure 1.14 An amplifier amplifies G times the input power P_{in}. The amplifier is an active component that consumes power.

The behaviour is reciprocal, as any other passive circuit, meaning that signals are split when travelling from the left to the right port and combined when travelling from the right to the left port.

1.2.11 Amplifier

An amplifier is a radiofrequency component with two ports that amplifies a radiofrequency signal. It is a unidirectional component, meaning that the signal is amplified from one port to another but not the other way around (Figure 1.14).

Among many parameters, two parameters mainly characterise an amplifier: gain and noise figure. The gain represents G times the outpower, which is amplified with respect to the input power. Noise Figure F shows how the signal-to-noise ratio (S/N) deteriorates. The amplifier magnifies not only the useful signal and noise but also adds noise, thus worsening the S/N. In this sense, the lower the F, the better, so the signal-to-noise ratio does not deteriorate. For example, an amplifier with an $F = 1$ dB deteriorates the output S/N by 1 dB with respect to the input S/N.

An amplifier needs to be powered; therefore, consumption is another parameter to consider.

1.2.12 Transceivers

A transceiver is a radiofrequency component in charge of transmitting and receiving a radiofrequency signal to an antenna through a radio frequency front-end module (Figure 1.15). The radiofrequency signal carries the information

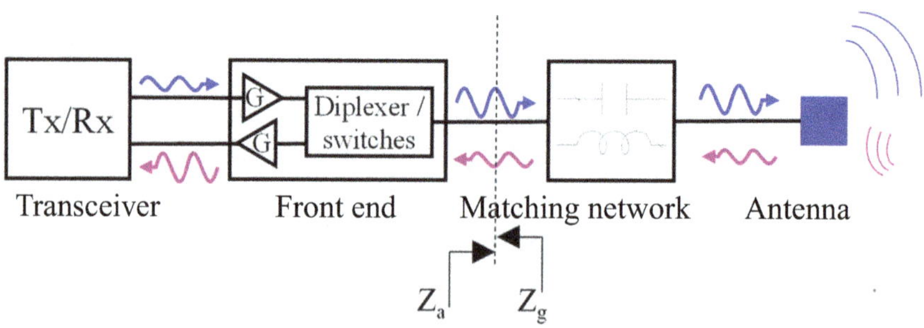

Figure 1.15 Basic scheme of a transceiver connected to a front end, to a matching network and antenna. Z_g is the input impedance at the input/output port of the front end (typically 50 Ω), and Z_a is the input impedance of the antenna system. A transceiver is connected to baseband components.

transmitted/received to/from space by the antenna. The purpose of the antenna and a matching network is to transfer the output power of the transceiver to space and vice versa. A transceiver adjusts the signal frequency from intermediate frequencies of a baseband signal to carrier frequencies (those to be transmitted to space) and vice versa in a process called *up-conversion* and *down-conversion* [15]. Transceivers are active devices and thus require power to operate.

The front-end part connected to a transceiver is in charge of amplifying the radiofrequency signal. From the antenna system perspective, its impedance (Z_g) matters (typically $Z_g = 50\ \Omega$). The internal impedance of the transceiver is crucial for the antenna system to efficiently extract power during transmission and deliver power during reception. This is achieved under impedance-matching conditions. To achieve this condition, $Z_a = Z_g^*$. For some antenna designs, Z_a can be designed in such a way that $Z_a = Z_g^* = 50\ \Omega$ or similar in the frequency bands of operation. This route requires antenna engineering since the geometry of the antenna must be designed to achieve a Z_a as close as 50 Ω. The approach is radically different with antenna booster technology, as will be evident across the chapters. Instead of a complex antenna shape, a very simple element called an antenna booster is used. Said element is in charge of exciting radiating currents in the ground plane of the device. However, its impedance is mainly reactive. Thus, a matching network transforms the impedance to a Z_a close to Z_g^* to achieve good impedance matching. From this rationale, antenna boosters need a matching network, which may be a drawback. However, it is the other way around. The matching network provides simplicity and flexibility to the design process because the frequency bands of operation are simply and systematically obtained thanks to the design of a matching network. Thus, it is unnecessary to customise an antenna for each new situation requiring other frequency bands. It is important to emphasise that the design of a matching network (either single or multiband) is addressed through matching network synthesis (Section 1.5), where matching networks are rapidly obtained by software force.

Although the impedance at the input/output port (Z_g) of the front end is usually 50 Ω, there are front ends with a complex impedance different from 50 Ω. Hence, the matching network allows impedance matching to such complex impedances (typically by using conjugate matching).

Besides the protocols handled by a transceiver, sensitivity is of paramount importance. Sensitivity is the minimum incoming power a receiver (the RX part of the transceiver) can correctly process. The lesser the sensitivity value, the better. Sensitivities are in the order of −100 dBm, whereas, in some radios, it can go down to −165 dBm [16].

1.3 The role of an antenna in a telecommunication system

An antenna is an indispensable component in a wireless device since it has the mission to efficiently radiate information to space and the other way around (Figure 1.16). When sending/receiving such information to/from space, the antenna

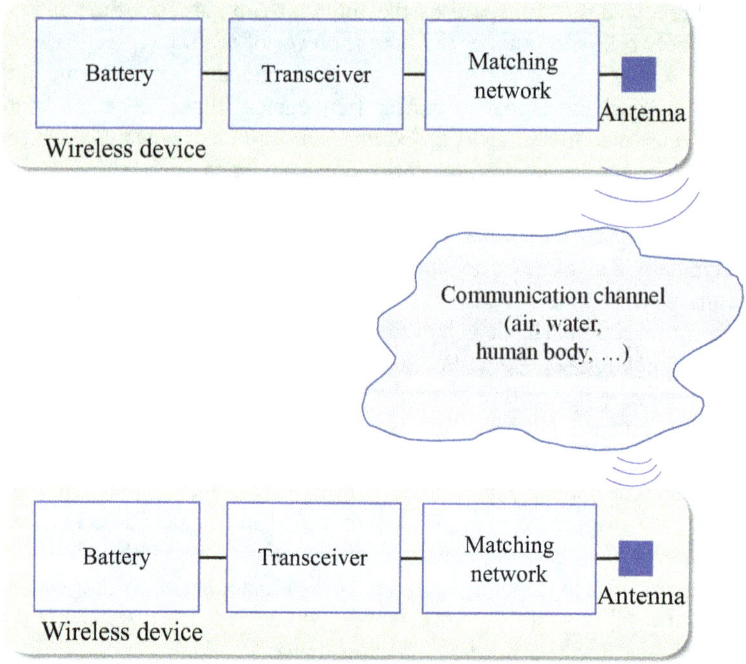

Figure 1.16 The role of an antenna is to radiate information into space and receive it from space efficiently. To simplify the drawing, the transceiver includes the front end (Figure 1.15).

has a directional behaviour. In some applications, such as point-to-point, such information must be directed towards a particular location; thus, a very directional antenna is needed. However, in those situations where the direction of the incoming signal is unknown or the orientation of the wireless device is arbitrary, a low directivity pattern is required.

Besides being a radiating device, an antenna must capture the power from the transmitter and radiate into space, as well as capture the incoming signal from space and transmit it to the receiver. This can be achieved if the antenna impedance matches that of the transceiver (Section 1.2.2). Furthermore, if the wireless device operates at several frequency bands (e.g., two frequency bands from $[f_1, f_2]$ and $[f_3, f_4]$), such impedance-matching conditions must hold across the whole set of frequencies. Said antenna impedance can be achieved by adequately designing the antenna geometry and/or adding a matching network.

1.4 Basic antenna parameters

This section details the most relevant antenna parameters to then relate those to a communication link.

1.4.1 Impedance and efficiency

From a circuit point of view, an antenna in its simplest form is a one-port passive device and, therefore, has an input impedance Z_a, which is the ratio of voltage to current at the input port. The impedance of an antenna has a real part $R(\omega)$ and an imaginary part $X(\omega)$, both depending on frequency f (note that $\omega = 2\pi f$). The antenna is resonant at a frequency of f_0 if the imaginary part of the input impedance at f_0 is zero. An antenna generally has many resonances, which we call modes or resonant frequencies. The impedance Z_a of an antenna can be expressed as:

$$\begin{aligned} Z_a &= R(\omega) + jX(\omega) \\ Z_a &= R_r(\omega) + R_\Omega(\omega) + jX(\omega) \end{aligned} \tag{1.3}$$

where R_r is the radiation resistance, representing the power that the antenna radiates into free space. R_r is defined as the resistance value that would dissipate the same power as that radiated by the antenna.

R_Ω is the loss resistance. It represents losses due to materials that make up the antenna, such as conductors and dielectrics. An antenna must present an $R_\Omega \ll R_r$ to be an effective radiator. However, R_r is typically small in electrically small antennas, where antenna dimensions are much smaller than the operating wavelength. Consequently, designing small, efficient antennas becomes an engineering challenge.

One can associate the power dissipated by an antenna P_a with power dissipated as radiated to space (P_r) and the other wasted as heat (P_Ω). Therefore, the primary purpose is to have an effective antenna where most of the power delivered to the antenna from a source such as a transceiver is emitted into space, implying $P_r \gg P_\Omega$ and hence, $R_r \gg R_\Omega$:

$$P_r = I^2 R_r \tag{1.4}$$

$$P_\Omega = I^2 R_\Omega \tag{1.5}$$

$$P_a = P_r + P_\Omega = I^2 (R_r + R_\Omega) \tag{1.6}$$

where I is the current at the antenna port.

Let us assume a lossless antenna $R_\Omega = 0$, with $X = 0$, so $Z_a = R_r$. This implies that all power delivered to the antenna will be radiated to space. But the big question is, how much power is delivered to the antenna? Let us assume a transceiver modelled by a voltage generator V_g and an internal impedance Z_g (typically 50 Ω) (Figure 1.17).

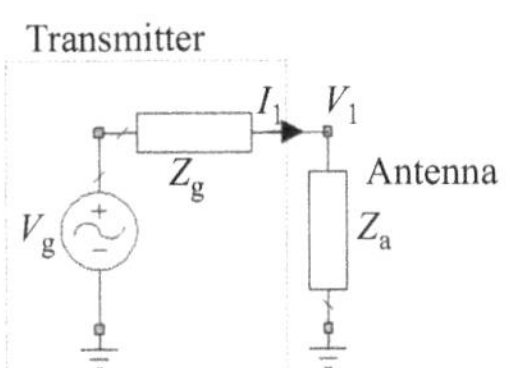

Figure 1.17 A transmitter modelled by a signal source V_g and an internal impedance Z_g connected to an antenna having an impedance Z_a

From basic Kirchhoff laws, we get the following:

$$P_a = \left| \frac{V_g}{Z_g + Z_a} \right|^2 \cdot \Re\{Z_a\} \tag{1.7}$$

If we consider $Z_a = R_r$, then:

$$P_a = \left| \frac{V_g}{Z_g + R_r} \right|^2 \cdot R_r \tag{1.8}$$

Assuming $V_g = 1$ V and $Z_g = 50\ \Omega$, what value of R_r maximises P_a? The answer can be solved analytically by $\frac{\partial P_a}{\partial R_r} = 0$ or in a more straightforward way by plotting P_a as a function of R_r (Figure 1.18).

It is shown that $R_r = Z_g$ maximises the power delivered to R_r. In a more general manner, the condition that antenna impedance Z_a should meet to maximise the power delivered to the antenna is $Z_a = Z_g^*$ (Figure 1.15). This condition is the so-called conjugate matching. From a practical perspective, most of the transceivers and amplifiers connected to the antenna feature a 50 Ω impedance. Therefore, the antenna should present 50 Ω to maximise the power delivered to the antenna so that

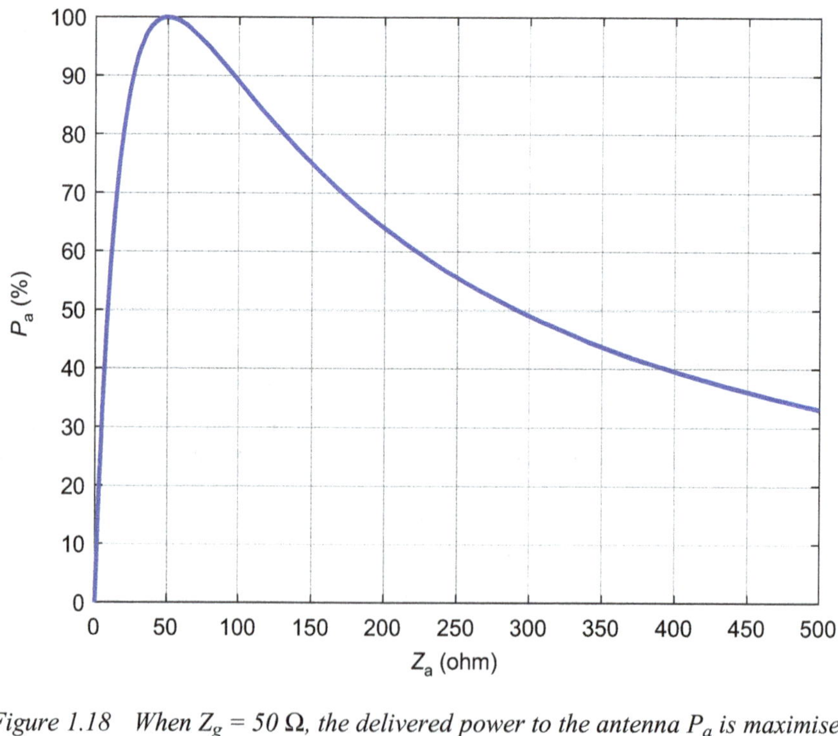

Figure 1.18 When $Z_g = 50\ \Omega$, the delivered power to the antenna P_a is maximised when $Z_a = Z_g^* = 50\ \Omega$. $P_a(\%) = P_a/P_{a\ max} \cdot 100$

the power radiated into space can be maximised, too. This condition holds both for transmission and reception. In effect, when the antenna works as a receiver, it has an equivalent generator representing the captured voltage from the incoming electromagnetic wave reaching the antenna with an impedance Z_a connected to a receiver with internal impedance Z_g. The result is the same circuit, but now the antenna is modelled as a signal generator.

Note that when $Z_a = Z_g^* = 50\ \Omega$, the power delivered to the antenna is maximum, being the available power from the generator P_i. Expressed as a function of Z_g, the available power P_i is:

$$P_i = P_a = \frac{V_g^2}{4Z_g} \tag{1.9}$$

For other values of Z_a different than 50 Ω, $P_a < P_i$.

How can we characterise the power delivered to an antenna with a basic equation once its impedance Z_a is known by simulation or measurements? This is addressed by the well-known reflection coefficient S_{11} as follows:

$$P_a = P_i \cdot \left(1 - |S_{11}|^2\right) \tag{1.10}$$

$$S_{11} = \frac{Z_a - Z_g^*}{Z_a + Z_g} \tag{1.11}$$

where P_a is the power delivered to the antenna, P_i is the power delivered by the generator, Z_a is the antenna impedance including both real and imaginary parts, Z_g is the impedance of the generator representing the transceiver connected to the antenna, and S_{11} is the reflection coefficient, which is a complex value having $|S_{11}| \leq 1$. S_{11} is easily computed through electromagnetic simulation or can be tested with a vector network analyser (Section 1.6). For multiport antennas or multi-antenna systems, coupling between ports and antennas S_{ij} appears in Equation (1.10), which is introduced in more detail in the following chapters (Chapter 6). In the end, less power will be delivered to the antenna if the ports are coupled.

As an example, if $Z_g = 50\ \Omega$, $Z_a = 300\ \Omega$, $S_{11} = 5/7 = -2.9$ dB or standing wave ratio (SWR) = 6 (Table 1.3), resulting from Equation (1.10), $P_a = 0.51 \cdot P_i$, meaning that only 51% is delivered to the antenna, representing half of the power lost. This is not a good scenario for a device since, in many applications, the front end will increase power to provide sufficient radiated power and, thus, will drain the battery faster, which is critical for battery-powered wireless devices.

Now, let us consider $Z_a = 50\ \Omega$, but in this case, $R_r = 25\ \Omega$ and $R_\Omega = 25\ \Omega$. The power delivered to the antenna is 100% of P_i, but how much power of the power delivered to the antenna is radiated to space? Using Equations (1.4)–(1.6) it results that $P_r/P_a = 50\%$. This represents half of the power lost in heating due to the antenna materials and any surrounding materials introducing losses. The ratio P_r/P_a is called radiation efficiency, representing the relative power portion that is radiated

Table 1.3 $(1 - |S_{11}|^2)$ *represents power delivered to the antenna over the available power from a source versus SWR and* S_{11}

SWR	S_{11} [dB]	Ratio of power delivered to the antenna over the available power: $1 - \|S_{11}\|^2$ [%]	Mismatch loss $1 - \|S_{11}\|^2$ [dB]
6	−2.9	49.0	−3.1
3	−6.0	75.0	−1.2
2	−9.5	88.9	−0.5
1.5	−14.0	96.0	−0.2
1.1	−26.4	99.8	−0.01

to space (P_r) with respect to the delivered power (P_a):

$$\frac{P_r}{P_a} = \eta_r = \frac{R_r}{R_r + R_\Omega} \tag{1.12}$$

In summary, the radiated power to space depends upon how much power from the generator (P_i) is delivered to the antenna (P_a) and then, of this P_a, how much power is radiated to space (P_r). Combining both contributions, we get:

$$P_r = P_i \cdot \eta_r \cdot \left(1 - |S_{11}|^2\right) \tag{1.13}$$

where $\eta_r \cdot \left(1 - |S_{11}|^2\right)$ is called total efficiency η_t:

$$\eta_t = \frac{P_r}{P_i} = \eta_r \cdot \left(1 - |S_{11}|^2\right) \tag{1.14}$$

Reflection coefficient and total efficiency

S_{11} informs on how much power from the generator is delivered to the matching network and antenna. However, since the matching network, the antenna, and nearby components (e.g., plastic covers, human body) may introduce losses, total efficiency is needed to take into account how much power is lost due to mismatching and losses.

Equation (1.14) is the most practical equation since it considers the radiated power to space in relation to the available power from a generator. For example, if a transceiver delivers P_i = 23 dBm to an antenna embedded in a wireless device having a η_t = 50%, the radiated power to space will be 20 dBm. Total efficiency can be computed with an electromagnetic simulator (Section 1.5) or measured using an anechoic chamber (Section 1.6).

Consider now an antenna having at a particular frequency a $Z_a = 50\ \Omega - j200\Omega$, and assume a lossless case, that is, $R_\Omega = 0$, so $\eta_r = 100\%$. What is the power delivered to the antenna? What is the power delivered to the space?

Since $\eta_r = 100\%$, $P_r = P_a$ (Equation (1.12)). Now, let us calculate P_a using Equations (1.10) and (1.11):

$$S_{11} = \frac{50 - j200 - 50}{50 - j200 + 50} = 0.8 - 0.4j \tag{1.15}$$

$$P_a = P_i \cdot \left(1 - |S_{11}|^2\right) = 0.2 \cdot P_i \tag{1.16}$$

Only 20% of the available power is delivered to the antenna and radiated to space. This severe mismatch can be compensated by removing the reactive component of the antenna using a matching network. In this case, the matching network can be a series inductor with a reactance of $X = 200$. Adding this series component to the antenna, $Z_a = 50\ \Omega$ and thus $S_{11} = 0$. Since $\eta_r = 100\%$, all the available power P_i from the generator will be radiated to space.

The term η_r in Equation (1.14) can include extra losses rather than the losses introduced by the antenna (Equation (1.12)), such as the losses of the matching network, the losses due to the materials of the wireless device (e.g., the cover), the human body. In Chapter 3, Equation (1.14) is revisited, considering the losses of the matching network.

Classic antenna engineering to design small and multiband antennas relies on shaping the geometry of an antenna to achieve a good S_{11} (good matching) across the frequency bands of interest while ensuring competitive radiation efficiency. However, Antenna Booster Technology, as profoundly analysed in the book, relies on a different approach. In essence, it utilises matching networks to shape the S_{11} response of a small, non-resonant element known as an antenna booster, which is responsible for inducing current in the ground plane of a wireless device to ensure high radiation efficiency. As shown in the following chapters, this is much more flexible than designing an antenna to achieve behaviour at the frequency bands of operation.

Impedance matching

When the antenna system is matched to the impedance of the transceiver, the power delivered to the antenna system is maximum, and the other way around, the incoming power from space captured by the antenna is maximised. Otherwise, power is lost due to reflections. During transmission, it indicates that not all the power generated is transferred to the antenna, and during reception, not all the available incoming power from space is captured by the antenna.

Radiation efficiency depends on the antenna size in terms of wavelength (size/λ). In general terms, an antenna with a small size in terms of wavelength will present less radiation efficiency and, thus, less total efficiency than a second antenna with a larger size. For example, given an antenna of 3 cm operating at 900 MHz and 5 GHz, the antenna size is $\sim\lambda/11$ and $\sim\lambda/2$ at 900 MHz and 5 GHz, respectively. Total efficiency will usually be smaller at 900 MHz since the antenna is smaller in wavelength than 5 GHz. The same applies when the antenna is embedded in a wireless device. In that case, the size depends not only on the antenna but also on the ground plane of the wireless device size since the ground plane takes part in the radiation process, as seen throughout the book. For example, given a small antenna on a 50 mm $\times$ 50 mm ground plane, the size of the ground plane is $\sim\lambda/6.6$ and $\sim\lambda$ at 900 MHz and 5 GHz, respectively. As before, it will be more likely to achieve higher total efficiency at 5 GHz because the radiating structure (the antenna and the ground plane of the wireless device) is bigger at 5 GHz than at 900 MHz.

1.4.2 Matching networks to improve total efficiency

As treated in the previous section, a matching network plays a vital role in matching the impedance of an antenna to a generator to transfer most of the available power from the generator to the antenna (and vice versa).

A matching network can comprise passive lumped components (capacitors and inductors) (Figure 1.19). Note that resistors are not included since they will increase losses of the antenna system, reducing the radiation efficiency (Equation (1.12)).

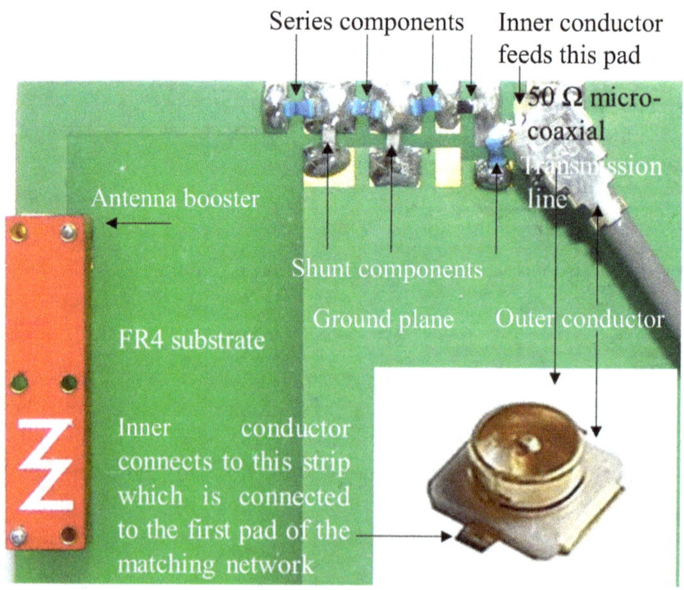

Figure 1.19 A matching network comprising SMD inductors (in blue) and capacitors. In red, an example of a 12 mm × 3 mm × 2.4 (height) antenna booster.

For example, let us consider an antenna with R_r = 10 Ω, R_Ω = 0, and X = 0. Since R_Ω = 0 (ideal case), radiation efficiency η_r = 100%. Let us assume the antenna is connected to a radio frequency source with Z_g = 50 Ω (Equation (1.11)). Without any matching network, the total efficiency will be $\eta_t = 100\% \cdot (1 - (4/6)^2) =$ 55%. If we now add a series resistor of 40 Ω, the Z_a = 50 Ω and thus S_{11} = 0. But most of the delivered power (80%) to the antenna will be dissipated in the 40 Ω resistor (Equations (1.4)–(1.6)), and only 20% will be radiated to space. As observed, although S_{11} = 0 (perfect match), η_t = 20% only compared to the 55% without matching. In conclusion, resistive elements must not be used in the matching network if we want to maximise the radiated/received power to/from space. This is extended to the losses of the components of the matching network. Those components should be selected to provide low losses, achieved using high-quality factors (Q). Now, if we design a matching network with low losses, the 10 Ω can be matched to 50 Ω, so S_{11} = 0, and as a result, we can get a total efficiency close to 100%, assuming low losses in the matching network.

Special care should be taken when selecting lumped components for a matching network to ensure minimum dissipation losses in the matching network. By selecting high-Q components, losses are minimised. For example, Figure 1.20 intentionally represents a lossy component (the one in green colour) where Q is

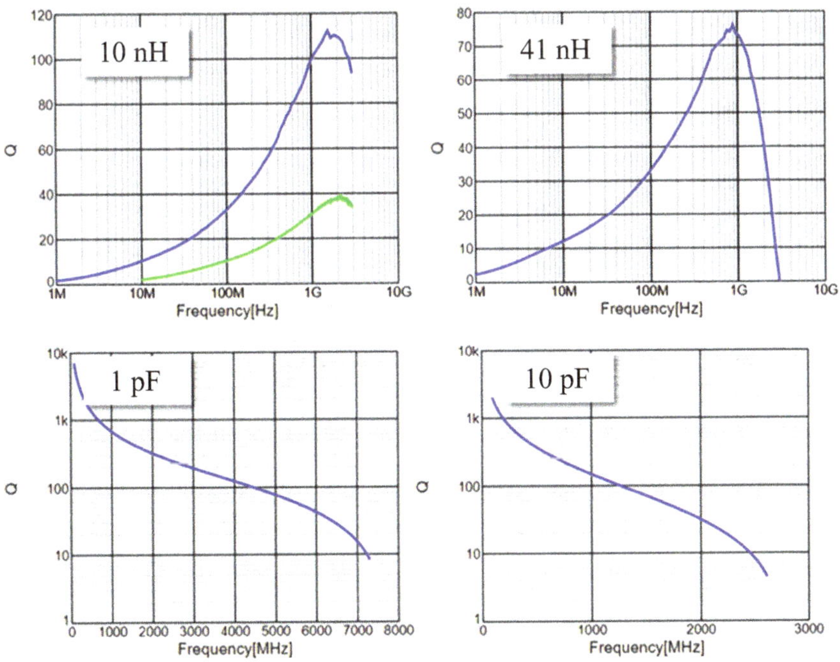

Figure 1.20 Quality factor Q for several capacitors and inductors as a function of frequency. Data obtained for SMD 0402 size components. From Murata datasheets.

much lower than the Q of the component with the blue line. The frequency range of operation can also be determined from the Q versus frequency response. For example, the 10 nH component may be used from hundreds of MHz (~600 MHz) up to 3 GHz. Beyond 3 GHz, Q starts to drop until it reaches the SRF (Self Resonant Frequency), where the component not only has a very low Q but will not behave as an inductor but as a capacitor due to parasitic effects. It is interesting to observe the same trend for 41 nH, but in this case, the component Q decreases beyond 1 GHz. This leads us to conclude that the larger the inductance, lower the maximum operating frequency will be. Fortunately, as discussed in this book, designing a matching network for antenna boosters is based on a matching network synthesis procedure where an optimisation algorithm discards those components introducing losses and selects those to minimise matching network losses.

Regarding capacitors, a similar trend occurs. For the 1 pF case, Q is larger than 100 for 4.5 GHz, whereas it is only 1.2 GHz for the 10 pF case. Again, this issue of selecting the best components is automated and requires no human intervention in designing wireless devices embedding antenna boosters.

In some cases where it is challenging to provide matching across all frequency bands of operation, tunable components are added to the matching network, resulting in a reconfigurable matching network. This reconfigurable network is driven by the intelligence of a transceiver that sets up the matching network to a particular state to minimise mismatch losses and ensures that most of the available power is delivered to the antenna and then radiated to space.

Quality factor *Q*

The higher the Q of an inductor and capacitor, the lesser the losses. Therefore, when designing a matching network with inductors and capacitors, Q should be as high as possible.

Finally, a matching network can comprise distributed elements such as transmission lines. For wireless devices, this may be suitable for high frequencies (above ~5 GHz) where the size of the transmission lines can be small. A quarter-wave transmission line typically used for matching purposes has a length of $\lambda_g/4$, where λ_g is the wavelength of the electromagnetic wave in the waveguide. For example, a $\lambda_g/4$ microstrip line at 5 GHz printed on a substrate with $\varepsilon_r = 4$ equals ~8 mm long. A combination of distributed and lumped elements can also gain design flexibility. However, simplicity and size should also be added to the design equation.

In Chapters 4 and 5, we will deal with matching networks to provide matching impedance for antenna boosters, either for single-band or multiband operation.

1.4.3 Bandwidth

According to the IEEE definition [7], bandwidth is the range of frequencies within which the antenna performance, with respect to some characteristic, conforms to a specified standard. In the present book, the characteristic will be S_{11}, and the selected reference will be $S_{11} = -6$ dB or SWR = 3, representing a maximum delivered power of 75% and reflected power of 25%, which is still acceptable for wireless devices where power is limited to a few watts (Table 1.3). For other applications, such as antennas for base stations transmitting tens of watts, an $S_{11} = -6$ dB is not accepted since 25% of the reflected power may damage the radiofrequency front end. Bandwidth for a particular S_{11}/SWR is defined as (Equation (1.17)):

$$BW = \frac{f_2 - f_1}{\frac{f_2+f_1}{2}} \cdot 100 \tag{1.17}$$

where f_2 and f_1 present a $S_{11}(f_2) = S_{11}(f_1)$ equal to the reference S_{11}. For example, the bandwidth for the S_{11} response given in Figure 1.21 is 21.3% and 47.2% at the first and second frequency regions considering a reference $S_{11} = -6$ dB.

It is important to distinguish between the single band and multiband. In the present book, a frequency band will refer to a set of frequencies for providing operation at a frequency region from f_1 to f_2: $[f_1, f_2]$. Multiband refers to a new frequency region $[f_3, f_4]$ that is non-continuous to $[f_1, f_2]$. The example shown in Figure 1.21 illustrates the S_{11} response of a multiband antenna, a first frequency band or frequency region from 0.782 GHz to 0.969 GHz and a second frequency region from 1.68 GHz to 2.72 GHz. The first frequency region may cover several operating bands; for example, 880 MHz–960 MHz is an operating band for cellular communications. An antenna with only operation at 0.782 GHz–0.969 GHz is not referred to

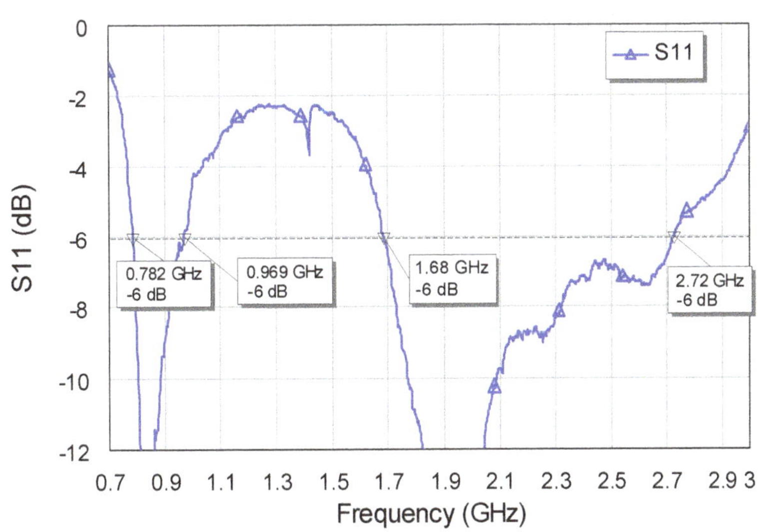

Figure 1.21 Bandwidth is typically calculated at a given S_{11} level. For an $S_{11} = -6$ dB bandwidth, the first and second frequency regions present 21.3% and 47.3%, respectively.

as a multiband since all operating bands belong to the same frequency region. However, suppose an operating band is allocated in a second frequency region. Then, the antenna receives the dual-band name in case it has two operating bands, one at $[f_1, f_2]$ and the other at $[f_3, f_4]$. As the response in Figure 1.21 has several operating bands at $[f_1, f_2]$ and $[f_3, f_4]$, the classification of multiband applies here.

Bandwidth is proportional to the electrical size of an antenna. It is, therefore, easier to obtain more bandwidth for an electrically large antenna than for an electrically small antenna. For example, the bandwidth of a $\lambda/2$ dipole is larger than that of a $\lambda/20$ dipole. For Antenna Booster Technology, we will show in Chapter 3 that bandwidth mainly depends on the electrical size of the ground plane of the wireless device. As a result, obtaining bandwidth for electrically small devices is more challenging than for larger ones. For instance, the ground plane of a wireless device #1 is 120 mm $\times$ 60 mm, and for a second device #2, it is 60 mm $\times$ 60 mm. If the frequency of operation is 900 MHz, the first and second device lengths are 0.36λ and 0.18λ, respectively. As a result, device #1 will have more bandwidth. Also, we will learn that there is an optimum device size that maximises bandwidth.

1.4.4 Radiation pattern, directivity, and gain

Besides being a microwave component able to capture the power from the generator, an antenna is also a device that transmits or receives electromagnetic waves into/from space. Such transmission/reception depends on direction so that the antenna can transmit/receive more power in a given direction while less in others. This directionality information is represented by the radiation pattern, which shows the intensity of the transmitted/received electromagnetic signal in a given direction. Radiation patterns may be represented in three- and two-dimensional charts following polar or cartesian coordinates (Figure 1.22).

When a radiation pattern presents a plane with equal intensity for any direction, the radiation pattern is called omnidirectional. Omnidirectionality is interesting for those situations where the wireless device does not know where the incoming signal may come from. Ideally, to transmit/receive the same in any direction, the radiation pattern should be a sphere receiving the name of the isotropic radiation pattern. From a practical perspective, omnidirectional patterns are the most typical for embedded antennas in wireless devices.

A figure of merit to characterise radiation patterns is called directivity. Directivity is the ratio between the power density radiated by an antenna in a given direction over the power density if the antenna were isotropic [17]. Mathematically:

$$D(\theta, \varphi) = \frac{\wp(\theta, \varphi)}{\frac{P_r}{4\pi r^2}} \tag{1.18}$$

Habitually, directivity is referred to as maximum directivity, therefore:

$$D = \frac{\wp_{\max}(\theta, \varphi)}{\frac{P_r}{4\pi r^2}} \tag{1.19}$$

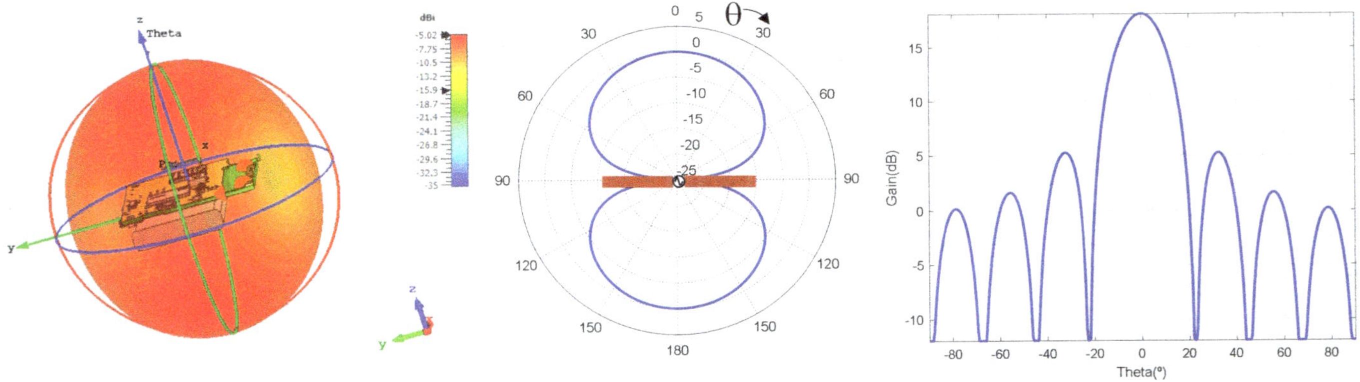

Figure 1.22 Representation of a radiation pattern in 3D, polar, and Cartesians coordinates using dB scale

where $\wp_{\max}(\theta,\varphi)$ is the power density (W/m^2) radiated by the antenna in the (θ, φ) direction at a distance r from the antenna, and P_r is the radiated power. Although D seems to depend on r, D does not depend on r once the observation point is far from the antenna in the so-called far-field region of the antenna. For example, suppose an antenna presents an isotropic pattern. In that case, the power density is $\wp_{\max}(\theta,\varphi) = \frac{P_r}{4\pi r^2}$, being the radiated power at a distance r over the area of an imaginary sphere of radius r. As a result, the directivity of an isotropic antenna is $D = 1$, being the minimum directivity possible. If an antenna concentrates power in a given direction, its directivity increases.

Let us analyse the following didactic example. For the first case, let us imagine our antenna radiates eight sweets for eight children equally placed around the antenna, forming a circle and spaced one to each other 45° (Figure 1.23). One antenna radiates those eight sweets uniformly distributed on a given plane, so each child receives only one sweet. A second antenna radiates those eight sweets in a single direction, so one kid will receive eight sweets, whereas the others will receive no sweets. The first antenna has a directivity of $D = 1$, whereas the second is $D = 8$. Both antennas radiate eight sweets, but the second antenna concentrates such transmission in one direction, and thus, the receiver (the kid) receives eight times more power. Note that this is a simplification of reality since not all the

Figure 1.23 (Left) The antenna with an omnidirectional radiation pattern transmits one sweet per kid; (right) an antenna with directional radiation transmits all sweets to a single kid. In both cases, the transmitted power is the same: eight sweets. However, the received power for the right case is more significant because of a larger directivity antenna.

power can be transmitted in a unique direction. It can be very narrow, though, in the case of high directivity antennas.

Directivity is given in decibels, abbreviated dB, which is obtained by:

$$D_{\mathrm{dB}} = 10 \cdot \log D \tag{1.20}$$

Directivity for embedded antennas in a wireless device features low directivity, typically less than 5 dB. Large antennas, such as antenna arrays for base stations, may achieve directivities around 18 dB. Large parabolic reflectors for radio-astronomy applications present large directivities of about 70 dB (Figure 1.24).

From a mathematical perspective, directivity is computed using the radiation pattern data [17].

For Antenna Booster Technology, we will learn that radiation patterns depend on the ground plane size of the device. Since devices are comparable to half of the wavelength at the frequencies of operation, radiation patterns will feature omni-directional patterns with low directivities of around 3 dB.

Following the example of sweets discussed above. Let us imagine that when sending those eight sweets, one is lost. For instance, in the case of the kid receiving eight sweets in the original case, only seven sweets will be received. In this case, we do not use directivity but gain. Gain is the same as directivity, but adding the radiation efficiency η_r:

$$G = \eta_r \cdot D \quad G_{\mathrm{dB}} = 10 \cdot \log \eta_r + 10 \cdot \log D = \eta_{r\mathrm{dB}} + D_{\mathrm{dB}} \tag{1.21}$$

Besides, mismatch loss can be added to the gain equation as well, resulting in the concept of realised gain:

$$G_{\mathrm{realised}} = \eta_r \cdot \left(1 - |S_{11}|^2\right) \cdot D = \eta_t \cdot D \tag{1.22}$$

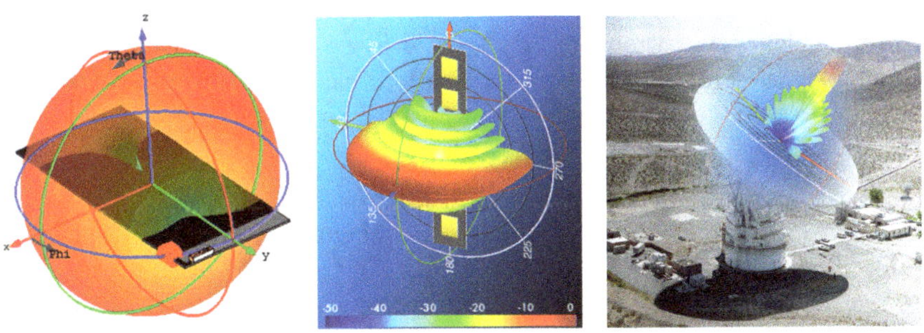

Figure 1.24 (Left) A radiation pattern for an embedded antenna in a wireless device featuring a quasi-isotropic pattern (low directivity); (middle) a directional radiation pattern of an antenna array of microstrip patch antennas for cellular communications; (right) a high directivity beam of a parabolic reflector for radio astronomy

Directivity is related to the radiation pattern, and gain G depends on the radiation pattern and the intensity of the radiated/received electromagnetic signal. Realised gain equals directivity in a perfect scenario having perfect matching ($S_{11} = 0$) and $\eta_r = 100\%$. Thus, the gain is a valuable parameter for power budgeting and range calculations (Section 1.4.7). For example, two antennas may have identical radiation patterns and, as a result, the same directivity D. However, one antenna is $\eta_t = 30\%$, whereas the other is $\eta_t = 60\%$. In this last case, the antenna presents more gain and, hence, will be able to provide more range and ensure more battery life.

Radiation patterns
Low directivity radiation patterns (quasi-isotropic) are suitable to transmit/receive in any direction which is effective when the position of the other device is unknown. This quasi-isotropic pattern allows a wireless device to be robust with respect to the incoming signal whose direction is unknown.

1.4.5 Polarisation

Acoustic waves, such as sound waves, are entirely defined by their field intensity and direction of propagation. However, electromagnetic waves need an extra parameter called polarisation. Polarisation indicates the trace of the radiated electric field over time. Such trace can be linear, which is known as linear polarisation. Linear polarisation can be vertical, horizontal, or any direction between vertical and horizontal. It can also be circular, which is referred to as circular polarisation. In other situations, the electric field trace follows an ellipse, receiving the name of elliptical polarisation. A receiving antenna should match the polarisation of the transmitting antenna to extract the maximum power from the incoming electromagnetic wave. For example, suppose the transmitting antenna radiates a linear vertical electromagnetic wave in the direction of the maximum radiation; then, the receiver wireless device should present a vertically polarised antenna to capture the incoming signal (Figure 1.25). Any rotation will incur less power as mathematically expressed in the polarisation loss factor (PLF) equation:

$$PLF = |\widehat{e}_t \cdot \widehat{e}_r|^2 \quad PLF_{\text{dB}} = 10 \cdot \log PLF \tag{1.23}$$

where $\widehat{e}_t$ and $\widehat{e}_r$ are the unitary vectors representing the polarisation of the transmitting and receiving antenna, respectively, and $(\cdot)$ is the dot product. For example, if a transmitting antenna presents linear horizontal polarisation $\widehat{e}_t = \widehat{x}$ and the receiver antenna presents the same linear polarisation $\widehat{e}_r = \widehat{x}$, $PLF = 1$ represents the best situation. But if the receiver antenna is vertically polarised $\widehat{e}_r = \widehat{y}$, then $PLF = 0$ results in zero received power. For intermediate cases, PLF results in $PLF = |\cos \alpha|^2$ α being the angle between the electric vectors defining the polarisation of the transmitting and receiving antenna. For circular polarisation, math is a bit different yet straightforward [17]. In a nutshell, if a transmitting antenna has

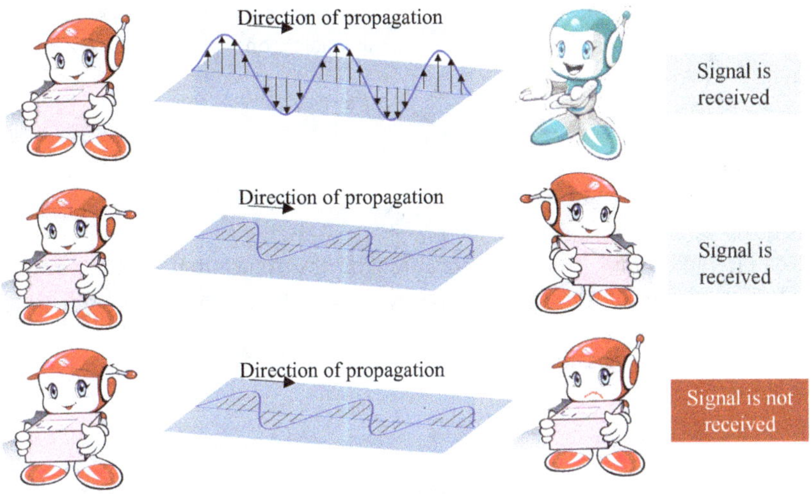

Figure 1.25 *(Top) The antenna transmits a vertical polarised electromagnetic wave that can be received by the receiving antenna on the right (PLF = 1); (middle) the same situation but with a horizontally polarised wave (PLF = 1); (bottom) antenna at the receiver side cannot capture the incoming wave since the antenna is only sensible to receive a vertically polarised wave but not a horizontally polarised wave (PLF = 0)*

circular polarisation and the receiver is linear, $PLF_{\mathrm{dB}} = -3$ dB, meaning that half of the power is not captured. Ideally, the antenna in the receiver should also be circularly polarised to seize the maximum power. However, this should not be taken as a general conclusion. For example, in many wireless devices, circular polarisation is achieved with a microstrip patch antenna [18]. The reason is that it is simple to generate two perpendicular out-of-phase radiated fields to result in a circular polarised wave. In contrast, achieving linear polarisation is easier as it requires only one mode, which can be achieved by a monopole antenna (see for example Figure 1.4). The radiation pattern for the microstrip patch antenna is typically more directive than that of a monopole antenna. Therefore, less signal is captured if the received signal to the circular polarised microstrip patch antenna does not impinge through its maximum. However, since the radiation pattern of a monopole is omnidirectional, a signal is received equally from all directions. As a result, although for a point-to-point radio link, the circular polarised antenna can extract more power from the incoming circular polarised wave, when the incoming waves may impinge the antennas from different angles, a linear polarised monopole is more robust [19]. This may be the case of a receiving device capturing waves from satellites moving around the sky since the incoming wave impinges the antenna at the receiver at different angles.

It should be outlined that circular polarisation presents two states: right hand and left hand [17]. For the present book, the basic concepts introduced here are enough.

Polarisation
Maximum power can be received if the polarisation of the antenna system matches the polarisation of the incoming signal. Being easy to attain in point-to-point radio links where the position of the transmitter and incoming polarisation are known, it becomes more difficult in multi-path situations where the incoming signal may be polarised differently with respect to the initial state of polarisation of the transmitter.

1.4.6 Total Radiated Power (TRP)/Total Isotropic Power (TIS)

Ideally, to obtain the transmitted power to space, given the output power of a generator, total efficiency is enough. For example, suppose a generator or a transceiver of a wireless device is connected to an antenna. In that case, the transmitted power to space, or radiated power P_r, can be estimated using Equation (1.13), where P_i is the output power of the transceiver and $\eta_r \cdot (1 - |S_{11}|^2)$ is the total efficiency. Total efficiency can be measured considering the embedded antenna in the wireless device and all the components and casings of the device. Therefore, if P_i is known, P_r can be calculated. However, P_i is not ideal, and the power accuracy may be around ± 2 dB. On the other hand, the output power may drop depending on the mismatch between the impedance of the transceiver and the input impedance of the antenna. Then, to correctly characterise P_r, a test called TRP (Total Radiated Power) measures P_r when a transceiver feeds the antenna. By performing a 3D measure, TRP can be computed. Therefore, TRP considers not only the antenna performance in terms of radiation efficiency and mismatch but also the quality of the transceiver connected to the antenna and its relation to the antenna.

On the other hand, TIS (Total Isotropic Power) considers noise to evaluate the performance of the device in terms of received power. As an acoustic simile, one person (the transmitter) speaks to another (the receiver) in a noiseless environment. The transmitter decreases the acoustic power until the receiver does not hear the information correctly. That power level is the sensitivity. Suppose the experiment is repeated now, but the transmitter and receiver are in an environment with more people. In that case, the power by which the receiver will begin to have trouble deciphering the message will be greater. In this case, the sensitivity is worse. From this, it follows that the sensitivity not only depends on the embedded antenna but also on the noise of the environment, such as those produced by the electronics of the wireless device. Thus, a low TIS is convenient since the wireless device can receive very little power yet can correctly process the information.

1.4.7 Friis equation

Now that we understand that the radiated power into space depends on the output power of a transceiver P_i and total efficiency, the question is: how much power P_L at a given frequency f will be received by a device including an antenna spaced at a distance r from the first device (Figure 1.26).

The equation that relates the received power P_L by a second antenna to the outpower of a transceiver P_i is known as the Friis equation (Equation (1.24)) named after Harald T. Friis (1883–1976) [20] (Figure 1.27). Without entering into details on its derivation ([21] can be consulted for further information), it is essential to note that radiated power into space following the scheme shown in Figure 1.26 is $P_r = P_i \cdot \eta_{r1} \cdot (1 - |S_{11}|^2)$. However, if we focus on a particular direction, the power density is equal to P_r over the area of the sphere multiplied by the directivity D of the antenna in such direction. When such power density arrives at the antenna at the receiver at a distance of r, it is captured depending on its total efficiency and directivity, which equals the realised gain of the antenna in the receiver.

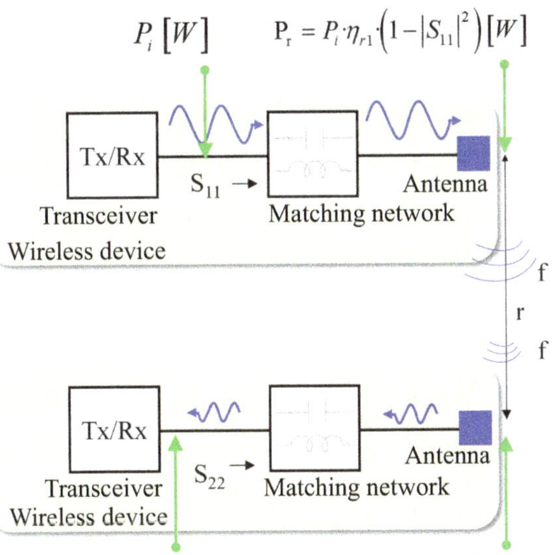

$$P_L = \frac{P_i \cdot \eta_{r1} \cdot \left(1-|S_{11}|^2\right) \cdot D_1}{4\pi r^2} \cdot D_2 \cdot \eta_{r2} \cdot \left(1-|S_{22}|^2\right) \cdot \left(\frac{\lambda^2}{4\pi}\right) \cdot PLF\,[W] = P_i \cdot \left(\frac{\lambda}{4\pi r}\right)^2 \cdot G_{1\,realised} \cdot G_{2\,realised} \cdot PLF\,[W]$$

$$\frac{P_i \cdot \eta_{r1} \cdot \left(1-|S_{11}|^2\right) \cdot D_1}{4\pi r^2}\left[\frac{W}{m^2}\right] = \frac{P_i \cdot G_{1\,realised}}{4\pi r^2}\left[\frac{W}{m^2}\right]$$

Figure 1.26 Key antenna parameters for power budget calculations. D_1 and D_2 are the directivities of antenna #1 and #2, respectively.

All these considerations result in the Friis equation (Equation (1.24)):

$$P_L = P_i \cdot \left(\frac{\lambda}{4\pi r}\right)^2 \cdot G_{1\,\text{realised}} \cdot G_{2\,\text{realised}} \cdot PLF = P_i \cdot \left(\frac{300}{4\pi r \cdot f(\text{MHz})}\right)^2 \cdot G_{1\,\text{realised}} \cdot G_{2\,\text{realised}} \cdot PLF$$
$$G_{1\,\text{realised}} = D_1 \cdot \eta_{t1} = D_1 \cdot \eta_{r1} \cdot \left(1 - |S_{11}|^2\right) \quad G_{2\,\text{realised}} = D_2 \cdot \eta_{t2} = D_2 \cdot \eta_{r2} \cdot \left(1 - |S_{22}|^2\right) \tag{1.24}$$

where λ is the wavelength of operation, $\lambda = c/f$, where c is the speed of the electromagnetic wave in the medium (in free space ~300,000 km/s), and f is the frequency of operation; $G_{1\ \text{realised}}$ and $G_{2\ \text{realised}}$ are the realised gains of antenna 1 and 2, respectively, *PLF* is the polarisation loss factor. Note that G_1 and G_2 both include the total efficiency of each antenna. Friis equation is valid when the second antenna is in the far field ($r > 2L^2/\lambda$, where L is the maximum size of the transmitting antenna) with respect to the first antenna, which is the typical situation for most radio links.

Several conclusions can be drawn from the Friis equation:

- If distance r is doubled, the received power decreases by 6 dB (four times).
- The total efficiency η_t has a direct impact on the received power. If the total efficiency of the receiver antenna decreases two times, the same applies to the received power.

Figure 1.27 Harald T. Friis (1883–1976), a Danish-American engineer and a pioneer in the field of radio wave propagation thanks to the Friis equations in terms of power and noise

- Equation (1.24) reflects the received power in a given direction. If the receiver antenna is very directional, it will receive more power in the direction where it presents the maximum gain, but the received power will be low if the incoming signal impinges on the antenna in a direction where the antenna presents a low gain. Hence, in scenarios where the direction of the incoming signal is uncertain, opting for a low-directivity antenna—a type characterised by a radiation pattern resembling a sphere—is advantageous for ensuring consistent received power, irrespective of the direction of the incoming electromagnetic wave.
- Given all parameters fixed, if the frequency f of the radio link increases twice, the received power decreases by 6 dB. Thus, as far as range is concerned, low frequencies are preferred. However, total efficiency depends on frequency (Section 1.4.1). It is more challenging to obtain high efficiency at those frequencies where the device is electrically small, which occurs at the lowest frequencies of operation. As a result, a trade-off exists. Besides, depending on bandwidth requirements, large frequencies may easily provide more bandwidth than those working at low frequencies.

To further understand the impact of the radiation pattern of an antenna in the Friis equation, let us imagine two scenarios, one with an omnidirectional antenna having a realised gain of 0 dB and a second with a directional antenna with 5 dB where all remaining parameters are the same: distance and frequency (Figure 1.28). For the first scenario, no matter the position of the transmitter, the received power P_L at the receiver will be the same (Figure 1.28, top left and right). For the second scenario, when the transmitter sends a radio signal to the second antenna through the direction of maximum gain, the received power will be 5 dB more. However, if the receiving device moves from its original position, the received power will decrease and, eventually, become very weak (Figure 1.28, bottom right), jeopardising the quality of the communication in terms of lost data or, even worse, communication drop. As a result, if the receiving device has an omnidirectional radiation pattern, it is more robust to position changes with respect to the transmitter.

Diving into another example, a transceiver's output power is $P_i = 23$ dBm. A first transmitting antenna is connected to a said transceiver with a directivity of $D = 3$ dB and a total efficiency of 50% at 900 MHz. A second receiver antenna spaced 1 km from the transmitting antenna has a $D = 3$ dB and a total efficiency of 50%. Assuming perfect polarisation alignment ($PLF = 1$), the received power is −68.5 dBm. If we repeat the exercise considering 1800 MHz, the receiver power is now −74.5 dBm. As observed, when working at higher frequencies, the received power decreases. However, it is easier to achieve larger total efficiencies at high frequencies than at low frequencies since the total efficiency of an embedded antenna in a wireless device depends on the electrical size of the device. Therefore, more range can be achieved if the total efficiency is acceptable when working at low frequencies.

Friis equation can be upgraded by adding additional loss terms due to propagation in buildings, multi-propagation, rain, etc. [21].

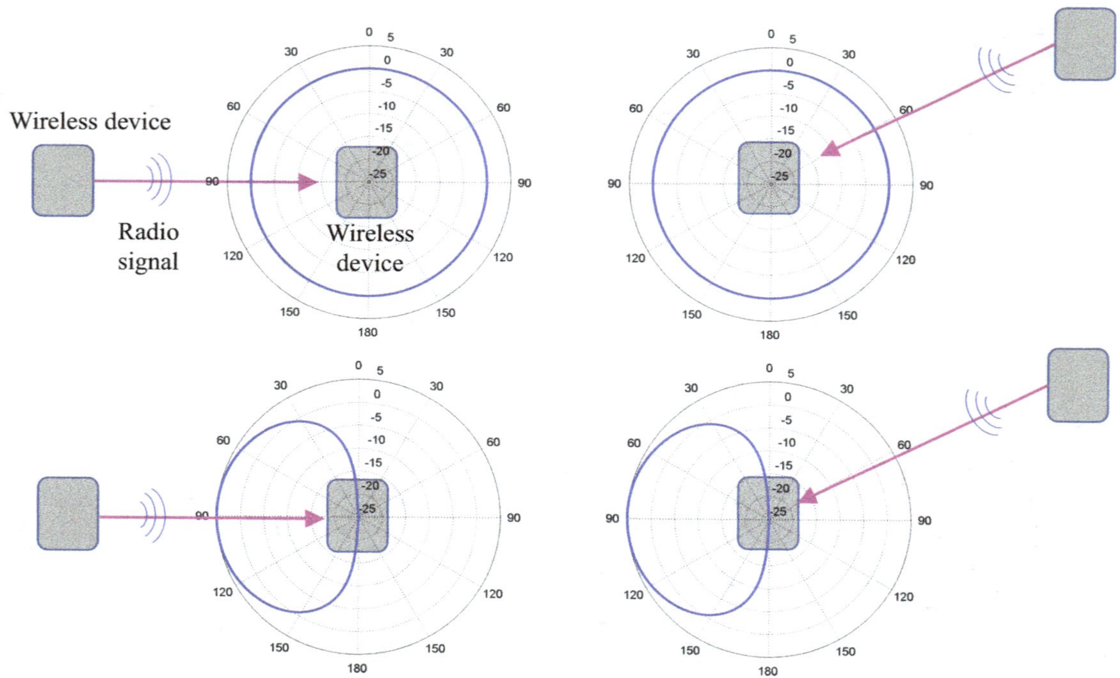

Figure 1.28 *(Top) For a device having an antenna with an omnidirectional pattern, the received signal has the same strength independently of the angle of arrival of the incoming radio signal; (bottom) when the antenna is directional, it is stronger in some directions while weak in others. Note that the radiation pattern represents realised gain where* $G = 0$ *dB is for the omnidirectional pattern and* $G = 5$ *dB is for the directional pattern. However, in other directions off maximum, the gain is weaker, and therefore, the received signal becomes very weak for the directional pattern case.*

High gain
High directive/gain radiation patterns imply strong transmitted/received signals in some directions but weaker in others. This is preferred in point-to-point radio links where the position of the wireless devices is known.

1.4.8 Impact of antenna parameters on a device performance

It is also interesting to analyse the range using the Friis equation. Let S be the sensitivity of a transceiver embedded in a wireless device. If the received power is less than sensitivity S, the quality of communication is poor or not tolerable (Section 1.2.12). From Friis Equation (1.24):

$$r \leq \frac{\lambda}{4\pi}\sqrt{\frac{P_i \cdot G_1 \cdot G_2 \cdot PLF}{S}} \quad r \leq \frac{\lambda}{4\pi}\sqrt{\frac{P_i \cdot D_1 \cdot \eta_{t1} \cdot D_2 \cdot \eta_{t2} \cdot PLF}{S}} \tag{1.25}$$

For example, if the output power of a transceiver at 900 MHz is $P_i = 23$ dBm, the directivities of both the transmitting and receiving antennas $D_1 = D_2 = 3$ dB, total efficiencies $\eta_{t1} = \eta_{t2} = 50\%$, $PLF = 1$, and the sensitivity $S = -100$ dBm, then the maximum r or range to guarantee that received power $P_L > S$ is $r_{\text{max}} \approx 37$ km. This ideal situation has no obstacles and no reflection on the ground. In this sense, Friis can be modified, including such effects, giving more realistic numbers [21]. However, as it is, it reflects the impact on the range of every single antenna parameter and frequency of operation. For instance, improving the total efficiency of the receiver η_{t2} two times increases the range by 41%.

It is essential to outline that directivity D, either of the antenna at the transmitter or the receiver, increases range. However, this increment in range in one direction decreases the range in other directions, as discussed in Figure 1.28. As a result, antenna radiation patterns should be quasi-isotropic ($D \sim 0$ dB) to obtain range for any direction.

As far as the battery life is concerned, total efficiency has a direct impact. According to [22], around 57% of the power consumption comes from the radio-communication module. If the signal strength is weak, the transmitted power of the module should be increased, resulting in high energy drainage. Thus, an efficient antenna at transmission frequencies will maximise battery life. In simple numbers, if the total efficiency of an antenna decreases by 3 dB (two times), the battery life will reduce a bit less than two times.

dB, dBi, dBW, dBm
dB refers to a ratio between two magnitudes in terms of power (Figure 1.29). For example, the gain G of an amplifier is the ratio of output power over the input power. In the dB scale, such gain is expressed as $G_{\text{dB}} = 10{\cdot}\log G$.

Figure 1.29 Alexander Graham Bell (1847–1992) was a Scottish inventor. The term dB, or decibel, is bestowed to his honour [23].

S_{11} is the ratio between reflected voltage and incoming voltage. Therefore, $S_{11}\text{dB} = 20\cdot\log|S_{11}|$.

Total efficiency η_t is the ratio of the radiated power over the incoming power to an antenna. Hence, $\eta_{t\text{dB}} = 10\cdot\log\eta_t$. For example, $\eta_t = 50\%$ equals −3 dB, meaning that the radiated power is 3 dB less than the incoming power to an antenna.

Gain or directivity of antennas is usually given in dB: $D_{\text{dB}} = 10\cdot\log D$. For example, a linear half-wavelength dipole $D = 1.6$ then $D_{\text{dB}} = 2.1$ dB. Sometimes also referred to as 2.1 dBi since it is compared to an isotropic radiator.

When dealing with absolute power, dBW or dBm are used, where both are magnitude power, the first referred to as W and the second to mW. $P_{\text{dBW}} = 10\cdot\log P$ and $P_{\text{dBm}} = 10\cdot\log(P/1\text{ mW})$. For example, 1 W is 0 dBW and 1 mW equals −30 dBW or 0 dBm, and 1 nW is −90 dBW or −60 dBm.

Note that dB can be added to dBW or dBm but dBW or dBm cannot be added between them. For example: if $P_i = 1$ mW enters an amplifier of $G =$ 10 dB gain, the output is $P_o = P_i\cdot G$. In dB, equals to $10\cdot\log(P_o) = 10\cdot\log(P_i\cdot G)$. Owing to log product properties, this results in $P_o(\text{dBW}) = P_i(\text{dBW}) + G(\text{dB})$ or $P_o(\text{dBm}) = P_i(\text{dBm}) + G(\text{dB})$. However, if we have two signals with 1 W, which in dB results in 0 dBW, the result must first be added in a linear scale and then converted to a log scale.

1.4.9 Frequency scaling

In the antenna and microwave/wireless engineering world, components are not big or small compared to a physical size but to their relation to the wavelength. For example, a half wavelength ($\lambda/2$) dipole antenna at 900 MHz is 16.6 cm long, and another $\lambda/2$ dipole antenna at 10 GHz is only 1.5 cm. Is the second dipole antenna small? Is the second dipole antenna smaller than the first dipole antenna? No, definitely not. Both antennas are half-wavelength dipoles; thus, neither the first nor the second are small ($L \ll \lambda$). Besides, both provide the same performance since both antennas are half-wavelength dipoles. Most parameters remain the same, such as directivity, bandwidth, and polarisation. However, efficiency may be different since losses are frequency-dependent. It is clear that the second antenna is physically smaller than the first antenna. Still, no wireless component should be compared in terms of physical size but in terms of electrical size. Consider another example. A ground plane for a wireless device is 120 mm × 120 mm with an antenna operating at 900 MHz. A second ground plane is 45 mm × 45 mm with an antenna running at 2400 MHz. Is the second ground plane smaller than the 120 mm one? No, it is not. To answer the question, we should calculate the size of the ground planes in terms of the wavelength as follows: 120 mm = $x_1 \cdot \lambda$ at 900 MHz, resulting in $x_1 = 0.36$, that is, 120 mm is 0.36λ at 900 MHz. Repeating the same process for the 45 mm × 45 mm ground plane at 2400 MHz results in the same 0.36λ.

Frequency scaling

A wireless device operating at a frequency f_1 can be scaled to another frequency f_2, keeping the same performance (bandwidth, radiation patterns, polarisation, directivity, gain) if dimensions are multiplied by f_1/f_2. Efficiency may change since losses in the conductor and substrates of antennas are frequency-dependent.

1.4.10 How small is small?

As discussed in Section 1.4.9, *small* should always be compared to the operating wavelength. An antenna, a radiating system (antenna including the ground plane), or a microwave component is small when its size is small compared to the wavelength. According to IEEE, an electrically small antenna is "an antenna whose dimensions are such that it can be contained within a sphere whose diameter is small compared to a wavelength at the frequency of operation" [7]. But how small is small? A generally accepted definition is "an antenna whose maximum dimension $2a$ is less than $\lambda/2\pi$, referred to as radiansphere" [24–28] (Figure 1.30).

Since the wavenumber $k = 2\pi/\lambda$, a small antenna satisfies $ka < 1$, which is equivalent to:

$$a < \frac{300}{2\pi \cdot f(\text{MHz})} \tag{1.26}$$

If we approximate the size of the wireless device L by $2a$, then a small wireless device satisfies:

$$L < \frac{300}{\pi \cdot f(\text{MHz})} \quad L < \frac{\lambda}{\pi} \tag{1.27}$$

When an antenna system becomes electrically smaller, its bandwidth decreases. This bandwidth reduction of small antennas has been first studied primarily by Chu [26] and Wheeler [27]. Chu derived an expression relating the antenna quality factor with the space the antenna fills: the radiansphere (Figure 1.30). Chu established a fundamental lossless limitation given by Equation (1.28), where λ is the operating wavelength, and a is the radius of the radiansphere. The diameter ($2a$) of the radiansphere is the largest antenna dimension, including the ground plane.

$$Q_{\min} = \eta_r \cdot \left(\frac{1}{(ka)^3} + \frac{1}{ka}\right) \tag{1.28}$$

Since bandwidth (BW) is a more practical figure of merit than Q, Equation (1.28) is transformed into:

$$\begin{aligned} BW_{\max} &= \frac{1}{Q_{\min}} \cdot \frac{SWR - 1}{\sqrt{SWR}} \\ BW_{\max} &= \frac{1}{\eta_r} \cdot \left(\frac{1}{(ka)^3} + \frac{1}{ka}\right)^{-1} \cdot \frac{SWR - 1}{\sqrt{SWR}} \end{aligned} \tag{1.29}$$

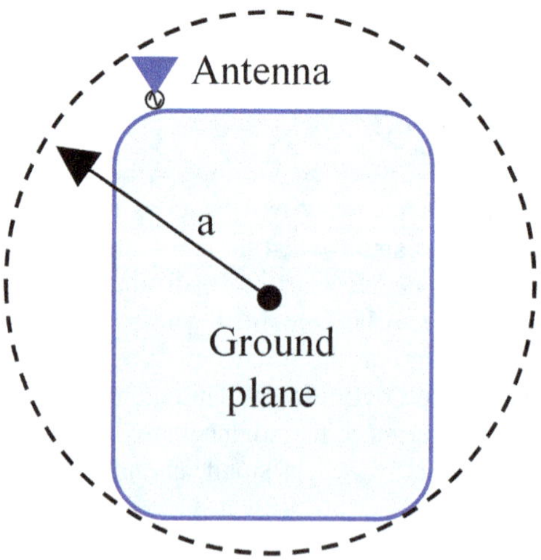

Figure 1.30 a is the radius of the smallest enclosing sphere, including the radiating structure (antenna and the ground plane)

where η_r is the radiation efficiency, and SWR is the reference of matching impedance at which bandwidth is computed. As observed, bandwidth can be increased if losses are increased. This is not a recommended route since it impacts range, consumes more power, and drains the battery of the wireless device faster.

The main conclusion derived from Equation (1.29) is that bandwidth is proportional to electrical size: the smaller the electrical size, the smaller the bandwidth (Figure 1.31). For example, given an antenna on a 120 mm × 60 mm ground plane and the same antenna on a 60 mm × 60 mm ground plane, both operating at 900 MHz, the bandwidth is larger for the first case since the electrical size is larger (BW = 20.0% and BW = 4.4% for the 120 mm × 60 mm and 60 mm × 60 mm cases, respectively. Note: an antenna booster has been considered).

Equation (1.29) is a fundamental limit that can be reached effectively using the radiansphere. In the literature, there are many different approaches to reaching the said limit [29]. In the present book, such a limit is difficult to achieve since antenna systems are 2D structures containing low-profile antenna elements and ground planes. Therefore, the Chu limit is helpful in the book as a reference to understand why obtaining bandwidth is more challenging when the antenna structure is small in terms of wavelength. In Chapter 4, the fundamental limit is revisited when considering broadband matching techniques that are able to increase the inherent antenna bandwidth.

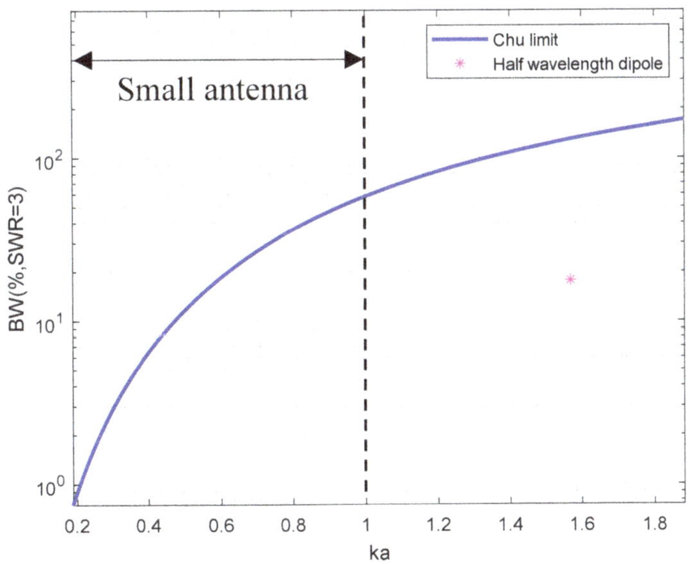

Figure 1.31 Bandwidth limit for a lossless antenna (η_r = 100%) using Equation (1.29)

Bandwidth and efficiency as a function of antenna system size
Bandwidth and total efficiency depend on the electrical size of the antenna system (the ground plane and the antenna element). The larger the electrical size, the larger the bandwidth and total efficiency and vice versa.

Besides bandwidth dependence on the antenna system volume, radiation efficiency η_r also depends on such an electrical volume. Using basic antenna theory, the radiation resistance R_r of a small linear dipole of length $2a$ is [30]:

$$R_r = 320 \cdot \pi^2 \cdot \left(\frac{a}{\lambda}\right)^2 \tag{1.30}$$

As shown, a decrease in a relative to the wavelength results in a decrease in R_r. According to Equation (1.30), radiation efficiency η_r becomes smaller as the electrical size decreases since ohmic losses R_Ω as a linear dependency on electrical size. With Equation (1.12), and assuming $a \ll \lambda$, radiation efficiency depends on $\frac{a}{\lambda}$. Therefore, with $a \ll \lambda$, Equation (1.29) is proportional to $\left(\frac{a}{\lambda}\right)^2$. If the size a is comparable to or larger than λ, the radiation efficiency is close to 100%. Once again, employing Equation (1.29) leads to the same conclusion: bandwidth depends on the antenna size. However, in this case, with a linear dependency $\propto \frac{a}{\lambda}$.

When considering antennas embedded in wireless devices, the smaller the size of the antenna system (antenna and the ground plane), the smaller the efficiency. However, this should not lead to the conclusion that more bandwidth can be achieved since radiation efficiency appears in the denominator in Equation (1.29). The explanation is that volume has more impact for the above reason, so the bandwidth decreases with the electrical size.

For example, at 600 MHz, the computed radiation efficiencies of an antenna element on a 120 mm × 60 mm and 40 mm × 40 mm ground plane are 88% and 53%, respectively. Computing the bandwidth for each case results in 2.8% (120 mm × 60 mm) and 0.75% (40 mm × 40 mm), showing that the one with a larger electrical size has more bandwidth.

Radiation efficiency can increase bandwidth for a given antenna when adding losses. For example, if one adds losses due to a human body, bandwidth may increase, but naturally, this is not a good feature since the total efficiency of the antenna system decreases (Figure 4.17, Chapter 4). Let us use a simple example. Consider an antenna whose impedance is modelled by a series RLC circuit where $R = 25\ \Omega$, $L = 8$ nH, and $C = 1$ pF. Using Equation (1.3), the antenna impedance is $Z_a = R + j\omega L - j/(\omega C)$. Besides, we assume a lossless antenna, where $R_r = R$ and $R_\Omega = 0$. In this situation, using two lossless components for the matching network, the bandwidth is BW ($SWR \leq 3$) = 28.6% with a total efficiency of 100%. If instead of the said matching network, we use a series resistance of 25 Ω, the $BW = 64.0\%$. In this second case, the radiation efficiency has decreased from 100% to 50% because half of the power delivered to the antenna system is dissipated in the added 25 Ω. Using Equation (1.29), the bandwidth is increased by $1/\eta_r$, resulting in two times when radiation efficiency is reduced from 100% to 50%. This theoretical value agrees with the calculations using the circuit model. In this case, bandwidth has increased, but with the negative impact of drastically reducing the losses. In the following chapters, a mechanism to enhance the inherent bandwidth of an antenna booster system will be proposed with broadband matching networks comprising inductors and capacitors, which will improve bandwidth while introducing few losses since the quality factor of the components will be high compared to the quality factor of the antenna system.

Finally, radiation patterns become almost the same for those frequencies where the antenna system (ground plane and antenna) is small in terms of wavelength. For example, a linear dipole has almost the same radiation pattern with a directivity of $D = 1.7$ dBi for any frequency where the dipole is smaller than half of the wavelength. In this regard, the antenna booster systems analysed in this book feature similar directivities across the whole frequency range with radiation patterns able to transmit or receive a signal with practically the same intensity along any direction. More details on radiation patterns will be given in Chapters 4 and 5.

1.5 Simulation tools

When designing a wireless device, two domains are involved. First, the virtual domain where the embedded antenna in a wireless device is simulated. And second, the physical domain where those simulated parameters are validated. The advantage of operating in the virtual world is that many experiments can be carried out until the performance is streamlined. This simplifies the fabrication of prototypes, making the design process cost- and time-effective.

For the virtual domain, two types of simulation tools are discussed. First, 3D electromagnetic simulators. Second, circuit simulators. 3D electromagnetic simulators allow us to obtain the antenna parameters (Section 1.4), given the 3D structure of a wireless device, including its constitutive materials. For example, a simulator of this category can simulate an entire wireless device, including the antenna, the PCB, the battery, covers, and even the effect of a human body in the case of wearable devices.

Why is a circuit simulator needed if a 3D electromagnetic simulator can simulate antenna parameters such as those described in Section 1.4 (S_{11}, total efficiency, radiation patterns, etc.)? The reason is that a circuit simulator can take the circuital information of the 3D electromagnetic simulator to make other procedures. One of those procedures is used in this book. A 3D electromagnetic simulator obtains the S_{11} of a wireless device embedding an antenna booster. With such S_{11}, a circuit simulator can synthesise a matching network including real components such as capacitors, inductors, transmission lines, and active devices such as switches and digitally tunable capacitors. Once the matching network is obtained, it can be embedded in the 3D simulator to simulate the wireless device with the components included in the matching network to get the total efficiency.

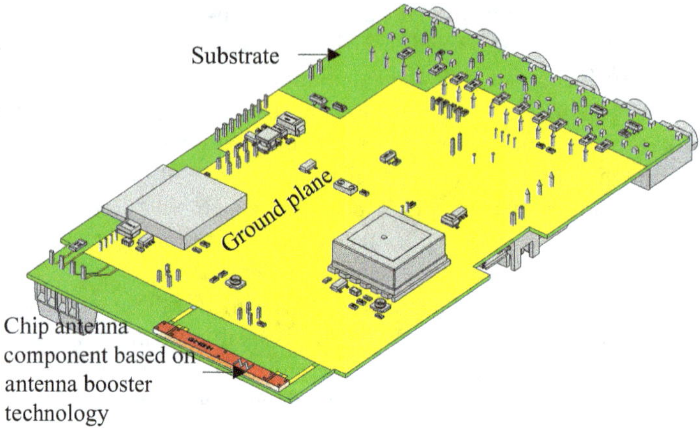

Figure 1.32 Electromagnetic simulators can obtain the antenna impedance, radiation and total efficiency, directivity, gain, polarisation, and radiation patterns considering the antenna embedded in a wireless device [34]

1.5.1 3D electromagnetic simulators

To obtain the input impedance, the S_{11}, the radiation efficiency, radiation patterns, directivity, gain, and polarisation discussed in the previous sections, the antenna embedded in a wireless device should be simulated using a full-wave simulator, also known as an electromagnetic simulator. Said simulator numerically solves the so-called Maxwell equations given a wireless system considering its geometry and constitutive materials [31]. Electromagnetic simulators solve the Maxwell equations by discretising the equations. Such discretisation involves intensive numerical computation that implies iterative methods such as the FDTD (Finite Difference Time Domain) [32] or matrix methods like MoM (Method of Moments) [33].

EM simulators can simulate not only an antenna on a ground plane but also much more complex scenarios considering the entire device with the casing, battery, and other components (Figure 1.32). Besides, more sophisticated environments, for example, the human body, can also be considered in the case of wearable devices.

1.5.2 Circuit simulators

Circuit simulators mainly work with impedance data (S-parameters), requiring simpler calculations than EM simulators solving Maxwell equations. In essence, a circuit simulator, as in the present book, helps treat the impedance of an embedded antenna on a ground plane to design matching networks for operation at several frequencies. Therefore, circuit simulators deal with impedances of antennas, matching networks either with lumped components or distributed elements, filters, amplifiers, and other circuit components (Figures 1.33 and 1.34).

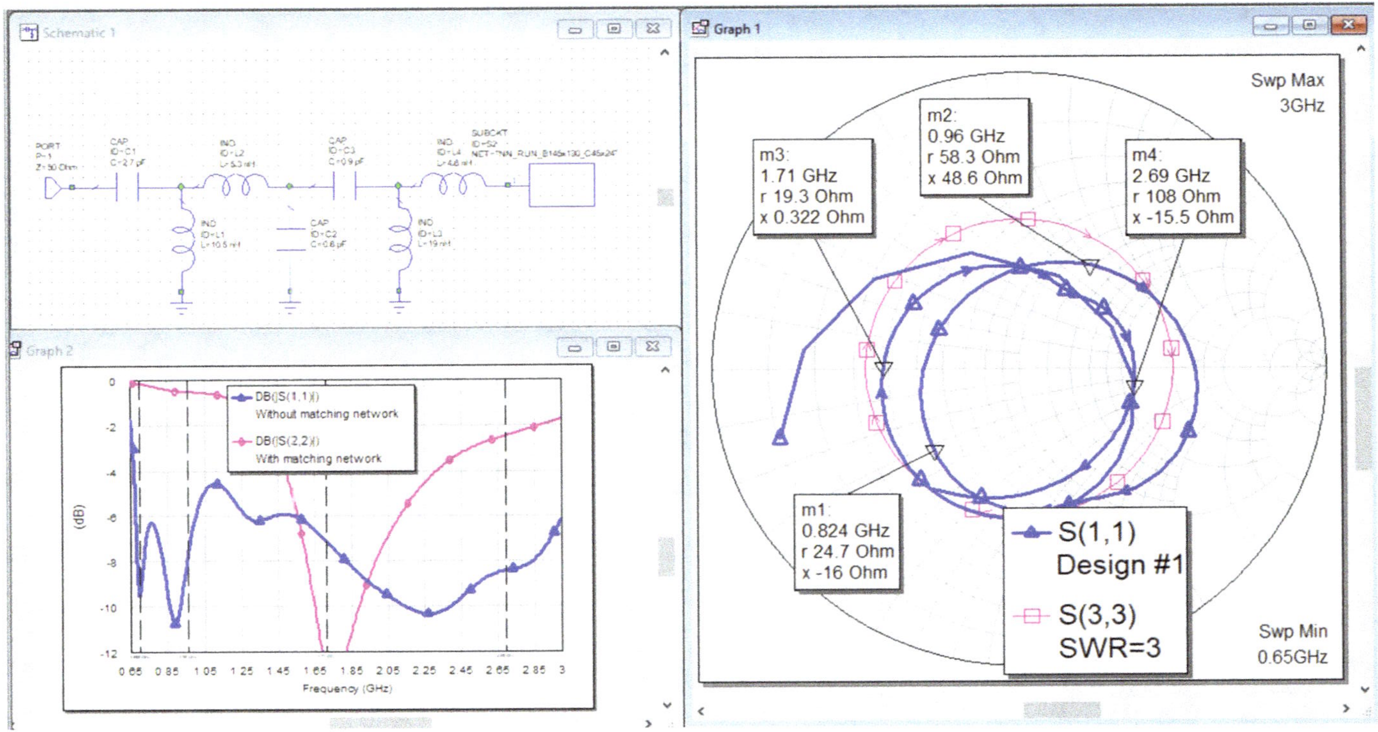

Figure 1.33 Circuit simulators allow optimising the S_{11} of an antenna, including a matching network (example using Microwave Office) [35]

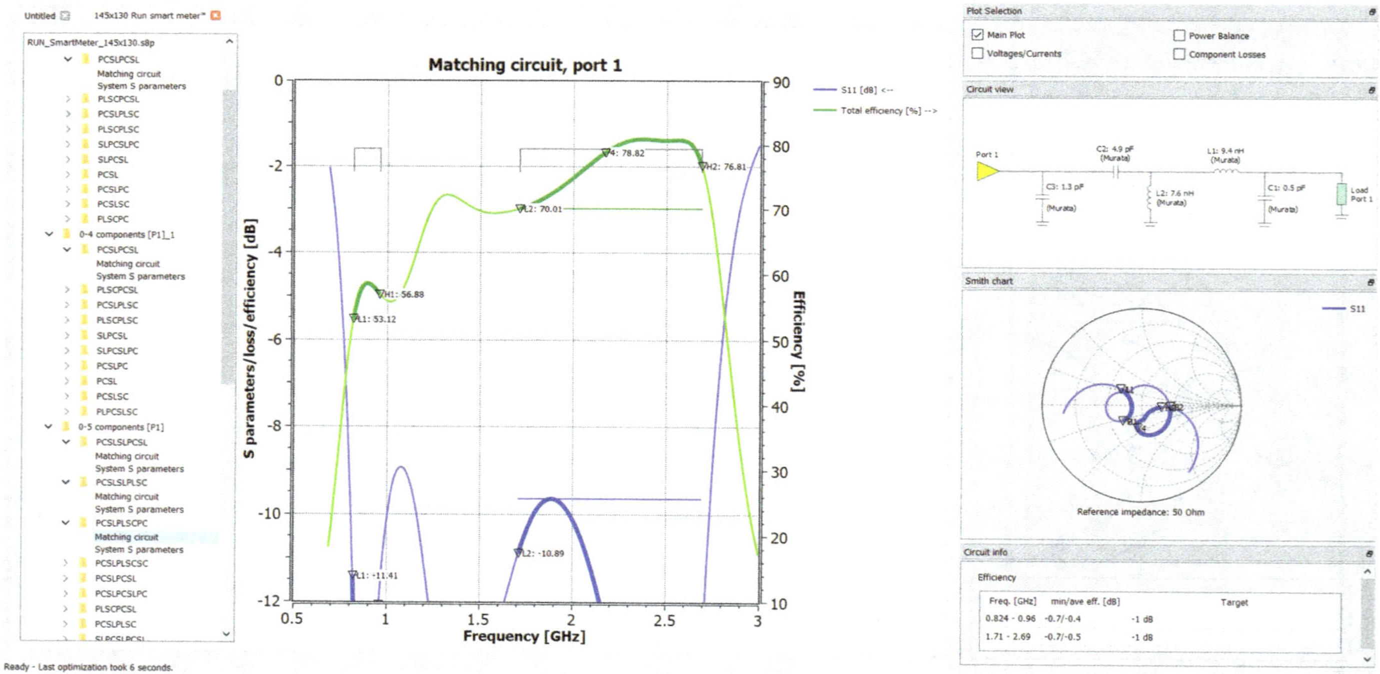

Figure 1.34 Circuit simulator allows matching network synthesis given the impedance of an antenna booster (example using Optenni Lab) [36]

Circuit simulators may also consume computation for optimisation or synthesis problems, as widely employed in the book to synthesise matching networks automatically. For example, given an antenna impedance Z_a as a function of frequency, a circuit simulator can synthesise a matching network to match Z_a at several frequency bands under some given targets, such as S_{11} or total efficiency across the band (Figures 1.33 and 1.34). Let us suppose a matching network with seven components (Figure 1.33) combining inductors and capacitors, where each may vary between a minimum and maximum value. The number of possible combinations is extremely large (~trillions), and thus, it is prohibitive to use brute force to test each candidate. In this case, optimisation algorithms find a solution by testing much less than the whole space of combinations. If there is no solution satisfying the targets, the simulator provides solutions that are closer to the target specifications. However, these computation problems are less computationally hungry than full-wave simulations.

To further connect a 3D electromagnetic simulator with a circuit simulator, let us imagine the following example: design an antenna for operating at 900 MHz and 1800 MHz with a 5% bandwidth at each band. For the sake of discussion, let us imagine the linear monopole antenna as shown in Figure 1.4. First, a 3D electromagnetic simulator is needed to model the ground plane and the vertical monopole, including the electrical properties of the materials, such as copper conductors for the ground plane and monopole. Also, we can consider in the simulation that the conductive ground plane is printed over a thin layer of dielectric substrate with a certain permittivity and loss tangent (Section 1.2.5). Finally, a radiofrequency generator is placed between the ground plane and the monopole to excite the antenna over the frequency range of interest, for example, from 700 MHz up to 2 GHz. Once the simulation is finished, S_{11}, bandwidth, radiation efficiency, total efficiency, polarisation, directivity, gain, realised gain, and radiation patterns can be obtained. However, what if the antenna is not correctly tuned to operate at 900 MHz and 1800 MHz? One possible route is to change the antenna geometry to achieve the desired response. This requires antenna engineering skills on how to properly create an antenna to match the target specification. Another possible route is to tune the antenna not by changing the geometry but by including a matching network. In this situation, a circuit simulator enters the scene. The impedance information (S_{11}), which is basically a vector including the S_{11} as a function of frequency, is transferred to the circuit simulator. The circuit simulator is now able to run the matching network design by either using human engineering skills or synthesis procedures to finally obtain the matching network design. Once the matching network is obtained, it is included in the 3D electromagnetic simulator to compute radiation parameters like radiation patterns.

1.6 Measurement systems

Once the simulation phase is closed, a prototype is built to validate the simulated data. The same simulated parameters are now tested in the physical domain. The most typical instrumentation is a vector network analyser to test the S-parameters of the embedded antenna in the wireless device. Afterwards, measuring the most relevant radiation parameters, such as total efficiency and radiation patterns, is convenient. The total efficiency can be tested either in an anechoic or a reverberation chamber, whereas radiation patterns are tested in an anechoic chamber. These measurements are passive since the radiofrequency signal is introduced from an external source. Finally, a radio communicator becomes an essential tool to test TRP and TIS to validate the device under network conditions. These measurements are active tests, as the radiofrequency signal is emitted by the transceiver within the wireless device, eliminating the need for external transmission lines to be connected to the device.

In some particular scenarios, such as wearable devices, it is required to test the wireless device on a human/animal/thing body. In this regard, phantoms are also discussed next.

1.6.1 Vector network analyser

A vector network analyser (VNA) is a microwave measurement system that measures the S_{11} or SWR of an antenna (Figure 1.35). Such measurements are done by injecting a radiofrequency signal into the antenna and measuring the reflected

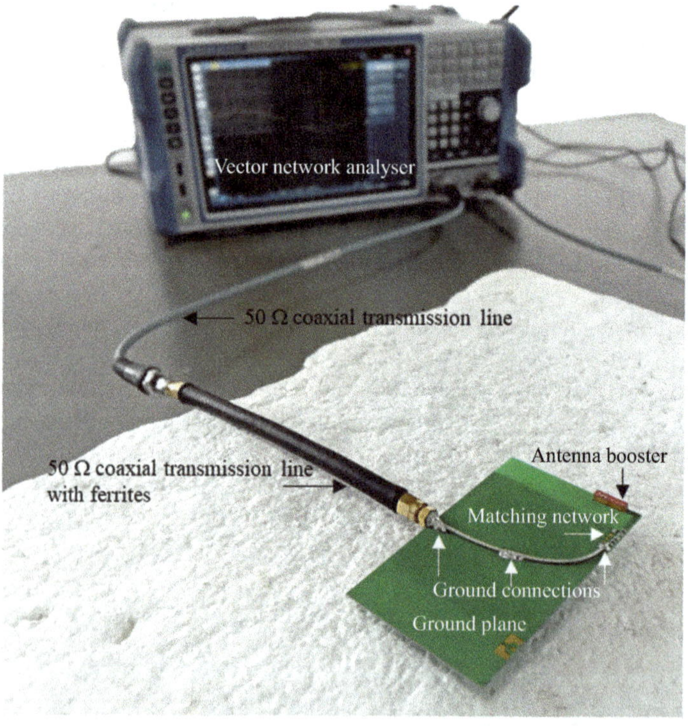

Figure 1.35 A vector network analyser for S_{11} testing

signal [37]. A VNA typically has two ports; thus, besides being able to measure S_{11}, coupling between two antennas can be tested by measuring the S_{21}. Port 2 is then able to measure the S_{22}. The measurement of the S-parameters requires attaching a transmission line (typically 50 Ω of characteristic impedance) to the wireless device. This allows testing the S_{11}, for example, including an antenna element on a PCB, considering all components such as the battery, chips, covers of the device, etc. Once the S-parameters are acceptable (e.g., $S_{11} \leq -6$ dB), the next step is to test radiation patterns and total efficiency with an anechoic chamber.

To test both S_{11} and radiation parameters such as patterns and efficiency, a 50 Ω transmission line (e.g., micro-coaxial) needs to be integrated. One end feeds the antenna or the matching network first if the antenna includes a matching circuit. The other end of the transmission line typically ends with an SMA connector. Such a position is chosen so as not to affect the impedance of the antenna system. A practical tip is to place it at the centre of the edge of the ground plane. Moreover, to further minimise the impact, ferrites may also be included in a portion of the transmission line reaching the SMA (Figure 1.35).

1.6.2 Anechoic chamber

Anechoic comes from the Greek meaning "without echo". Therefore, an anechoic chamber is a room with no echo. No echo means that the electromagnetic waves radiated by the device under test, or device under test (DUT), do not reflect back when they reach the walls of the chamber. To achieve no echo, the chamber is included with pyramidal absorbers (Figure 1.36). This avoids any reflected wave back to the

Figure 1.36 An anechoic chamber where the DUT is placed on a rotatory platform in the middle of the gate. Several receiving antennas along the gate electronically sample the power radiated by the DUT [38]

receiving antenna or antennas in charge of measuring the transmitted wave by the DUT that may interfere with the measurement. Besides the absorbers, the full chamber is covered with a conductive sheet (Faraday cage) to impede external electromagnetic waves from entering the chamber, interfering with the measurement system.

An anechoic is needed to measure the radiation patterns, total efficiency, directivity, gain, and polarisation of an antenna. The DUT emits an electromagnetic wave into space, which is then captured by a receiver antenna. All previous parameters can be tested when the receiving antenna changes its position along a sphere surrounding the DUT. To accelerate the measuring time, several receiving antennas are located along an arch for sophisticated anechoic chambers (Figure 1.36). In the middle of the arch, the DUT is placed on a turntable that rotates while the receiving antennas along the arch measure the transmitted power by the DUT. This way, when the DUT has turned 180°, the transmitted power from the DUT throughout an entire sphere has been acquired. From this data, radiation patterns, gain, total efficiency, polarisation, etc., can be obtained.

1.6.3 Reverberation chamber

Unlike an anechoic chamber, a reverberation chamber is fully reflective and creates a field with many angles of arrival at the DUT simultaneously, i.e., many echoes but no direct signal path [39]. As the so-called mode stirrers are moved, signals will combine in different ways, and over a stirring sequence, all angles of arrival will be equally probable (Figure 1.37). Hence, parameters like total efficiency can be extracted as a direct result of a measurement sequence. However, no radiation patterns can be obtained with this approach, including those linked to the radiation pattern, such as directivity, gain, and polarisation.

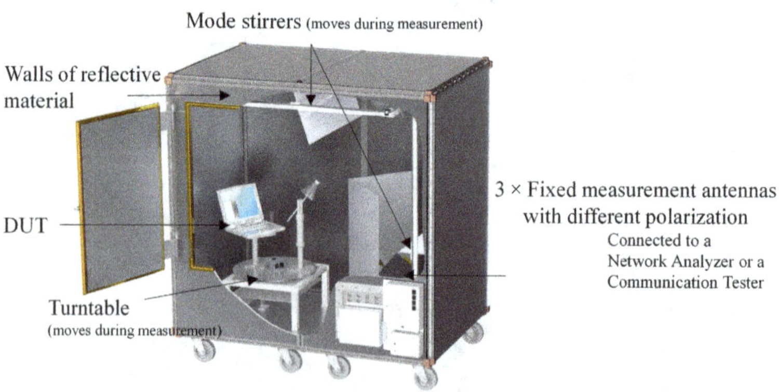

Figure 1.37 A reverberation chamber with a DUT, including an antenna on a turntable. Three antennas inside the chamber generate front waves impinging the DUT from several directions. The stirrers help to create an environment with multiple wavefronts at the DUT [39].

Vector network analyser/anechoic and reverberation chamber
To measure S_{11} a VNA is needed. However, since S_{11} does not inform if the antenna system is efficient, an anechoic or reverberation chamber is needed to measure total efficiency.

1.6.4 Radio communicator

In the development phase of wireless devices, it becomes essential to test the device using real network conditions without any transmission line, as was the case with the vector network analyser. This is the case of testing TRP and TIS. A radio communicator is a device able to monitor a wireless device to transmit a certain amount of power. Such power is then recollected to compute TRP as the ratio of the received power over the transmitted power by the front end. For TIS, the radio communicator transmits a radiofrequency signal according to the communication protocol under test in the wireless device to determine the minimum acceptable power the device can correctly process. TRP and TIS can be tested in an anechoic or reverberation chamber [39].

1.6.5 Phantoms

Wireless devices must perform well under real conditions, such as those for wearable devices. However, it is not feasible to rely on performance measurement using a human head and handgrip of a real person, as each person's head and hand would differ from one to another. A focus group was needed to investigate how to come up with a standardised model of a human head and hand. One such organisation is the CTIA – The Wireless Association [40]. A sub-group in this organisation was created to develop a set of standardised heads and hands to obtain consistent and reliable performance measurements in a controlled lab environment.

The phantom head model, called SAM (Specific Anthropomorphic Mannequin), was first introduced in 2002 (Figure 1.38, left). The material inside the plastic shell

Figure 1.38 Phantom head and hands for testing wireless devices embedding antennas in a reliable manner

has specific electrical properties, i.e., relative permittivity and conductivity, that are modelled closely to the real human head. Modelling the hand was more complicated, and it took longer to get to the final set of phantom hands (Figure 1.38, right).

In this regard, phantoms are used to ensure a stable set-up for testing wireless devices close to human or animal bodies.

1.7 Antennas for wireless devices

Antennas for wireless devices have evolved from classic half-wavelength dipoles and quarter-wavelength monopoles, transitioning from external to internal configurations. Small and multiband antennas for wireless devices can be classified as follows (Figure 1.39):

- L-monopole: a basic monopole antenna for single-band or single-frequency region operation with an L-shape to fit in the device.
- IFA (Inverted-F antenna): a monopole with a short circuit to provide impedance matching. Also for single-band or single-frequency region operation.
- Multi-branch monopole: like the L-monopole but with a complex shape to create a multiband antenna.
- PIFA (Planar Inverted-F Antenna) is the multi-branch monopole, but the antenna is on the ground plane. Usually, it has less bandwidth than the multi-branch monopole. If multiband operation is required, the antenna volume should be larger than its non-grounded versions, such as the multi-branch monopole. Another on-ground antenna is the microstrip patch antenna. A microstrip is typically loaded with a substrate between the patch and the ground plane to miniaturise the antenna. However, such loading impacts

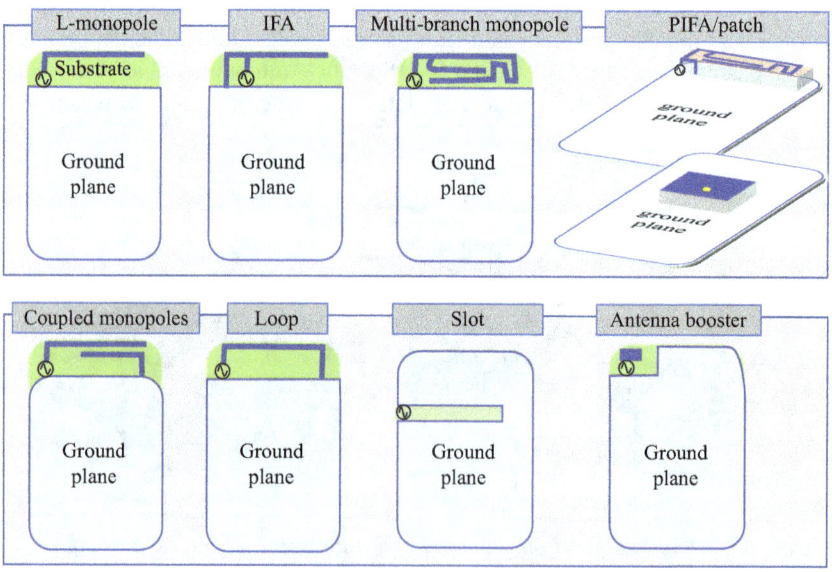

Figure 1.39 From classic antennas to Antenna Booster Technology

bandwidth and makes this antenna usable for narrow-band applications. For broadband operation, the height of the patch must be increased [41].

Having both the PIFA and the microstrip patch on a ground plane underneath makes them more robust to detuning or absorption due to nearby components such as conductive bodies or the human body, for example.

- Coupled monopoles comprise at least a parasitic element, a grounded conductive portion coupled to the driven element to enlarge bandwidth.
- Loop antenna: as the monopole, the antenna has the form of a loop where the end of the conductive strip is connected to the ground of the PCB.
- Slot antenna: a slot on the ground plane that works as an antenna. It has the advantage of being a planar structure. However, if the antenna size is large, it complicates the integration of electronics since no component can overlap the slot area at any layer of the PCB.

All these types can be implemented as printed elements on the PCB in both 2D and 3D configurations. Typically, the bandwidth of a 2D version is narrower compared to what a 3D structure can achieve [42].

All previous designs share the common feature that the frequency response is determined by the antenna geometry. Achieving a single-band antenna may require a relatively simple geometry, such as the L-monopole. However, for applications requiring multiple bands, the antenna geometry must become more complex, as seen in designs like the multi-branch monopole of PIFA. In the case of the multi-branch monopole, it typically adopts a volumetric structure to achieve sufficient bandwidth and efficiency (Figure 1.40).

Achieving impedance matching solely through antenna design can be challenging in certain situations. Consequently, integrating a matching network provides an additional degree of freedom for allocating frequency bands of operation. As observed, depending on the frequency bands of operation, a different antenna design must be utilised, making the number of possible designs huge. For example, if two devices need one antenna for single-band operation at [f_1, f_2] and another device needs operation at [f_3, f_4] and [f_5, f_6], a different antenna is needed.

A radically different approach appears with the introduction of Antenna Booster Technology [43–48]:

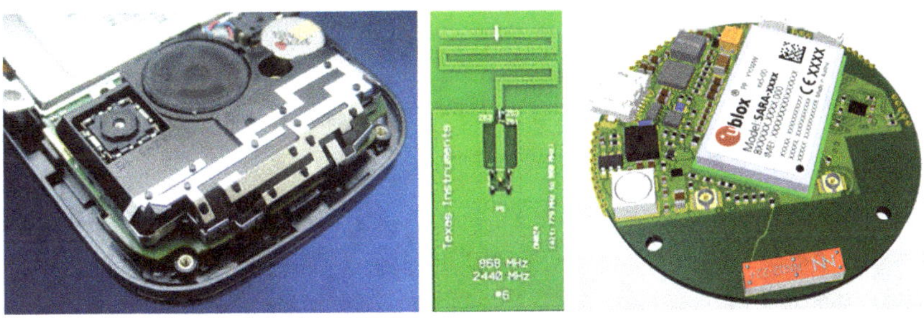

Figure 1.40 (Left) 3D multi-branch monopole; (middle) printed monopole [43]; (right) antenna booster on a device [44]. Figures not to scale.

- Antenna Booster Technology: relies on a simple conductive non-resonant element called antenna booster, tiny in size ($\sim\lambda/30$ and even less) that, when properly placed on a PCB, excites radiating currents on the ground plane, making the radiating structure as big as the wireless device. The addition of a matching network gives the flexibility to select the frequencies of operation, from single band to multiband. Therefore, the same antenna booster can be used for any frequency band of operation and any device size. An antenna booster typically has no ground underneath as the L-monopole, IFA, etc. However, an antenna booster can be on the ground plane as the PIFA/patch, where the bandwidth is less than the non-ground counterpart.

As far as manufacturability is concerned, several methods are utilised (Figure 1.41). Stamped metal comprises the shape of the metal antenna manually stamped on a plastic carrier. It requires moulding, making it only cost-competitive for very high volumes. MID (Molded Interconnect Device) and LDS (Laser Direct Structure) can afford much more complex shapes than stamped metal and allow the integration of electronics in the same device [49]. Like stamped metal, MID and LDS are suitable for high volumes. LTCC stands for low-temperature cofired ceramics. This technology allows the embedding of different types of components from SMD, integrated circuits, conductors, and vias [50].

Printed on the substrate of the device is low-cost, but due to its 2D profile, the size of the antenna should be increased [42]. In the case of multiband antennas, the geometry becomes complex and significant since it cannot benefit from a 3D volumetric antenna to gain bandwidth. FPC (Flexible Printed Circuit) follows the principle of a printed antenna but on a different substrate, usually very thin, which

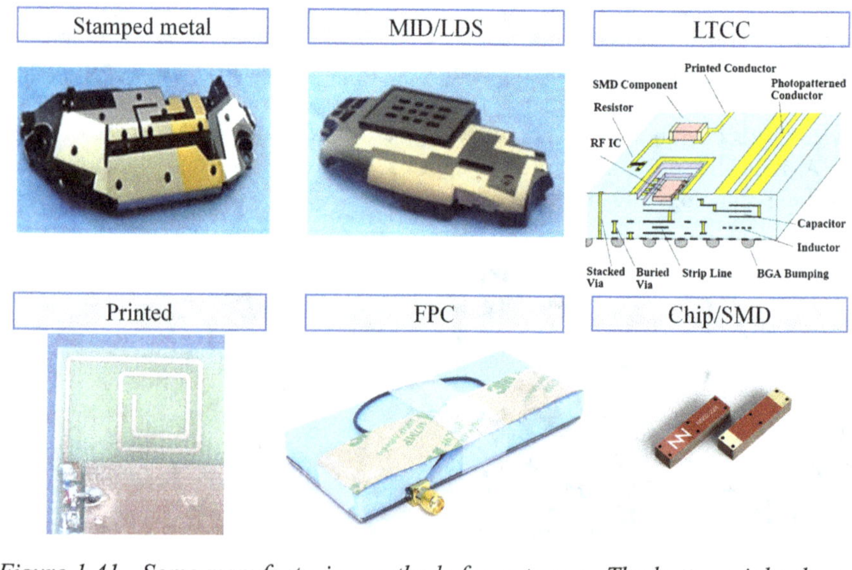

Figure 1.41 Some manufacturing methods for antennas. The bottom right shows an antenna booster product (12 mm × 3 mm × 2.4 mm) [38].

can be assembled to the device through a micro-coaxial transmission line. Due to the slim profile, the antenna size is big for multiband antennas. Also, the performance is dependent on the position of the antenna on the PCB, making its performance unpredictable. SMD, or chip antennas, are off-the-shelf antennas using low-cost substrates such as epoxy glass, where conductive materials can be printed on the top and bottom layers. Also, a ceramic substrate allows the printing of conducting shapes at all faces but is less cost-competitive than glass epoxy materials such as FR4 (Table 1.2) [51].

SMD antennas can be efficiently assembled onto the PCB using a pick-and-place machine, ensuring consistent repeatability. Antenna booster products shown in the book are within this chip category.

1.8 Classic antenna approach versus Antenna Booster Technology

To design small and multiband antennas, the classic approach copes with a sophisticated and artisan design of the antenna geometry. As a result, a diverse range of techniques exists to attain design flexibility. These methods encompass shaping the geometry of an antenna alongside incorporating supplementary elements such as short circuits and parasitic elements (Figure 1.39). Consequently, designers need to tackle the project with a high level of expertise in antenna engineering, making the process difficult for other radiofrequency experts. On the other hand, once the antenna geometry has been designed to meet the frequency requirements, it is hard to reuse the same design for other frequency bands. Moreover, when the antenna needs to be electrically small (less than the basic $\lambda/4$ monopole, as shown in Figure 1.4), the length of the conductor that makes up the antenna must increase. For example, let us imagine an $H_0 = \lambda/4$ monopole, such as the one illustrated in Figure 1.42. Now, if we want to miniaturise the antenna by a factor of 3.4 [52], that is, to reduce the size of the antenna by 3.4 ($H_1 = H_0/3.4$) while providing the same frequency of operation, the length of the conductor should be increased five times H_0, that is, when making the antenna smaller, the

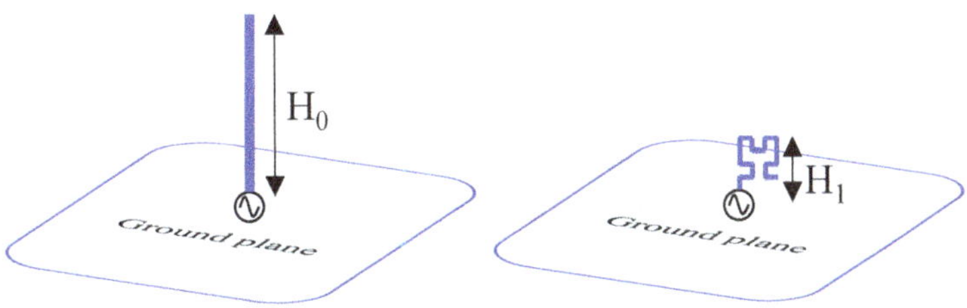

Figure 1.42 A straight monopole antenna with length H_0 and a monopole antenna smaller in size both have the same resonant frequency. The length of the conductive wire of the small antenna is five times larger than the length of the straight monopole for $H_1 = H_0/3.4$ [52].

length of the antenna should be increased. This has an impact on the radiation efficiency since longer lengths result in an increase of the R_Ω.

As observed, making an antenna small is not just bending a linear conductor into a small area but also needs to have more conductor trace than the unfolded version. This results in more losses since more conductor is required and since the conductive wire needs to be packed into a small space, the width of the conductor must be reduced, increasing losses as well. Finally, although the resonant frequencies of antennas H_1 and H_0 are the same, the input impedance of H_1 will differ from that of H_0, and thus, efforts to match the antenna are necessary. As observed from this discussion, making an antenna small is a challenging task. If we now add that the antenna needs to be multiband, the design effort increases vastly.

In effect, if more frequency bands are needed, the design of the antenna is sophisticated (Figure 1.43) by either adding more conductive branches to the antenna or parasitic elements and even short circuits. In the end, how do we know that a particular geometry is the best for the design under consideration?

Antenna Booster Technology is a different approach to designing radiating systems and, more importantly, developing wireless devices that need an antenna element. First, an antenna booster is a non-resonant element at most frequencies of operation using a very simple and electrically small conductive structure such as a conductive parallelepiped. Then, thanks to a matching network, the designer can freely tune the system at any frequency of operation, from single band to multiband. For example, a simple matching network with less than three lumped components may be enough to provide a single-band operation. For multiband operation, matching networks with more components are needed. In all situations, the same antenna booster may be used. This means that a single antenna booster can manage any frequency band of operation.

If the ground plane of the wireless device changes in size, it impacts the antenna impedance. With Antenna Booster Technology, the detuning is again compensated by changing the matching networks, which is simpler and faster than designing a new antenna geometry from scratch, as occurs following the classic antenna approach. On top of that, the size of an antenna booster is typically ten times smaller than conventional self-resonant antennas [53], featuring sizes less than $\lambda/20$.

While the design of classic small and multiband antennas based on complex geometry is affordable by antenna experts, the design of wireless devices embedding antenna boosters is easily achievable by the whole community of microwave or radiofrequency experts dealing with their daily activities with matching networks for amplifiers, filters, and the like. Besides, the matching network design for an antenna booster is carried out with software. In effect, the design of those matching networks is easily and rapidly obtained by matching network synthesis using optimisation algorithms available in most microwave circuit simulators. Chapters 4 and 5 illustrate how easy and fast the design of either single-band or multiband antenna systems with antenna boosters is.

It is essential to revisit the scalability principle introduced in Section 1.4.9. Let us take the multi-branch design for multiband operation (Figure 1.43). If the

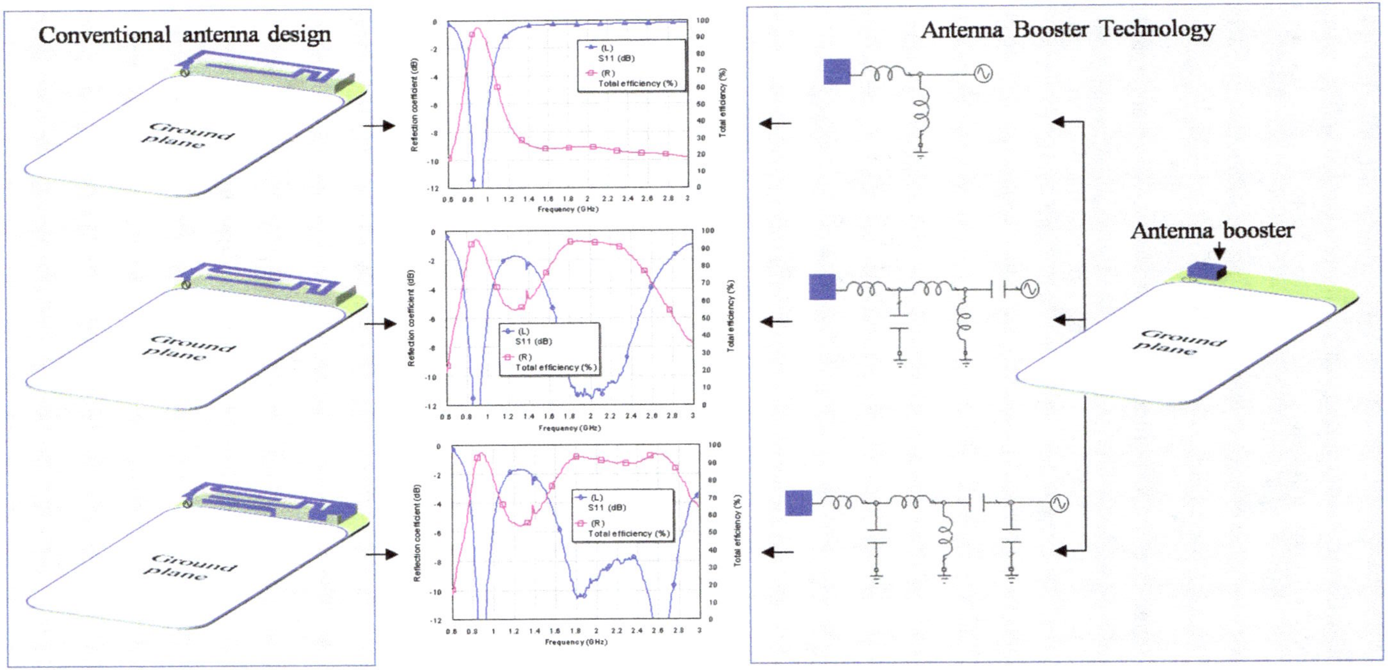

Figure 1.43 From classical antenna design to Antenna Booster Technology

frequency of operation for device #1 is 824–960 MHz and 1710–2690 MHz and for device #2 is 2.4–2.5 GHz and 5.9–7.2 GHz, the multiband multi-branch monopole will be different owing to the scalability principle of resonant antennas. However, for the antenna booster, there is no need to change the antenna booster but only the design of a matching network, which is more straightforward, faster, and more cost-effective than designing an antenna from scratch. Also, multi-branch monopoles for operation at devices #1 and #2 cannot be reused for future devices since the frequency bands must be readjusted. Nevertheless, antenna booster technology is effective across present and future protocols, as the operating frequency is determined not by the booster itself, but by the design of the matching network. Therefore, an antenna booster is ready to adopt any future protocol from day zero since the combination of an antenna booster and an appropriate matching network is flexible enough to operate at any new frequency band to appear in the future.

The essential concepts of Antenna Booster Technology

- The architecture comprises three main actors: an antenna booster, a matching network, and a ground plane.
- An antenna booster is a non-resonant element, in particular at the lowest frequency of operation.
- It is tiny in size: $\sim\lambda/30$, typically less than $\lambda/20$. For example, an antenna booster at 900 MHz can be as small as 12 mm and even smaller. Compared to self-resonant antennas, an antenna booster is typically ten times smaller.
- Its position on the ground plane of a wireless device enables the excitation of radiating current in the ground plane. Normally, the corner provides better bandwidth and efficiency. This makes the ground plane the main radiating plane, allowing antenna boosters to be very small. As such, an antenna booster is not considered an antenna, but an excitation means for the ground plane of a wireless device to become the antenna.
- An antenna booster can be an off-the-shelf SMD component, simplifying its integration into a PCB thanks to a pick-and-place assembly process. However, antenna boosters can adopt any other form, including being printed on the PCB, on a flexible substrate, etc. However, 3D SMD parts are preferred to obtain more bandwidth and efficiency.
- The frequency of operation is easily adjusted by a matching network. The design of the matching network allows single to multiband operation. Therefore, an antenna booster can be reused for any frequency band of operation and any PCB size since only the matching networks need to be designed.
- Matching networks for antenna boosters can be automatically designed by matching network synthesis software tools, simplifying it not only for

antenna experts but across a huge variety of engineers: wireless, radio-frequency, and microwave engineers.

- Matching networks can consist of lumped SMD components like inductors and capacitors, as well as reconfigurable elements, to enhance versatility and achieve a broader range of operational bands.
- The size of the ground plane of the device is key for obtaining enough bandwidth and efficiency. If the device is small in terms of the operating wavelength (length$\sim <\lambda/\pi$), obtaining bandwidth and efficiency will be more challenging. The size of the ground plane also determines the radiation patterns, being generally quasi-isotropic due to the size of ground planes typically smaller than one wavelength at the lowest frequency of operation.

It is interesting to use an analogy to better understand the concept of an antenna booster (Figure 1.44). It is well-known that a drumstick excites the drum membrane to create a sound wave. Therefore, the element in charge of radiating a sound wave is the drum. The stick matters in the sense that its position is relevant. For example, the centre of the membrane is preferred to make the membrane vibrate at its maximum amplitude, and thus, maximum radiation is achieved. The size of the stick is not as relevant as the position of the stick and the size of the drum. In this sense, the antenna booster acts like a drumstick. Its correct position on the device makes the ground plane vibrate and, hence, creates electromagnetic waves. The antenna booster is equivalent to the stick, and the ground plane is equivalent to the drum. As a result, the antenna booster is not considered an antenna but an exciter means to boost the ground plane to become the main player in the radiation process. That is the reason that other possible names are *ground plane booster*, *radiation booster*, *antennaless*, or *Virtual Antenna®*. To unify the wording into a single concept, antenna booster is adopted through the book as well as in many scientific publications.

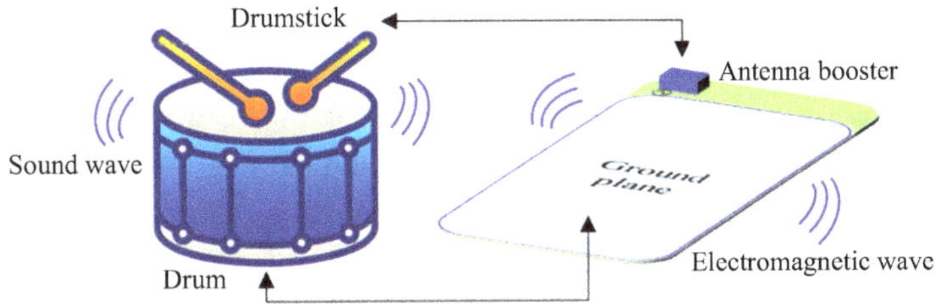

Figure 1.44 A drumstick boosts the membrane to create sound waves the same way an antenna booster boosts currents on the ground plane to create electromagnetic waves

1.9 Ready to master Antenna Booster Technology?

The following chapters will take you to the next level in designing wireless devices that embed antenna boosters.

Chapter 2 reviews the state-of-the-art techniques to design small and multi-band antennas, including antenna boosters.

Chapter 3 is crucial to understanding Antenna Booster Technology since it deals with the conception, the architecture, and the basic steps to embed an antenna booster in a wireless device. The chapter also discusses the physics to understand how and why an antenna booster can excite radiating currents in the ground plane. The impact of antenna booster size and position on bandwidth is deeply studied. Furthermore, the role of the ground plane is investigated.

Chapter 4 discusses matching networks in the design of single-band antenna systems with antenna boosters. The impact of the quality factor and tolerance is also analysed. Broadband matching networks are presented as a simple technique to enhance the inherent bandwidth of the antenna system.

Chapter 5 deep dives into the design of multiband systems. It also introduces a fully automatic design method where multiband matching networks for antenna boosters are synthesised by software without any human intervention in the design process.

Chapter 6 introduces how antenna boosters are also helpful for multi-radio applications. The chapter shows how a single antenna booster can operate several different radio protocols without the need to add one antenna booster per radio protocol.

Chapter 7 deals with reconfigurable matching networks in contrast to previous chapters, where the designs are based on passive matching networks comprising capacitors and inductors. Several reconfigurable architectures based on DTC and RF switches are compared. The chapter discusses reconfigurable matching networks in comparison to passive matching networks.

Chapter 8 drives the design of wireless devices embedding antenna boosters close to conductive bodies and ways to mitigate the impact of a closely spaced conductor.

Chapter 9 brings to a practical arena how to deal with the layout of a PCB to host a matching network. The design methods presented in previous chapters are revisited, considering the layout effect.

Chapter 10 compares antenna boosters with other antenna techniques, such as flexible printed antennas, self-resonant antennas, and microstrip patches.

Chapter 11 summarises the benefits of Antenna Booster Technology.

1.10 Conclusions

Table 1.4 answers the objectives given at the beginning of the chapter. It also points to relevant figures and equations to consolidate the concepts learnt.

Table 1.4 Key aspects to take away

Topic	Review and take away
Wavelength	Size in antenna engineering is always compared to the operating wavelength (Equation (1.1)). Typical antennas for wireless devices feature size that is comparable to $\lambda/4$. On the contrary, antenna boosters feature small electrical dimensions $\sim\lambda/30$.
Antenna	A key RF component to ensure connectivity, battery life or power consumption, and range in a wireless device.
Matching network	A matching network is a circuit system that connects a radio-frequency source to an antenna booster to maximise the power radiated to space and vice versa. It is needed in antenna boosters due to the non-resonant nature of an antenna booster, which does not match the typical 50 Ω internal impedance of sources such as a transceiver. In the antenna booster world, matching networks comprise lumped passive components such as inductors and capacitors and active components such as digitally tunable capacitors and switches (Figure 1.6).
Quality factor of a lumped component	It determines the losses of components such as those existing in a matching network. The higher the Q, the fewer losses the matching network introduces.
Materials	Materials play an essential role. On the one hand, dielectrics are mainly employed as casings. On the other hand, conductors are used in antenna elements and ground planes (Tables 1.1 and 1.2).
Transmission lines	A transmission line helps to guide an electromagnetic wave from one point of the circuit to another. It may be printed or take the form of a flexible structure (Figure 1.9). In an antenna booster system, transmission lines are employed to connect a radio-frequency source to the matching network and the matching network to the feeding line. To minimise losses, the length of the transmission line should be minimised. There are some situations where such length cannot be short because the matching network is far away from a transceiver. In that case, the matching network can be connected to the input/output of the transceiver with a printed or flexible transmission line. That may be the case of a MIMO antenna booster system where antenna boosters are spread at different PCB locations, and thus, they have to be connected to the transceiver with long transmission lines. Losses should be taken into consideration to minimise the impact on total efficiency.
Filter and Multiplexer	Filters reject or allow the signal to pass from one port to the other port of the filter (Figure 1.12). Multiplexers allow several frequencies to be combined into a single port and the other way around (Figure 1.13). Both components help design multi-radio antenna booster radiating systems, as Chapter 6 shows.
Amplifier	A one-way component that amplifies a signal. It amplifies weak signals to be above the sensitivity level of a receiver (Figure 1.14).
Resonant and non-resonant	Small-multiband antennas are typically resonant or close to the circle of $SWR = 3$. Therefore, a matching network is not needed to provide operation. When needed, they allow a fine-tuning of the impedance to match it better. On the contrary, antenna boosters are non-resonant elements with poor impedance matching, which is

(Continues)

Table 1.4 (Continued)

Topic	Review and take away
	easily compensated with a matching network. This makes an antenna booster a tiny element comparable to the wavelength ($\sim\lambda/30$). This results in a typical ten times less volume compared to resonant antennas.
Reflection coefficient/ S_{11}/SWR	Tells how much power we can capture from a transceiver to the antenna system and vice versa, how much power from the incoming electromagnetic wave we can deliver to the transceiver. Since the power contains both useful and lost, it is not enough to know how much power is radiated. Thus, total efficiency knowledge is needed.
Total efficiency	Total efficiency is one of the key parameters to determine the performance of an antenna system within a wireless device, showing how much power is radiated into space over the input power to the antenna system (Equation (1.14)). Total efficiency depends on the matching of the antenna system and the losses due to the antenna system, which comprises the antenna, the losses introduced by the components of the matching network, materials of the wireless device, and even a lossy body (e.g., human, animal, thing) that can absorb part of the radiated power.
Radiation pattern	How the antenna system distributes power into space and vice versa. In those scenarios where the direction of the incoming signal is unknown, quasi-isotropic radiation patterns are preferred (Figure 1.28). In the antenna booster arena, the radiation patterns are mainly determined by the size of the device because the ground plane is the main radiation actor.
Polarisation	The trace of the electric component of the electromagnetic wave over time. Radio communication between a transmitter and a receiver is maximised if both radiation pattern maximums are aligned and the polarisation matches (Figure 1.25).
TRP/TIS	Besides total efficiency, which is a parameter tested including a transmission line in the wireless device to excite the antenna booster system with a power source, TRP represents the same feature of radiated power, but instead, the power comes directly from the real source in the device such as a transceiver. This is a more realistic figure of merit since it considers the entire radio-frequency chain. However, it depends on the total efficiency, which is the first step to optimising the transmitted/received power to/from space. TIS is the minimum power the device can receive to process the received data correctly.
Power budget	Equation (1.24) reflects how much power is received as a function of the frequency, the distance between a transmitter and a receiver, and the realised gains of both the transmitter and receiver. Essential numbers: • Improving total efficiency by 3 dB implies increasing the range by 1.41 times. • Decreasing the frequency of operation twice while keeping other parameters the same increases the range by two.

(Continues)

Table 1.4 (Continued)

Topic	Review and take away
Antenna types	The vast majority of small-multiband antennas for wireless devices can be roughly classified as monopoles, loops, and PIFA (Figure 1.39), where several frequency bands of operations are achieved by shaping the geometry of the antenna. On the contrary, antenna booster technology relies on a small non-resonant element, the antenna booster, that, when strategically located on a ground plane of a PCB, can excite radiating currents to radiate effectively. Thanks to a matching network, performance is customised by the designer, who freely determines the frequency of operation. This is simpler and faster since the matching network design can be automated thanks to matching network synthesis tools, as explained in Chapters 4 and 5. Besides the technical classification, antennas for wireless devices can also be classified depending on their mechanical nature or fabrication process (Figure 1.41). Antenna booster technology belongs to the set of off-the-shelf chip components easily embedded into a wireless device thanks to a robotised pick-and-place assembly process ensuring mounting quality when building wireless devices at a massive industrial level.
Electromagnetic simulation tools	Given the whole 3D structure of the wireless device, including the constitutive materials (conductors and dielectrics) (Figure 1.32), electromagnetic full-wave simulators allow for obtaining the main parameters such as impedance, total efficiency, and radiation patterns. It simplifies the fabrication of prototypes since parametric analysis in the simulation domain can screen the best candidates. Therefore, it saves time and cost of building prototypes. The core behind this is based on Maxwell equations (Figure 1.3). Besides full-wave simulators, circuit simulators deal with circuital parameters such as impedance (Figures 1.33 and 1.34). Hence, they help design matching networks for antenna boosters.
Measurement tools	Physical measurements are needed to validate the designs obtained in the simulation domain. To test S_{11}, a vector network analyser is required (Figure 1.35). 3D measures involving an anechoic chamber with a 3D acquisition system (Figure 1.36). A reverberation chamber (Figure 1.37) can also be employed to test total efficiency. To test radiation patterns, an anechoic chamber is necessary. Besides, to analyse the impact of the human body, phantom elements are available (Figure 1.38).
Antenna booster design approach	While conventional small-multiband antenna design mainly relies on shaping the geometry of the antenna, ending with complex shapes depending on the frequency bands to cover, antenna booster technology, on the contrary, is based on an antenna booster that can be used for any frequency band thanks to a matching network (Figure 1.43). An antenna booster can operate from single band to multiband for any ground plane size. The only customisation needed is the matching network design, which is flexible, simple, and fast, thanks to matching network synthesis tools available in circuit simulators.

It is also advisable to refresh all essentials included in the grey boxes distributed along the chapter.

1.11 Exercises

A 3D electromagnetic software and a circuit simulator, including matching network synthesis, are needed to solve the following exercises.

Assume the internal impedance of a radio frequency generator is 50 Ω when needed.

1.11.1 Exercise 1

What is the electrical size of a 100 mm × 60 mm ground plane at 900 MHz? And at 1800 MHz? And at 2.5 GHz? Which are electrically small based on $k{\cdot}a < 1$? What is the electrical size of a 12 mm × 3 mm × 2.4 mm (height) antenna booster at 800 MHz? Consider 12 mm as the maximum dimension for the calculation.

1.11.2 Exercise 2

An antenna presents a total efficiency of $\eta_t = 30\%$ at a given frequency f_0. The S_{11} at f_0 is −3 dB. What will the total efficiency be if we match the antenna to $S_{11} = -6$ dB? And at $S_{11} = -10$ dB? Assume a lossless matching network.

1.11.3 Exercise 3

An antenna presents an impedance at a given frequency of $Z_a = 50 - j300\ \Omega$. Assuming a radiation efficiency of $\eta_r = 70\%$, what is the total efficiency of the antenna? What will the total efficiency be if we use a series inductor to cancel the $-j300\ \Omega$? Obtain the value of such an inductor if the frequency of operation is $f = 900$ MHz.

1.11.4 Exercise 4

Be an antenna with an impedance $Z_a = 25\ \Omega$ and a radiation efficiency of 90%. What is the total efficiency? What will the S_{11} be if we add a series $R = 25\ \Omega$? What will the total efficiency be in this case? Based on the results, analyse if S_{11} is enough to determine how much power is radiated to space. Analyse also if it is convenient to include resistive elements in a matching network.

1.11.5 Exercise 5

An antenna #1 presents a total efficiency of 70% and a directivity of $D = 3$ dB. A second antenna #2, shows a total efficiency of 70% and a $D = 10$ dB. Which antenna radiates more power into space? What if the total efficiency of antenna #2 is 35% and $D = 6$ dB?

1.11.6 Exercise 6

Calculate the received power in dBm for an antenna embedded in a wireless device for the following conditions: a transceiver with an output power of $P_i = 23$ dBm connected to an antenna with a total efficiency of $\eta_t = 80\%$ and directivity of $D = 3$ dB. The receiver is situated at a 1 km distance with a receiver antenna with $D = 3$ dB and a total efficiency of 50%. Consider first $f = 900$ MHz. Repeat the exercise with $f = 1800$ MHz. Consider first $PLF = 1$. Then, consider that the incoming linear polarised wave impinging the linearly polarised antenna at the receiver is rotated 30°. Repeat with 60°.

1.11.7 Exercise 7

For the previous exercise 1.11.6, the receiver is now located on top of a metallic container, impacting its total efficiency, dropping from 50% to 5%. Calculate the received power for $f = 900$ MHz. Compare the results concerning the previous case.

1.11.8 Exercise 8

Simulate a half-wavelength dipole at 900 MHz. Consider the dipole a flat conductor of 2 mm width and 35 μm conductor thickness. Compute the S_{11}, radiation, and total efficiency for a wide frequency range (e.g., 600 MHz to 3 GHz). Compute the radiation pattern at 900 MHz, directivity, and polarisation. Compare the radiation patterns at 900 MHz and at those higher frequencies where the S_{11} of the antenna presents a good match.

1.11.9 Exercise 9

Repeat the previous exercise 1.11.8, but design the half-wavelength dipole at 1800 MHz. Consider the dipole a flat conductor of 1 mm width and 35 μm conductor thickness. Compare bandwidth, radiation patterns, and directivity.

1.11.10 Exercise 10

Simulate a straight quarter-wavelength monopole on an infinite ground plane for having its first resonance at 900 MHz. Consider the wire as a flat rectangular, having a 2 mm width and 35 μm conductor thickness. Compute the radiation and total efficiency and discuss the results when the antenna resonates (f_0) and for off-resonance ($f_1 = 0.5{\cdot}f_0$, and $f_1 = 0.1{\cdot}f_0$).

1.11.11 Exercise 11

Simulate a straight quarter-wavelength monopole at 900 MHz on a 120 mm × 60 mm ground plane, but in this case, not vertical to the ground plane but on the same plane. Feed it at the corner of the ground plane (Figure 1.45). Change the straight monopole to an L-type. Compare the results. Change the size of the ground plane to 60 mm × 60 mm. Analyse the results.

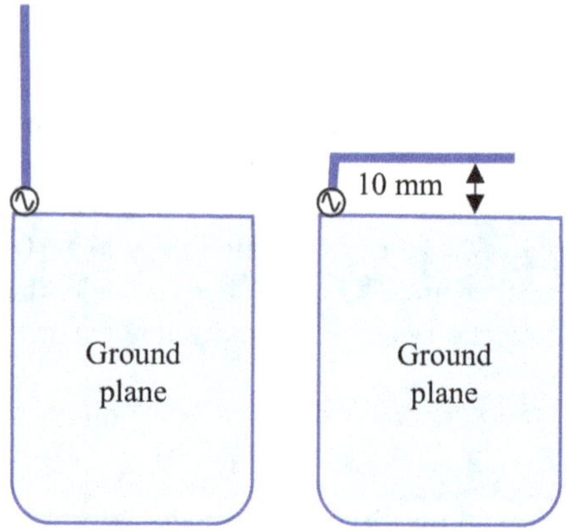

Figure 1.45 A straight monopole coplanar to the ground plane and an L-shaped monopole

Figure 1.46 Wireless device embedding two antenna booster elements (the red chip-like components on the right) [38,54]

1.11.12 Exercise 12

For the wireless device of Figure 1.46 incorporating antenna booster technology, the largest element (30 mm × 3 mm × 1 mm) is employed for cellular communications while the smallest element (7 mm × 3 mm × 2 mm) is employed for satellite communications. Identify the matching networks, the transmission lines to connect with the transceiver, the amplifier in the satellite path, and the transceiver. Why does the matching network for the cellular case comprise more lumped components? Considering a top and a bottom ground plane, recognise the kind of transmission line. Identify the clearance area.

If the lowest frequency of operation for the satellite antenna booster is 1.56 GHz, determine its electrical size.

References

[1] J. Ding, M. Nemati, C. Ranaweera, and J. Choi, "IoT Connectivity Technologies and Applications: A Survey," *IEEE Access*, vol. 8, pp. 67646–67673, 2020, doi: 10.1109/ACCESS.2020.2985932.

[2] A. Al-Fuqaha, M. Guizani, M. Mohammadi, M. Aledhari, and M. Ayyash, "Internet of Things: A Survey on Enabling Technologies, Protocols, and Applications," *IEEE Communications Surveys & Tutorials*, vol. 17, no. 4, pp. 2347–2376, Fourthquarter 2015, doi: 10.1109/COMST.2015.2444095.

[3] DGT, www.dgt.es. https://www.dgt.es/muevete-con-seguridad/tecnologia-e-innovacion-en-carretera/Dispositivos-de-presenalizacion-V16/

[4] Hella. https://www.hella-v16.com/

[5] Kepar. https://kepar.es/en/

[6] K. L. Lueth, "State of the IoT 2020: 12 billion IoT connections, surpassing non-IoT for the first time," *IoT Analytics*, 2021.

[7] "IEEE Standard Definitions of Terms for Antennas," *Revision of IEEE* Std 145-1983.

[8] D. L. Sengupta and T. K. Sarkar, "Maxwell, Hertz, the Maxwellians, and the Early History of Electromagnetic Waves," *IEEE Antennas and Propagation Magazine*, vol. 45, no. 2, pp. 13–19, 2003.

[9] T. K. Sarkar, R. J. Mailloux, A. A. Oliner, M. Salazar, and D. L. Sengupta, "*History of Wireless*," Wiley-Interscience, Kai Chang, Series Editor, 2006.

[10] "IEEE Standard Definitions of Planar Transmission Lines," *Revision of IEEE Std* 1004-1987.

[11] T. Homasai, C. Phongcharoenpanich, and W. Sriseubsai, "Bio-degradable material realization for antenna fabrication in wireless communications," *Matec Web of Conferences* 192, 01035, 2018.

[12] mangOH® yellow device. https://mangoh.io/mangoh-yellow

[13] H. F. Pues and A. R. Van de Capelle, "An Impedance-matching Technique for Increasing the Bandwidth of Microstrip Antennas," *IEEE Transactions on Antennas and Propagation*, vol. 37, no. 11, pp. 1345–1354, 1989, doi 10. 1109/8.43553.

[14] L. Miller, *"RF Filter Technologies for Dummies,"* Wiley, 2015.

[15] U. L. Rohde and D. P. Newkirk, *"RF/Microwave Circuit Design for Wireless Applications,"* Wiley, 2000.

[16] Y. Sun, "*Wireless Communication Circuits and Systems,*" ISBN: 0852964439, Institution of Electrical Engineers, 2004.

[17] J. Anguera and A. Pérez, "Teoría de Antenas," *La Salle, Universidad Ramon Llull*, 2008.

[18] D. M. Pozar and D. H. Schaubert, *"Microstrip Antennas: The Analysis and Design of Microstrip Antennas and Arrays,"* Wiley-IEEE Press, May 1995, ISBN: 978-0-780-31078-0.

[19] J. Gui, J. L. Leiva, A. Andújar, J. Groot, J. L. Pijoan, and J. Anguera, "Linearly Polarized Antenna Boosters Versus Circularly Polarized

Microstrip Patch Antennas for GPS Reception in IoT Devices," *Energies*, vol. 15, 9623, 2022. https://doi.org/10.3390/en15249623

[20] H. T. Friis, "A Note on a Simple Transmission Formula," *IRE Proc.* Vol. 34, no. 5, pp. 254–256, 1946.

[21] R. S. Saunders, *"Antennas and Propagation for Wireless Communication Systems,"* Wiley, 2nd Edition, 2007.

[22] P. Pramanik, N. Sinhababu, B. Mukherjee, *et al.* "Power Consumption Analysis, Measurement, Management, and Issues: A State-of-the-Art Review of Smartphone Battery and Energy Usage," *IEEE Access*, vol. 8, pp. 182113–182172, 2019.

[23] J. E. Brittain, "Electrical Engineering Hall of Fame: Alexander Graham Bell," in *Proceedings of the IEEE*, vol. 93, no. 2, pp. 467–470, 2005, doi: 10.1109/JPROC.2004.840443.

[24] H. A. Wheeler, "Fundamental Limitation of Small Antennas," *Proceedings of the IRE*, vol. 35, pp. 1479–1484, 1947.

[25] R. C. Hansen, "Fundamental Limitation in Antennas," *Proceedings of the IEEE*, vol. 69, no. 2, pp. 170–182, 1981.

[26] L. J. Chu, "Physical Limitations on Omni-directional Antennas," *Journal of Applied Physics*, vol. 19, pp. 1163–1175, 1948.

[27] H. A. Wheeler, "The Radiansphere Around a Small Antenna," *Proceedings of the IRE*, vol. 47, pp. 1325–1331, 1959.

[28] J. L. Volakis, C. C. Chen, K. Fujimoto, *Small Antennas: Miniaturization Techniques and Applications*, McGraw-Hill, 2019.

[29] D. F. Sievenpiper *et al.*, "Experimental Validation of Performance Limits and Design Guidelines for Small Antennas," *IEEE Transactions on Antennas and Propagation*, vol. 60, no. 1, pp. 8–19, 2012, doi: 10.1109/TAP.2011.2167938.

[30] W. L. Stutzman and G. A. Thiele, *Antenna Theory and Design*. 3rd Edition, Wiley, May 2012.

[31] G. A. E. Vandenbosch and A. Vasylchenko, *A Practical Guide to 3D Electromagnetic Software Tools*, INTECH Open Access Publisher, 2011.

[32] A. Taflove and S. C. Hagness, *Computational Electrodynamics: The Finite-Difference Time-Domain Method*. Artech House, Norwood 2005.

[33] Roger F. Harrington, *Field Computation by Moment Methods*. Wiley-IEEE Press, 1993.

[34] Nuvathings company. www.nuvathings.com

[35] D. Vye, "Network Synthesis Wizard Automates Interactive Matching-Circuit Design," *Microwave Journal*, pp. 96–102, 2018.

[36] J. Rahola, "Optimization of Matching Circuits for Antennas," *European Conference on Antennas and Propagation*, EuCAP 2011, Rome, April 2011.

[37] D. M. Pozar, *Microwave Engineering*, Hoboken, NJ: Wiley, 2012.

[38] Ignion company. www.ignion.io

[39] J. Anguera, A. Andújar, M.C. Huynh, C. Orlenius, C. Picher, and C. Puente, "Advances in Antenna Technology for Wireless Handheld Devices,"

International Journal on Antennas and Propagation, Volume 2013, Article ID 838364.

[40] www.ctia.org

[41] G. Kumar and K. P. Ray, *"Broadband Microstrip Antennas,"* Artech House, 2003.

[42] C. Deng, Y. Li, Z. Zhang, and Z. Feng, "Planar Printed Multi-Resonant Antenna for Octa-Band WWAN/LTE Mobile Handset," in *IEEE Antennas and Wireless Propagation Letters*, vol. 14, pp. 1734–1737, 2015, doi: 10.1109/LAWP.2015.2421335.

[43] Texas Instruments. https://www.ti.com/

[44] https://cactusiot.com/

[45] J. Anguera, A. Andújar, C. Puente, and J. Mumbrú, "Antennaless Wireless Device". *Pat. Application* WO2010/015365, Feb. 2010.

[46] J. Anguera, A. Andújar, C. Puente, and J. Mumbrú, "Antennaless Wireless Device Capable of Operation in Multiple Frequency Regions". *Pat. Application* WO 2010/015364 A2, Feb. 2010.

[47] J. Anguera, A. Andújar, C. Puente, and J. Mumbrú, "Antennaless Wireless Device," *Patent US.* 8,203,492 B2, Jume 2012.

[48] J. Anguera, A. Andújar, C. Puente, and J. Mumbrú, "Antennaless Wireless Device Capable of Operation in Multiple Frequency Regions". *Patent US.* 8,237,615 B2, Aug. 2012.

[49] C. Rowell and E. Y. Lam, "Mobile-Phone Antenna Design," *IEEE Antennas and Propagation Magazine*, vol. 54, no. 4, pp. 14–34, 2012, DOI: 10.1109/MAP.2012.6309152.

[50] Ubaid Ullah, Mohd Fadzil Ain, Nor Muzlifah Mahyuddin, Mohamadariff Othman, Zainal Arifin Ahmad, Mohd Zaid Abdullah, and Arjuna Marzuki, "Antenna in LTCC Technologies: A Review and the Current State of the Art," *IEEE Antennas and Propagation Magazine*, vol. 57, no. 2, pp. 241–260, 2015.

[51] *Electronic Materials Handbook: Packaging*. ASM International, 1989.

[52] J. Anguera, C. Puente, E. Martínez, and E. Rozan, "The Fractal Hilbert Monopole: A Two-dimensional Wire". *Microwave and Optical Technology Letters*, vol. 36, no. 2, pp. 102–104, 2003.

[53] J. Anguera, A. Andújar, G. Mestre, J. Rahola, and J. Juntunen, "Design of Multiband Antenna Systems for Wireless Devices Using Antenna Boosters," *IEEE Microwave Magazine*, pp. 102–114, 2019.

[54] Cactus company. https://cactusiot.com/

Chapter 2

State of the art

This chapter reviews strategies to design small and multiband antennas for wireless devices. Familiarising yourself with different techniques to design small and multiband antennas is essential.

The chapter has the following objectives:

1. Familiarise yourself with small and multiband antenna designs based on adding intelligence to the antenna geometry.
2. Understand the role of matching networks combined with antennas.
3. Understand how several antennas can be combined to enhance the frequency band of operation.
4. Learn how the ground plane plays a significant role in the radiation process.
5. Know metal rim antennas.
6. Acquaint yourself with the role of tunable radio frequency components in designing reconfigurable multiband antennas.
7. Introduction to Antenna Booster Technology.

2.1 Introduction

This chapter reviews small and multiband antenna technologies for wireless devices and introduces Antenna Booster Technology (Figure 2.1). Section 2.2 introduces small and multiband antennas based on geometry shaping. Additionally, the document introduces antenna configurations utilising both a driven and parasitic monopole, along with arrangements comprising PIFA (Printed Inverted-F Antenna) designs in conjunction with ground plane slots. The role of matching networks for bandwidth enhancement is discussed in Section 2.3. The concept of using several radiators distributed along a ground plane is shown in Section 2.4. Besides adding intelligence to the antenna geometry, other techniques put intelligence in the ground plane, as presented in Section 2.5. Metal enclosures envelop certain wireless devices. In that case, since having an antenna inside the enclosure would block the radiation, the enclosure includes an antenna with a technique known as metal rim antennas, as discussed in Section 2.6. Section 2.7 outlines the utilisation of tunable elements such as PIN diodes, varactors, DTC (Digitally Tunable Capacitors), BST (Barium Strontium Titanate) capacitors, and RF-switches as a method to expand bandwidth when passive antenna systems have reached their limit. Finally, Antenna Booster

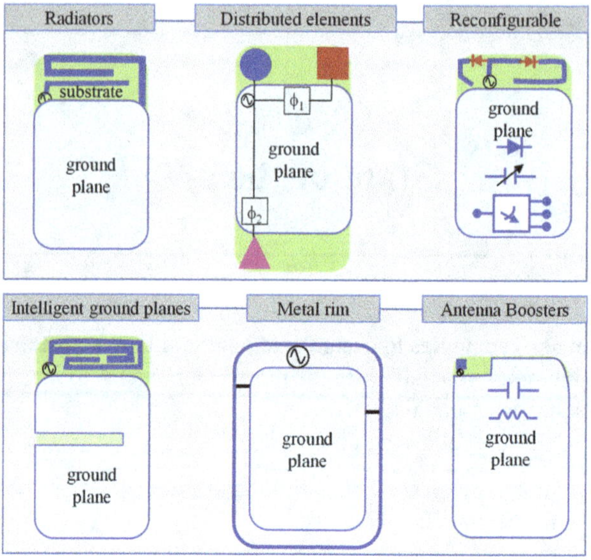

Figure 2.1 Small and multiband techniques classification: (top left) based on shaping the geometry of an antenna; (top middle) distributing several antenna elements along the ground plane; (top right) adding tunable elements to the antenna to electronically tune the frequency response; (bottom left) adding intelligence on the ground plane by introducing slots; (bottom middle) making use of a metal rim as an antenna; (bottom right) antenna booster based on non-resonant electrically small elements to boost currents on the ground plane and a matching network to select the frequency bands of operation

Technology is introduced as a radically different approach than previous ones. In effect, the techniques discussed in Sections 2.2 to 2.7 rely on the use of complex antennas, where the frequency bands are determined by designing the geometry of the antenna. The design process is often artisanal and cumbersome and it takes many iterations until the final design. However, Antenna Booster Technology relies on a non-resonant, electrically small element ($\sim\lambda/30$) called an antenna booster (Section 2.8). The correct location of an antenna booster on the ground plane of a wireless device boosts currents on the ground plane, making it the main radiator. In this manner, the size of an antenna booster can be ten times smaller than conventional antennas. Thanks to a matching network, the designer can easily shape the frequency response of the device from a single band to a multiband. As a result, embedding an antenna booster in a wireless device is faster and easier, occupies less space, and provides flexibility across the several stages of the device design, from conception to mass production. This flexibility is due to the fact that the frequency bands of operation can be easily tuned by modifying a matching network, which is simpler and more cost/time effective than adjusting the geometry of the antenna.

Conclusions are drawn in Section 2.9. Then, a collection of exercises is proposed in Section 2.10 for becoming familiar with the design of small and multiband antennas based on shaping the geometry of the antenna. This will help better understand the advantages of antenna boosters in designing devices for single and multiband operation, as explained in the following chapters. Finally, references for further reading are included in the References section.

2.2 Radiators

Internal antennas like patch/PIFAs and monopoles are common designs for handsets [1–7]. For PIFAs, several well-known techniques provide dual-band or multiband operations, such as inserting slits in the radiating path or using slotted ground planes. This fact increases the complexity of the design and makes their integration into slim platforms difficult since, to guarantee good performance, the PIFA antenna must be arranged at a certain height with respect to the ground plane, hence occupying a considerable volume (>1,000 mm^3). Monopole antennas are an alternative design to provide multiband operation on slim platforms, mainly due to their low-profile characteristics [8]. In this section, two kinds of radiators are briefly reviewed. The first one employs monopole antennas. The second radiator combines a PIFA with a slot to make a modular design because the number of bands is controlled independently with each radiator.

2.2.1 Fractal-inspired monopoles

The space-filling properties of some fractal-inspired curves have helped to design small antennas [9–17]. In this regard, a monopole antenna following a Hilbert pattern reduces the resonant frequency as the number of iterations increases (Figure 2.2). The straight monopole resonates at 799 MHz, whereas the 5th iteration

Figure 2.2 Several iterations of the Hilbert curve to generate a monopole: From the straight monopole (7 cm) to the 5th iteration. The feed point is shown at the base. All antennas are mounted vertically on an 80 cm × 80 cm ground plane [9].

of the Hilbert-base antenna resonates at 71.7 MHz. This results in a size reduction of about ten times, which helps design small antennas, as shown next.

Following the space-filling properties inspired by the Hilbert fractal, a small antenna for dual-band performance at 2.5 GHz and 5 GHz shows a total volume of only 7 mm × 3 mm × 2 mm. This represents a maximum dimension of $\sim\lambda/18$ at 2.4 GHz, smaller than a classical $\lambda/4$ monopole antenna (Figure 2.3).

Besides being small in size, the antenna presents a competitive performance for the 2.5 GHz and 5 GHz bands, both in terms of VSWR (Voltage Standing Wave Ratio) and total efficiency (Figure 2.4). This makes the antenna convenient for

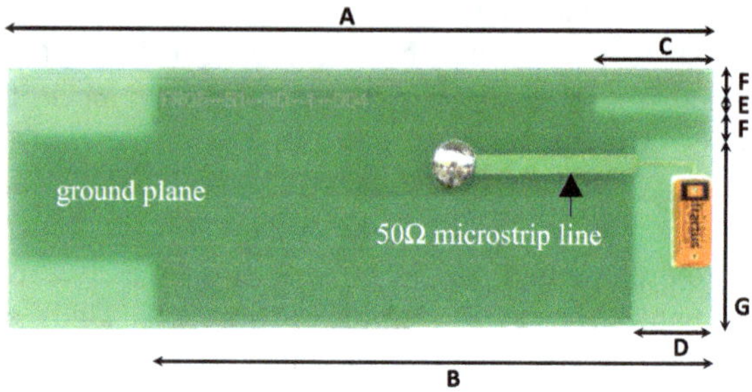

Figure 2.3 A tiny antenna inspired by the Hilbert curve. The antenna operates at 2.5 GHz and 5 GHz bands within a reduced space of 7 mm × 3 mm × 2 mm. A = 55.4, B = 44.2, C = 9.2, D = 6.0, E = 0.0, F = 2.4, G = 14.3, all cotes in mm. The clearance area is D × G.

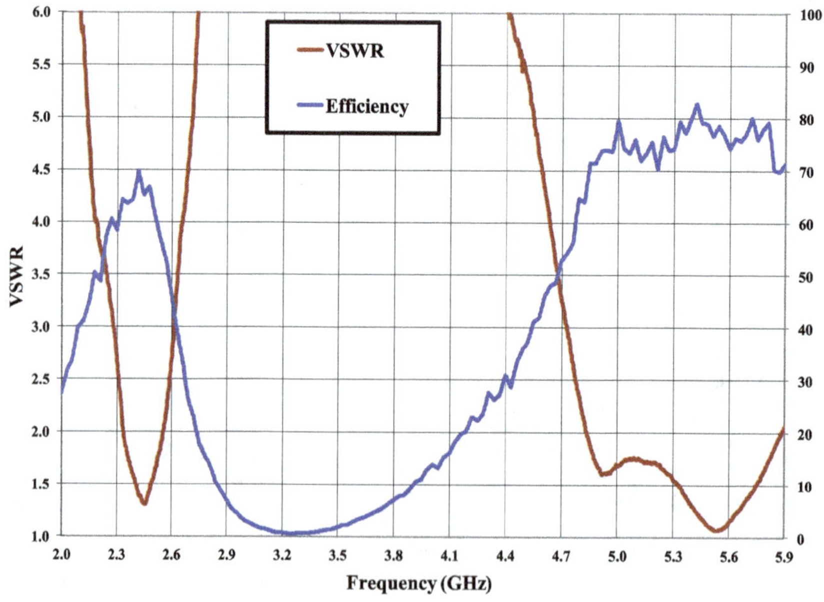

Figure 2.4 Measured VSWR and total efficiency for the antenna shown in Figure 2.3

being integrated into wireless devices such as smartphones, headsets, and smartwatches. However, due to its self-resonant nature, it is difficult to provide optimal performance at other frequencies, as shown in Chapter 5.

2.2.2 Coupled monopoles

The impact of monopole antennas in wireless handheld devices has increased thanks to their low-profile characteristics that simplify their integration into wireless platforms. Many designs have appeared in the literature and industry intending to cover the largest number of frequency bands possible without reducing the antenna performance [18–24].

A multiband behaviour (824–960 MHz and 1710–2170 MHz) is obtained with a technique using parasitic elements coupled to a driven element. At the same time, the proposal maximises the space on the PCB (Printed Circuit Board) to integrate other cellular components [25,26]. The proposed antenna also has a planar profile, which is attractive for slim platforms (Figure 2.5). The driven element is located closer to the ground plane, separated at a distance from the parasitic elements. The ground plane area located at the right side of the antenna provides a valuable space to integrate electronic components, such as sensors. On the other hand, the design considers the most critical variables when defining the operating frequency ranges. These variables are the element lengths and the gap between them, which determine their coupling effect. Furthermore, the location of the elements determines the correct performance, especially at the low-frequency bands (824–960 MHz).

Coupling between the driven and a parasitic element allows the apparition of an impedance loop in the Smith chart. By properly controlling the coupling between both elements, the performance can be wideband or multiband. Electrical models give physical insight into the coupling mechanism [26]. In this case, the first parasitic element is tightly coupled to the driven element to obtain two

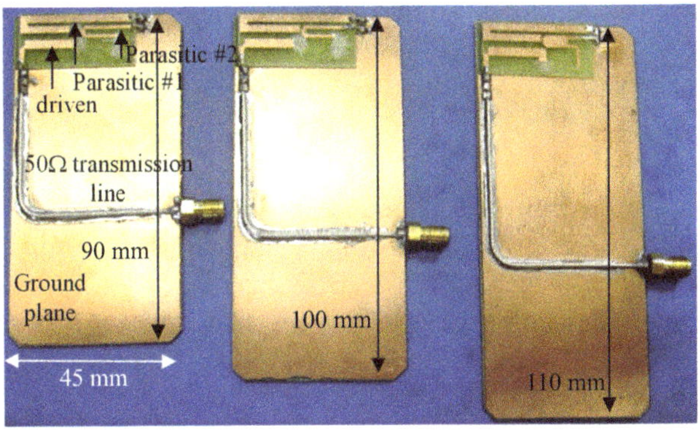

Figure 2.5 Coupled monopoles on a ground plane of 45 mm × 90 mm, 100 mm, and 110 mm. The driven element is fed through a 50 Ω transmission line. Ground clearance is 35 mm × 15 mm. Ground planes are printed on a thin substrate 1 mm thick, $\varepsilon_r = 4.15$, tan$\delta = 0.017$ [26].

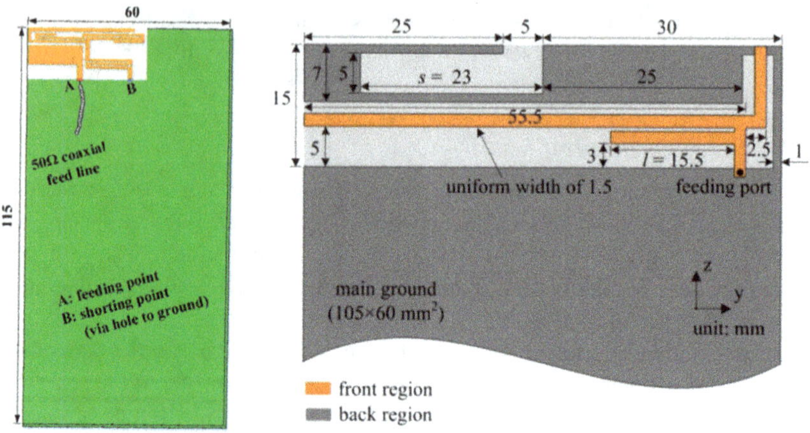

Figure 2.6 *(Left) A driven monopole coupled to a parasitic element with a complex geometry on a 35 mm × 15 mm ground clearance [27]; (right) a 60 mm × 15 mm × 0.8 mm-driven monopole coupled to a parasitic element [29]*

separated bands (Figure 2.5). Another parasitic element (parasitic #2, Figure 2.5) is weakly coupled to the driven element to get a wide bandwidth at the upper-frequency region. Interestingly, a comparable effect is observed in microstrip antennas consisting of both driven and parasitic components. Therefore, using electric models helps understand the behaviour of the impedance performance of antennas.

Many other architectures based on a driven and parasitic element, as the one explained above, have appeared in the literature (Figure 2.6) [27–32]. While achieving satisfactory multiband performance, the structure lacks flexibility when ground clearance is reduced. This limitation arises from the challenge of compactly packing either the driven or parasitic components into a small space, necessitating complex geometries [27].

2.2.3 Monopoles and loops

Besides driven and parasitic monopoles, as explained in Section 2.2.3, other antenna structures relying on sophisticated geometries are proposed, such as monopoles and loops (Figure 2.7) [33–36]. In order to achieve performance within a primary frequency range (∼698–960 MHz) and a secondary frequency range (∼1710–2690 MHz), antenna geometry is customarily crafted in an artisanal manner, often through an intuitive and iterative process to fine-tune the operating frequency bands. In some cases, where the antenna design is not sufficient to provide impedance matching, a matching network is needed to fine-tune the frequency bands of operation. Overall, it requires a laborious design in the simulation and physical domains, where the antenna needs to be adjusted several times to obtain the final design. Besides, if the size of the ground

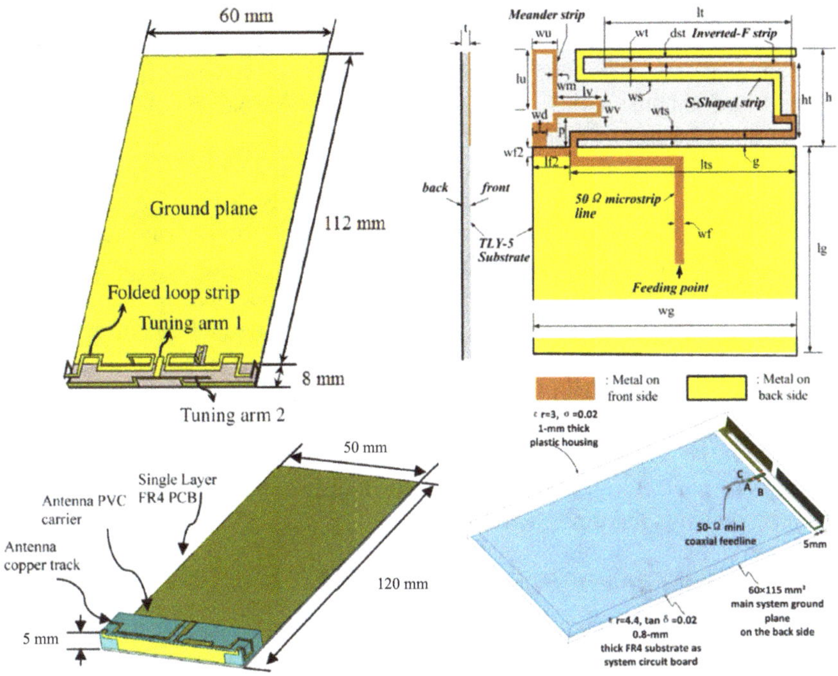

Figure 2.7 (Top left) A 60 mm × 8 mm × 5 mm (height) loop antenna including lumped components at the feeding port [33]; (top right) a 42 mm × 15 mm × 0.5 mm monopole antenna [34]; (bottom left) a 50 mm × 13 mm × 5 mm (height) loop antenna [35]; (bottom right) a 60 mm × 5 mm × 5 mm monopole antenna, including a matching network [36]

clearance needs to be reduced, the antenna must be changed, starting the whole process over again [37].

It is paramount to consider that an antenna may need to be tuned across several stages of a device design: from the conception of the design, where only a ground plane is considered, to the design phase, where all electronics components are considered, and even in certification. In the case of those devices that need to fulfil a specific certification, like TRP (Total Radiated Power), the problem may be solved by adding or modifying the matching network. However, if it is not enough, the antenna geometry should be changed, which is almost impossible at this stage when the antenna is ready to enter its mass production phase.

In the case of printed antennas [27–29,34], an additional problem arises. In effect, if the antenna needs to be tuned across several stages of the design of a wireless device, each adjustment will require the manufacture of a new PCB, making the process cumbersome. For single-band antennas, if the design has more bandwidth than required, this problem is alleviated since the antenna has enough bandwidth to absorb any frequency deviation. However, for multiband antenna design, this becomes more critical since it is difficult to have an antenna with much more bandwidth at all frequency bands. This obeys physical reasons since the space devoted to the antenna is tiny; therefore, achieving more bandwidth to absorb any detuning is impossible in many circumstances.

From a design standpoint, another drawback of complex geometric designs is that their tuning necessitates expertise in antenna engineering, making adjustments challenging for radio frequency professionals such as microwave or wireless engineers who may lack specialised antenna expertise.

Finally, another drawback is the antenna volume required to achieve either single-band or multiband performance while keeping a small electrical size. In [37], several antennas were analysed based on complex geometries like the ones shown here. In such a reference, the criterion was calculating antenna volume for achieving performance across 824 MHz–960 MHz and 1710 MHz–2690 MHz. The findings give an average volume of 971mm^3. As explained in the forthcoming chapters, Antenna Booster Technology can achieve the same performance by 86.4 mm^3, which is ten times less volume [37].

2.2.4 *Combination of PIFA and slots*

PIFA and slots have been widely studied in the literature [1,38,39]. The PIFA needs a 3D volume to radiate efficiently, whereas the slot antenna can be completely flat. However, due to the ground plane, the space underneath the antenna cannot be reused to place other electronic components (batteries, shieldings, sensors, etc.) since they would significantly affect the antenna's performance. To combine the benefits of PIFAs and slot antennas (planar structures), a concept that combines a PIFA with a slot antenna is discussed here. Other combinations, such as monopole and slot antennas using a self-complementary structure, have been proposed in [40].

An illustration of how the concept works is shown next [41,42]. Figure 2.8(a) depicts a slot in a 100 mm × 40 mm ground plane. In this case, the slot is excited around 1900 MHz, resulting in a $\lambda/4$ slot antenna. The obtained bandwidth covers 1.71–2.17 GHz at $SWR \leq 3$. Figure 2.8(b) shows a 900 MHz PIFA on the same ground plane. The feeding mechanism is in the same position to excite the previous slot. Both designs are combined, that is, the PIFA and the slot share the same feeding mechanism (Figure 2.8(c)). The antenna combines both reflection coefficients (Figure 2.8(d)). To increase the bandwidth at the second band, the slot width may be increased [43].

Since the PIFA has only one branch, the space can be reused to allocate more branches to increase the number of frequency bands [42]. For this technique, it can be concluded that:

(a) Number of bands = number of PIFA bands + number of slot bands.
(b) Bands can be adjusted independently due to the PIFA and the slot.

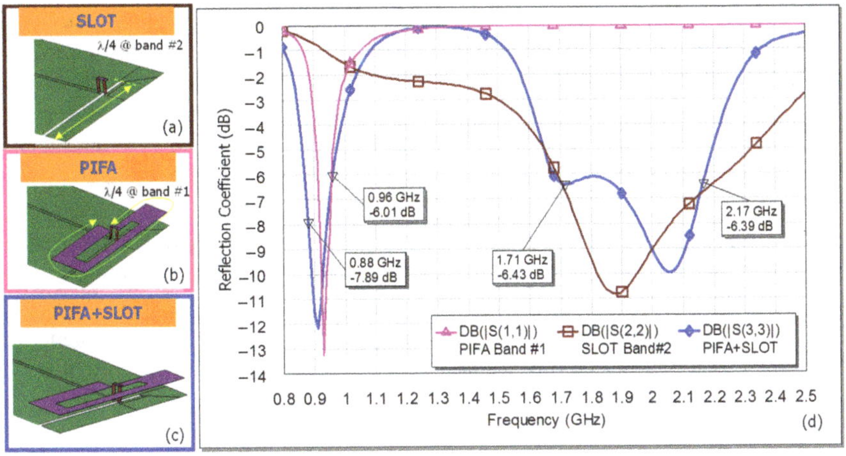

Figure 2.8 *Sequence showing the antenna concept. (a) A slot on the ground plane is tuned at 1900 MHz (band#2); (b) PIFA is tuned at 900 MHz (band#1); (c) parallel excitation of both antennas (PIFA+Slot); (d) reflection coefficient of the antenna system. The ground plane is 100 mm × 40 mm for all cases.*

This concept relies on the parallel excitation of a PIFA slot, which proves particularly beneficial in crafting multiband wireless devices. Here, the number of frequency bands is determined by the addition of bands from each radiator. Furthermore, these bands can be independently controlled, introducing an additional level of design flexibility.

Thanks to the slot radiator, the PIFA volume can be reused to add more bands. With this structure, an extra band centred at 2.6 GHz has been added to finally design a multiband prototype including 0.88–0.96 GHz and 1.71–2.17 GHz, and 2.5–2.6 GHz [41]. The antenna volume is 39 mm × 11 mm × 2 mm (858 mm^3).

Shaping the antenna geometry
Shaping the geometry of an antenna is a common way to design small/multiband antennas. The design process is laborious in both the simulation and prototyping stages since several iterations are needed to reach the final antenna.

2.3 Matching networks

In combination with antenna techniques, matching networks play a significant role not only in tuning the frequency band location but also in providing greater bandwidth [44–48]. A technique consisting of a simple circuit is discussed to enhance the bandwidth of a simple antenna by a factor of about 2.45 times for $SWR = 3$ [45,46].

Matching networks using lumped components are widely included in many commercial handset devices. In some situations, a matching network helps fine-tune the operating bands. Here, a technique for broadening the inherent bandwidth of an antenna is reviewed. The method consists of adding an LC shunt circuit that allows the creation of an impedance loop of the proper size to be inscribed inside the circle of a given target SWR [46].

A circuit analysis shows that the bandwidth of an antenna featuring an input impedance similar to that produced by an RLC series circuit around the central operating frequency can be improved by a theoretical factor of 2.45 regarding an $SWR = 3$ [46]. To demonstrate the potential of this technique, a single L-shaped monopole featuring an RLC series input impedance along the central frequency of operation is matched with a broadband matching network (Figure 2.9). Bandwidth and efficiency measurements demonstrate that this single element provides total efficiencies of ~55% from 1.7 GHz to 2.7 GHz (Figure 2.10).

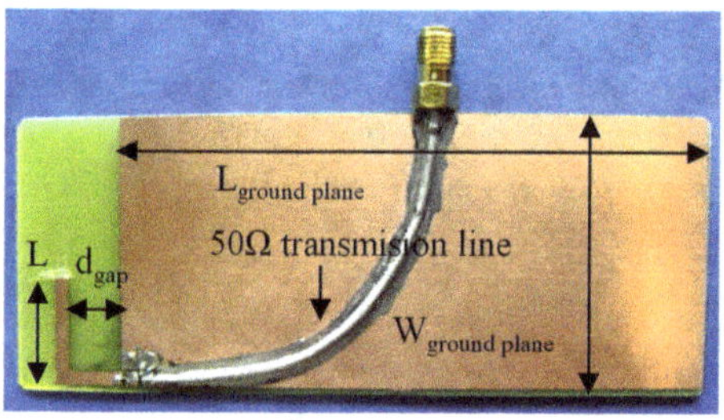

Figure 2.9 L-shaped monopole printed on a ground plane. The dimensions of the monopole antenna are $L = 23$ mm with a strip width of 2 mm, and it is located on the shorter edge of a PCB at a distance $d_{gap} = 4$ mm from the ground plane. The ground plane dimensions are $L_{ground\ plane} =$ 90 mm and $W_{ground\ plane} = 40$ mm.

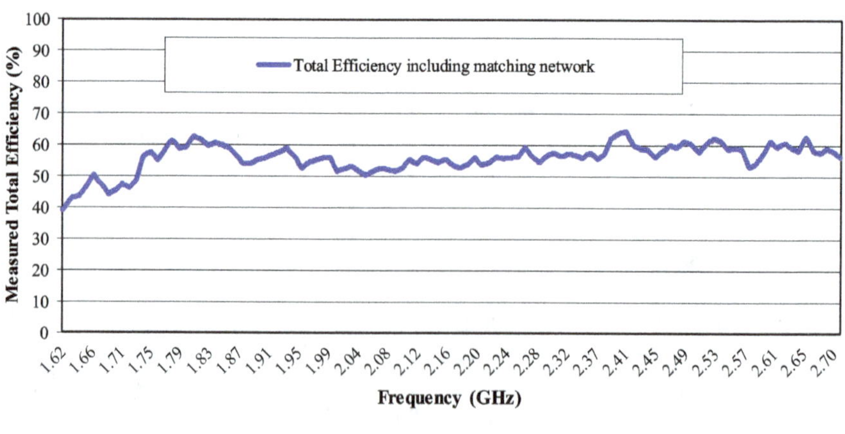

Figure 2.10 Measured total efficiency of the L-monopole shown in Figure 2.9. Broadband matching network consisting of a shunt $L = 3.3$ nH and a shunt $C = 1.3$ pF.

Measured radiation patterns are stable across the frequency range of operation, being omnidirectional and having a minimum along the long axis of the PCB. Measured directivities range from 2.8 dB to 4.4 dB. As a result, a bandwidth enhancement of at least one-half of Fano's limit [49] is achieved with a simple two-stage matching network. A concrete illustration involves enhancing a monopole, originally possessing a 14.2% bandwidth at $SWR \leq 3$, to attain a bandwidth of 52.4% at $SWR \leq 3$, accompanied by an average measured total efficiency of 56.5%.

Matching networks, such as the proposed broadband matching network, allow increasing the bandwidth of the antenna element without the necessity of increasing the antenna size. Broadbanding applied to antenna boosters is analysed in Chapter 4.

2.4 Distributed antennas

The physical principle to increase the bandwidth relies on combining several single monopole antennas distributed along the ground plane of a wireless device by applying a proper phase difference between them [50] (Figure 2.11). In addition to broadening the bandwidth when compared to a standalone monopole antenna, the distributed architecture ensures the antenna system remains resilient to human interaction, such as handloading, which can cause detuning and absorption of radiated power [51–53].

The method suggests an antenna design comprising an array of compact monopoles strategically positioned on a ground plane to boost bandwidth and offer resilience against human loading effects, particularly finger loading effects. (Figure 2.11). Other techniques are reported in [54–57].

It is well known from microwave theory that an array of in-phase antennas presents the same return loss at the input port of the feeding system as the return loss of a single antenna. However, if a phase delay is introduced, for example, to achieve a certain beam-tilt, the bandwidth may be enhanced at the input port due to the non-constructive sum of all the reflections coming from each radiator. This principle of array theory is applied here to obtain not only a broadband antenna but also a more insensitive system to finger-loading effects than the one using a single element.

The proposed system is entirely passive, which is considered advantageous in terms of simplicity and battery consumption.

Electric models have been applied to give physical insight into the broad banding mechanism of distributed antenna systems [53].

A prototype with a single monopole, another prototype with two monopoles, and a third one integrating three small monopoles combined in a single port is built and measured to demonstrate the effectiveness of the proposal (Figure 2.11) [52]. The bandwidth ($SWR \leq 3$) for the system with three monopoles is broader than that attained by the other prototypes. The bandwidth is 15.6% (0.827–0.967 GHz), 23.6% (0.772–0.979 GHz), and over 34.0% (below 0.7 GHz to 0.979 GHz) for the single, two, and three antenna cases, respectively. However, it should be considered that the array with three antennas also operates from 700 MHz to 824 MHz, where neither the array of two antennas nor the single antenna presents a good reflection coefficient (Figure 2.12).

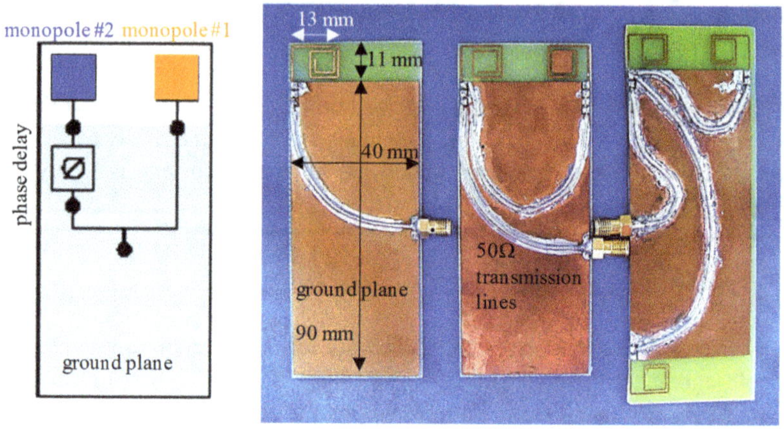

Figure 2.11 (Left) Illustration of a distributed antenna system having two elements placed at different locations of a handset device. (Right) Single monopole; distributed antenna system comprising two monopoles; distributed antenna system of three monopoles. The ground plane, in all cases, presents dimensions of 90 mm × 40 mm and is printed on an FR4 substrate 1 mm thick. The footprint for the monopoles is 13 mm × 11 mm.

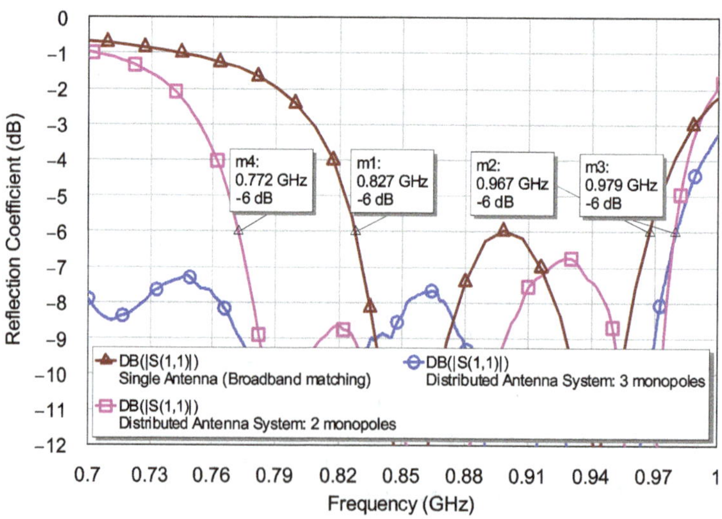

Figure 2.12 Measured reflection coefficient for the prototypes in Figure 2.11

To determine the robustness of the human loading impact, a hand phantom is used. The hand phantom is filled with liquids emulating the electromagnetic properties of the human hand at the frequencies of interest [53,58]. Different experiments with the finger located 1 mm from the antenna have been carried out considering three distinct positions: left, middle, and right. The phantom palm is 20 mm spaced from the ground plane to characterise a realistic scenario when the user is holding the device. For the three monopoles, the same scheme is utilised (the bottom monopole does not suffer from the finger-loading effect).

For the single monopole antenna, the finger in the left position is critical since the finger covers the antenna, whereas the finger is further away from the left position. This effect is appreciated in the measured results. For the right case, the total efficiency reaches an average efficiency of 23%, whereas, for the left position, it is not more than 3% (Figure 2.13) [53].

Regarding the distributed system of two monopoles, total efficiency is better than that of the single monopole for all cases except for the right position, where the single monopole does not suffer the finger effect since it is further away from the antenna. However, in the best case (finger right), average total efficiencies for the single monopole and the distributed system of two monopoles remain comparable. The benefits of the distributed antenna system of two monopoles become apparent for the other cases (finger middle and left), where the efficiency is clearly above the efficiency of the single monopole. Thus, it is possible to conclude that the distributed antenna system of two monopoles offers better performance than that of a single monopole regarding the presence of the hand phantom.

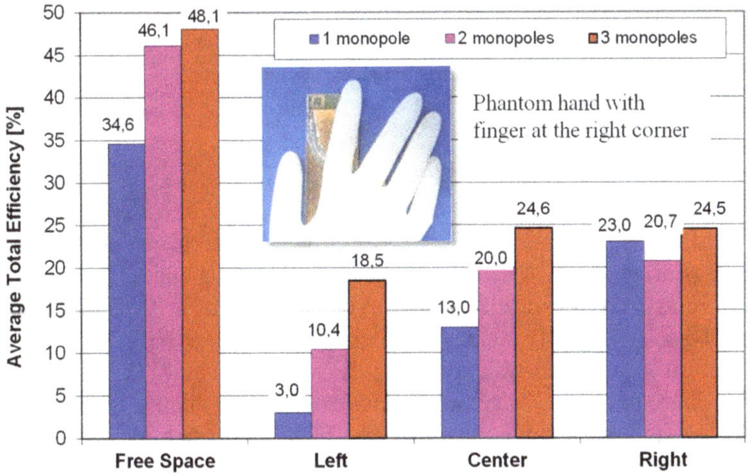

Figure 2.13 Measured total efficiency for the single and distributed antenna system with two and three antenna elements. Left, centre, and right represent the location of the finger above the PCB. The phantom hand included a liquid with $\varepsilon_r' = 41.1$ and $\sigma = 0.88$ S/m at 835 MHz.

A distributed handset antenna system using three small monopoles features enough bandwidth to cover communication protocols in the range of 686–970 MHz. This concept uses an array of monopoles with a proper phase shift to improve the bandwidth compared with a single antenna element. Moreover, the proposed system is robust to the finger effect because, when the finger blocks one antenna element, there are still two more elements that efficiently contribute to the radiation. Finally, it should be emphasised that the proposed distributed system is entirely passive and advantageous in terms of simplicity and battery consumption.

2.5 Intelligence in the ground plane

The efforts on the antenna design have been mainly addressed to the antenna geometry and not to the ground plane since its relevance in the radiation process was underestimated. Accordingly, the antenna element was typically a self-resonant element that provided efficient radiation independently from the ground plane structure. Nevertheless, the ground plane is relevant since several studies have demonstrated its vital contribution to the radiation properties [59–72]. It has been shown that a rectangular ground plane length of 0.4λ optimises bandwidth and efficiency [6].

Hence, the dimensions of the ground plane predominantly dictate the antenna design, constrained by the size of the wireless device. Another significant limitation is the antenna height, which should be small enough to allow slim devices. Therefore, new techniques are needed to attain maximum performance with an antenna that occupies the smallest possible space. Three strategies to manipulate the ground plane are revisited:

- Use of a slot to lengthen the ground plane
- Use of a conductive strip to lengthen the ground plane
- Use of traps to electrically reduce the ground plane

2.5.1 Lengthen the ground plane by using slots

To effectively enlarge the ground plane, slots can be used. The idea is illustrated in Figure 2.14, where the slot is used to tune the ground plane mode (enlarging the current path) at the low-frequency range (900 MHz) while placed underneath the antenna area to act as a parasitic element at higher frequencies (1800–2100 MHz).

Prototypes of three PIFA antennas, namely, a dual-band PIFA without slots, a quad-band PIFA with one slot, and the proposed hexa-band PIFA with multiple slots on the ground plane, have been constructed and studied (Figure 2.15) [70]. The electromagnetic simulation software IE3D was used for optimising the design.

In this concept, a slotted ground plane is used to improve the bandwidth in both low- and high-frequency regions without increasing the volume of the antenna. On the one hand, at low frequencies, the slot is below resonance but forces the ground plane mode to be excited to increase bandwidth; on the other hand, the slots are

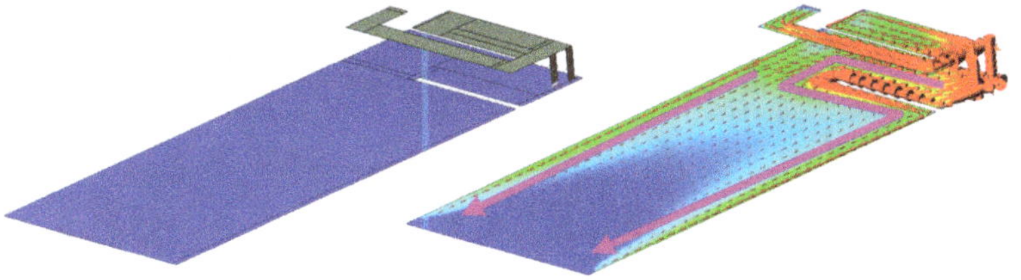

Figure 2.14 Introducing slots in the ground plane for electrically lengthening the current path. Continuous arrows are a qualitative representation of the main current distribution for 900 MHz, distributed along the long edges of the 100 mm × 40 mm ground plane.

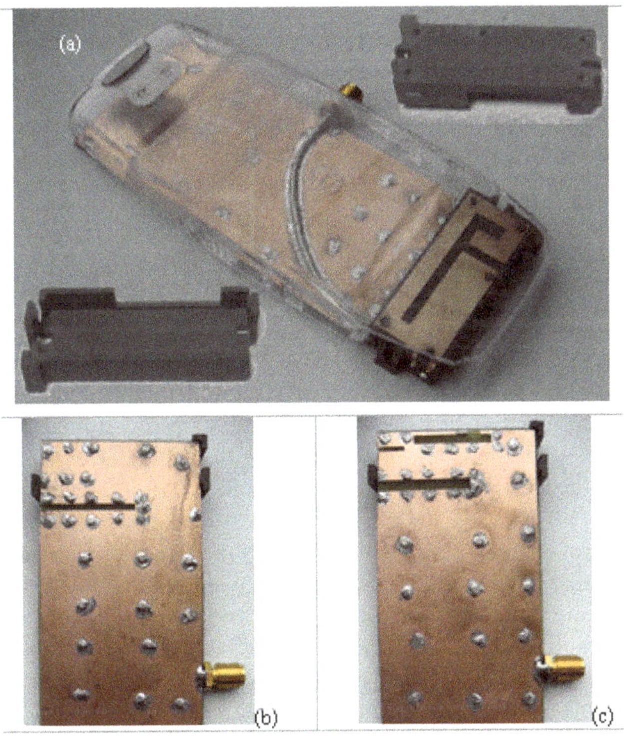

Figure 2.15 Manufactured antenna prototypes: (a) Dual-band PIFA and rear view of the (b) Quad-band PIFA and (c) the Hexa-band PIFA. In (a) the carrier to attach the metal plate and the plastic cover is also shown.

comparable to $\lambda/4$ at high frequencies and, therefore, they enhance the bandwidth (Figure 2.16). This solution does not directly excite the slots, as in the case with PIFA explained in Section 2.2 but by electromagnetic coupling, with the PIFA as the driven element.

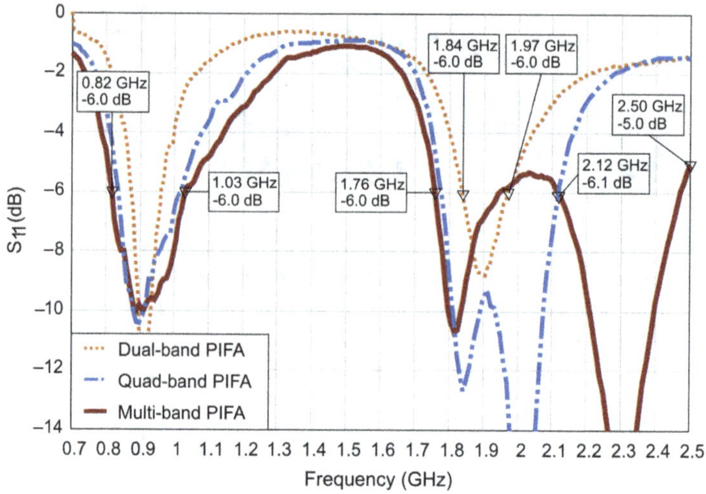

Figure 2.16 Measured reflection coefficient for the three studied prototypes. The proposed multiband design with slots enhances the bandwidth of the PIFA without slots (dual-band case).

The placement of a component (speaker) over the slot (without any metallic contact between the speaker and the ground plane) does not affect the antenna's performance at low frequencies. However, it is critical at high frequencies when the component is close to the open edge of the slot [70]. The effect is minimised at the centre and the short end of the slot. Also, SAR (Specific Absorption Rate) has been evaluated for this concept and the ones using slots in the ground plane. Results show that this concept presents a similar SAR to that of the PIFA on the bare PCB with the advantage that more bands are covered with the slotted ground plane solution [71].

This new procedure has been compared with the same design without the slots. Results show that the bandwidth and, consequently, the total efficiency are improved, obtaining a radiator useful for multiband wireless applications.

This technique enhances the bandwidth of a low-profile PIFA of only 3.2 mm in height inside a cellular phone (Figure 2.17). The slots are applied to all ground layers of the PCB.

2.5.2 Lengthen the ground plane by using conductive strips

As discussed above, the ground plane plays an essential role in the electromagnetic behaviour of an embedded antenna. The following technique integrates a conductive strip on the ground plane to effectively produce an electromagnetic enlargement capable of tuning the resonant frequency of the fundamental mode to lower values close to 900 MHz (Figure 2.18). A strip at the opposite edge of the antenna location makes the ground plane larger. Such a strip is designed to tune the ground plane mode [73,74]. As a result, bandwidth and efficiency have increased.

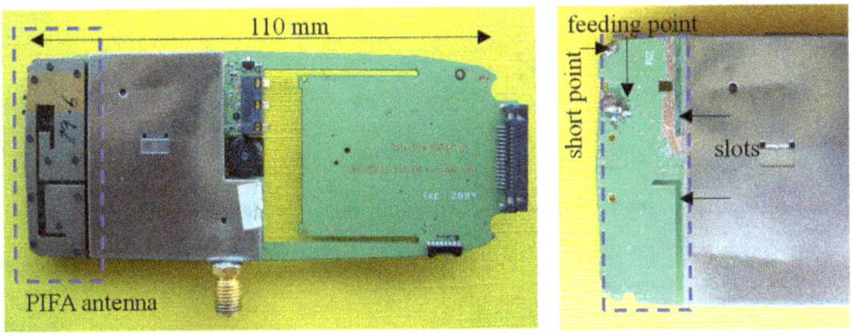

Figure 2.17 PIFA antenna with a slotted ground plane. PIFA height with respect to the ground plane is 3.2 mm.

Figure 2.18 Conductive strip on a real device to enlarge the ground plane. An inductor connecting the strip to the ground plane helps to shorten the length of the strip [73].

Inductive and dielectric loading enables the reduction of the strip length. Physical insight is given by electrical models [73] and using radar cross-section analysis [75].

Other authors have used the strip to mitigate the hand-loading effect [76]. In [77], a mechanism to control near electrical and magnetic fields is beneficial for hearing-aid compatibility.

Four case studies are selected to give a better perspective of the efficiency improvement (Figure 2.19): handset phone without strip, with the strip, with the strip length having 48 mm and 23 mm with the respective loading inductor for the last two cases. On the one hand, it is clearly shown how the efficiency is improved at the low-frequency region. The unloaded and inductively loaded strips of 48 mm length perform very similarly, demonstrating the benefit of inductance loading. The 23 mm case improves the efficiency peak, but the efficiency drops at 960 MHz. In

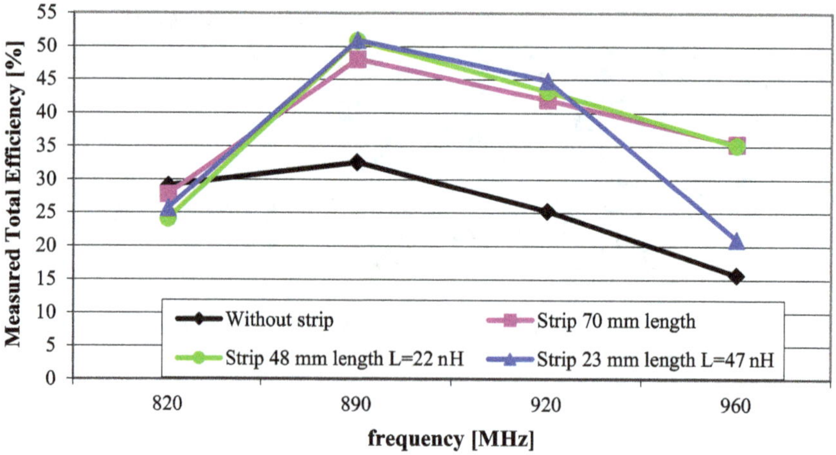

Figure 2.19 Measured total efficiency without the strip and the strip considering loading inductors for the case shown in Figure 2.18. The case having L of 22 nH and 48 mm length effectively enhances the efficiency across the 820–960 MHz band.

summary, the strip with a 48 mm length improves the efficiency across the band. In particular, the improvement at 960 MHz is very significant: 3.5 dB. On the other hand, the strip does not alter the performance in the high-frequency region [73].

In conclusion, this technique is helpful to improve the bandwidth and efficiency at the low-frequency region where the ground plane is smaller than 0.4λ, which is approximately the optimum length to excite the fundamental mode of the ground plane and, thus, maximise the bandwidth and efficiency.

The role of a ground plane
The ground plane of a wireless device is a crucial element in performance in terms of efficiency, bandwidth, and radiation patterns.

2.5.3 *Reducing the ground plane using stubs*

In some platforms, for example, clamshell-type handsets, the ground plane is large in the open position. Moreover, if the antenna is placed at one edge instead of the hinge, it may excite a particular mode that results in a radiation pattern with many lobes and a minimum in the horizontal plane. In this regard, the present technique reduces the ground plane's electrical length by adding a trap (Figure 2.20) [78]. In [79], the method of using traps increases the bandwidth in the high-frequency region. In effect, at this frequency, a typical length of a bar-type handset of 100 mm is $0.63\lambda_0$ at 1900 MHz, which is larger than $0.4\lambda_0$. Therefore, the strip forces the ground plane to be $0.4\lambda_0$ in length at such frequencies. Similar effects can be obtained by introducing a slot in the ground plane [80].

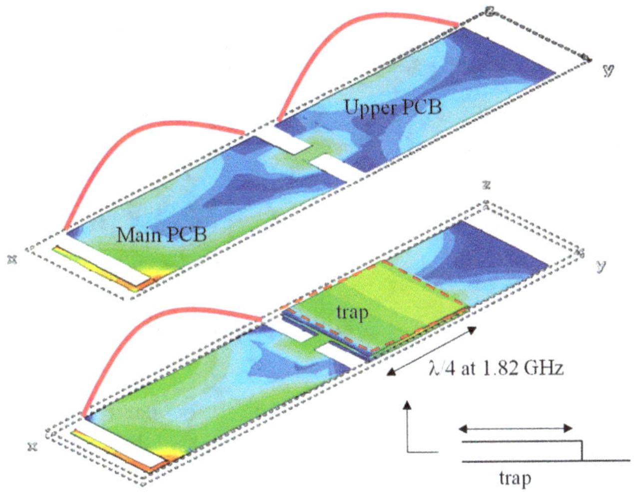

Figure 2.20 Simulated current distribution at 1.82 GHz without and with a shortening mechanism consisting of a short circuit plate of λ/4 at 1.82 GHz. The continuous line is a qualitative approach to the currents on the ground plane. For the top case, a current mode having two sinusoids is supported, causing a multi-lobe pattern. For the bottom case, the current in the upper PCB board has been mitigated due to the trap.

When the antenna is placed at one edge of a clamshell platform, the radiation in the horizontal plane does not present maximum radiation due to a multi-lobe pattern. By adding the trap, which is a short-ended $\lambda/4$ stub at the central frequency of operation, the current is blocked due to the high impedance of the stub. In this way, the current is minimised. As a result, the higher-order mode has been removed at the ground plane, supporting now a fundamental mode that radiates with a maximum in the horizontal plane. This technique can also be applied to other wireless devices to manipulate currents on the ground plane to improve bandwidth, control the radiation pattern, or enhance isolation between antennas.

2.6 Metal-rimmed antennas

Smartphone companies consider metal-rimmed antennas to add aesthetics to wireless devices. Besides, metal is a better dissipator than plastic casings. However, from the antenna perspective, a metal prevents the antenna from radiating. Therefore, to overcome the blocking effect, metal rim antennas rely on exciting the rim (Figure 2.21) [81–89]. Some metal rim antennas make use of the slot-antenna concept. In effect, a metal rim is spaced from the ground plane by a gap. Such a gap acts as an antenna by adding short-point and feeding lines [81]. Other strategies break the metal rim into several parts; thus, the broken parts function as monopole antennas [89].

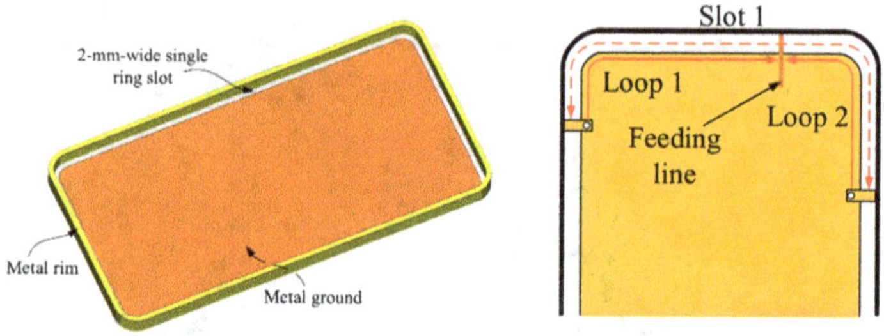

Figure 2.21 Metal rim antenna [81]

The design of metal rim antennas depends on the device design since the rim is part of the device. Therefore, any change altering the structure of the rim can determine the antenna performance.

2.7 Reconfigurability

The number of frequency bands depends on the antenna geometry and size, as well as the ground plane size. In those situations where many frequency bands need to be covered or where the ground plane of the wireless device is small and, therefore, challenging to obtain bandwidth, reconfigurability plays a key role. Reconfigurability involves the incorporation of adjustable radiofrequency components, such as PIN diodes, varactors, DTC (Digitally Tunable Capacitors), BST (Barium Strontium Titanate) capacitors, or micro switches [90–100]. In addition, reconfigurability plays a vital role in dynamic scenarios where the device is not always in the same place or interacts with a human body. In this scenario, the environment can change the input impedance with respect to a free-space scenario. Hence, a reconfigurable architecture can recover impedance matches.

Reconfigurable components have other applications, such as control of the radiation pattern and polarisation. However, in the present book, the focus on this review section, as well as in Chapter 7, is devoted to providing flexibility in tuning the frequency bands of operation to achieve acceptable matching impedance and total efficiency values in a larger bandwidth than that of a passive matching network (Figure 2.22).

PIN diodes allow you to connect and disconnect parts of the antenna to tune the frequency of operation. In Figure 2.23, two PIN diodes are included along with a monopole antenna. Depending on the on-off state of the diodes, the antenna is longer or shorter and, as a result, operates at a different frequency band. This concept can be applied to other types of antennas, such as loops (Figure 2.23 top right), IFA, PIFA, etc. The feeding mechanism should be compact in order not to degrade the performance of the antenna.

DTC is a tunable component that provides an array of capacitors. In Figure 2.23, bottom left, an IFA is loaded at its end with a DTC. Depending on the

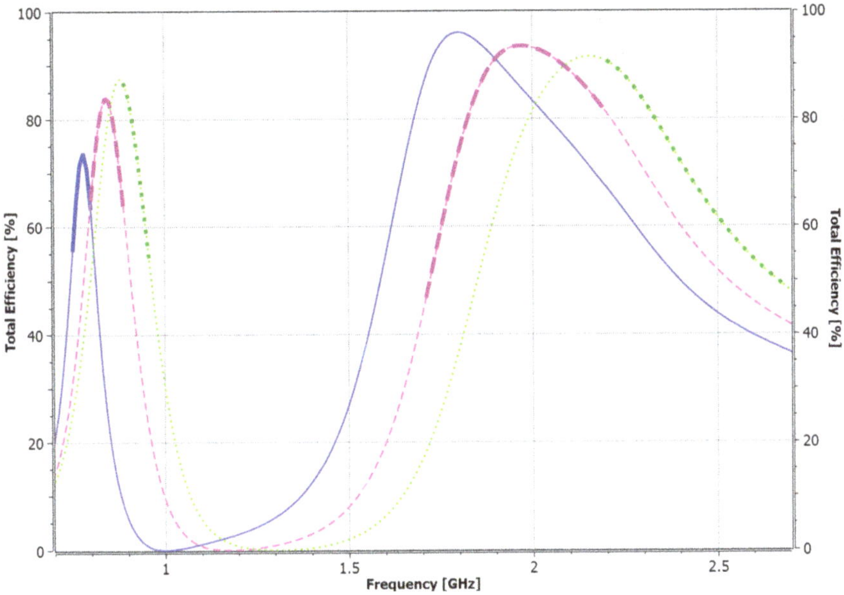

Figure 2.22 *Tuning the total efficiency response using three different states of a tunable component. In this example, the three states correspond to a series DTC for three different values in a matching network for an antenna booster.*

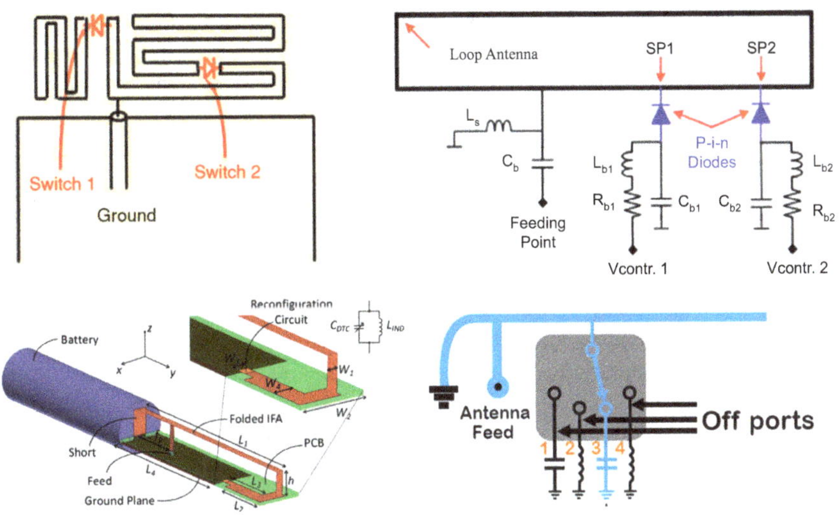

Figure 2.23 *(Top left) A PIN-diode monopole antenna [90]; (top right) a PIN-diode loop antenna [91]; (bottom left) a monopole antenna loaded with a DTC at its end [92]; (bottom right) an IFA (Inverted-F antenna) loaded with a switch with several lumped components [93]*

value of the capacitor, the antenna is tuned at a different frequency. While a DTC provides a discrete number of capacitors, varactors, on the contrary, provide a continuous set of capacitors depending on the reverse voltage applied to a varactor diode, which may range from low voltages close to zero to more than ten volts. In a standard solution, a DTC is controlled through interfaces such as GPIO (General Purpose Input Output) and RFFE (Radio Frequency Front Ent), which can easily engage with radio transceivers without additional electronics to control the tunable component. Similarly, a BST capacitor is a tunable capacitor where the capacitance can be changed as a function of an applied constant voltage [94–96]. This constant voltage may not be directly available in a wireless device, and therefore, a charge pump voltage conversion is required to control the BST capacitor [96].

A similar approach to DTC is based on micro-RF switches where a common input is connected to an output (Figure 2.23, bottom right). Each output can be loaded with lumped elements such as inductors, capacitors, or even open or short circuits. As in DTCs, the micro-RF switches can also be controlled through GPIO and RFFE interfaces.

Reconfigurability
In those situations where an antenna does not provide enough impedance bandwidth, reconfigurability adds a degree of freedom to expand the frequency bands.

2.8 Antenna Booster Technology

The previous sections have presented antenna techniques that rely on antenna engineering principles, where the geometry of the antenna plays a major role in determining the frequency bands of operation. Although those techniques result in competitive designs, their design is laborious, cumbersome, and requires advanced antenna engineering skills. A radically different approach is introduced with Antenna Booster Technology, where a very simple and electrically small conductive element, called an antenna booster, allows a device designer to select the frequency bands of operation not by shaping a complex antenna geometry but by designing a matching network [37,101–118]. J. Anguera and A. Andújar invented Antenna Booster Technology in 2008, and it first appeared in the literature through two patent applications in 2010 [101,102], which were granted some years later [103,104]. Because of the non-resonant nature of the antenna booster, a designer can easily select the frequency bands of operation since network synthesis tools can address the matching network design, as Chapters 4 and 5 explain. Besides being small and multiband, an antenna booster can be multi-rat (multi-radio access technology), supporting several wireless radio protocols within the same antenna booster chip (Chapter 6).

An antenna booster may be an off-the-shelf component ready to be used from the beginning of a design project. As a result, designing wireless devices with antenna boosters is simpler, easier, and faster than those designs relying on complex antenna shapes. Consequently, this technology is more accessible to wireless microwave engineers since the antenna system design looks more like a microwave design than an antenna design.

This technology employs the ground plane of a wireless device as the main radiator. The ground plane booster, or antenna booster, is in charge of adequately boosting the efficient radiation modes of a ground plane. The proper location of an antenna booster in the ground plane of a wireless device, together with a matching network architecture, allows from single band to multiband operation with small antenna booster dimensions (ten times smaller than self-resonant antennas), thus making it attractive for being embedded in wireless devices (Figure 2.24).

A wireless device employing antenna booster technology is advantageous as it makes integrating the radiating structure into the wireless device easier (Figure 2.25). The volume freed up by the absence of prominent antenna

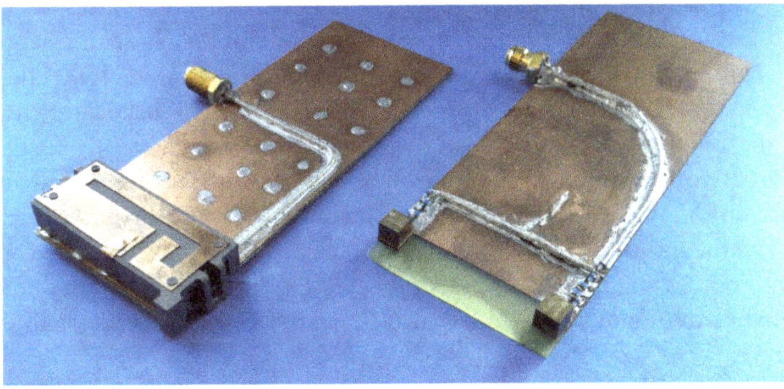

Figure 2.24 Comparison of a PIFA antenna (left) and a solution based on antenna boosters for operation at 824–960 MHz and 1710–2170 MHz on a 100 mm × 40 mm ground plane. The volume of the PIFA is 4600 mm^3 whereas the antenna booster-based solution is only 250 mm^3.

Figure 2.25 A wireless device embedding an antenna booster located at the short edge of the PCB

elements enables smaller and thinner devices or even the adoption of radically new form factors. Furthermore, by eliminating precisely the element that requires customisation, a standard solution is obtained, which only requires minor adjustments (if any) in the matching network to be implemented in different wireless devices.

The antenna booster architecture is simple: an antenna booster, a matching network, and a ground plane. An antenna booster features a small electrical size ($\sim\lambda/30$), and its correct location in a PCB, including a ground plane, enables the excitation of currents to make the ground plane the main radiating element. Ground clearance is needed to be free of any component or conductive traces. Since the input impedance of an antenna booster is non-resonant due to its small electrical size, it is not matched to the transceiver. Hence, a matching network permits the shaping of the frequency bands of operation, either single-band or multiband, to achieve impedance matching (Figure 2.26). A matching network may take the form of a combination of lumped elements (inductors and capacitors) (Chapters 4–6) and also include reconfigurable/tunable elements to gain flexibility in those situations where a passive matching network is not enough to achieve the desired frequency bands of operation (Chapter 7). Designing a matching network is much faster than designing a new antenna, and besides, such matching networks can be obtained through software thanks to matching network synthesis tools. And last but not least, the ground plane is vital in providing enough bandwidth and efficiency. It is more challenging for small ground planes to obtain bandwidth and efficiency than for larger ground planes (Chapter 3).

The steps to embed an antenna booster into a wireless device are simple: first, place the antenna booster in the device. The corners are preferred since they provide the best bandwidth and efficiency, but depending on the bandwidth requirements, other positions work as well. The impact of the antenna booster position is studied in Chapter 3.

Second, design a matching network. A matching network with two lumped components is typically enough for a single-band operation. For a multiband case, the number increases and can reach seven, for instance. More lumped components are typically rare since they may impact tolerances and efficiency. If the frequency range is above approximately 5 GHz, matching networks using distributed elements (transmission lines) may feature a compact size depending on the relative permittivity of the substrate.

The third step consists of checking the response in SWR/S_{11} with a vector network analyser and testing total efficiency to ensure the device embedding the antenna booster and the matching network are radiating properly. If the design has been correctly modelled in the electromagnetic simulation stage, little or no adjustment is needed in the physical prototype. These steps are detailed in Chapter 9.

A significant advantage is the flexibility of the antenna booster technology to adjust the frequency response at any stage of a design project. This becomes

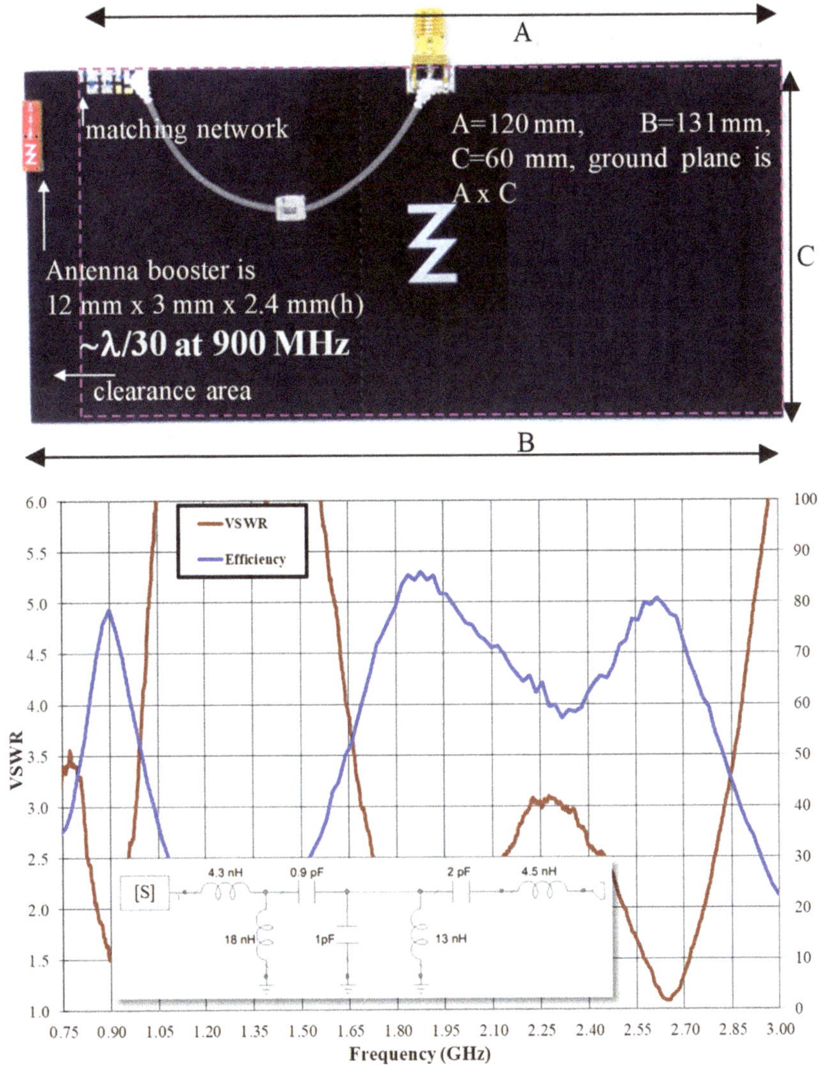

Figure 2.26 *A 12 mm × 3 mm × 2.4 mm (height) antenna booster on a 120 mm × 60 mm ground plane provides operation at 824–960 MHz and 1710–2690 MHz thanks to the proper location of the antenna booster and a matching network. Results show measured VSWR and total efficiency.*

essential when a project is in the certification stage before starting mass production. If certain radiation parameters, for example, the TRP, do not meet the specs, fine-tuning the matching network to optimise performance at a particular frequency band can save the day.

Antenna Booster Technology

- Antenna booster technology relies on electrically small antenna boosters ($\sim\lambda/20$) that are in charge of boosting currents on the ground plane of a wireless device. Its correct location on a ground plane and a matching network allows the device designer to easily operate at single or multi-band frequencies in a simple and fast manner.
- Antenna booster technology comprises three basic ingredients: an antenna booster, a matching network, and a ground plane.
- An antenna booster does not resonate, especially at the lowest frequency of operation. The frequency bands of operation are not determined by a sophisticated antenna geometry, as in classic antenna design, but by the antenna booster with a matching network. A matching allows for a single-band to multiband operation. Software facilitating the design of wireless devices can address matching networks.
- Besides being small and multiband, antenna boosters may support different wireless radio protocols with the same antenna chip with independent ports for each radio protocol.
- Antenna boosters operate in any frequency band, and thus, the same element can be used in a given device.
- Antenna boosters can take the form of off-the-shelf components such as SMD (Surface Mounted Component) and, therefore, can be included in a wireless design from the very beginning, simplifying the design process.
- The pick-and-place assembly process can automatically integrate antenna boosters into the PCB of a wireless device due to their SMD nature.
- The ground plane size of the wireless devices mainly determines bandwidth, efficiency, and radiation patterns.

2.9 Conclusions

The continuous apparition of wireless communications systems with new platforms makes antenna design a challenge since not only are more antennas needed to operate at new frequency bands, but also the antennas require multiband operation and a small size to be integrated into wireless devices.

Antenna design requires a high level of expertise in antenna engineering, electromagnetics, microwave theory, and tones of imagination. However, the antenna booster technology relies on a simple non-resonant element to properly excite radiating current on the ground plane. In addition, a matching network provides the flexibility to achieve operation at the desired frequency band. This is a faster and easier method than designing a new antenna, and moreover, the matching network can be totally addressed by synthesis algorithms, so no human intervention is needed. Naturally, physical limitations also restrict the bandwidth that can be achieved for antenna booster technology, which mainly depends on the size of the ground plane with respect to the wavelength of operation.

Table 2.1 answers the objectives given at the beginning of the chapter. It also points to relevant figures and equations to consolidate the concepts learnt. It is also advisable to refresh all essentials included in the grey boxes distributed throughout the chapter.

Table 2.1 Key aspects to take away

Topic	Review and take away
Antenna design	Common antenna designs, both small and multiband, rely on shaping the geometry of the antenna. To increase the frequency bands, the geometry is shaped to create resonances around the frequency bands of operations (Figure 2.3). This method stands for self-resonant structures that require expert skills in designing the antenna. Besides complex geometries, other techniques use parasitic elements coupled to a driven element (Figures 2.5 and 2.6), as well as a combination of radiators of different natures, such as monopoles/PIFAs and slots (Figure 2.8).
Distributed antenna elements	Other routes employ distributed antenna elements where bandwidth is increased thanks to the contribution of several elements where the phase delay plays a significant role in increasing the bandwidth (Figures 2.11 and 2.12). Besides providing more bandwidth than a single element, the distributed solution is more robust to detuning when one external element perturbs any of the antenna elements (Figure 2.13). This technique needs a transmission line to connect the different antenna elements, which must be appropriately adjusted to achieve the bandwidth increment.
Ground plane intelligence	Since the ground plane of a wireless device is key in the radiating process, several techniques add intelligence to the ground plane to increase the bandwidth. One technique consists of adding slots in the ground plane. The purpose of the slot or slots is to increase the current path to emulate a larger ground plane close to the length where the bandwidth is maximised (Figures 2.14 and 2.15). Such length is typically $\sim 0.4\lambda$ for a rectangular ground plane. Furthermore, the slot can be reused as a parasitic element to improve the bandwidth as well. A larger ground plane can also be emulated by adding a conductive strip at the opposite edge of the antenna location (Figure 2.18). The length of the strip can be reduced by loading a series inductor between the strip and the ground plane. This technique is an alternative to the slotted ground plane since embedding slots is not always possible due to interference with other components, like batteries, that short-circuit the slot effect. These techniques are convenient when the ground plane is small in terms of the wavelength ($\sim\lambda/\pi$ at the lowest frequency of operation).
Metal rimmed antennas	Certain wireless devices (e.g., smartphones) include a metallic casing instead of plastic. In that situation, the metal rim can be reused as part of the antenna radiating process (Figure 2.21).

(Continues)

Table 2.1 (Continued)

Topic	Review and take away
Reconfigurability	When a passive solution cannot cover the frequency bands of operation, reconfigurability is an alternative. Reconfigurable antennas comprise tunable components such as digitally tunable capacitors, switches, diodes, etc., where the antenna's response can be adjusted beyond the frequencies covered by the non-tunable antenna (Figure 2.23).
Antenna Booster Technology	Ground plane boosters, or antenna boosters, are tiny components able to excite radiating currents in the ground plane of a wireless device (Figure 2.26). Its location on the ground plane is essential to excite currents effectively and, as a result, provide competitive efficiency values. An antenna booster is not an antenna but an exciter of the current in the ground plane, which is the crucial element in the radiating process. The impedance of an antenna booster is non-resonant, at least at the lower frequency bands of operation. Far from being an issue, this is an advantage since the frequencies of operation are easily obtained by the design of a matching network. As a result, antenna boosters have the following features: 1. Typically, they are smaller than self-resonant antennas. 2. The same antenna booster can be reused for any ground plane size (i.e., any form factor) since the impact of the ground plane size on the impedance is compensated with a matching network. 3. It is easy to embed in a PCB of a wireless device thanks to its pick-and-place nature. 4. Able to operate at any frequency band by modifying the matching network design and not the antenna booster. 5. Enables fast, easy, and automated design of wireless devices embedding antenna boosters since the design of the matching network is synthesised by smart algorithms. This is a radically different approach compared with the self-resonant antenna design, where antenna expert skills are a must. Antenna boosters can also be multi-radio, meaning that with a single chip, several radios are supported without the need to have as many antenna booster chips as radios.

2.10 Exercises

A 3D electromagnetic software and a circuit simulator are needed to solve the following exercises.

2.10.1 Exercise 1

Consider the coupled monopoles in [26]. Simulate the S_{11} for the ground planes given in the reference. Reduce the length of the ground plane to 70 mm length.

Where is the bandwidth more affected? Why? Compute current distribution at 900 MHz, 2 GHz, as well as total efficiency. Does the ground plane have any impact on bandwidth?

2.10.2 Exercise 2

Consider the coupled monopoles in [26]. Simulate the S_{11} for the 110 mm × 45 mm ground planes. Compare S_{11} and total efficiency when the loss tangent of the substrate is $\tan\delta = 0.01$, $\tan\delta = 0.05$, and $\tan\delta = 0.1$.

2.10.3 Exercise 3

Consider the coupled monopoles in [26]. Using the frequency scaling principle explained in Chapter 1, modify the design to operate at 2.4–2.5 GHz and 5–7 GHz. Note the time it takes to achieve an $S_{11} < -6$ dB across both frequency regions. Use a 50 mm length ground plane.

2.10.4 Exercise 4

Simulate the printed antenna given in [27]. Change substrate thickness to ±20% and analyse the robustness of the performance. Repeat by increasing/reducing the permittivity of the substrate by ±10%.

2.10.5 Exercise 5

How should the design from exercise 2.10.1 be changed to cover 698–960 MHz and 1710–2690 MHz? Note the time it takes till obtaining the antenna with an $S_{11} < -6$ dB.

2.10.6 Exercise 6

Simulate the driven-parasitic monopole of [28]. How should the design be changed to increase bandwidth in the low-frequency range to include from 698 MHz to 960 MHz while keeping the high-frequency region? Note the time it takes to obtain the final design.

2.10.7 Exercise 7

Simulate the PIFA antenna given in [67] without any slot in the ground plane. Include the slot in the ground plane following the teachings in [67]. Evaluate the bandwidth and total efficiency with and without slots. Analyse the effect of moving the slot position as well as its length.

2.10.8 Exercise 8

Considering the antenna in [33], obtain the S_{11} and total efficiency. Include a cover considering a thin dielectric layer having the same size as the whole PCB with

1 mm thick and $\varepsilon_r = 3.2$ and $\tan\delta = 0.02$. Place such dielectric touching the antenna on the top. Evaluate the impact on S_{11} and efficiency. Following the teachings in [35], tune the antenna to recover the original performance without the enclosure.

2.10.9 Exercise 9

Repeat exercise 2.10.5 with the antenna design shown in [35]. Compute S_{11}, total efficiency, directivity versus frequency, and radiation patterns for several frequencies with the operating bands.

2.10.10 Exercise 10

Select ten papers disclosing small and multiband antenna designs for 698–960 MHz and 1710–2690 MHz. As a reference, search for the ten most cited papers in the IEEE database (IEEE Xplore). Compute the volume for each antenna case considering the volume of the smallest parallelepiped, including the antenna. Calculate the average volume.

References

[1] J. Anguera, A. Andújar, M. C. Huynh, C. Orlenius, C. Picher, and C. Puente, "Advances in Antenna Technology for Wireless Handheld Devices," *International Journal on Antennas and Propagation*, vol. 2013, Article ID 838364, 2013.

[2] Taga and K. Tsunekawa "Performance Analysis of a Built-in Planar Inverted-F Antenna for 800MHz Band Portable Radio Units," *IEEE Trans. Selected Areas in Communication*, vol. SAC-5, no. 5, pp. 921–929, 1987.

[3] C.R. Rowell and R.D. Murch, "A Compact PIFA Suitable for Dual-Frequency 900/1800-MHz Operation," *IEEE Transactions on Antennas and Propagation*, vol. 46, no. 4, 1998.

[4] D. Manteuffel, A. Bahr, and I. Wolff, "Investigation on Integrated Antennas for GSM Mobile Phones," *Millennium Conference on Antennas & Propagation*, ESA, AP2000, Davos, Switzerland, April 2000.

[5] C. Puente, C. Borja, J. Anguera, and J. Soler, "Multilevel Antennas," *Patent US* 7,505,007 B2, March 2009.

[6] K. L. Wong, "*Planar Antennas for Wireless Communications,*" Wiley Inter-Science, 2003.

[7] M. Martínez-Vázquez, O. Litschke, M. Geissler, D. Heberling, A. M. Martínez-González, and D. Sánchez-Hernández, "Integrated Planar Multiband Antennas for Personal Communication Handsets," *IEEE Transactions on Antennas and Propagation*, vol. 54, no. 2, 2006.

[8] C. Lin, K. L. Wong, "Printed Monopole Slot Antenna for Internal Multiband Mobile Phone Antenna," *IEEE Transactions on Antennas and Propagation*, vol. 55, no. 12, pp. 3690–3697, 2007.

[9] J. Anguera, C. Puente, E. Martínez, E. Rozan, "The Fractal Hilbert Monopole: A Two-dimensional Wire," *Microw. Optical Technol. Lett.*, 36, 102–104, 2003.

[10] J. Anguera, C. Puente, J. Soler, "Miniature Monopole Antenna Based on the Fractal Hilbert Curve," *IEEE Antennas Propag. Soc. Int. Symp.*, 4, 546–549, 2002.

[11] M. Z. Azad, M. Ali, "A miniaturized Hilbert PIFA for Dual-band Mobile Wireless Applications," *IEEE Antennas Wireless Propagation Lett.*, 4, 59–62, 2005.

[12] R. Azaro, F. Viani, L. Lizzi, E. Zeni, A. Massa, A. "A Monopolar Quad-Band Antenna Based on a Hilbert Self-Affine Prefractal Geometry," *IEEE Antennas Wirel. Propag. Lett.*, 8, 177–180, 2009.

[13] C. Puente, E. Rozan, J. Anguera, "Space-Filling Miniature Antennas," *Patent US* 7,202,822, 10 April 2007.

[14] I. Sanz, J. Anguera, A. Andújar, C. Puente, C. Borja, "The Hilbert Monopole Revisited," in *European Conference on Antennas and Propagation*; EuCAP: Barcelona, Spain, 2010.

[15] I. S. Bangi and J.S. Sivia, "Minkowski and Hilbert Curves-based Hybrid Fractal Antenna for Wireless Applications," *AEU-Int. J. Electron. Commun.*, vol. 85, pp. 159–168 2018.

[16] S. A. Saputro, J. Y. Chung, "Hilbert Curve Fractal Antenna for Dual On-and Off-Body Communication," *Progress in Electromagnetics Research*, vol. 58, pp. 81–88, 2016.

[17] M. M. Roushdy, H. F. Hammad, "Inkjet Printed Wearable Hilbert Monopole Fractal Antenna Optimized for BAN Systems," in *Proceedings of the 2016 33rd National Radio Science Conference* (NRSC), Aswan, Egypt, 22–25 January 2016; pp. 102–108.

[18] K.L. Wong, G.Y. Lee, and T.W. Chiou, "A Low-profile Planar Monopole Antenna for Multiband Operation of Mobile Handsets," *IEEE Transactions on Antennas and Propagation*, vol. 51, no. 1, pp. 121–125, 2003.

[19] K.L. Wong and C.H. Huang, "Printed Loop Antenna with a Perpendicular Feed for Penta-band Mobile Phone Application," *IEEE Transactions on Antennas and Propagation*, vol. 56, no. 7, pp. 2138–2141, 2008.

[20] K. L. Wong and S. C. Chen, "Printed Single-Strip Monopole using a Chip Inductor for Penta-Band WWAN Operation in the Mobile Phone," *IEEE Transactions on Antennas and Propagation*, vol. 58, no. 3, pp. 1011–1014, 2010.

[21] H. Kanj and S.M. Ali, "Compact Multiband Folded 3-D Monopole Antenna," *IEEE Antennas and Wireless Propagation Letters*, vol. 8, pp. 185–188, 2009.

[22] J. Ma, Y. Z. Yin, J.L. Guo, and Y.H. Huang, "Miniature Printed Octaband Monopole Antenna for Mobile Phones," *IEEE Antennas and Wireless Propagation Letters*, vol. 9, pp. 1033–1036, 2010.

[23] H.W. Hsieh, Y.C. Lee, K.K. Tiong, and J.S. Sun, "Design of a Multiband Antenna for Mobile Handset Operations," *IEEE Antennas and Wireless Propagation Letters*, vol. 8, pp. 200–203, 2009.

[24] C.T Lee and K.L. Wong "Planar Monopole with a Coupling Feed and an Inductive Shorting Strip for LTE/GSM/UMTS Operation in the Mobile Phone," *IEEE Transactions on Antennas and Propagation*, vol. 58, no. 7, pp. 2479–2483, 2010.

[25] C. Puente, J. Anguera, J. Soler, and A. Condes, "Coupled multiband antennas," *US Patent* 7,315,289 B2, Jan. 2008.

[26] S. Risco, J. Anguera, A. Andújar, A. Pérez, and C. Puente, "Coupled Monopole Antenna Design for Multiband Handset Devices," *Microwave and Optical Technology Letters*, vol. 52, no. 10, pp. 359–364, 2010.

[27] P. Huang and L. Huang, "Design of an Antenna with Ground Plane Slots for LTE/WWAN Mobile Phones," *Progress in Electromagnetics Research Letters*, vol. 83, pp. 15–22, 2019.doi:10.2528/PIERL19012701.

[28] A. Chen, M. Sun, Z. Zhang and X. Fu, "Planar Monopole Antenna with a Parasitic Shorted Strip for Multistandard Handheld Terminals," *IEEE Access*, vol. 8, pp. 51647–51652, 2020, DOI:10.1109/ACCESS.2020.2979483.

[29] C. Deng, Y. Li, Z. Zhang and Z. Feng, "Planar Printed Multi-Resonant Antenna for Octa-Band WWAN/LTE Mobile Handset," *IEEE Antennas and Wireless Propagation Letters*, vol. 14, pp. 1734–1737, 2015, doi:10.1109/LAWP.2015.2421335.

[30] D. Huang, Z. Du and Y. Wang, "An Octa-band Monopole Antenna with a Small Nonground Portion Height for LTE/WLAN Mobile Phones," *IEEE Transactions on Antennas and Propagation*, vol. 65, no. 2, pp. 878–882, 2017, DOI:10.1109/TAP.2016.2632530.

[31] X. Wang, Y. Wu, W. Wang and A. A. Kishk, "A Simple Multi-Broadband Planar Antenna for LTE/GSM/UMTS and WLAN/WiMAX Mobile Handset Applications," *IEEE Access*, vol. 6, pp. 74453–74461, 2018, DOI:10.1109/ACCESS.2018.2883308.

[32] C. Yang, Y. Jung and C. W. Jung, "Octaband Internal Antenna for 4G Mobile Handset," *IEEE Antennas and Wireless Propagation Letters*, vol. 10, pp. 817–819, 2011, DOI: 10.1109/LAWP.2011.2164049.

[33] D. Wu, S. W. Cheung and T. I. Yuk, "A Compact and Low-Profile Loop Antenna with Multiband Operation for Ultra-Thin Smartphones," *IEEE Transactions on Antennas and Propagation*, vol. 63, no. 6, pp. 2745–2750, 2015, DOI:10.1109/TAP.2015.2412962.

[34] T. Zhang, R. Li, G. Jin, G. Wei and M. M. Tentzeris, "A Novel Multiband Planar Antenna for GSM/UMTS/LTE/Zigbee/RFID Mobile Devices," *IEEE Transactions on Antennas and Propagation*, vol. 59, no. 11, pp. 4209–4214, 2011, DOI: 10.1109/TAP.2011.2164201.

[35] M. Zheng, H. Wang and Y. Hao, "Internal Hexa-Band Folded Monopole/Dipole/Loop Antenna with Four Resonances for Mobile Device," *IEEE*

Transactions on Antennas and Propagation, vol. 60, no. 6, pp. 2880–2885, 2012, DOI: 10.1109/TAP.2012.2194687.

[36] Zhong-Xiang Chen, Yong-Ling Ban, Zhi Chen, Kai Kang, and Joshua Le-Wei Li, "Two-Strip Narrow-Frame Monopole Antenna with a Capacitor Loaded for Hepta-Band Smartphone Applications," *Progress in Electromagnetics Research*, vol. 145, pp. 31–38, 2014. doi:10.2528/PIER13112808.

[37] J. Anguera, A. Andújar, G. Mestre, J. Rahola, and J. Juntunen, "Design of Multiband Antenna Systems for Wireless Devices Using Antenna Boosters," *IEEE Microwave Magazine*, Dec. pp. 102–114, 2019.

[38] C. Lin and K.L. Wong, "Printed Monopole Slot Antenna for Internal Multiband Mobile Phone Antenna," *IEEE Transaction on Antennas and Propagation*, vol. 55, no. 2, pp. 3690–3697, 2007.

[39] C.H. Wu and K.L. Wong "Hexa-band Internal Printed Slot Antenna for Mobile Phone Application," *Microwave and Optical Technology Letters*, vol. 50, pp. 35–38, 2008.

[40] N. Takemura, "Inverted-FL Antenna with Self-Complementary Structure," *IEEE Transactions on Antennas and Propagation*, vol. 57, no. 10, pp. 3029–3034, 2009.

[41] J. Anguera, I. Sanz, J. Mumbrú, and C. Puente, "Multi-Band Handset Antenna with a Parallel Excitation of PIFA and Slot Radiators," *IEEE Transactions on Antennas and Propagation*, vol. 58, no. 2, pp. 348–356, 2010.

[42] J. Anguera and C. Puente, "Shaped Ground Plane for Radio Apparatus," *Patent US* 7,932,863 B2, April 2011.

[43] S. Kumar, L. Shafai, and N. Jacob, "Investigation of Wide-band Microstrip Slot Antenna," *IEEE Transactions on Antennas and Propagation*, vol. 52, no. 3, pp. 865–872, 2004.

[44] J.S. Lee, G.C. Kang, B. Jung, W. Jung, M.J. Park, S.H. Han, F.J Harackiewicz, and B. Lee, "Triple Band Internal Antenna Using Matching Circuits," *IEEE Antennas and Propagation Society International Symposium*, 2005, vol. 1A, pp. 442–445 vol. 1A, 3–8, 2005.

[45] J. Anguera, C. Puente, C. Borja, G. Font, and J. Soler "A Systematic Method to Design Single-patch Broadband Microstrip Patch Antennas," *Microwave and Optical Technology Letters*, vol. 31, no. 3, pp. 185–188, 2001.

[46] A. Andújar, J. Anguera, and C. Puente, "A Systematic Method to Design Broadband Matching Networks," *European Conference on Antennas and Propagation* – EuCAP 2010, Barcelona, Spain, 2010.

[47] Y. Li, B. Derat, D. Pasquet, and J.C. Bolomey, "Matching Limits for a Dual-Band Mobile Phone Antenna," *2007 International Symposium on Microwave, Antenna, Propagation and EMC Technologies for Wireless Communications*, vol., pp. 656–659, 16–17 Aug. 2007.

[48] Y. Li, T. Cantin, B. Derat, D. Pasquet, and J.C. Bolomey, "Application of Resonant Matching Circuits for Simultaneously Enhancing the Bandwidth of Multi-band Mobile Phones," *International Workshop Antenna Technology:*

Small and Smart Antennas Metamaterials and Applications, 2007. IWAT '07. 2007, pp. 479–482.

[49] R.M. Fano, "Theoretical Limitations on the Broad-Band Matching of Arbitrary Impedances," *Journal of the Franklin Institute*, vol. 249, pp. 57–83, 1950, and pp. 139–154, 1950.

[50] J. Anguera and C. Puente, "Distributed Antenna System Robust to Human Loading Effects," *Patent US* 9,007,275 B2, April 2015.

[51] J. Anguera, A. Camps, A. Andújar, and C. Puente, "Enhancing the Robustness of Handset Antennas to Finger Loading Effects," *Electronics Letters*, vol. 45, no. 15, pp. 770–771, 2009.

[52] J. Anguera, A. Andújar, Y. Cobo, C. Picher, and C. Puente, "Handset Antenna Array to Mitigate the Finger Loading Effect," *European Conference on Antennas and Propagation*, EuCAP 2011, pp. 611–614, Rome, Italy.

[53] A. Andújar, J. Anguera, Y. Cobo, C. Picher, "Distributed Antenna Systems for Wireless Handheld Devices Robust to Hand Loading," *IEEE Transactions on Antennas and Propagation*, vol. 60, no. 10, pp. 4830–4837, 2012.

[54] C.H. Li, E. Ofli, N. Chavannes, and N. Kuster, "Effects of Hand Phantom on Mobile Phone Antenna Performance," *IEEE Transactions on antennas and propagation*, vol. 57, no. 9, pp. 2763–2770, 2009.

[55] M. Pelosi, O. Franek, M. B. Knudsen, G. F. Pedersen, and J. B. Andersen, "Antenna Proximity Effects for Talk and Data Modes in Mobile Phones," *IEEE Antenna and Propagation Magazine*, vol. 52, no. 3 pp. 15–27, 2010.

[56] J. Ilvonen, O. Kivekäs, J. Holopainen, R. Valkonen, K. Rasilainen, and P. Vainikainen, "Mobile Terminal Antenna Performance with the User's Hand: Effect of Antenna Dimensioning and Location," *IEEE Antennas and wireless propagation letters*, vol. 10, pp. 772–775, 2011.

[57] W. Yu, S. Yang, C.-L. Tang, and D. Tu, "Accurate Simulation of the Radiation Performance of a Mobile Slide Phone in a Hand-Head Position," *IEEE Antennas and Propagation Magazine*, vol. 52, pp. 168–177, 2010.

[58] "Basic Standard for the Measurement of Specific Absorption Rate related to human exposure to Electromagnetic fields from mobile phones (300 MHz–3 GHz," *CENELEC-European Committee for Electrotechnical Standardization Std.* EN 50 361, Jul. 2001.

[59] T. Y. Wu and K. L. Wong, "On the Impedance Bandwidth of a Planar Inverted-F Antenna for Mobile Handsets," *Microwave Opt. Tech. Lett.* vol. 32, pp. 249–251, 2002.

[60] M.C. Huynh and W. Stutzman, "Ground Plane Effects on Planar Inverted-F Antenna (PIFA) Performance," *IEE Proc. Microw. Antennas Propag.*, vol. 150, no. 4, 2003.

[61] K. L. Wong, J. S. Kuo, and T. W. Chiou, "Compact Microstrip Antennas with Slots Loaded in the Ground Plane," *11th International Conference on Antennas and Propagation*, No. 480, April 2001.

[62] P. Vainikainen, J. Ollikainen, O. Kivekäs, and I. Kelander, "Resonator-Based Analysis of the Combination of Mobile Handset Antenna and

Chassis," *IEEE Transactions on Antennas and Propagation*, vol. 50, no. 10, pp. 1433–1444, 2002.

[63] R. Hossa, A. Byndas, and M. E. Bialkowski, "Improvement of Compact Terminal Antenna Performance by Incorporating Open-End Slots in Ground Plane," *IEEE Microwave and Wireless Components Letters*, vol. 14, no. 6, 2004.

[64] A. Byndas, R. Hossa, M.E. Bialkowski, and P. Kabacik, "Investigations into Operation of Single Multi-Layer Configuration of Planar Inverted-F Antenna," *IEEE Antennas and Propagation Magazine*, vol. 49, no. 4, pp. 22–33, 2007.

[65] M. F. Abedin and M. Ali, "Modifying the Ground Plane and Its Effect on Planar Inverted-F Antennas (PIFAs) for Mobile Phone Handsets," *IEEE Antennas and Wireless Propagation Letters*, vol. 2, pp. 226–229, 2003.

[66] B. Sanz-Izquierdo, J. Batchelor, and R. Langley, "Multiband Printed PIFA Antenna with Ground Plane Capacitive Resonator," *Electronics Letters* vol. 40, no. 22, 2004.

[67] J. Anguera, I. Sanz, A. Sanz, T. Condes, D. Gala, C. Puente, and J. Soler, "Enhancing the Performance of Handset Antennas by Means of Groundplane Design," *IEEE International Workshop on Antenna Technology: Small Antennas and Novel Metamaterials* (iWAT 2006). New York, USA, March 2006.

[68] M. Cabedo, E. Antonino, V. Rodrigo, and C. Suárez, "Análisis Modal de un Plano de Masa Radiante Doblado y con una Ranura para Terminales Móviles," *Proceedings of the XXI National Symposium* URSI '06. Oviedo, Spain, Set. 2006.

[69] J. Anguera, I. Sanz, A. Sanz, T. Condes, C. Puente, and J. Soler, "Multiband PIFA Handset Antenna by Means of Groundplane Design," *IEEE Antennas and Propagation Society International Symposium*, Albuquerque, New Mexico, USA, July 2006.

[70] A. Cabedo, J. Anguera, C. Picher, M. Ribó, and C. Puente, "Multi-Band Handset Antenna Combining a PIFA, Slots, and Ground Plane Modes," *IEEE Transactions on Antennas and Propagation*, vol. 57, no. 9, pp. 2526–2533, 2009.

[71] C. Picher, J. Anguera, A. Andújar, C. Puente, and S. Kahng, "Analysis of the Human Head Interaction in Handset Antennas with Slotted Ground Planes," *IEEE Antennas and Propagation Magazine*, vol. 54, no. 2, pp. 36–56, 2012.

[72] C. Picher, J. Anguera, A. Cabedo, C. Puente, and S. Kahng, "Multiband Handset Antenna Using Slots on the Ground Plane: Considerations to Facilitate the Integration of the Feeding Transmission Line," *Progress in Electromagnetics Research C*, vol. 7, 95–109, 2009.

[73] J. Anguera, A. Andújar, and C. Puente, "A Mechanism to Electrically Enlarge the Ground Plane of Handset Antennas: A Bandwidth Enhancement Technique," *Microwave and Optical Technology Letters*, vol. 53, no. 7, pp. 1512–1517, 2011.

[74] J. Anguera and T. Condes, "Antenna Set, Portable Wireless Device, and Use of a Conductive Element for Tuning the Ground-plane of the Antenna Set," *Patent US* 7,903,034 B2, March 2011.

[75] J. Anguera and A. Andújar, "Ground Plane Contribution in Wireless Handheld Devices using Radar Cross Section Analysis," *Progress in Electromagnetics Research M*, vol. 26, pp. 101–114, 2012.

[76] J. M. Jung, S. J. Kim, K. H. Kong, J. S. Lee, B. Lee, "Designing Ground Plane to Reduce Hand Effects on Mobile Handsets," IEEE Antennas and Propagation Society International Symposium, Honolulu, Hawaii, USA, June 2007.

[77] J. Holopainen, J. Ilvonen, O. Kivekäs, R. Valkonen, C. Icheln, and P. Vainikainen, "Near-Field Control of Handset Antennas Based on Inverted-Top Wavetraps: Focus on Hearing-Aid Compatibility," *IEEE Antennas and Wireless Prop. Lett.* vol. 8, pp. 592–595, 2009.

[78] C. Puente and J. Anguera, "Handset with Electromagnetic Bra," *Patent US* 7,456,792 B2, Nov. 2008.

[79] P. Lindberg and E. Öjefors, "A Bandwidth Enhancement Technique for Mobile Handset Antennas Using Wavetraps," *IEEE Transactions on Antennas and Propagation*, vol. 54, no. 8, 2006.

[80] C.T. Lee and K.L Wong "Internal WWAN Clamshell Mobile Phone Antenna Using a Current Trap for Reduced Ground Plane Effects," *IEEE Transactions on Antennas and Propagation*, vol. 57, no. 10, pp. 3303–3308, 2009.

[81] Q. Chen *et al.*, "Single Ring Slot-Based Antennas for Metal-Rimmed 4G/5G Smartphones," *IEEE Transactions on Antennas and Propagation*, vol. 67, no. 3, pp. 1476–1487, 2019, DOI:10.1109/TAP.2018.2883686.

[82] Y. Liu, Y. Zhou, G. Liu and S. Gong, "Heptaband Inverted-F Antenna for Metal-Rimmed Mobile Phone Applications," *IEEE Antennas and Wireless Propagation Letters*, vol. 15, pp. 996–999, 2016, DOI:10.1109/LAWP.2015.2489763.

[83] Z. Xu, Q. Zhou, Y. Ban and S. S. Ang, "Hepta-Band Coupled-Fed Loop Antenna For LTE/WWAN Unbroken Metal-Rimmed Smartphone Applications," *IEEE Antennas and Wireless Propagation Letters*, vol. 17, no. 2, pp. 311–314, 2018, DOI:10.1109/LAWP.2017.2787863.

[84] B. Yuan, Y. Cao, G. Wang and B. Cui, "Slot Antenna for Metal-Rimmed Mobile Handsets," *IEEE Antennas and Wireless Propagation Letters*, vol. 11, pp. 1334–1337, 2012, DOI: 10.1109/LAWP.2012.2227663.

[85] Y. Liu, J. Zhang, A. Ren, H. Wang and C.-Y.-D. Sim, "TCM-Based Hepta-Band Antenna with Small Clearance for Metal-Rimmed Mobile Phone Applications," *IEEE Antennas and Wireless Propagation Letters*, vol. 18, no. 4, pp. 717–721, 2019, DOI:10.1109/LAWP.2019.2901808.

[86] P. Qiu and Q. Feng, "Low-Profile Compact Antenna for Octa-Band Metal-Rimmed Mobile Phone Applications," *IEEE Transactions on Antennas and Propagation*, vol. 68, no. 1, pp. 54–61, 2020, DOI:10.1109/TAP.2019.2935135.

[87] Y. Yang, Z. Zhao, W. Yang, Z. Nie and Q. Liu, "Compact Multimode Monopole Antenna for Metal-Rimmed Mobile Phones," *IEEE Transactions*

on Antennas and Propagation, vol. 65, no. 5, pp. 2297–2304, 2017, DOI:10.1109/TAP.2017.2679059.

[88] J. Lian, Y. Ban, Y. Yang, L. Zhang, C. Sim and K. Kang, "Hybrid Multi-Mode Narrow-Frame Antenna for WWAN/LTE Metal-Rimmed Smartphone Applications," *IEEE Access*, vol. 4, pp. 3991–3998, 2016, DOI:10.1109/ACCESS.2016.2593538.

[89] Xiao-Yan Zhang, Xinxing Zhong, Chuanyun Wang, and Zhiwei Liu, "A Tunable Multiband LTE Antenna for Metal-Rimmed Smartphone Applications," *Progress in Electromagnetics Research Letters*, vol. 60, pp. 89–94, 2016. doi:10.2528/PIERL16041701.

[90] Songnan Yang, Chunna Zhang, H. K. Pan, A. E. Fathy and V. K. Nair, "Frequency-Reconfigurable Antennas for Multiradio Wireless Platforms," *IEEE Microwave Magazine*, vol. 10, no. 1, pp. 66–83, 2009, DOI:10.1109/MMM.2008.930677.

[91] Y. Li, Z. Zhang, J. Zheng, Z. Feng and M. F. Iskander, "A Compact Hepta-Band Loop-Inverted F Reconfigurable Antenna for Mobile Phone," *IEEE Transactions on Antennas and Propagation*, vol. 60, no. 1, pp. 389–392, 2012, DOI: 10.1109/TAP.2011.2167949.

[92] T. Houret, L. Lizzi, F. Ferrero, C. Danchesi and S. Boudaud, "DTC-Enabled Frequency-Tunable Inverted-F Antenna for IoT Applications," *IEEE Antennas and Wireless Propagation Letters*, vol. 19, no. 2, pp. 307–311, 2020, DOI:10.1109/LAWP.2019.2961114.

[93] Abhinay Kuchikulla, "How to Implement Aperture Tuning," Best Practices for 4G/5G Smartphones, e-guide, Qorvo, 2020.

[94] M. Rammal, L. Huitema, A. Crunteanu *et al.*, "BST Thin Film Capacitors Integrated within a Frequency Tunable Antenna," *2016 International Workshop on Antenna Technology (iWAT)*, 2016, pp. 44–47, doi: 10.1109/IWAT.2016.7434796.

[95] A. Cihangir, F. Ferrero, C. Luxey *et al.*, "Investigation of Tunable Matching Circuits for Multiband 4G Handsets," *The 8th European Conference on Antennas and Propagation (EuCAP 2014)*, 2014, pp. 601–604, doi: 10.1109/EuCAP.2014.6901830.

[96] A. Cardona, "Tunable BaSrTiO3 Applications for the RF Front End," *2011 IEEE MTT-S International Microwave Symposium*, 2011, pp. 1–4, doi: 10.1109/MWSYM.2011.5972714.

[97] A. Petosa, "An Overview of Tuning Techniques for Frequency-Agile Antennas," *IEEE Antennas and Propagation Magazine*, vol. 54, no. 5, pp. 271–296, 2012, doi: 10.1109/MAP.2012.6348178.

[98] S. Haydhah, E. Assouad, G. Forget, F. Ferrero, L. Lizzi and M. S. Sharawi, "Casing and Hand Effects on a Compact Robust Pattern Reconfigurable Sub-GHz IoT Antenna," *2022 IEEE International Symposium on Antennas and Propagation and USNC-URSI Radio Science Meeting (AP-S/URSI)*, 2022, pp. 1474–1475, doi: 10.1109/AP-S/USNC-URSI47032.2022.9886709.

[99] L. H. Trinh, F. Ferrero, "Multiband Frequency Tuneable Antennas for Selection Combining Strategy in White Space Applications," *Applied Science*, vol. 12, p. 11062, 2022. https://doi.org/10.3390/app122111062.

[100] L. A. Bronckers, A. Roc'h and A. B. Smolders, "A New Design Method for Frequency-Reconfigurable Antennas Using Multiple Tuning Components," *IEEE Transactions on Antennas and Propagation*, vol. 67, no. 12, pp. 7285–7295, 2019, DOI:10.1109/TAP.2019.2930204.

[101] J. Anguera, A. Andújar, C. Puente, J. Mumbrú, "Antennaless Wireless Device," *Pat. Application* WO2010/015365, Feb. 2010.

[102] J. Anguera, A. Andújar, C. Puente, J. Mumbrú, "Antennaless Wireless Device Capable of Operation in Multiple Frequency Regions," *Patent Application* WO 2010/015364 A2, Feb. 2010.

[103] J. Anguera, A. Andújar, C. Puente, J. Mumbrú, "Antennaless Wireless Device," *Patent US*. 8,203,492 B2 June 2012.

[104] J. Anguera, A. Andújar, C. Puente, J. Mumbrú, "Antennaless Wireless Device Capable of Operation in Multiple Frequency Regions," *Patent US*. 8,237,615 B2, Aug. 2012.

[105] J. Anguera and A. Andújar, "AntennaLess Wireless Device Comprising One or More Bodies," *Patent US* 9,147,929 B2, Sep. 2015.

[106] A. Andújar, J. Anguera, and C. Puente, "Ground Plane Boosters as a Compact Antenna Technology for Wireless Handheld Devices," *IEEE Transactions on Antennas and Propagation*, vol. 59, no. 5, pp. 1668–1677, 2011.

[107] A. Andújar, J. Anguera, C. Puente, and C. Picher, "Wireless Device Capable of Multiband MIMO Operation," *Patent US* 8,952,855 B2, Feb. 2015.

[108] A. Andújar and J. Anguera, "Compact Radiating Array for Wireless Handheld or Portable Devices," *Patent US* 9,577,325 B2, Feb. 2017.

[109] J. Anguera, C. Picher, A. Andújar, C. Puente, "Concentrated Antennaless Wireless Device Providing Operability in Multiple Frequency Regions," *Patent US* 9,379,443 B2, June 2016.

[110] A. Andújar and J. Anguera, "On the Radiofrequency System of Ground Plane Booster Antenna Technology," *Electronics Letters*, vol. 48, no. 14, pp. 815–817, 2012.

[111] A. Andújar and J. Anguera, "Multiband Coplanar Ground Plane Booster Antenna Technology," *Electronic Letters*, vol. 48, no. 21, pp. 1326–1328, 2012.

[112] J. Anguera, N. Toporcer, and A. Andújar, "Slim Bar Booster for Electronics Devices," *Patent US* 9,960,478 B2, May 2018.

[113] A. Andújar and J. Anguera, "Magnetic Boosters for Multi-band Operation," *Microwave and Optical Technology Letters*, vol. 55, no. 1, pp. 65–75, 2013.

[114] J. Anguera, A. Andújar, and C. García, "Multiband and Small Coplanar Antenna System for Wireless Handheld Devices," *IEEE Transactions on Antennas and Propagation*, vol. 61, no. 7, pp. 3782–3789, 2013.

[115] A. Andújar, J. Anguera, C. Picher, and C. Puente, "Ground Plane Booster Antenna Technology. Human Head Interaction: Functional and Biological

Analysis," *European Conference on Antennas and Propagation*, EuCAP 2012, Prague, Czech Republic.

[116] J. Anguera, C. Picher, A. Bujalance, and A. Andújar, "Ground Plane Booster Antenna Technology for Smartphones and Tablets," *Microwave and Optical Technology Letters*, vol. 58, no. 6, pp. 1289–1294, 2016.

[117] J. Anguera, A. Andújar, and C. Puente, "Antenna-Less Wireless: A Marriage Between Antenna and Microwave Engineering," *Microwave Journal (invited paper)*, vol. 60, no. 10, pp. 22–36, 2017.

[118] J. Anguera, A. Andújar, J. L. Leiva, O. Massó, J. Tonnesen, E. Rindalsholt, R. Brandsegg, Roberto Gaddi, "Reconfigurable Multiband Operation for Wireless Devices Embedding Antenna Boosters," *Electronics* vol. 10, no. 7: 808. 2021.

Chapter 3
Antenna boosters: fundamentals

This chapter introduces Antenna Booster Technology to lay the foundations for the following chapters on designing wireless devices with antenna boosters.

The chapter has the following objectives:

1. What is an antenna booster?
2. To understand the design of either single-band or multiband wireless with antenna boosters
3. The basic design steps
4. Why a small antenna booster can be efficient and provide bandwidth
5. To determine the impact of the position and size of an antenna booster on bandwidth and efficiency
6. The role of the ground plane size as well as the impact of ground clearance on bandwidth, efficiency, and radiation patterns.

3.1 Introduction

Smart sensors, smart meters, smart tracking, smart factories, smart agriculture, and all possible smart things you can imagine need to be connected. Wireless communications such as IoT enable connecting things with other things; consequently, a relevant player is required, which transmits and receives data to and from each thing: an antenna.

Wireless engineers working on developing all the new wireless devices are looking for the best antenna covering all desired frequency bands with the smallest footprint and the highest efficiency. On top of this, the antenna needs to be integrated into the design with other radiofrequency components such as modules, amplifiers, and filters. Therefore, why not treat it as another component?

Antenna Booster Technology enables one antenna component to cover any frequency band within a wide range (~0.4–12 GHz). The antenna booster is a tiny chip component that can be tuned to any frequency as needed (Figure 3.1). And because it is off-the-shelf, no customisation of the antenna will be required. Hence, the wireless architecture becomes predictable from day one and is the ideal solution for mass production.

Being surface-mount and chip-like, the antenna booster fits seamlessly in an electronic printed circuit board the same way any other electronic component (an amplifier, filter, or switch, to name a few) does, making the manufacture and design of the new generation of wireless devices simpler, faster, and more cost-effective. Antenna booster technology gets its name from the fact that an antenna booster, when strategically placed on a device's PCB (Printed Circuit Board), can boost currents on the ground plane of the PCB to make it an effective radiator [1–7]. Consequently, the size of an antenna booster is tiny, about ten times smaller, compared to other conventional multiband antennas (Figures 3.1 and 3.2). For example, the antenna boosters shown in Figure 3.1 have 5 mm × 5 mm × 5 mm and 12 mm × 3 mm × 2.4 mm, which, in terms of the wavelength, represent only $\lambda/66$ and $\lambda/30$, respectively, at 824 MHz referring to their longest dimension.

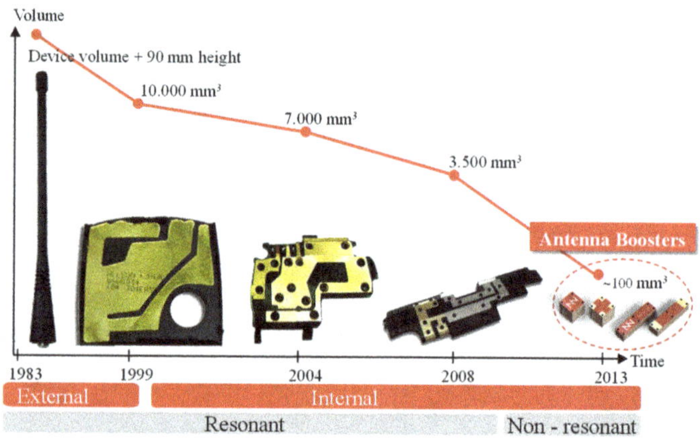

Figure 3.1 From external quarter-wavelength monopole antenna to internal custom antenna based on complex geometries to the tiny off-the-shelf antenna boosters. Antenna booster technology is based on replacing complex and usually, customised antenna designs with an off-the-shelf, miniature, and multiband component called an antenna booster.

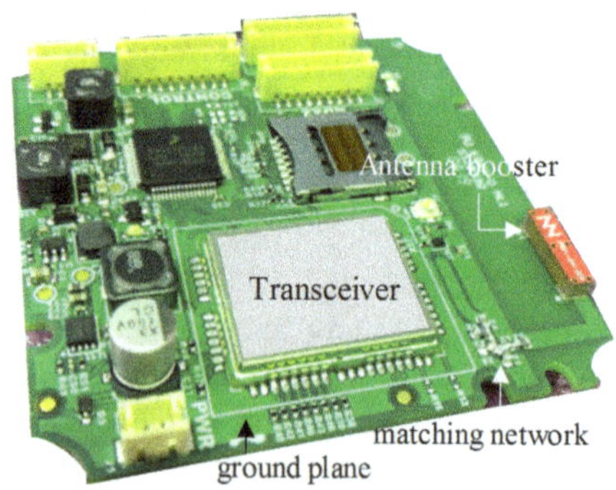

Figure 3.2 A 12 mm × 3 mm × 2.4 mm (height) antenna booster on a wireless device [8]

An antenna booster has a non-resonant impedance. Thus, a matching network comprising lumped components (inductors and capacitors) is combined with an antenna booster to achieve either single or multiband operation. The design of single and multiband matching networks is addressed in Chapters 4 and 5, respectively. In devices characterised by a small ground plane relative to the operating wavelength ($\sim 0.1\lambda$ at the lowest frequency of operation) or a high number of frequency bands, designs integrate reconfigurable matching networks with antenna boosters to function across multiple frequency bands [9–11]. The reason is that a passive matching network comprising inductors and capacitors cannot infinitely expand bandwidth. Hence, reconfigurability can provide additional frequency bands in those situations where a passive matching network has limitations.

The chapter is structured as follows (Figure 3.3). First, the history of the invention of the technology is presented in Section 3.2. The architecture of an

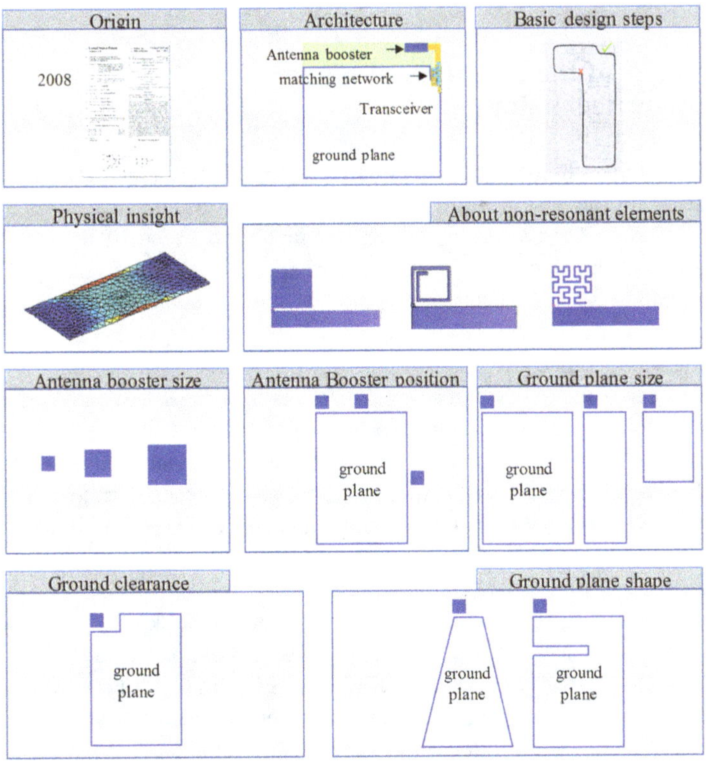

Figure 3.3 Chapter structure. Origins (Section 3.2), architecture (Section 3.3), the design steps (Section 3.4), physical insight (Sections 3.5–3.7), about non-resonant elements (Section 3.8), the impact of the antenna booster size (Section 3.9) and position (Section 3.10), on the ground plane size (Section 3.11), the ground plane clearance (Sections 3.12 and 3.13), ground plane of different shapes (Section 3.14).

antenna system with antenna boosters is discussed in Section 3.3. The basic steps to embed an antenna booster in a wireless device are detailed in Section 3.4. The physics behind the technology is studied in Section 3.5. Since antenna boosters rely on the excitation of currents on the ground plane of a wireless device, a deep dive into the modes supported by a conductive ground plane is explained in Section 3.6. The impedance nature of an antenna booster is analysed in Section 3.7. Since antenna boosters are non-resonant elements, a discussion between resonant and non-resonant elements is given in Section 3.8. The impact on impedance bandwidth as a function of the size of an antenna booster and its position on a ground plane is presented in Sections 3.9 and 3.10, respectively. Once the antenna booster size and position are studied, the following step is to understand the impact of the ground plane size and the ground clearance, as discussed in Sections 3.11–3.13. Ground planes of different forms are examined in Section 3.14. Conclusions are drawn in Section 3.15. Exercises are proposed in Section 3.16. And finally, references are included in the References section.

3.2 History of the invention of antenna booster technology

The classic approach to designing small and multiband antennas for wireless devices relies on the design of a resonant antenna that usually features a complex geometry [12–23]. On the one hand, an antenna must be small to fit inside the reduced space of the device, and on the other hand, the antenna must cover several frequency bands, and in many cases, in different frequency regions such as 698–960 MHz and 1710–2690 MHz. As a result, antenna engineering put all efforts into shaping the geometry to achieve those goals (Figure 3.4).

Although the antenna design approach results in competitive antennas in terms of size and performance, the design process is not efficient since it requires a new design for each device. One antenna designed for one device cannot be reused in another. Besides, the process is cumbersome because it needs iterations both in the simulation and lab domains to obtain the antenna shape. Therefore, it takes time to end with the final product. The big question is: can we invent a new antenna technology for any device and frequency band?

After deep-diving into research, the authors of the book, Dr. J. Anguera and Dr. A. Andújar, invented a new technology called ground plane booster or antenna booster technology. In essence, the novel idea relies on the use of a very simple and small non-resonant component, an antenna booster that, when strategically placed on the ground plane of a wireless device, boosts currents on such a ground plane, making it an effective radiating means for transmitting and receiving electromagnetic waves. Since the ground planes of wireless devices are comparable to the wavelength, enough bandwidth and efficiency can be obtained when suitably exciting efficient modes in the ground plane thanks to an antenna booster. Here is where the name ground plane booster comes from, since it boosts the currents in the ground plane. From a commercial perspective, the technology is called Virtual

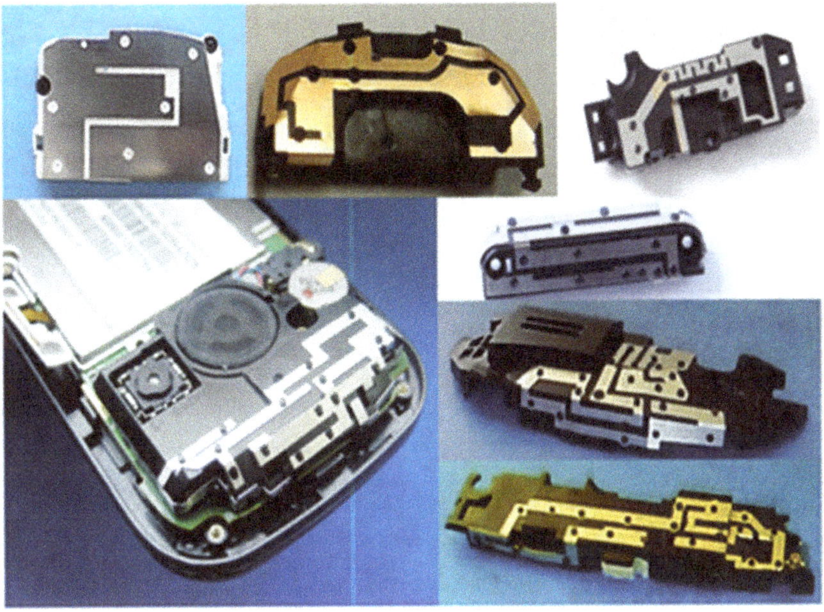

Figure 3.4 Customised small-multiband self-resonant antennas for mobile phones. Each antenna geometry is customised for a given device, making the design process slow and, therefore not efficient since a new design from scratch is needed for every device.

Antenna®, referring to the fact that the main radiator is the ground plane and the antenna booster is merely a boosting component of currents in the ground plane. A voltage probe, like a short monopole, excites parallelism in a rectangular waveguide, serving not as an antenna but as the mechanism to feed energy and excite modes within the waveguide.

In 2007, the authors of the book designed the first 5 mm × 5 mm × 5 mm antenna booster, representing only $\lambda/70$ at 824 MHz and being very small in terms of wavelength. With the addition of a single matching network with six lumped components, we showed it was possible to achieve multiband performance [24], with a volume ten times smaller than a typical PIFA (Printed Inverted-F Antenna) with similar performance [25].

Due to the novelty of the idea, we filed two patents in 2008, with the first-ever publication [1] in 2010. Those patent applications were granted in 2012 [2,3] (Figure 3.5). We did not stop there but kept researching, filing new patents, and publishing papers in prestigious scientific journals and conferences. Once the physical principles were understood, it was the right moment to translate the research into a product. In 2013, the first commercially available antenna booster was designed, featuring dimensions of 5 mm × 5 mm × 5 mm (Figure 3.1).

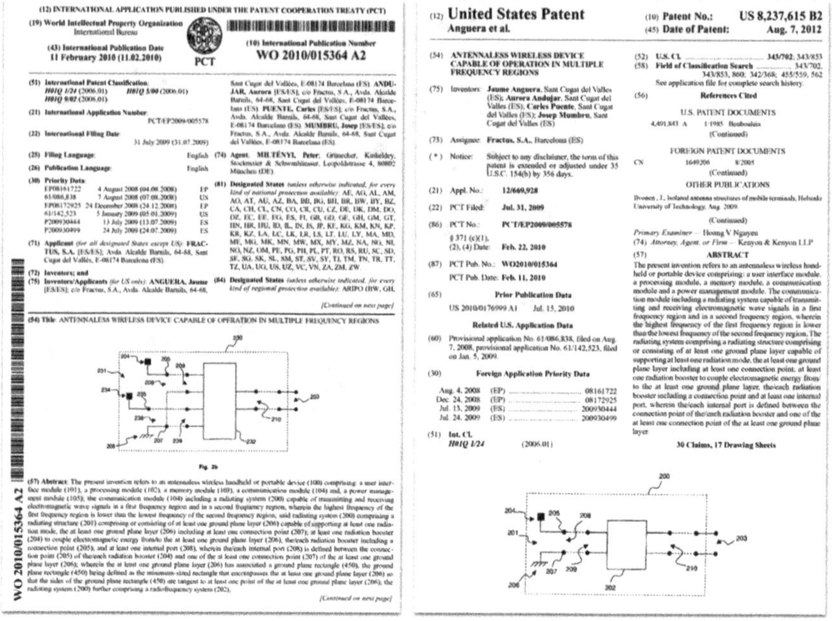

Figure 3.5 *First publication on antenna boosters in 2010 [1] and first granted patent in 2012 at the US Patent Office [3]*

The antenna booster answered the question: one antenna for any band and device. Customising a different antenna for a new device or new frequency bands was no longer necessary. With an antenna booster, it is possible to operate from 0.4 GHz up to 10.4 GHz thanks to the correct selection of the antenna booster, its location on the ground plane of a wireless device, and the design of a matching network. In fact, the design of wireless devices embedding antenna boosters looks more like a microwave problem than an antenna problem, where the antenna booster is treated as an impedance box. Moreover, the frequency of operation is not determined by the antenna but by the designers, who can freely shape the frequency band to operate, from single-band to multiband, with the design of a matching network (Figure 3.6) [5]. Besides, the matching network design is quickly addressed by matching network synthesis tools [25,26]. Therefore, this design procedure is faster and simpler than designing a complex antenna geometry from scratch. And being off-the-shelf and SMD (Surface Mounted Device) compatible, antenna boosters are ready to use, removing the need for manual intervention for assembly on a printed circuit board and thus making them convenient for mass production.

In 2015, Ignion, a technology company, was founded in Barcelona, Spain to empower the IoT ecosystem with antenna boosters [27]. Since then, Antenna Booster Technology has been adopted in many wireless IoT devices such as smart meters, trackers, modules, and wearables.

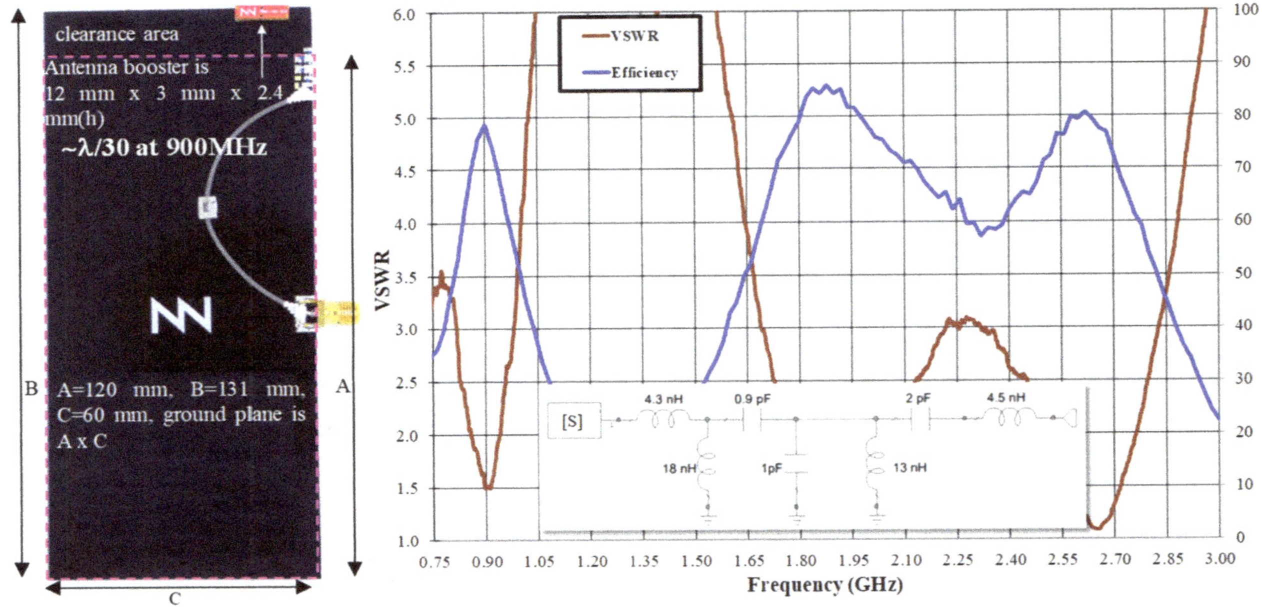

Figure 3.6 By the proper location of the antenna booster on the device and the design of a matching network, performance at any band can be obtained

3.3 Architecture of Antenna Booster Technology

The architecture of Antenna Booster Technology comprises three main elements: an antenna booster, a matching network, and a ground plane (Figure 3.7).

The antenna booster size and location on the device matter. The size of an antenna booster is tiny compared to conventional antennas. An antenna booster is typically about ten times smaller than self-resonant antennas, as discussed in Chapter 10. Some of them have a larger dimension of less than $\lambda/20$ and even less than $\lambda/30$. Being small, a minimum size is required to boost currents on the ground plane of a wireless device and, as a result, to achieve the required bandwidth. The size impact of an antenna booster is discussed in Section 3.9. As for the position, usually, corners are preferred to boost currents on the ground plane of a wireless device and, therefore, obtain the best bandwidth and efficiency. The impact of the antenna booster location is examined in Section 3.10.

Once the antenna booster type and location are determined, the matching network is another relevant element. The matching network allows an impedance match between the transceiver impedance and the antenna system to radiate/receive the maximum power possible. Matching networks for antenna boosters discussed in this book comprise lumped elements such as capacitors, inductors (Chapters 4–6), and an active component to create reconfigurable matching networks (Chapter 7). Unlike conventional antenna design, where the frequency bands of operation are obtained by a proper design of the antenna's geometry, the design is moved to the

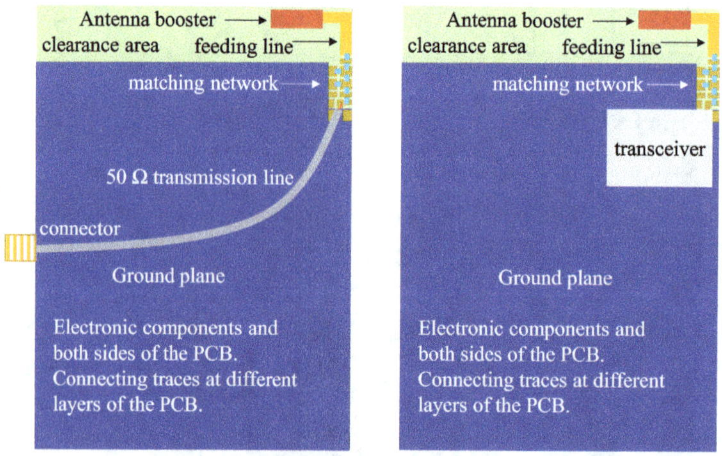

Figure 3.7 Architecture of antenna booster technology: the antenna booster, the matching network, and the ground plane. (Left) Configuration for lab testing, including a 50 Ω transmission line for testing S_{11}, total efficiency, radiation patterns, etc. (Right) Configuration of a real device where an input/output port of a transceiver is connected to the matching network of the antenna booster.

matching network for antenna booster technology. In this manner, the designer, not the antenna, determines the frequency bands of operation by freely designing a matching network. This procedure is easier and faster than creating a new antenna from scratch. Moreover, network synthesis tools address the design of a matching network automatically, as elaborated in Chapters 4 and 5. As observed, the design looks like a microwave problem rather than an antenna problem.

As far as design is concerned, the matching network is connected to a 50 Ω transmission line, such as a micro-coaxial line (Figure 3.7, left). The outer conductor of said coaxial line is connected to the ground plane in at least three points: at the connector side, in the middle of the transmission line, and at the end when connected to the matching network. This ensures that the transmission line is not disturbing the measurement of the S_{11}. The inner conductor of the transmission line is then connected to the first pad of the layout, hosting the matching network. This setup is recommended for testing the S_{11} and measuring efficiency and radiation patterns. Other transmission lines, such as CPW (CoPlanar Waveguide), can be used as well. Once the matching network is designed, the real device has a transceiver instead of said transmission line (Figure 3.7, right). The transceiver is in charge of transmitting and/or receiving the radiofrequency signal. To minimise losses and interferences, the matching network and the transceiver should be as close as possible since any transmission line connecting the matching network and the transceiver may add losses or capture noise.

The third relevant element is the ground plane of the wireless device. Since the antenna booster excites radiating modes in the ground plane, its electrical size matters. Thus, if the electrical size of the ground plane is small (maximum length $\sim 0.1\lambda$), it is a challenge to obtain bandwidth and efficiency. The impact of the ground plane on bandwidth and efficiency is deeply explained in Sections 3.11–3.13. From a practical perspective, a ground plane should be a continuous layer in at least one layer of the PCB. Usually, the top and bottom layers form the ground plane, both layers connected with vias (through-hole) to ensure good grounding connections; vias are approximately spaced one to each other $\lambda_m/20$, where λ_m is the wavelength in the medium $\lambda_m = \lambda_0/\sqrt{\varepsilon_r}$, λ_0 the wavelength in vacuum, and ε_r the relative permittivity of the substrate.

The architecture of antenna booster technology
The architecture of an antenna booster system comprises three elements: an antenna booster, a matching network between the antenna booster and a transceiver, and a ground plane.

3.4 Basic steps to embed an antenna booster in a wireless device

Antenna Booster Technology enables the integration of embedded antennas in wireless devices in three simple steps. Following these basic steps, wireless engineers designing wireless devices can easily embed antenna boosters.

3.4.1 Step 1

The first step is to place the antenna booster on the PCB of the device. As a recommended rule, the preferred positions are the corners of the ground plane of the device (Figure 3.8). In this situation, bandwidth and efficiency can be maximised, which is especially attractive for multiband applications such as devices operating at 698–960 MHz and 1710–2690 MHz. However, in other situations where bandwidth is not a constraint, such as in single band applications, for example, ISM – Industrial Scientific Medical (BW∼3%), GNSS – Global Navigation Satellite Systems (BW∼3%), Bluetooth (BW∼3%), other positions different than the corners can be used as well. A detailed analysis of the relevance of the antenna booster position in a wireless device is addressed in Sections 3.5 and 3.10. Not all corners behave in the same manner. For example, Figure 3.8 illustrates two corners: the first one with 270° free space (marked with ✓) and the other with 90° free space (marked with ×). The one marked with ✓ is generally preferred among the two cases, as demonstrated in Section 3.10. However, since ground plane size also impacts bandwidth and efficiency, such an × position may become preferred since it may be able to excite an efficiency mode on the ground plane.

It is important to emphasise that antenna location should be determined at the beginning of a device design project to optimise the performance of the antenna within the device. Otherwise, performance can be poor and difficult to improve in more advanced project phases. This means that the device performance should not consider the antenna alone: the performance depends on the antenna in combination with the device, more precisely with the ground plane of the device. Therefore, a correct selection of the antenna booster location will optimise radiofrequency performance in terms of transmitted and received power.

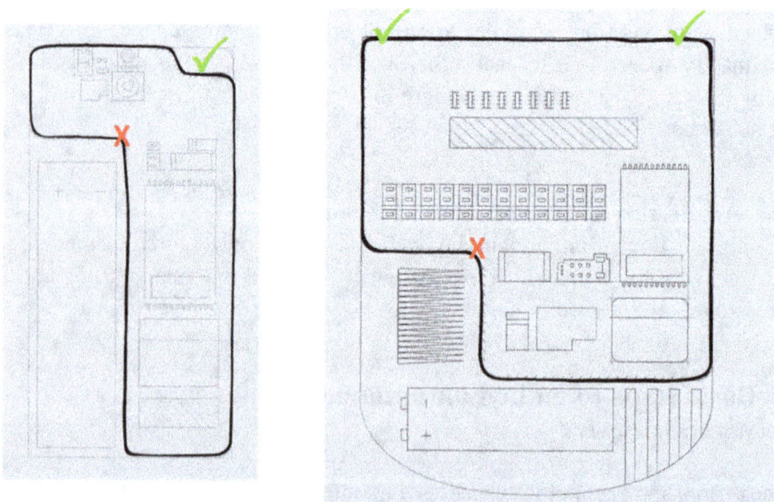

Figure 3.8 Illustrations showing the preferred (✓) and non-preferred locations (×) for an antenna booster

3.4.2 Step 2

The next step, step 2, is to design a matching network. Since the input impedance of an antenna booster is mainly reactive across the frequency bands of operation, a matching network is needed to maximise the radiated power into space and maximise the received power from space (Figure 3.9). This design flow differs from the traditional antenna design, where the antenna geometry is customised depending on the frequency bands of operation [24]. For Antenna Booster Technology, however, the only customisation needed is for the matching network, resulting in a faster and more accessible design flow. In effect, the design of a matching network can be fully automatised by using network synthesis tools available in most microwave circuit simulators used in the wireless industry [25,28,29].

Thanks to this design flow, an antenna booster can operate at any band and in any device by just designing a suitable matching network. For example, a simple L-type matching network can work if a wireless device needs operation at the 900 MHz band (Figure 3.10). However, if the device needs to operate at multiple

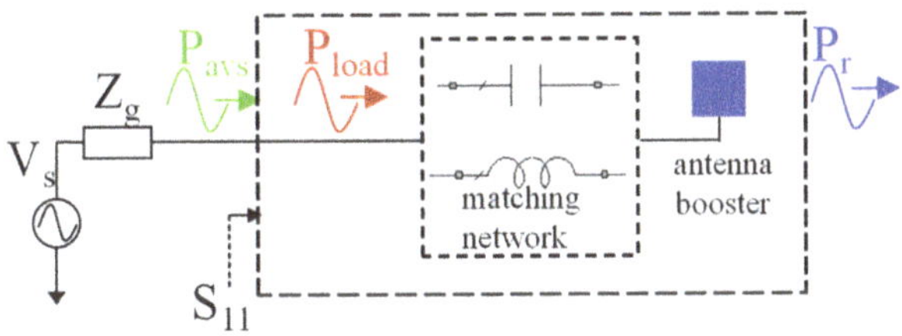

Figure 3.9 A matching network is designed and located between a radiofrequency module (represented by the source V_s and impedance Z_g) and the antenna booster to maximise radiated power P_r

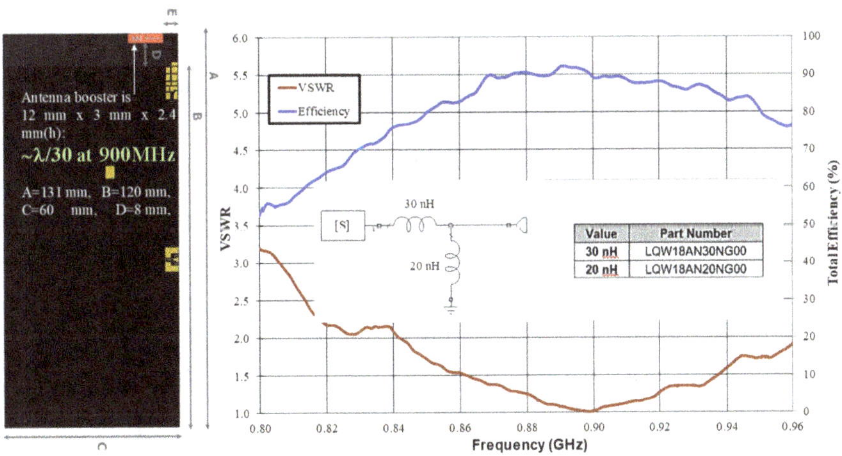

Figure 3.10 (Left) A 12 mm × 3 mm × 2.4 mm antenna booster with a matching network comprising two lumped SMD components for operation at 900 MHz. (Right) Measured VSWR and total efficiency.

frequency bands with two separate frequency regions, the same antenna booster as in the previous case is utilised but with a different matching network (Figure 3.6).

As seen from these two examples (Figures 3.6 and 3.10), the antenna booster is the same component for both cases, and the only part that changes is the matching network design for each case. Therefore, a new antenna design is unnecessary for each situation since the same antenna booster can be used for any band. Furthermore, if the PCB size changes, the same design flow applies; that is, the antenna booster can be the same, and only the matching network is designed to operate at the desired number of bands [30]. Consequently, this is an advantage for wireless engineers who need to embed an antenna in their devices since the antenna booster remains the same. Hence, there is no need to choose a different antenna to integrate into all their various devices. Also, this represents an economy of scale advantage since the same antenna booster can be integrated across other platforms.

3.4.3 Step 3

The third step in the design flow is to test the device. Once the matching network has been implemented into the device's PCB, VSWR (Voltage Standing Wave Ratio) and efficiency must be tested. VSWR can be tested with a vector network analyser (VNA), providing information about VSWR (or S_{11}). By measuring VSWR (see examples in Figures 3.6 and 3.10), the wireless designer knows how well the matching network and the antenna booster behave. Usually, the VSWR results of less than three are preferred across the frequency bands of operation.

Once the VSWR achieves a specified target, total efficiency (η_t) must be tested, which is carried out with the device inside an anechoic chamber (Figure 3.11). Total efficiency is the ratio between the power radiated into space (P_r) and the available

Figure 3.11 Anechoic chamber for testing total efficiency – the full 3D pattern can be measured thanks to a mechanical rotation of the device under test in the middle of the arc while several probes measure the vertical plane thanks to electronic scanning

power of the radiofrequency module (P_{avs}) (Figure 3.9). Although VSWR measurement provides a good sense of how well the antenna system is performing, total efficiency will inform how much power from the module is radiated into space and how much is lost in nearby components, materials, and components of the matching network. Total efficiency is, therefore, a relevant figure of merit to be sure that the device will be competitive in the market. Moreover, when total efficiency is measured, taking the whole device parts (the PCB including the antenna booster, the matching network, as well as battery, displays, casing, etc.) provides information about the TRP (Total Radiated Power), which is utilised in many wireless device certifications. TRP is linked to total efficiency, as shown in Equation (3.1):

$$TRP(dBm) = P_{avs}(dBm) + 10 \cdot \log \eta_t \tag{3.1}$$

where P_{avs} is the nominal output power from the radiofrequency module.

For example, consider that we have a radiofrequency module with a nominal output power of 23 dBm. Let us suppose the TRP should be above 18 dBm to certify the product. This means that the total efficiency must be above 31.6%. This is, however, a theoretical limit since the output power of a transceiver may vary due to tolerances, and instead of transmitting the nominal power, it may transmit 1 dB or 2 dB less. This is not an antenna responsibility, but the antenna can help if the total efficiency is higher than the theoretical limit. A second issue may appear if the antenna impedance is not well-matched to the transceiver. In that case, the output power P_{avs} may decrease [31]. Therefore, total efficiency and good matching are vital in providing a competitive performance to the wireless device in terms of range and battery life. This reveals that total efficiency is a parameter of paramount relevance. In this sense, Antenna Booster Technology can easily optimise total efficiency in a product certification since if any adjustment needs to be made, this can be addressed by adjusting the matching network. Changing the antenna geometry in this situation will put the project at risk since a new design takes weeks of work, which is critical at this stage. On the contrary, Antenna Booster Technology has the flexibility to simply adjust the matching network, which is faster and easier, and, in the end, this is an efficient process that is very convenient for wireless design engineers developing IoT devices.

The role of matching networks

A matching network typically comprises inductors and capacitors. The matching network is located between the transceiver and the antenna booster. It allows extracting the maximum power from the transceiver and delivering it to the antenna booster and then to the space, and vice versa, it maximises the received power from space to be delivered to the transceiver. Wireless devices embedding antenna boosters provide from single frequency band to multi-band operation in combination with a matching network. The design of matching networks is fully addressed by software, making the design process fast and easy.

3.5 The physics behind antenna booster technology

The term "ground plane booster" or "antenna booster" describes an electrically small, non-resonant element with limited standalone radiation capabilities that enhances currents and excites efficient modes within the ground plane of a wireless device. By exciting such modes, operation can be achieved for any wireless device size and frequency band by designing a matching network. Therefore, the complex geometries of small and multiband antennas are simplified with the very simple structure of the ground plane booster and moving the intelligence to the matching network. This makes ground-plane boosters ten times smaller in size while being flexible enough to operate on any radiocommunication protocol of a wireless device.

An antenna booster should not be considered an antenna but an element that boosts currents on the ground plane. In fact, an antenna booster is, per se, a poor radiating element. In effect, the performance is poor if the antenna booster is isolated from the wireless device. The following analysis considers a typical 12 mm × 3 mm × 2.4 mm (height) antenna booster in a dipole configuration, a monopole configuration on an infinite ground plane, and the antenna booster in a wireless device having a 120 mm × 60 mm ground plane (Figure 3.12). Antenna boosters are modelled as a conductive 3D parallelepiped with a conductivity $\sigma = 5.9 \times 10^7$ S/m.

Computed input impedance, where all cases present a reactive behaviour for frequencies below 1 GHz (Figure 3.13). Besides, for the dipole and monopole configuration, the real part at 900 MHz is low compared (less than 1 Ω) to the 38.3 Ω of the antenna booster on the PCB.

Bandwidth is computed according to Equations (3.2) and (3.3) [32], where it clearly shows a very poor bandwidth for frequencies below 3 GHz (Figure 3.14) for the monopole and dipole configurations. At the sub-1 GHz region, bandwidth is below 0.2%, making it impractical for most common radio systems, such as those in wireless communications. Note that Equations (3.2) and (3.3) allow bandwidth computation without needing to have the antenna matched. Based on the input impedance of the antenna, a bandwidth estimation can be easily computed. Other

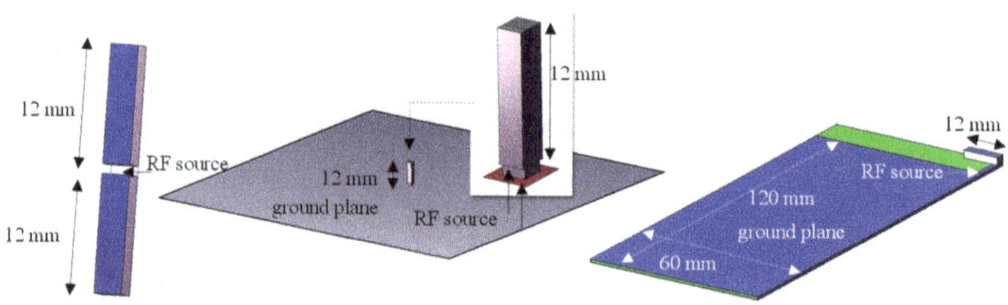

Figure 3.12 Antenna booster in a dipole and monopole configuration and an antenna booster on a 120 mm × 60 mm ground plane

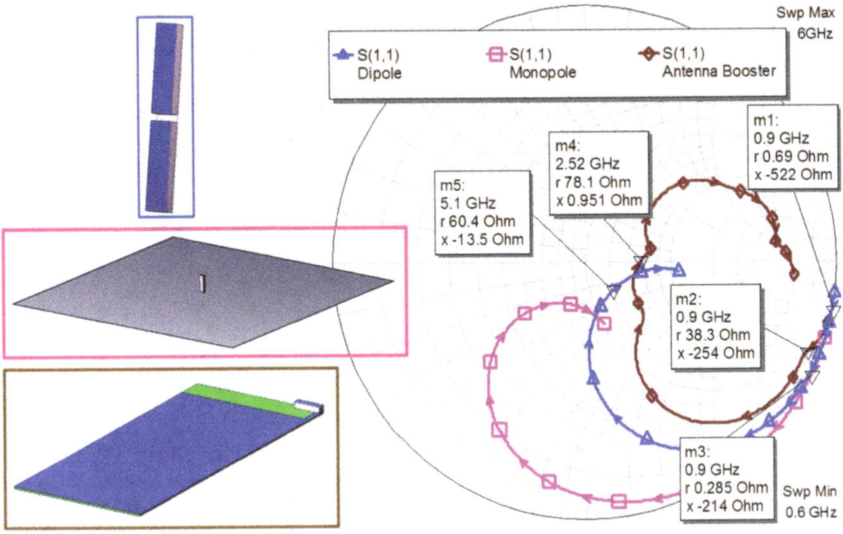

Figure 3.13 Impedance for the antenna booster on a 120 mm × 60 mm ground plane and of an antenna booster in a dipole and monopole configuration

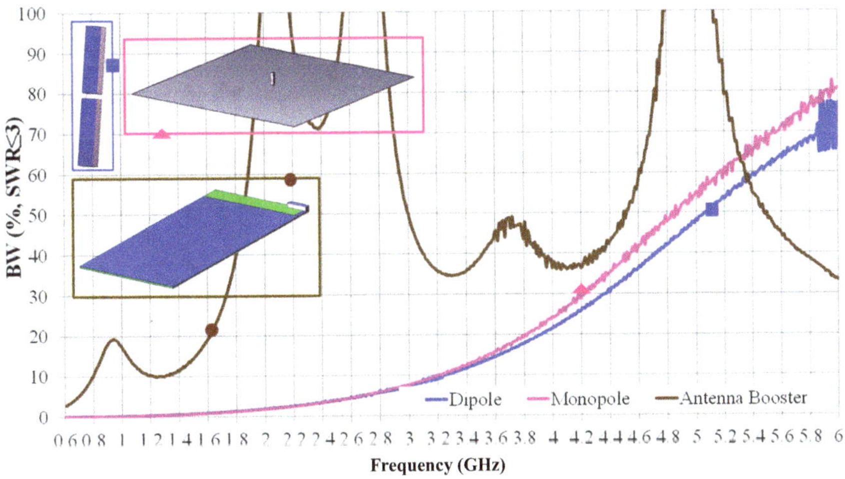

Figure 3.14 Computed bandwidth for the antenna booster on a dipole and monopole configuration and antenna booster. Not to scale. See the exact size in Figure 3.12.

approaches to calculating this bandwidth are also known as potential bandwidths [33]. The name potential bandwidth is also adopted here. The wording *potential* refers to the bandwidth one could expect when matching the impedance with a simple matching network with two components (inductor/capacitor). Bandwidth estimation with Equations (3.2) and (3.3) may diverge for certain frequencies. In that situation, the bandwidth is large but smaller than what Equations (3.2) and (3.3) predict. A more accurate bandwidth can be obtained by matching the particular frequency with a two-component matching network. Nevertheless, Equations (3.2) and (3.3) are employed here, sufficient to compare several scenarios.

$$Q_a(\omega) = \frac{\omega}{2R(\omega)}\sqrt{\left[\frac{dR(\omega)}{d\omega}\right]^2 + \left[\frac{dX(\omega)}{d\omega} + \left|\frac{X(\omega)}{\omega}\right|\right]^2} \tag{3.2}$$

$$BW = \frac{f_2 - f_1}{f_0} = \frac{SWR - 1}{Q_a \cdot \sqrt{SWR}} \tag{3.3}$$

Besides bandwidth, efficiency should also be considered when embedding antenna boosters in a wireless device. Radiation efficiencies have been computed for the antenna booster connected to a ground plane, dipole, and monopole configuration (Figure 3.15). Although the electrical size of both the antenna booster in a dipole or monopole configuration is small in terms of the wavelength at 900 MHz ($\sim\lambda/30$ at 900 MHz for the monopole configuration), the radiation efficiency is above 95%. Therefore, although the bandwidth is small, one may consider using an antenna booster in a dipole configuration for a narrow band application since a matching network can match the reactive impedance to 50 Ω. However, as shown in Figure 3.13, the real part of the input impedance at 900 MHz is very low, only 0.69 Ω. This low impedance is comparable to the resistance of a real lumped component (inductor or capacitor) of a matching network that matches the reactive impedance to 50 Ω. Therefore, most of the power will be dissipated in the matching network. This means that the quality factor of the components of the matching network (Q_m) should be larger than the Q of the antenna booster plus ground plane (Q_a) to ensure that the matching network is introducing low losses (Equation (3.4)

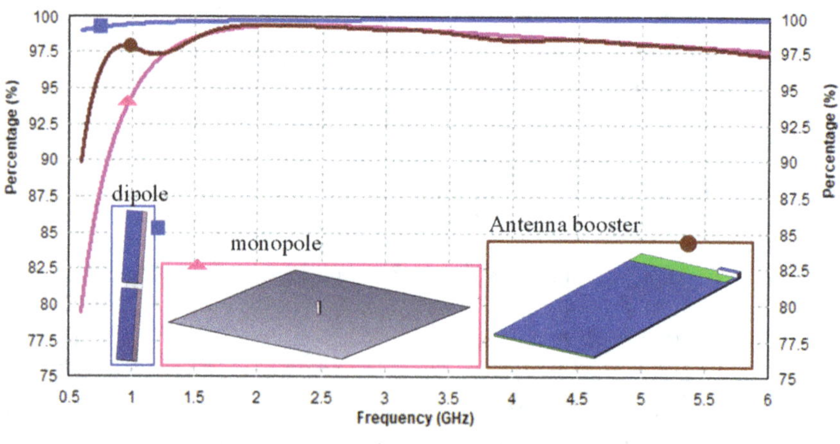

Figure 3.15 Simulated radiation efficiency. Conductors are modelled with $\sigma = 5.9 \times 10^7$ S/m.

[34]). Equation (3.4) is derived considering that the Q_m of the components of a matching network is the same.

$$\eta_t = \eta_r \cdot \eta_m \cdot \left(1 - |S_{11}|^2\right) = \frac{\eta_r \cdot \left(1 - |S_{11}|^2\right)}{1 + \frac{Q_a}{Q_m}} \qquad \eta_m = \frac{1}{1 + \frac{Q_a}{Q_m}} \tag{3.4}$$

Note that Equation (3.4) is more specific than Equation (1.14). In this situation, the losses due to the matching network components are explicitly included to better understand the impact. As shown next, the losses due to a matching network do not depend on Q_m but on the relationship between Q_m over Q_a. For example, in those situations where Q_a is low, the Q_m of the components is not as critical as far as losses are concerned. However, if the Q_a is high (small ground plane sizes, as shown next), the Q_m should be carefully selected to avoid excessive losses.

The term η_r considers the radiation efficiency of the antenna booster system and any other lossy mechanism (rather than the matching network already considered with η_m). For example, losses due to nearby components of the device, a human/animal body may absorb power.

The following discussion deals with the importance of the ratio Q_a/Q_m.

Building upon the previous scenario, given the high Q of the dipole configuration at 930 MHz (with Q_a = 700), the Q_m of the lumped components must surpass this value significantly. However, achieving such high Q_m values is challenging with real components, especially with inductors having Q_m values around 100 and 200.

To reflect the importance of the Q_m, two simple matching networks to bring the impedance at 930 MHz to 50 Ω are designed. The first has a component Q_m of 700, and the second has a Q_m of 70 (Table 3.1). As noted, while the initial radiation efficiency of the dipole configuration at 930 MHz was 99%, it decreases to 49.5% when the Q_m for the lumped inductors is 700, and further plummets to only 2.8% for Q_m = 100, which represents the typical Q of lumped SMD components with current technology.

This fact reflects that an antenna booster is not an antenna per se since it has very poor bandwidth and total efficiency performance. However, the situation changes drastically when an antenna booster is connected to a ground plane. Its correct location ensures not only competitive bandwidths but also, due to a lower Q of the antenna system (the combination of the ground plane and the antenna booster), total efficiencies are high. For validation purposes, the same matching network with $Q_m = 100$ is adjusted for the antenna booster on the PCB, resulting in a total efficiency of 90%. Note that the Q_a for a 12 mm × 3 mm × 2.4 mm (height) antenna booster on a 120 mm × 60 mm ground plane is $Q_a = 6$ at 930 MHz. Therefore, as long as lumped components feature a $Q_m \gg Q_a$, losses due to the matching network will be insignificant. Equation (3.4) gives a practical design rule: if $Q_m > 10 \cdot Q_a$, losses due to the matching network is 0.41 dB. This topic is revisited in Chapters 4 and 5, dealing with the design of both single-band and multiband matching networks, where matching networks comprise both inductors and capacitors featuring different Q_m.

Table 3.1 Impact of the total efficiency as a function of the relation of Q_m with respect to Q_a at 930 MHz

<table>
<tr><th></th><th>Matching network with lumped components with finite Q</th><th>Q_m</th><th>Losses of the matching network</th><th>Total efficiency</th></tr>
<tr><td>Dipole configuration, Q_a = 700</td><td>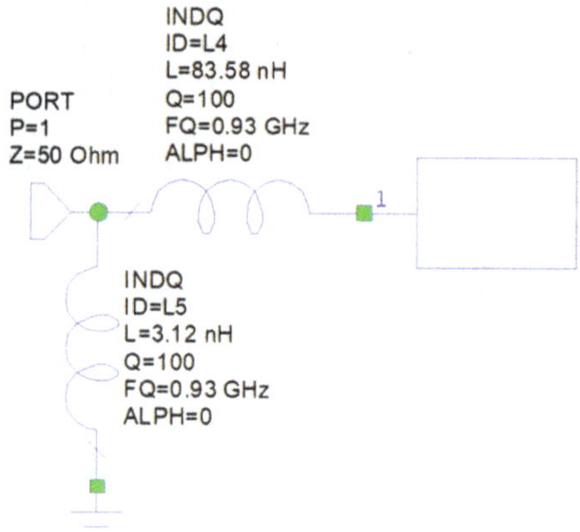
</td><td>100</td><td>8.9 dB</td><td>2.8%</td></tr>
<tr><td></td><td>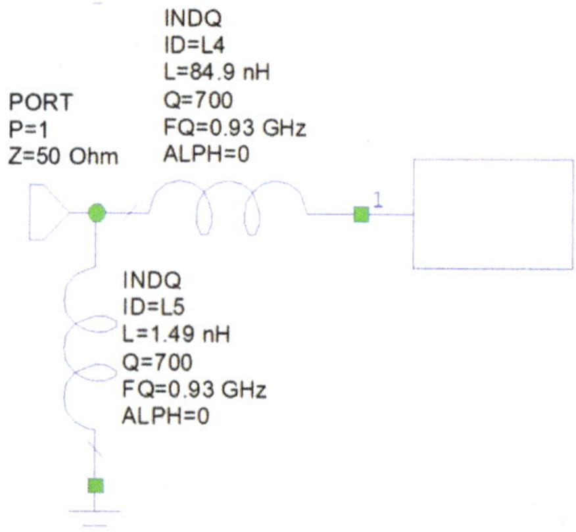
</td><td>700</td><td>3 dB</td><td>49.5%</td></tr>
<tr><td>Antenna booster on PCB, Q_a = 6</td><td>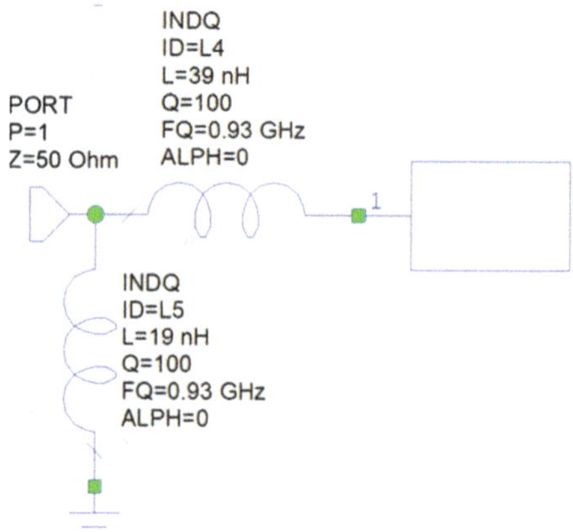
</td><td>100</td><td>0.26 dB</td><td>90.0%</td></tr>
</table>

Quality factor of components (Q_m) and radiating system (Q_a)
The quality factor of the components (Q_m) in the matching network should be as high as possible to minimise dissipation losses and be larger than the Q_a of the antenna booster, including the ground plane. As a practical rule, if $Q_m > 10 \cdot Q_a$, losses introduced by the matching network are less than 0.5 dB.

3.6 Modes on a ground plane

To understand how an antenna booster can effectively excite currents on a ground plane depending on the antenna booster location, a characteristic mode analysis is presented. Characteristic modes theory [35–38] becomes a valuable tool for understanding the ground plane contribution to bandwidth and radiation. The characteristic modes theory was initially introduced by Garbacz in 1968 and later refined by Harrington and Mautz in 1971. The theory allows for determining the wave modes supported by a particular conducting body as a function of its shape and size [36,38]. This fact provides a significant physical insight into the radiation properties of a particular conducting object.

The theory is based on the impedance matrix ($\widehat{Z}$), which relates a surface current density (J_s) to the tangential components of the electrical field on the surface S of the conducting body (Equation (3.5)).

$$E_{\tan} = \widehat{Z} \cdot J_s = (\widehat{R} + j\widehat{X}) \cdot J_s \tag{3.5}$$

Note that $\widehat{R}$ and $\widehat{X}$ represent the real and imaginary parts of the impedance matrix $\widehat{Z}$, respectively. The reciprocity theorem states that if $\widehat{Z}$ is a linear symmetric operator, $\widehat{R}$ and $\widehat{X}$, it will also be a real and symmetric operator. Accordingly, all the eigenvalues are real, and all the eigen currents are real over the surface in which they are defined. This problem formulation, developed in detail in [35–38], leads via the generalised eigenvalue problem to real eigenvalues λ_n and real eigenfunctions or eigen currents $J_{s,n}$, which present orthogonal properties (Equation (3.6)). This formulation can be reduced to a matrix form (Equation (3.7)) that can be solved using standard algorithms for eigenvalues problems.

The solution of Equation (3.7) provides the corresponding eigenvalues associated with each characteristic mode, and their knowledge offers relevant physical insight into their radiation properties. A characteristic mode is in resonance when its corresponding eigenvalue is zero. In contrast, a mode contributes to storing electric energy or magnetic energy when its eigenvalue is lower than zero ($\lambda_n < 0$) and larger than zero ($\lambda_n > 0$), respectively.

$$\widehat{X} \cdot J_{s,n} = \lambda_n \cdot \widehat{R} \cdot J_{s,n} \tag{3.6}$$

$$[X] \cdot \vec{J}_n = \lambda_n \cdot [R] \cdot \vec{J}_n \tag{3.7}$$

Similarly, the electric fields produced by these eigen currents can be derived from Equation (3.5) according to Equation (3.8). The electric fields produced by two currents J_m and J_n are also equiphasal and present orthogonal properties, which become apparent from the Poynting theorem (Equation (3.9)).

$$E_n(\vec{J}_n) = Z(\vec{J}_n) = R(\vec{J}_n) + jX(\vec{J}_n) = R(\vec{J}_n) \cdot (1 + j\lambda_n) \tag{3.8}$$

$$\begin{aligned} P(J_m, J_n) &= \oiint_{S'} \vec{E}_m \times \vec{H}_n^* ds + j\omega \iiint_{\tau'} (\mu \vec{H}_m \cdot \vec{H}_n^* - \varepsilon \vec{E}_m \cdot \vec{E}_n^*) d\tau \\ &= (1 + j\lambda_n)\delta_{mn} \end{aligned} \tag{3.9}$$

Due to the orthogonal properties of both electric fields and surface currents, each characteristic mode radiates power independently from the other. Thus, they can be used as a basis for expanding the total current J on the surface of the conducting body, according to Equation (3.10).

$$J = \sum_n \frac{V_n^i \cdot J_n}{1 + j\lambda_n} \tag{3.10}$$

It is worth noticing that characteristic modes are independent of the source excitation (V_n^i). Nevertheless, this source excitation (V_n^i) influences the total current on the surface of the conducting body (J) and determines whether a characteristic mode will be adequately excited. The entire current surface is directly proportional to the term, $\frac{1}{1+j\lambda_n}$, and its modulus is the so-called modal significance (MS) (Equation (3.11)). This term represents the normalised amplitude of the characteristic modes as well as an alternative representation of their electromagnetic properties. In this sense, the more effectively the characteristic mode contributes to the radiation process, the closer the modal significance approaches one.

$$MS_n = \left| \frac{1}{1 + j\lambda_n} \right| \tag{3.11}$$

Characteristic mode analysis is now applied to understand how an antenna booster can excite currents on a ground plane and how the position of the antenna booster is relevant to exciting such modes.

Be a conductive 120 mm × 50 mm rectangle (a ground plane of a wireless device). Eigen currents and eigenvalues for the first modes are computed using an MoM (Method of Moments) code in MATLAB® [39] (Figure 3.16). From the eigenvalues, the modal significance is calculated (Figure 3.17). From the results (Figures 3.16 and 3.17), it is observed that below 1.1 GHz, there is only a dominant mode or fundamental mode. The current of this eigenmode 1 has a maximum current at the centre of the long edges and a minimum at the corners. This information can also be read using the electrical field, where the field is maximum at the corners of the ground plane, whereas it has a null at the centre of the long edges. For other modes, the distribution changes (Figure 3.16).

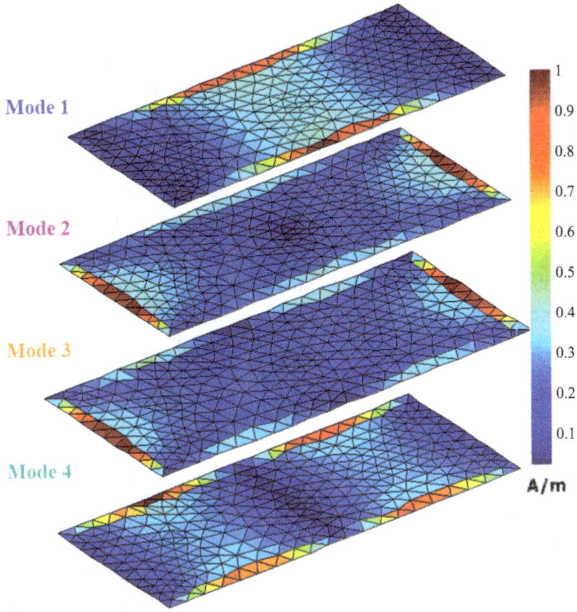

Figure 3.16 *Eigen currents associated with the characteristic modes J_1, J_2, J_3, J_4 of a rectangular 120 mm × 50 mm ground plane*

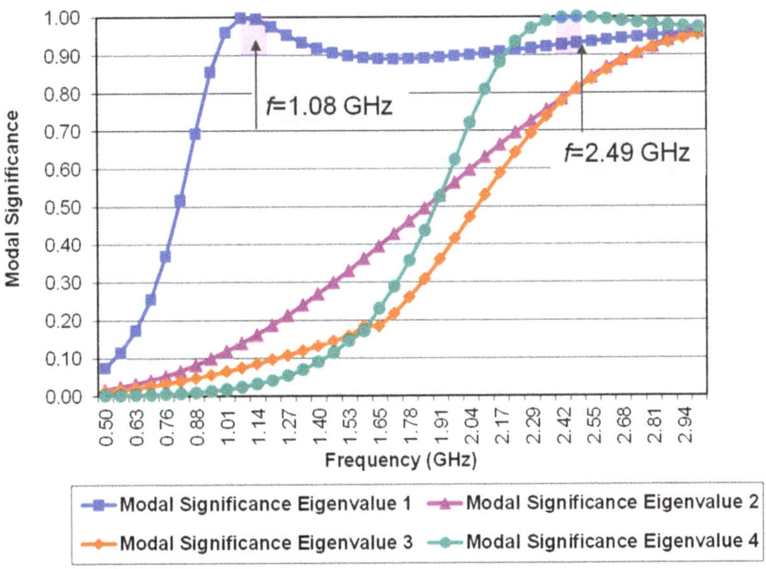

Figure 3.17 *Modal significance associated to the characteristic modes J_1, J_2, J_3, J_4 of a rectangular 120 mm × 50 mm ground plane*

Several conclusions can be derived from this data:

- There is the first maximum of modal significance at 1.08 GHz. Since the ground plane length is 120 mm, this size corresponds to 0.43λ. Therefore, as a practical rule, if a device needs to operate at a particular frequency region, if the length is 0.43λ at the lowest frequency of operation, the modal significance is close to 1. In return, the best bandwidth and efficiency will be obtained. This number changes a bit when considering the antenna booster, as shown in the following sections, depending on the ground plane width. As a reasonable range, the optimum length fits within 0.36λ and 0.43λ, and thus, 0.4λ can be considered as an optimum length.
- For frequencies below the first maximum, modal significance eigenmodes tend to be zero. This means that currents will be excited with little intensity, and in return, bandwidth and efficiency will decrease as the ground plane length decreases in terms of the wavelength. For example, suppose one wireless device has a maximum length of 0.3λ, and another has only 0.1λ. In that case, this last one will present less bandwidth and efficiency, provided both use the same antenna booster.
- For ground planes large in terms of wavelength, the density of modes increases, with modal significance close to one. Therefore, it is easier to obtain bandwidth compared to electrically small ground planes.

A complementary manner of determining the size of a ground plane to excite the first mode better uses radar cross-section (RCS) analysis [40]. In this situation, a plane wave with an E-field aligned with the longest dimension of a rectangular ground plane impinges from the normal direction.

Using a commercial MoM code (IE3D), RCS is computed versus frequency. RCS is normalised in all cases to its maximum value. Therefore, RCS ranges from 1 to 0, where the maximum value 1 indicates the maximum backscattering, which coincides with the better excitation of the metal plate, whereas 0 means transparency, i.e., no scattering is produced by the metal plate (Figure 3.18).

To validate the method, the first characteristic mode (fundamental mode) has been computed for several platforms. The modal significance parameter becomes practical for comparison purposes.

As [36] suggests, one can compute RCS by summing all characteristic modes supported by a structure. In the present case, the proposed method directly calculates RCS to predict the fundamental frequency at which the ground plane can be excited. Hence, only the first characteristic mode is used.

Both RCS and MS have been compared for the three platforms, concluding that the prediction of the resonant frequency of the fundamental mode is in good agreement (difference less than 6.3% in the worst case – Figure 3.18). It should be pointed out that RCS and MS agree from the low-frequency region up to the resonance of the fundamental mode. Beyond that, as more modes are excited, RCS and modal significance diverge. The reason is that some higher modes cannot be excited when the plane wave impinges on the metal plate from the normal direction. Nevertheless, as far as the fundamental mode is concerned, the RCS is a

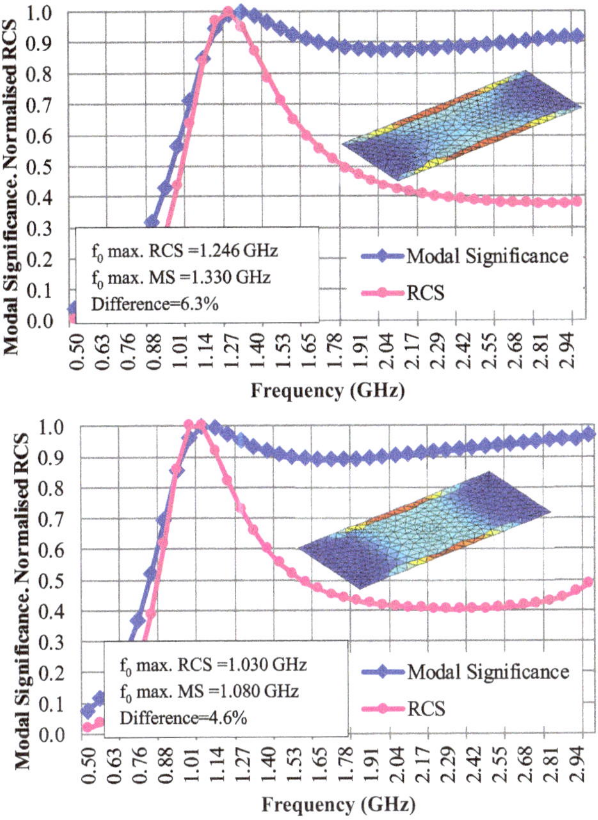

Figure 3.18 Comparison between normalised RCS and modal significance; (top) 100 mm × 40 mm ground plane; bottom) 120 mm × 50 mm ground plane

simple and fast method to compute the resonant frequency at which the ground plane of a wireless device can be better excited.

Regarding the obtained results, it is observed that the resonance of the fundamental mode decreases when the ground plane length is larger. For example, for the 100 mm × 40 mm ground plane (Figure 3.18(a)), the resonance based on the RCS method occurs at 1.246 GHz, while it is 1.030 GHz for the 120 mm × 50 mm ground plane (Figure 3.18(b)). This means that an antenna operating in the frequency region around 900 MHz would produce better performance in terms of bandwidth and efficiency when integrated into the second case since the RCS indicates that the ground plane resonance is placed at lower frequencies, as shown in Figure 3.19. It is interesting to note further that 100 mm length results in 0.39λ at 1.17 GHz, which is aligned with the experimental results to optimise bandwidth for a PIFA antenna demonstrated in [41].

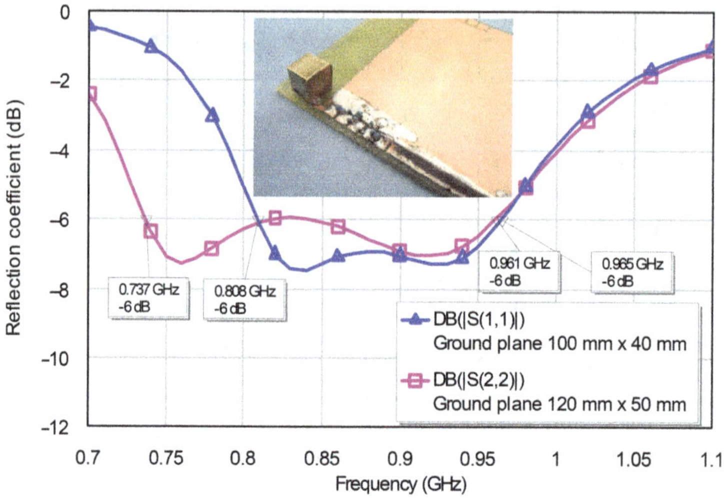

Figure 3.19 Measured reflection coefficient for a 5 mm × 5 mm × 5 mm antenna booster with two different ground plane lengths. Both cases comprise a matching network.

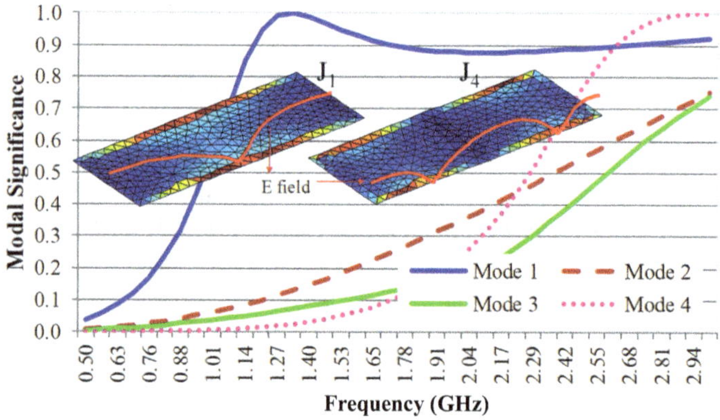

Figure 3.20 Modal significance for the first four modes for a 100 mm × 40 mm ground plane. Eigen current associated with mode 1 and mode 4.

The question now is: how can we effectively excite the first mode? If we analyse the eigen current for the first mode J_1, we observe a maximum E-field at the corners of the ground plane, whereas there is a null at the middle of the long edge (Figure 3.20). Therefore, the suitable position for the antenna booster to excite J_1 is at the corner of the ground plane, whereas if we place it at the centre of the long edge, the mode will be weakly excited. At this point, we can consider an antenna booster, not an antenna but an electrical probe able to boost currents on the ground plane. There is a direct analogy with microstrip patch antennas. For example, if we consider a square microstrip patch operating at the fundamental mode TM_{10}, the electrical field distribution is null along the middle line parallel to the radiating apertures. Therefore, if we place a voltage probe at such a point, the TM_{10} will not be excited. But if we put it at the edge, the TM_{10} will be excited, and thus, the antenna will radiate. The feeding probe to excite the cavity of a microstrip patch antenna is not considered an antenna but a probe. A similar situation occurs for an antenna booster; depending on the location of the antenna booster, a particular mode can be excited. Note that the antenna booster is connected to the ground plane with a voltage source. We called this type of antenna booster an *electric antenna booster* since it is placed at a point with a maximum E-field or at least at a position that does not have a weak intensity of the E-field.

What if the antenna booster is placed at the corner of the rectangular ground plane under consideration or the centre of the long edge? To answer this question, a 5 mm × 5 mm × 5 mm booster is situated in the corner, and in a second scenario, it is located at the centre of the long edge (Figure 3.21).

The potential bandwidth is calculated by computing the input impedance and using Equations (3.2) and (3.3). At low frequencies (f< 1.1 GHz), where J_1 dominates, the bandwidth obtained for the booster at the corner is much larger than the case with the booster at the centre of the long edge. This confirms that the antenna booster at the corner excites mode 1; consequently, acceptable bandwidth is obtained. For monopole antennas, the same conclusions can be found in [42].

In the same way, for the antenna booster at the centre of the long edge, mode J_1 is not excited because the E-field is null, and thus, poor bandwidth (<1%) is obtained. Regarding radiation efficiency, the antenna booster at the corner again provides much better efficiency than the centre position at low frequencies, being practically the same beyond 1.7 GHz (Figure 3.22).

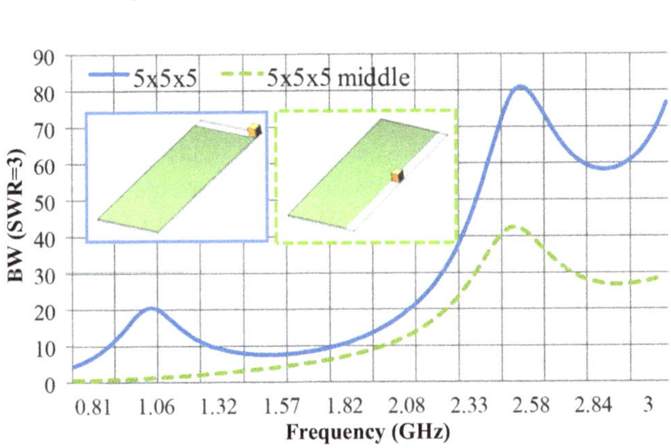

Figure 3.21 Impact on bandwidth as a function of the antenna booster position. Antenna boosters are 5 mm × 5 mm × 5 mm, ground planes are 100 mm × 40 mm. Bandwidth is computed according to Equations (3.2) and (3.3).

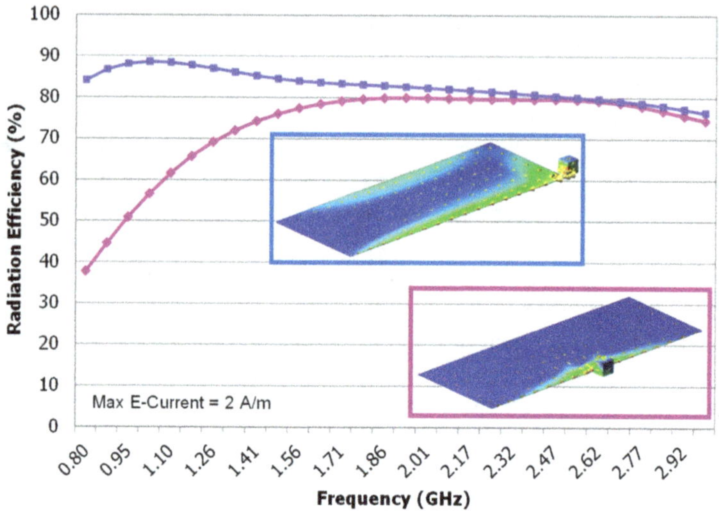

Figure 3.22 Simulated current distribution and radiation efficiency of a 5 mm × 5 mm × 5 mm antenna booster on a 100 mm × 40 mm ground plane

It is worth comparing the frequency having the maxim of modal significance for J_1 (1.28 GHz) with the frequency having the maximum potential bandwidth when adding an antenna booster for the 100 mm × 40 mm ground plane (1.07 GHz). They are close, but not the same. The reason is that the antenna booster excites not only J_1 but other modes as the frequency increases. For example, as frequency increases, a second mode starts to be excited, with the currents aligned with the short 40 mm edge. Therefore, since there is a contribution of at least two modes, the bandwidth is lower than when mainly J_1 is excited. This means that if we reduce the width of the ground plane, J_1 is excited with less impact of other higher-order modes, and thus, more bandwidth should be obtained with a thinner ground plane. This hypothesis is proven in Section 3.11.

When frequency increases, J_2 (a mode having currents aligned with the short 40 mm edge) becomes predominant. Therefore, a maximum should appear close to 100/40 times the 1.07 GHz (the frequency having the maximum of J_1, which is proportional to 100 mm). This frequency results in 2.675 GHz. As observed, the bandwidth potential has a second region where bandwidth is maximised, close to such a frequency (Figure 3.21).

As far as high frequencies are concerned ($f > 1.5$ GHz), the antenna booster at the middle edge now presents more bandwidth than at low frequencies. The reason is that although the antenna booster cannot excite mode J_1 because it has a null E-field at the antenna booster position, other higher modes exist. For example, eigenmode J_4 features a maximum electrical field at the middle of the long edge. Thus, the antenna booster excites such a mode, resulting in better bandwidth than that at low frequencies ($f < 1.5$ GHz).

At this point, it is interesting to study if we can excite J_1 with an antenna booster at the middle of the long edge of the ground plane, where the current is maximum. The answer is yes. Let us think for a moment about the field of wave-guide cavities or dielectric resonator antennas. The E and H field distributions are computed to determine how to excite a particular mode. At a maximum of E-field for a given mode, a short monopole or voltage probe is employed to excite such mode, or a short loop or slot is placed at those points having a maximum of H-field. The same principle is applied here. At the centre of the long edge (Figure 3.20), the current is at its maximum, and so is the H-field. Therefore, instead of using an electric antenna booster, a new antenna booster type is introduced: a *magnetic antenna booster*. A magnetic antenna booster is placed at a point of the ground plane having a maximum H-field (large current) or at least at a point not having a null H-field of current [2,43] (Figure 3.23).

Antenna booster
An antenna booster is an electrically small ($\sim\lambda/30$ at the lowest frequency of operation) non-resonant element, typically at the lowest frequencies of operation, able to excite currents on a ground plane of a wireless device to transmit and/or receive electromagnetic signals.

As studied before, efficiency can be optimised by placing the antenna booster at the corner of the ground plane for electric antenna boosters, and the frequency bands can be shaped by designing a matching network. But what about radiation patterns? And polarisation? The currents on the ground plane primarily determine the radiation patterns, rather than the currents on the antenna booster. For example,

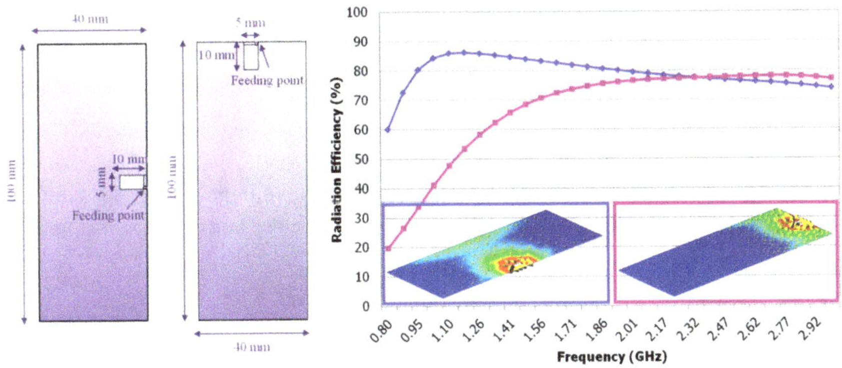

Figure 3.23 A magnetic booster (non-resonant slot element) at a maximum current for J_1. The 100 mm × 40 mm ground plane is printed over a 1 mm thick substrate (ε_r = 4.15, tanδ = 0.017). Currents are computed at 900 MHz.

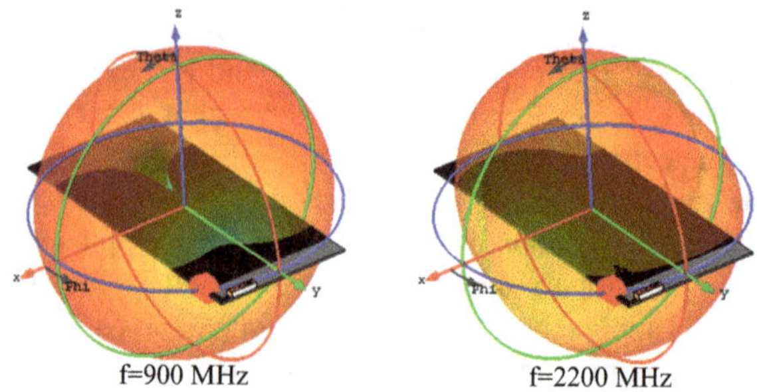

Figure 3.24 Simulated radiation patterns for a 12 mm × 3 mm × 2.4 mm antenna booster on a 120 mm × 60 mm ground plane. Directivity is 2.6 dB and 3.5 dB at 900 MHz and 2200 MHz, respectively

let us take a 120 mm × 60 mm ground plane operating at a mode where the currents are mainly aligned with the long dimension, similar to a half sinus as the one shown in Figure 3.20 (J_1 mode). For J_1, the radiation pattern will be like that of a half-wavelength dipole (Figure 3.24), that is, a maximum at $\varphi = 0°$ plane, two deeps at ($\varphi = 90°$, $\theta = 90°$) and ($\varphi = 270°$, $\theta = 90°$) directions. As for the polarisation, no matter the orientation of the antenna booster (vertical or horizontal), the polarisation is determined by the currents on the ground plane. In this case, polarisation will be linearly aligned with the y-axis. For higher frequencies where more modes appear, polarisation changes direction.

Antenna booster position
As a practical rule of thumb, placing an electrical antenna booster at the corner of a ground plane of a wireless device provides the best bandwidth and efficiency. For a magnetic booster, however, positions are complementary, and usually, the centre of an edge is a good option as long as it is not short in terms of the wavelength (edge comparable to 0.4λ).

Finally, since the ground plane is comparable to the wavelength, radiation patterns present low directivity (~3 dBi), which is convenient for wireless devices because the direction of the incoming signal is not known, and also, the orientation of the wireless device is random. We will dive deep into the analysis of radiation patterns in Section 3.11.2.

3.7 The impedance nature of an antenna booster

Antenna boosters, both electrical and magnetic, present a non-resonant input impedance, at least for the lowest frequency of operation. While an electrical booster presents a reactive capacitive impedance, a magnetic booster, on the

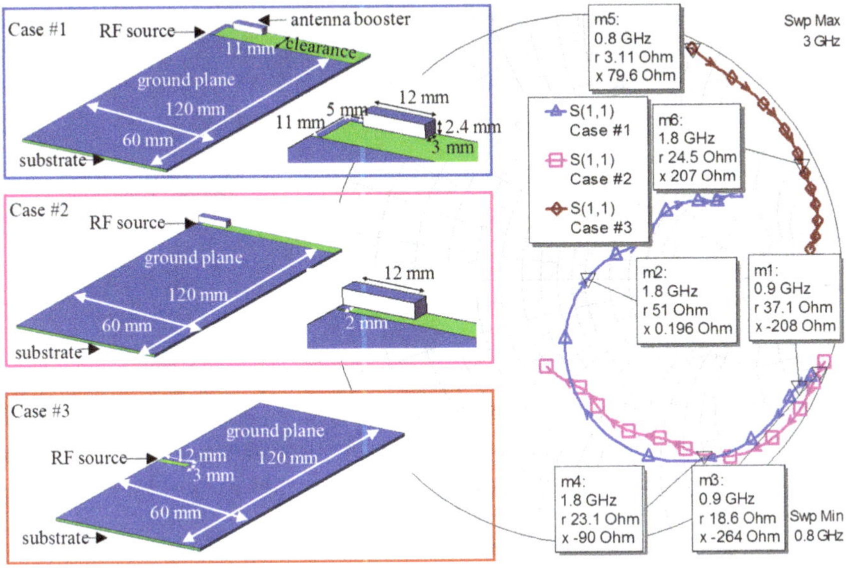

Figure 3.25 Input impedance for 12 mm × 3 mm × 2.4 mm (height) antenna boosters. (Top) With a feeding line 16 mm feeding line (BW = 20% at 900 MHz, SWR = 3); (middle) with a 2 mm feeding line, BW = 9.6%; (bottom) a magnetic antenna booster of 12 mm × 3 mm, BW = 8.5%.

contrary, features a reactive inductive impedance (Figure 3.25). However, the same design procedure applies to both types of antenna boosters since a matching network can match both types of input impedances. This situation differs from self-resonant antennas, where the antenna geometry is shaped to provide an input impedance matched at the frequency bands of interest. However, in lieu of being an advantage, the design of a self-resonant antenna requires an iterative process since the geometry needs to be carefully designed to create the frequency bands of operation. This iterative process occurs not only in the electromagnetic simulation domain but also in the prototyping phase, where the antenna geometry made of conductive traces must be adjusted, either enlarging or shortening parts of the antenna till the final design is achieved. To even add more drawbacks to the design process of a self-resonant antenna, once the antenna is designed, it is difficult to operate at other frequency bands [25]. A new design from scratch is then needed, resulting in a cumbersome design process. Besides, if the size of the device changes, the impedance changes, and therefore, the antenna needs tuning. However, all these drawbacks are overcome with antenna boosters since the response is shaped quickly and simply with a matching network, either in the simulation or prototyping domains, as explained in Section 3.4. And if new bands are added or the size of the device changes, there is no necessity to change the antenna booster, only the matching network.

For the electric antenna booster, the input impedance can change depending on the length of the feeding line (Figure 3.25). Depending on the size of the antenna booster and the feeding line, the input impedance is non-resonant across the whole frequency range. In other cases, the input impedance is non-resonant at the lowest frequency region. Spacing the antenna booster from the ground plane with a feeding line adds a degree of freedom to adjust the bandwidth, as shown in Section 3.13.

For a magnetic booster (Figure 3.25), the impedance is inductive rather than capacitive. The book addresses electric boosters. However, the same design process of adding a matching network is applied to a magnetic booster to achieve either a single or multiband performance.

3.8 Resonant antennas versus non-resonant elements

Usually, the most challenging situation for designing multiband-small antennas for wireless devices occurs at the lowest frequencies of operation ($f < 1$ GHz, where many communications standards are allocated). In this frequency region, the ground plane is small in terms of wavelength, and thus, it is challenging to obtain efficiency and bandwidth. This section is focused on determining which element becomes preferred for exciting efficient radiation modes of a typical ground plane of a wireless device. With this aim, four elements are analysed: a spiral, two fractal-inspired elements based on the Hilbert geometry (second (H2) and third iteration (H3)) [44], and an antenna booster with a square shape (Figure 3.26).

All elements are 15 mm × 15 mm. In the first experiment, the spiral and the Hilbert-based antennas are tuned to resonate at 900 MHz. In contrast, the square antenna booster has its first resonance at 1921 MHz, well above the frequency region of operation. Simulations are performed using MoM (IE3D software). The ground plane dimensions are 90 mm × 40 mm, printed on a thin FR4 substrate of 1 mm thick ($\varepsilon_r = 4.15$, $\tan\delta = 0.017$). For this ground plane dimension, the first characteristic mode (J_1) resonates at 1170 MHz. This resonance has been

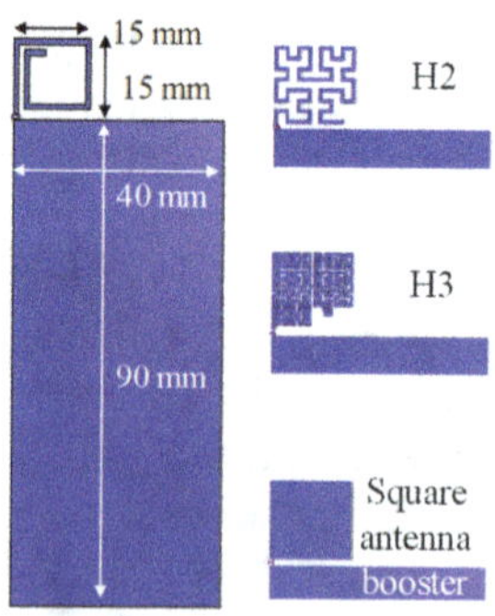

Figure 3.26 Geometry of the proposed structures (spiral, H2, H3, and square antenna booster) in a 90 mm × 40 mm ground plane etched in a 1 mm FR4 substrate $\varepsilon_r = 4.15$ $\tan\delta = 0.017$. For this case, spiral, H2, and H3 resonate at 900 MHz.

calculated using characteristic mode analysis, computing the zero crossing of the eigenvalue λ_1 [35–38]. For this reason, in the second experiment, the spiral, H2, and H3 are tuned by changing their geometry to resonate at 1170 MHz, i.e., at the same frequency as the eigenmode J_1. The basis for these two experiments is found in the electrical circuit modelling an antenna with a ground plane [45,46]. The previous circuit analysis demonstrates that the bandwidth of a radiating system is improved when the resonant frequency of the antenna and that of the ground plane are the same. At 900 MHz, the antenna can easily resonate but not the ground plane unless it has enough length (approximately 0.4λ, resulting in 133 mm at 900 MHz). Previous work in the literature [45–51] teaches that the resonant frequency of short ground planes (around 100 mm) shifts down to 900 MHz by introducing slots that force currents to travel a longer path, thus emulating a longer ground plane. In this section, the approach does not focus on manipulating the ground plane but rather on analysing the effect of the relationship between the resonant frequency of the antenna element and that of the ground plane. Moreover, the impact of the antenna geometry on the potential bandwidth is studied as well. To solve these questions, the proposed experiments analyse three antenna elements with different geometry (spiral, H2, and H3) having a resonance at three different frequencies:

1. 900 MHz, the centre frequency of the frequency region where some wireless protocols are allocated.
2. 1170 MHz, coinciding with the resonant frequency of the fundamental mode of a 90 mm × 40 mm ground plane (eigenmode J_1).
3. a frequency above the resonance frequency of J_1.

The behaviour of the previous proposals is further compared with a square antenna booster having a resonant frequency well above the resonance of the ground plane.

To compare the four elements, the quality factor (Q_a) is computed from input impedance with Equations (3.2) and (3.3) [32].

The advantage of this procedure lies in its simplicity since, through a single simulation of input impedance, bandwidth across a large frequency range can be obtained without the need to match the antenna at each frequency. This last procedure is undesirable since, usually, a matching can misrepresent the potential bandwidth. The validity of this method holds if the input impedance presents a behaviour like a series or parallel RLC circuit [32]. For those cases where the input impedance presents small loops, such as the ones caused by couplings between different radiators, the validity does not hold anymore.

3.8.1 Resonance of the element (f_r) equal to 900 MHz

The geometries of the spiral and the two Hilbert-based antennas are tuned to resonate at 900 MHz, whereas the square antenna booster, as indicated above, has its first resonance at 1921 MHz. Several conclusions are drawn (Figures 3.27 and 3.28):

(a) The maximum bandwidth for each resonant element is obtained approximately at the resonant frequency of the element.

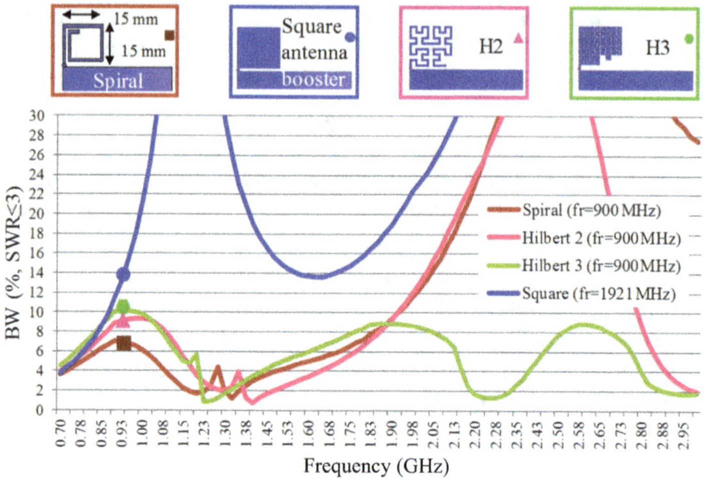

Figure 3.27 Bandwidths for the three resonant antennas at 900 MHz and the square antenna booster regarding a 90 mm × 40 mm ground plane (partially shown). f_r is the first resonant frequency

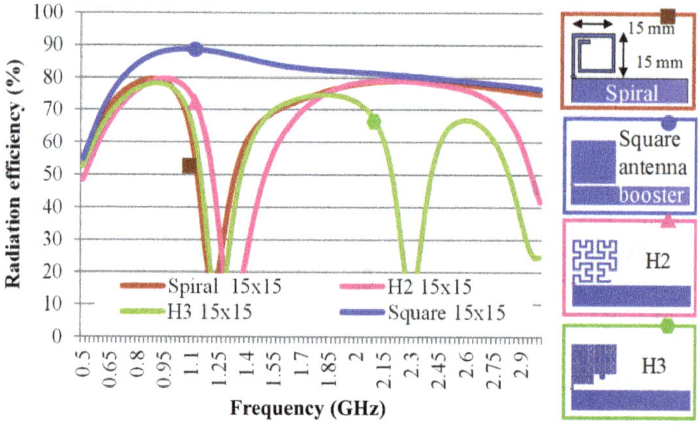

Figure 3.28 Radiation efficiency of the spiral, square antenna booster, and Hilbert elements on a 90 mm × 40 mm ground plane (partially shown)

(b) Resonant elements having a resonance below that of the ground plane do not effectively excite the ground plane at the resonant frequency of the eigenmode J_1 (1170 MHz) compared to the square element being non-resonant below 1.921 GHz.

(c) The maximum bandwidth value for the resonant cases is dependent on the geometry, with H3 being the best one (peak of 10%) while the spiral is the worst (peak of 7%).

(d) The square antenna booster has the maximum bandwidth at the resonance of the ground plane, reaching huge values (62%). It should be noted that the Q_a method is not accurate but is indicative that a large bandwidth can be obtained. For instance, the bandwidth is 44% using conjugate matching.
(e) The bandwidth for the square antenna booster is the best among the other cases using geometry-based radiators.
(f) The square antenna booster has slightly better bandwidth values near the resonant frequency of the resonant elements. The behaviour from 700 to 850 MHz is very similar to the Hilbert 3, with the last one slightly better.
(g) Radiation efficiency is better for the square antenna booster, not only at the non-resonant region ($f < 1.921$ GHz) but in the whole frequency range (Figure 3.28). At 900 MHz, radiation efficiency η_r = 87.3% for the square element, whereas it is approximately 79.0% for the spiral and Hilbert cases. Furthermore, the square element does not present any efficiency drop across the 0.7–3 GHz frequency range, whereas spirals and Hilbert antennas present some nulls. This means that the square element can provide high efficiency when using a matching network. However, spiral and Hilbert would give low efficiency at the efficiency drops.

From these conclusions, it is determined that a simple antenna booster provides better bandwidth and efficiency. Therefore, it can be used at any frequency band, either for single band or multiband operation, by adding a matching network. Migrating from a square 2D element to a 3D structure results in an antenna booster providing better bandwidth. For example, consider a 90 mm × 40 mm ground plane, a 5 mm × 5 mm square element has a 2.8% bandwidth at 900 MHz, whereas it is almost double, 4.7% for a 5 mm × 5 mm × 5 mm antenna booster, with a similar trend at higher frequencies. Therefore, 3D antenna boosters are preferred to provide more bandwidth within a small volume.

At this point, an interesting question to address is whether the square antenna booster is an efficient element or is the one that better excites the fundamental mode of the ground plane. To answer this question, the Q_a is computed for the element in a dipole configuration (Figure 3.29).

The spiral resonates at 900 MHz, while the square element resonates at upper frequencies (2.33 GHz), both featuring the same footprint. At the frequency of interest (900 MHz), both elements have a computed radiation efficiency of ~40%,

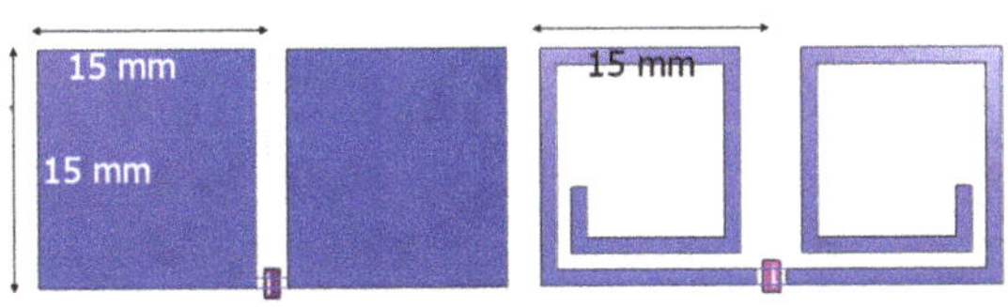

Figure 3.29 Dipole configuration for the square and spiral used to compute the Q_a of the element itself. A dipole arm corresponds to the same element used in Figure 3.26.

Table 3.2 Simulated results for Q_a

Element	**Q_a-dipole element only at 900 MHz (Figure 3.29)**	**Q_a-total at 900 MHz (element coplanar to the ground plane as in Figure 3.26)**
Spiral	80.5	20.1
Square	101.1	12.2

but the Q_a for the square is larger than the one of the spiral. Therefore, as a single element operating at 900 MHz, the spiral has better bandwidth than the square. However, as shown before, the situation changes completely when the ground plane is considered since the Q_Z of the square case in the ground plane of 90 mm × 40 mm is lower than the spiral case. This means that the element is not the main contributor to the bandwidth but the way the element excites a mode or modes in the ground plane. For this case, the Q_a of 12.2 for the square antenna booster with the ground plane represents a bandwidth of 11.5%, while the Q_a of 20.1 for the spiral with the ground plane represents a bandwidth of 7.0%, that is, 64% larger for the square (Table 3.2).

3.8.2 Resonance of the element (f_r) equal to the resonance of the ground plane (f_{GP})

The second experiment explores resonant elements having the first resonance equal to the resonance of the fundamental mode of the ground plane. Therefore, the spiral and Hilbert-based antennas have been geometrically tuned at 1170 MHz. At this frequency, the resonance of the ground plane is computed using characteristic mode analysis [35]. The footprint for all cases is maintained as in the previous experiments, except for H3, which, thanks to its space-filling geometry, is smaller (Figures 3.30 and 3.31). The potential bandwidth is computed using Equations (3.2) and (3.3). Several conclusions are drawn (Figures 3.30 and 3.31):

(a) for the resonant elements, the maximum bandwidth is obtained at f_r, corresponding to the resonance of the antenna element as well as to the ground plane resonance
(b) the square element also presents a maximum bandwidth at the resonant frequency of the ground plane mode
(c) the maximum bandwidth is obtained for the Hilbert antennas, with H3 having the largest peak (75%)
(d) the spiral antenna becomes the antenna element that attains the lower potential bandwidth
(e) the bandwidth at the 690–960 MHz frequency range is larger for the antennas having the resonance at the resonance of the ground plane (1170 MHz) in comparison with the case where the resonant elements have their resonance centred at 900 MHz (Figure 3.27). This is exemplified in Figure 3.31, where the H3 tuned at 900 MHz and 1170 MHz are compared. It is clearly shown that the bandwidth is greater when the resonance of the antenna matches the

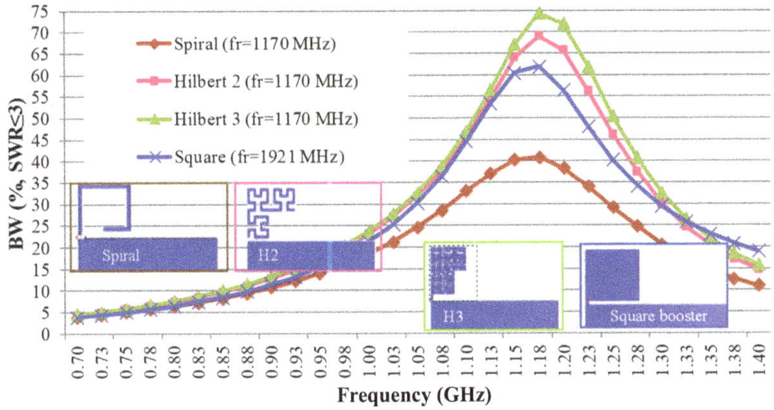

Figure 3.30 Bandwidth for the spiral, H2, H3, and square when the resonant frequency of the antenna elements (1170 MHz) becomes equal to the resonance of the ground plane mode (90 mm × 40 mm)

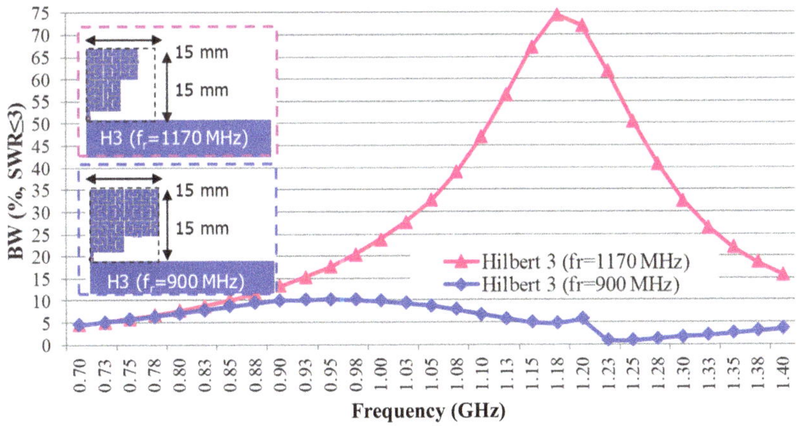

Figure 3.31 Comparisons of the bandwidth attained by an H3 antenna when it is designed to resonate at 900 MHz and at 1170 MHz, coinciding with the ground plane resonance (90 mm × 40 mm)

resonance of the ground plane mode. In particular, the bandwidth at 900 MHz for H3 resonating at 900 MHz is 10.0%, in contrast with the 13.2% for H3 resonating at 1170 MHz and with the advantage of having a smaller footprint.

The previous subsections illustrate essential conclusions. Contrary to what was initially believed, the antenna should not be designed to resonate at the centre frequency of the frequency region of interest but at the resonant frequency of the ground plane mode to maximise the potential bandwidth. The results further illustrate that this resonance could also be placed well above the resonant frequency of the fundamental mode of the ground plane without a significant loss of performance. In fact, the square antenna booster offers a larger bandwidth at 900 MHz than other antenna elements resonating at this frequency. This conclusion is wholly aligned with that obtained from the electrical model presented in [46], where the bandwidth was maximised if the resonant frequency of the antenna matched that of the ground plane. However, a new conclusion is derived here since if the resonant frequency of the element is placed above that of the ground plane, the bandwidth can be large enough and very similar to that obtained in those cases where both resonances become the same (Figures 3.30 and 3.31).

3.8.3 Resonance of the element (f_r) above to the resonance of the ground plane (f_{GP})

Finally, the spiral, H2, and H3 are geometrically tuned at a frequency above the resonance of the ground plane (the resonance of the eigenmode J_1, which is 1170 MHz). In this case, all the elements are 5 mm × 5 mm (Figure 3.32). The resonant frequencies for the spiral, H2, H3, and square are 2500 MHz, 3800 MHz, 2300 MHz, and 5800 MHz, respectively, well above 1170 MHz. For this experiment, all cases present practically the same bandwidth at the frequency region of 690–960 MHz, similar to the result shown in Figure 3.30. For example, at 900 MHz, bandwidth is 3.1% and 2.8% for H3 and square elements, respectively. However, the radiation efficiency is 61.5% for the H3 antenna and 66.2% for the square antenna booster, making the booster the better option.

Non-resonant nature of antenna boosters

The non-resonant nature of antenna boosters is attractive because they can provide better bandwidth than resonant elements and a flat efficiency over a wide frequency range, avoiding the null of the resonant elements (Figure 3.27). This enables operation at any frequency band thanks to a matching network that can result in single band (Chapter 4) or multiband operation (Chapter 5).

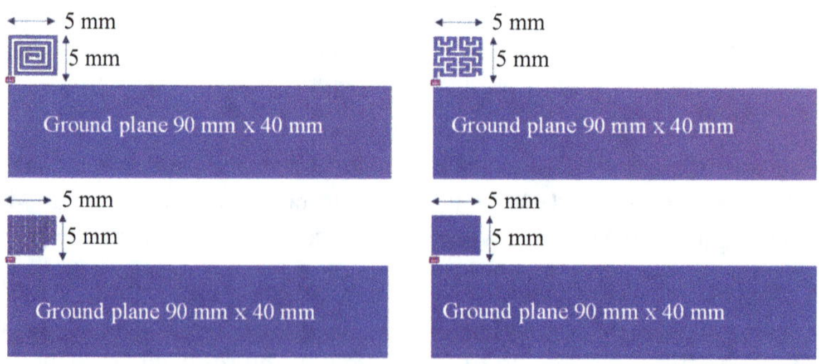

Figure 3.32 Small antenna elements on a 90 mm × 40 mm ground plane (partially shown)

3.9 Antenna booster size

The bandwidth of an antenna booster-based device depends on the ground plane size, as profoundly analysed in this chapter, Section 3.11. But before addressing the impact of the ground plane size, the question to solve is: how much does the size of an antenna booster help to modulate the bandwidth? Therefore, this section aims to determine the role of antenna booster size in bandwidth.

An antenna booster excites radiating currents on a ground plane, and therefore, its size plays a vital role in bandwidth, efficiency, and radiation patterns. Based on the characteristic mode analysis explained in Section 3.6, exciting a particular mode on the ground plane of a wireless device will determine the bandwidth. But what about the antenna booster? Does its size matter to excite those currents on the ground plane better? Let us address these questions by comparing several antenna elements, from antenna boosters to a resonant element: from a small antenna booster of 5 mm × 3 mm × 1 mm to a resonant element of 60 mm × 3 mm × 1 mm (f_0 = 0.91 GHz). If we consider a first frequency region of operation with the highest frequency of 960 MHz, the 60 mm × 3 mm × 1 mm is resonant within the band, whereas the others are not (Table 3.3).

In the first case, all are feeding the same 120 mm × 60 mm ground plane (Figure 3.33). Later, the ground plane is reduced to 90 mm × 60 mm and 60 mm × 60 mm. Bandwidth is computed according to Equations (3.2) and (3.3), where it is observed that at the low-frequency region (f< 1 GHz), the size of the element helps to modulate the bandwidth, where the larger the size, the larger the bandwidth. The bandwidth is not proportional to the element size at a second frequency region (f> 1.6 GHz). In fact, the bandwidth is larger for the small element and decreases as the element size increases.

Therefore, there is a trade-off for multiband applications requiring operation at frequencies below 1 GHz and beyond 1.6 GHz (Table 3.4). The elements between 15 mm and 30 mm in length are good choices.

For example, let us consider operation at 824–960 MHz, and 1710–2690 MHz are required. This requires a BW = 15.2% and a BW = 44.5% at the first and second frequency regions, respectively. By computing bandwidth at 900 MHz and 2200 MHz, considering an SWR = 3, it is shown that an element in the range of

Table 3.3 Simulated resonant frequency and electrical size for different elements on a 120 mm × 60 mm ground plane

Size (mm)	f_0 (GHz)	Electrical size at 900 MHz
5 × 3 × 1	3.7	$\lambda/67$
15 × 3 × 1	2.3	$\lambda/22$
25 × 3 × 1	1.57	$\lambda/13$
30 × 3 × 1	1.44	$\lambda/11$
40 × 3 × 1	1.23	$\lambda/8$
60 × 3 × 1	0.91	$\lambda/5$

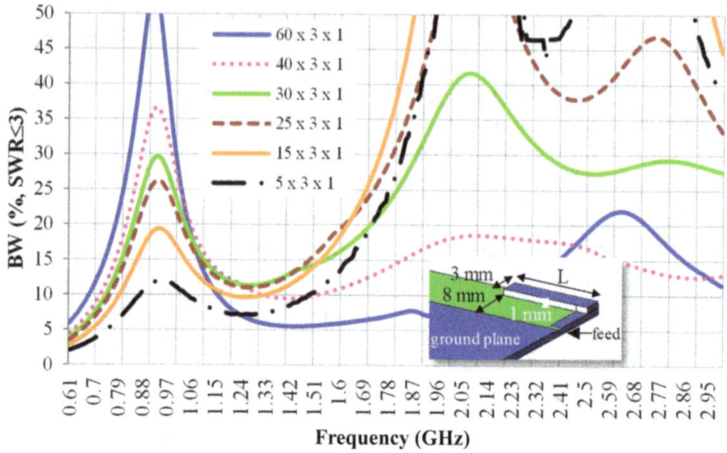

Figure 3.33 Bandwidth as a function of different element lengths for a 120 mm × 60 mm ground plane

Table 3.4 Bandwidth (SWR = 3) for several elements on a 120 mm × 60 mm ground plane at two different frequencies: 900 MHz and 2200 MHz

Size (mm)	BW(%) at 900 MHz	BW > 15.2%?	BW (%) at 2200 MHz	BW > 44.5%?
5 × 3 × 1	11.2	No	54.0	Yes
15 × 3 × 1	18.1	Yes	98.3	Yes
25 × 3 × 1	24.5	Yes	53.3	Yes
30 × 3 × 1	27.7	Yes	37.2	No
40 × 3 × 1	34.3	Yes	18.3	No
60 × 3 × 1	51.1	Yes	9.9	No

15 mm and 25 mm in length is appropriate for achieving enough bandwidth at both frequency regions once a multiband matching network is designed, as shown in Chapter 5.

It is interesting to observe that all cases present a first bandwidth maximum at approximately 930 MHz. This maximum is due to a first resonance of the 120 mm × 60 mm ground plane, resulting in a ground plane length of 0.372λ at 930 MHz.

Let us consider the same element but with a shorter ground plane, one of 90 mm × 60 mm and the other of 60 mm × 60 mm. For the 90 mm × 60 mm, there is a first bandwidth maximum at 1.2 GHz (Figure 3.34). This corresponds to a ground plane length of 0.360λ at 1.2 GHz, similar to the previous case. In this case, the best bandwidth coincides when the resonance of the element matches the resonance of the ground places, as discussed in Section 3.8.3, which is the 40 mm element featuring a resonance at 1.23 GHz (Table 3.3). This is again consistent

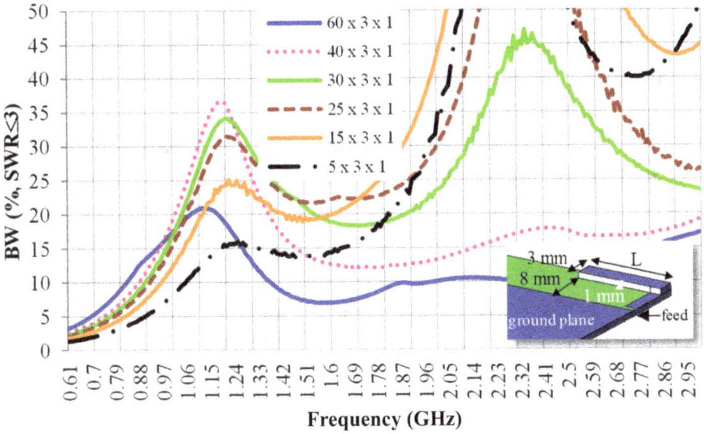

Figure 3.34 Bandwidth as a function of different element lengths for a 90 mm × 60 mm ground plane

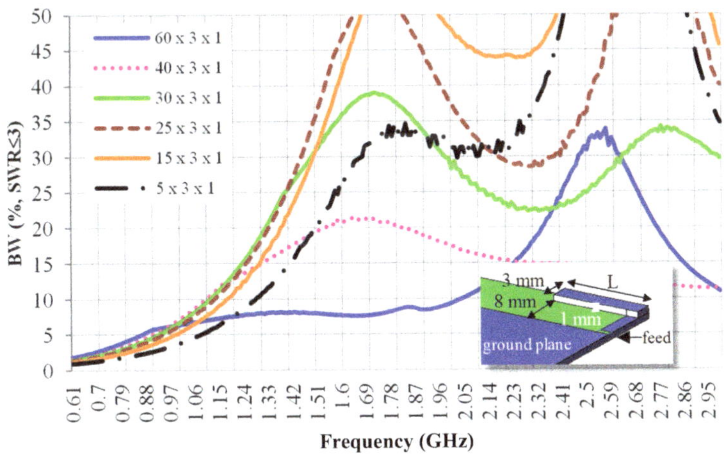

Figure 3.35 Bandwidth as a function of different element lengths for a 60 mm × 60 mm ground plane

with the 60 mm × 60 mm ground plane (Figure 3.35). The first bandwidth maximum appears at 1.82 GHz. In this case, since the element having the closest resonance to 1.82 GHz is the 15 mm element length (Table 3.3), the maximum bandwidth is achieved with such a size. In this case, the ground plane length maximising bandwidth is 0.364λ at 1.82 GHz.

All these results drive again to the conclusion that only sometimes large volume matters. Suppose the first resonance of the ground plane is within the frequency of operation. In that case, the best bandwidth at such a frequency region is achieved by having an element with the same resonant frequency. However, for frequencies below 1 GHz, the element size is large and thus not convenient to embed into the scarce space in wireless devices where other electronic components need to be integrated. Besides, the bandwidth at higher frequencies is low because of the second resonance (an anti-resonance).

Besides the previous multiband example at 824–960 MHz and 1710–2690 MHz, where an element between 15 mm and 25 mm in length is enough to cover both frequency regions, let us consider some more cases. Imagine we need to provide operation from 863 MHz to 928 MHz, a frequency range for some wireless protocols transmitting data from sensors. The required bandwidth is 7.3%. In this case, a 15 mm × 3 mm × 1 mm antenna booster is enough as long as the ground plane is at least 90 mm in length. Reducing the ground plane will reduce bandwidth. For example, if the ground plane length is 60 mm, the bandwidth with the 15 mm × 3 mm × 1 mm antenna booster is 4.1% (Figure 3.36). This bandwidth can be achieved with a simple matching network of two components. However, as discussed in Chapter 4, a broadband matching network can increase bandwidth by a factor of around 2.5 times. Therefore, it is possible to obtain a BW of ~10% for a 60 mm length-long ground plane with a 15 mm × 3 mm × 1 mm broadband matching network. Note that the resonant frequency of the 15 mm × 3 mm × 1 mm antenna

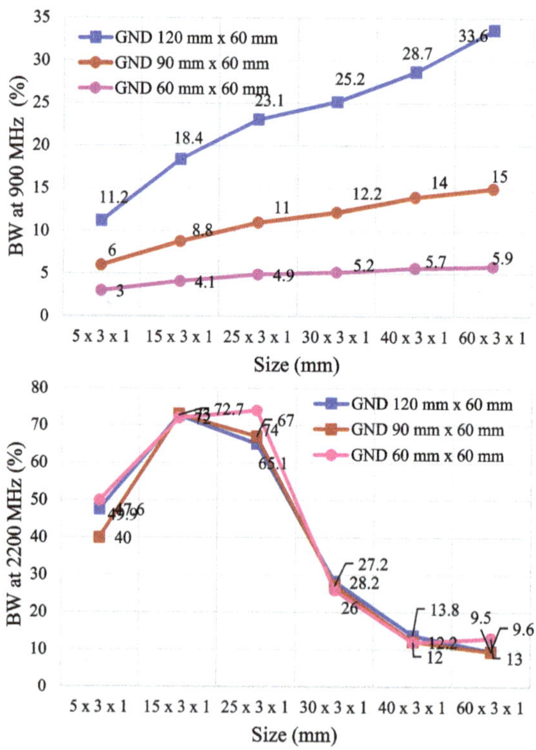

Figure 3.36 Potential bandwidth at 900 MHz (top) and 2200 MHz (bottom) for the different elements shown in Figures 3.33–3.35

booster is 2.3 GHz, which is beyond the frequency of operation (Table 3.3), resulting in a non-resonant element across the frequency band of operation. As observed, an antenna booster is enough to obtain the required bandwidth without using a large self-resonant element that would require a large footprint on the PCB. In this case, the antenna booster size at the lowest frequency of operation (863 MHz) is $\lambda/23$.

Let us consider another single-band example for satellite communication systems from 1561 MHz to 1606 MHz that require a 2.8% bandwidth. For this example, the size of the antenna booster can be reduced since a 5 mm × 3 mm × 1 mm antenna achieves a bandwidth close to the target. Not that the resonant frequency of a 5 mm × 3 mm × 1 mm antenna booster is 3.7 GHz (Table 3.3), showing the non-resonant nature of the antenna booster. In this case, the size of the antenna booster using the 5 mm length element is $\lambda/38$. Following the scalability concept explained in Chapter 1, the larger the frequency, the smaller the antenna booster can be. For example, using as a reference an antenna booster of $\lambda/30$, the size of an antenna booster at 3 GHz can be as small as 3.3 mm. Thus, a size between $\lambda/20$ and $\lambda/30$ is a good reference when choosing the antenna booster length.

It has been shown that antenna booster length helps modulate bandwidth. The principle, the larger, the better, does not apply in the antenna booster world. A last question to address is the impact on height. Does the height have a substantial effect on the bandwidth? Let us take the 15 mm × 3 mm × 1 mm antenna booster as a reference and compare the bandwidth with an antenna booster three times bigger in volume by increasing its height to 3 mm.

Finally, it is interesting to analyse the impact of antenna volume while keeping the length constant. The impact is studied for two boosters having three times more volume than a reference booster of 15 mm × 3 mm × 1 mm: a first booster with 15 mm × 3 mm × 3 mm (height) and another with 15 mm × 9 mm × 1 mm (Figure 3.37).

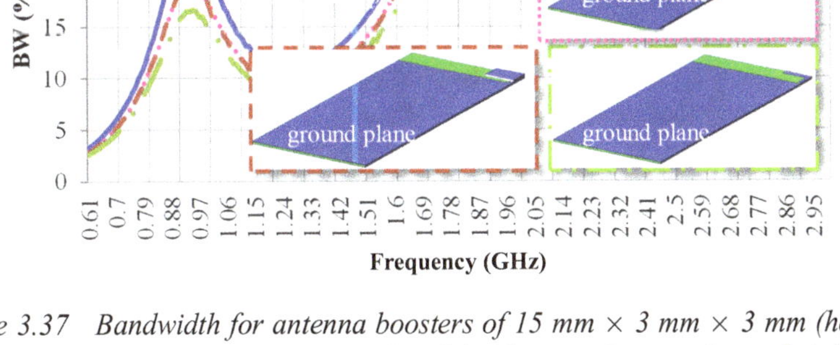

Figure 3.37 Bandwidth for antenna boosters of 15 mm × 3 mm × 3 mm (height), 15 mm × 3 mm × 1 mm (height), 15 mm × 9 mm × 1 mm (height), and 15 mm × 9 mm × 0 mm (height) on a 120 mm × 60 mm ground plane

The conclusions reflect that volume does not always matter. For example, the antenna booster with three times more volume due to its three times more height helps in getting more bandwidth in the low-frequency region (f< 1 GHz). However, the antenna booster, which has three times the volume due to its larger width, does not increase bandwidth but provides the same bandwidth. Therefore, to make a compact antenna booster, width is not relevant, and thus, the width can be kept thin. Finally, antenna booster height helps again to modulate bandwidth, in particular, well below the first resonance (f< 2.3 GHz) (note that a 15-mm-long antenna booster has its first resonant frequency at 2.3 GHz considering the feeding line of 8 mm shown in Figure 3.33). In this case, the 15 mm × 3 mm × 3 mm antenna booster provides at 900 MHz a bandwidth of BW = 22.4%, whereas the printed version (15 mm × 3 mm × 0 mm) obtains a BW = 15.4%. However, although the bandwidth is smaller, it may be enough to cover the required bandwidth.

Now that the role of the antenna booster size has been discussed, the impact of the position, size of the ground plane and ground clearance on bandwidth and efficiency are discussed in the following sections.

3.10 Antenna booster location

As shown in Section 3.6, the position of the antenna booster on the ground plane matters to excite a particular mode, and as a result, a better bandwidth can be obtained depending on its position. The objective of this section is to determine how bandwidth and efficiency depend on the position of an antenna booster for several positions along the edges of the ground plane of a wireless device.

The following analysis considers a 12 mm × 3 mm × 2.4 mm (height) antenna booster on a rectangular 120 mm × 60 mm ground plane, where the antenna booster is located at six different locations (Figure 3.38). The first resonant frequency occurs at 1.795 GHz. Therefore, if the minimum frequency of operation is located at 700 MHz, the electrical size is $\lambda/35$, which is very small and non-resonant. Naturally, if the minimum frequency of operation is the resonant frequency, the element is a resonant element with an electrical size of $\lambda/14$.

The bandwidth is computed with Equations (3.2) and (3.3) for a large frequency range from 0.6 GHz up to 6 GHz. In all cases, a dielectric sheet of $\varepsilon_r = 4.15$, $\tan\delta = 0.017$, and 1 mm thick is included, having the size of the minimum rectangle that includes the ground plane as well as the antenna booster. The said substrate has no significant impact on the bandwidth estimation, but still, it has been included to provide more realistic results since it is a typical substrate employed in many PCBs.

3.10.1 Impact of the position of the antenna booster

The results at the low-frequency region (frequencies below 1 GHz) conclude that the best position in terms of bandwidth is the corners, either C_Top or C_Side. This result is aligned with Figure 3.16, where the current of the fundamental mode (J_1) has a minimum at the corners of the ground plane and, therefore, a maximum voltage. Thus, the antenna booster best excites such a mode when it is located at a

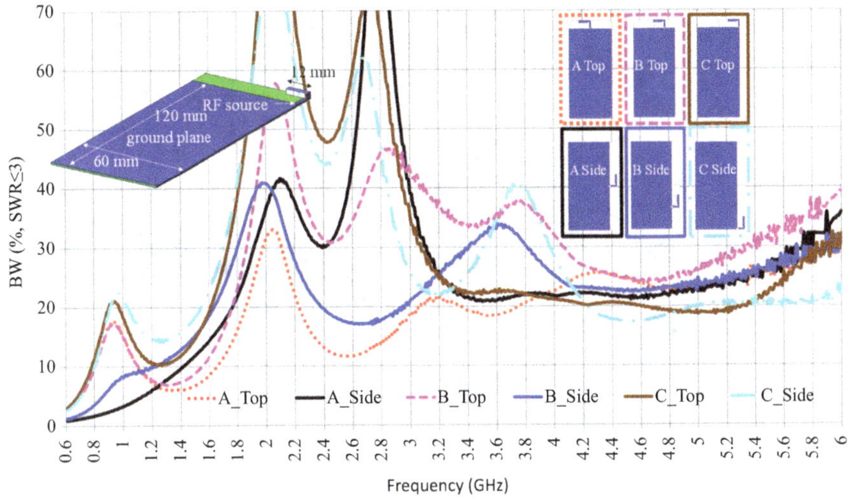

Figure 3.38 Computed bandwidth for a 12mm × 3 mm × 2.4 mm (height) antenna booster at different positions on a rectangular 120 mm × 60 mm ground plane

maximum voltage point. Following the same argument, the worst case occurs for A_side, where the antenna booster is situated in the centre of the long edge. Indeed, the voltage is null at such a point, and thus, little bandwidth can be obtained. As the booster is moved slightly from its null position (B_Side), bandwidth increases since the mode is better excited. In conclusion, for the low-frequency region, in this situation, the antenna booster at the corner gives the best results in terms of bandwidth as well as efficiency (Figure 3.39).

It is relevant to discuss the impact on efficiency based on Equation (3.4). A_side and C_top both have high radiation efficiency at 900 MHz ($\eta_r > 85\%$). However, C_top achieves a larger potential bandwidth because of its lower Q. In particular, Q is 5.9 and 47.9 for C_top and A_side, respectively. Based on Equation (3.4), losses of a matching network considering the same $Q_m = 80$ for the matching network components result in $\eta_m = -0.3$ dB and $\eta_m = -2.0$ dB for the C_top and A_side, respectively. This result agrees with a more accurate calculation using a circuit simulator including a matching network with real components: $\eta_m = -0.3$ dB and $\eta_m = -2.6$ dB, for the C_top and A_side, respectively. Therefore, although some antenna booster positions, such as A_side, may be interesting for narrowband applications, their high Q can lead to a lossy matching network since the Q_m of the components is comparable to Q_a. More examples are discussed in Chapter 4.

For a second frequency region (1.6–3 GHz), again, the corner cases provide better bandwidth. The reason is that the eigenmode with a longitudinal current distribution can be easily excited in the corner. Again, the centre of the short edge has a voltage minimum point, and therefore, a small bandwidth is obtained. Finally,

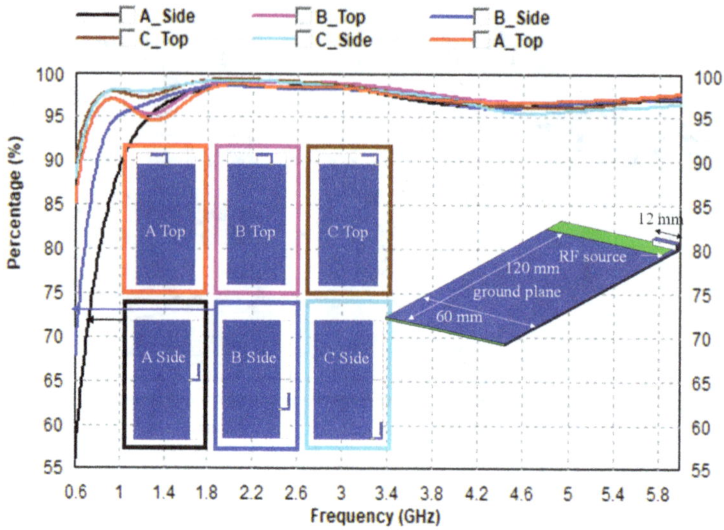

Figure 3.39 Simulated radiation efficiency for the cases illustrated in Figure 3.38

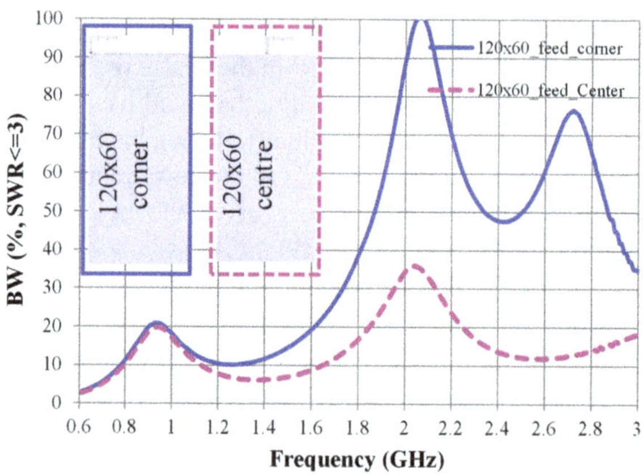

Figure 3.40 Computed bandwidth for the same 12 mm × 3 mm × 2.4 mm (height) antenna booster at two different positions on the 120 mm × 60 mm ground plane

at a higher frequency region (frequency larger than 3 GHz), all cases provide roughly similar performance.

Given the rectangular ground plane of 120 mm × 60 mm, the antenna booster at the centre of the short edge and corner provides similar bandwidth at the lower frequency region (f < 1 GHz) (Figure 3.40). In this case, both the corner and centre edge positions offer the same bandwidth because the fundamental mode supported by the rectangular ground plane presents a maximum electrical field (zero or close to zero current) at the short edges of such a plane (Figure 3.20). However, for frequencies above 1 GHz, bandwidth is much less at the second frequency region (Figure 3.40). This position is ineffective if one aims to cover several radio systems at this frequency region. For example, the bandwidth from 1.71 GHz up to 2.69 GHz represents approximately 44% bandwidth and, as clearly shown, cannot be achieved when the antenna booster is at the centre of the short edge (Figure 3.40).

Understanding the current distribution sheds light on the problem, confirming the eigenmode analysis (Figure 3.41). The current distribution is computed at 2.1 GHz, where it is observed that at the corner position, the current has a predominant horizontal effect, exciting the fundamental mode of the short 60 mm edge. However, at the centre position, such a mode is no longer excited based on the two opposite current directions aligned with the long axis of the ground plane. Since said mode has a weak modal significance, the bandwidth cannot be as large as the corner case, where the modal significance is 1.

In terms of input impedance, the corner case presents a compact impedance locus on the Smith chart, whereas the centre case is widely spread and, therefore, difficult to match across a wide frequency range (Figure 3.42). An example of a wireless device with an antenna booster element corner-fed to maximise bandwidth and efficiency is shown in Figure 3.43.

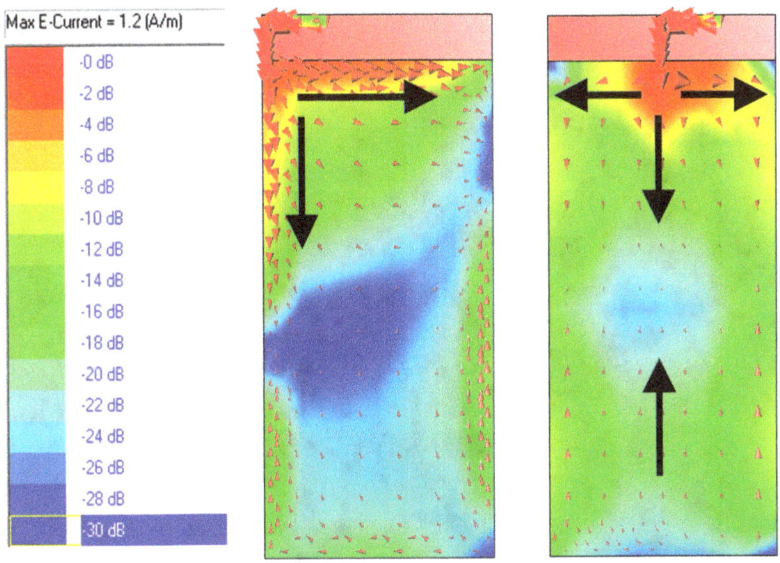

Figure 3.41 Simulated current distribution for a 12 mm × 3 mm × 2.4 mm (height) antenna booster at 2.1 GHz. Black arrows have been intentionally added to represent the main current intensity

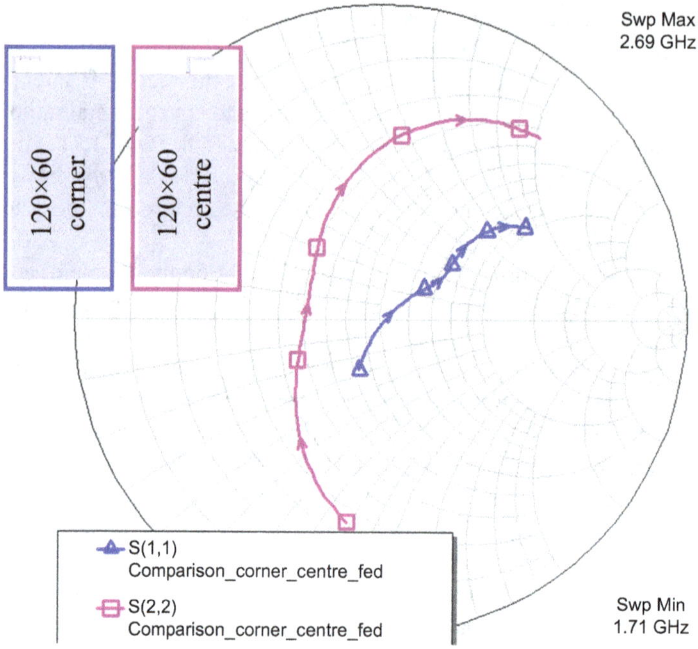

Figure 3.42 Simulated input impedance for a 12 mm × 3 mm × 2.4 mm (height) antenna booster from 1.71 GHz to 2.69 GHz

Figure 3.43 Wireless device with antenna booster elements (30 mm × 3 mm × 1 mm and 7 mm × 3 mm × 2 mm) fed at the corner of the ground plane

3.10.2 Position of the antenna booster for a small ground clearance

Now, we are considering a scenario akin to the one depicted in the prior analysis, but with attention given to a restricted ground clearance area. In some wireless devices, ground clearance is scarce since other electronic components, such as sensors, connecting lines, radiofrequency modules, and the like, should be integrated. Therefore, it is intriguing and of practical application to determine the best location of an antenna booster for a limited ground clearance area. In this regard, the same 120 mm × 60 mm ground plane size and 12 mm × 3 mm × 2.4 mm antenna booster have been considered for the analysis, but in this case, the antenna booster is embedded in a small portion of such a ground plane. The impact of the size of such a ground clearance portion is studied in Section 3.13.

Since the corner of the ground plane shows the best results regarding bandwidth, all examined cases consider a ground clearance area at the corner of the ground plane (Figure 3.44).

It is interesting to observe that corner situation 1.1 provides a bandwidth of 11.1% at 1 GHz, which is not the best at the lower frequency region since cases 2 and 2.1 provide a bandwidth at 1 GHz of 15.0% and 14.0%, respectively. It is interesting to validate this result by designing a matching network for cases 1.1 and 2.1 at the lower frequency region (Figure 3.45). Leaving apart the steps on how to design a matching network for antenna boosters that will be explained in Chapters 4 and 5, the simulation confirms that, as far as bandwidth at low frequencies is concerned, case 2.1 provides a bandwidth of 14.9%. In contrast, case 1.1 provides 12.0%, which is aligned with the potential bandwidth calculations (Figure 3.45).

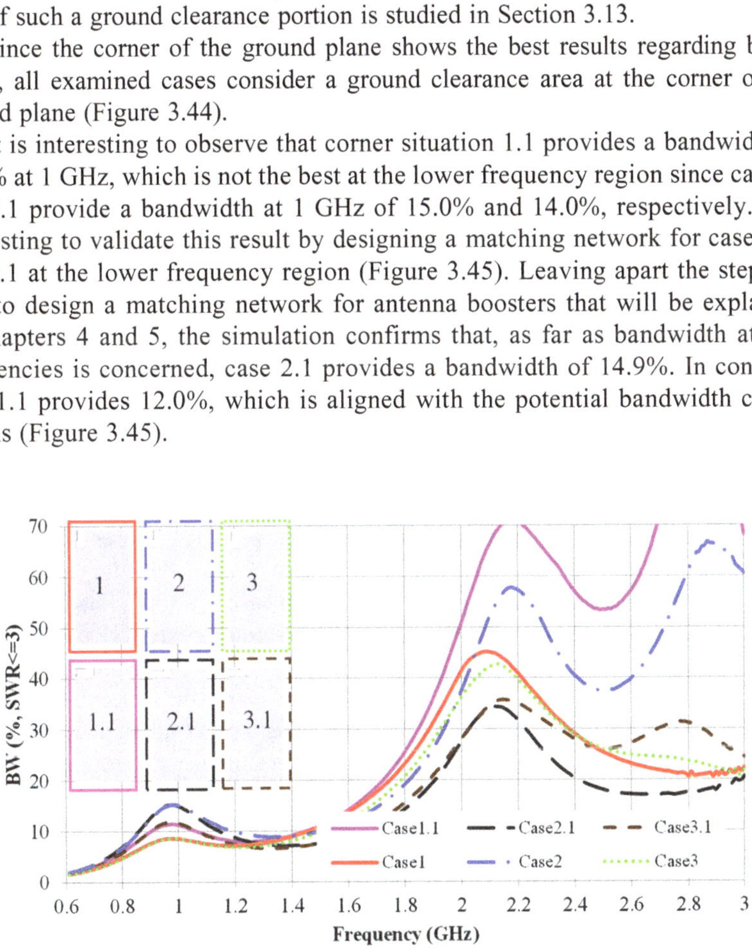

Figure 3.44 Computed bandwidth for a 12 mm × 3 mm × 2.4 mm (height) antenna booster at different positions on a 120 mm × 60 mm rectangular ground plane with a 15 mm × 11 mm ground clearance

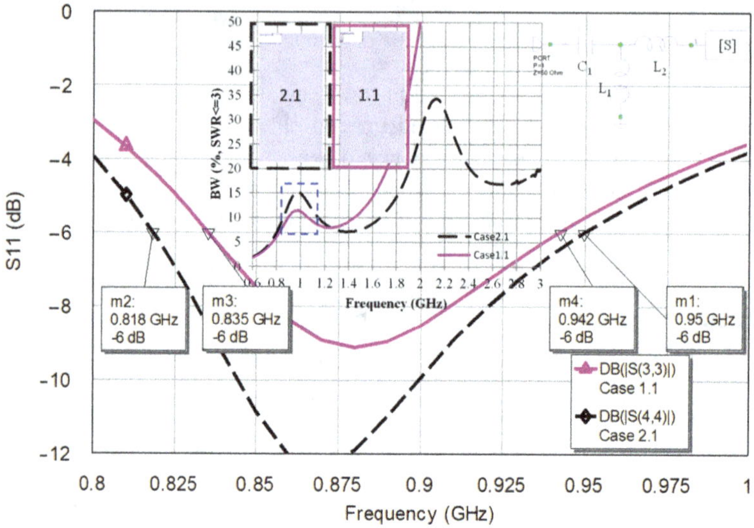

Figure 3.45 Simulated S_{11} using a matching circuit with the same topology for both cases. The dashed rectangle represents the potential bandwidth showing a larger bandwidth for case 2.1 than 1.1, as confirmed with the S_{11} using the matching network shown in the top right corner. [S] represents the S_{11} of cases 2.1 and 1.1 without any matching network. Matching network for case 2.1 is: $C_1 = 4.5$ pF, $L_1 = 8.6$ nH, $L_2 = 37$ nH; for case 1.1 is: $C_1 = 3.8$ pF, $L_1 = 8.9$ nH, $L_2 = 40$ nH.

Current simulation at 880 MHz for both cases 1.1 and 2.1 reveals an interesting distribution where case 2.1 better excites a longitudinal mode aligned with the long axis, corresponding to the fundamental mode, but case 1.1 excites such a longitudinal mode along one edge and part of the transversal mode (Figure 3.46).

While case 2.1 provides better bandwidth than case 1.1 at the lower frequency region, a different behaviour appears at a high-frequency region (1.6–3 GHz), where case 2.1 is worse than case 1.1 (Figure 3.44). Again, an explanation of why case 2.1 is now the worst can be found by analysing the current distribution (Figure 3.47). For case 2.1 (Figure 3.47, left), the current aligned with the horizontal axis (the shortest one) presents two out-of-phase currents. However, for case 1.1, the current is aligned along the horizontal axis, a kind of J_1 mode that provides more bandwidth. As a reminder, the J_1 mode is the fundamental one (Figure 3.16). Observe that this conclusion is consistent with the results obtained in Figure 3.41, where the antenna booster at the corner excites horizontal currents. In contrast, it excites out-of-phase currents when the antenna booster is in the middle of the edge. As a result, more bandwidth is achieved for the first situation, where currents are aligned along the edge as a J_1 mode (Figure 3.42).

Therefore, the antenna booster position behaves better for case 2.1 at a first frequency region centred at 900 MHz, but case 1.1 behaves better at a second

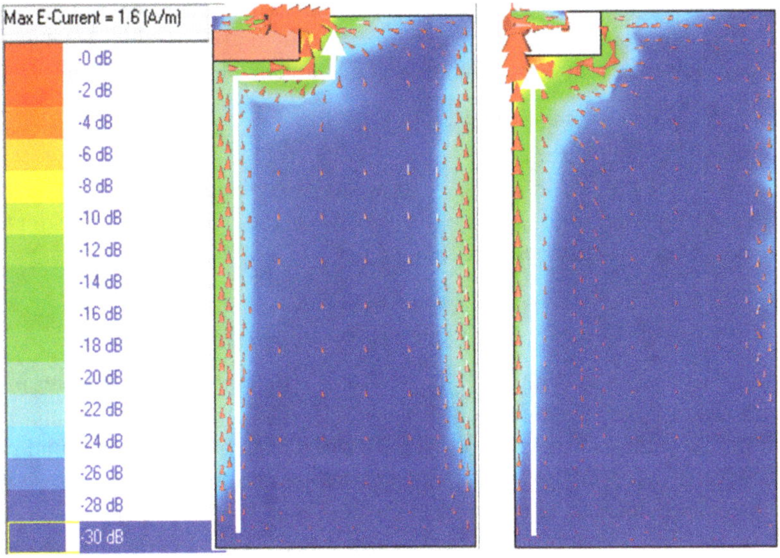

Figure 3.46 *Simulated current distribution at 880 MHz for cases 2.1 (left) and 1.1 (right). The white arrow has been intentionally added to show the main current distribution.*

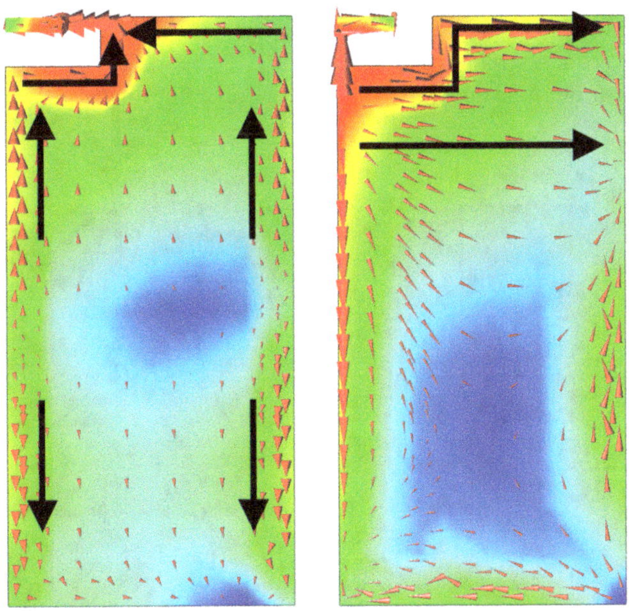

Figure 3.47 *Simulated current distribution at 2100 MHz for cases 2.1 (left) and 1.1 (right). Black arrows have intentionally been added.*

frequency region (1.6–3 GHz). Therefore, as far as multiband operation is concerned, both positions should be analysed depending on the required bandwidth at each frequency region to determine the suitable antenna booster location.

Antenna booster position
Antenna booster position on the ground plane may be adequate for one frequency region but not for another. Depending on the frequency bandwidth to achieve, a compromise needs to be taken.

3.10.3 On the position on a square ground plane

This section analyses a 12 mm × 3 mm × 2.4 mm (height) antenna booster in two different locations on a square ground plane of 120 mm × 120 mm (Figure 3.48). Trends are aligned with the previous results where the corner of the ground plane provides better results at the lower frequency region (frequencies less than 1 GHz), but a different behaviour is observed at the higher frequency region (1.6–3 GHz) where bandwidths are quite the same (Figure 3.48). At the low-frequency region, the achievable bandwidth at 900 MHz is 15.3% and 6.0% for the corner and centre positions, respectively, with similar radiation efficiencies over 90% for both cases at the whole frequency region.

Since the ground plane is square, both fundamental modes along the longitudinal and transversal directions have the same resonant frequency. Therefore, both can be excited with the same amplitude because modal significances are the same. In effect, both modes are excited when the antenna booster is located at the corner of the square ground plane (Figures 3.49 and 3.50). The eigen currents associated with the fundamental mode are the same but rotated 90° and have the

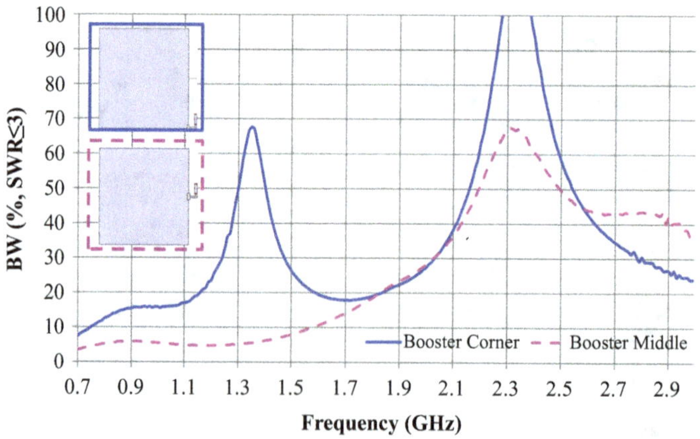

Figure 3.48 Computed bandwidth for a 12 mm × 3 mm × 2.4 mm antenna booster on a 120 mm × 120 mm square ground plane

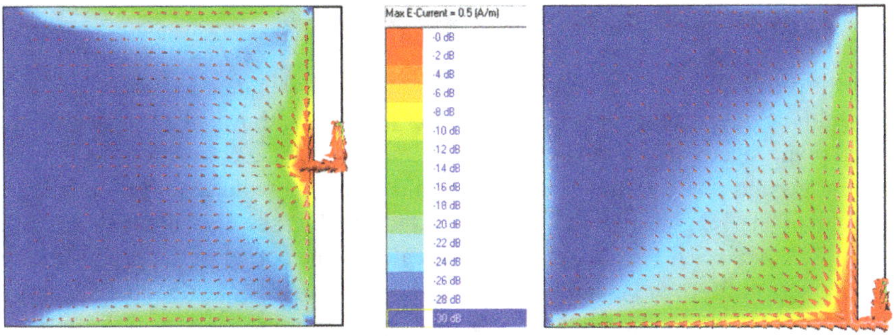

Figure 3.49 Simulated current distribution at 900 MHz for the cases shown in Figure 3.48

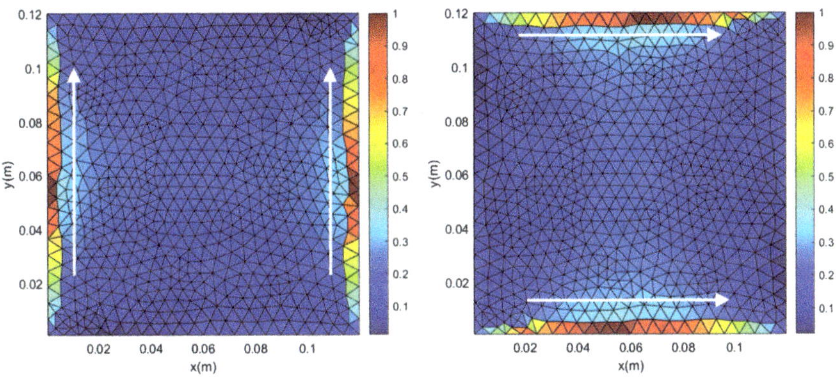

Figure 3.50 Eigen currents for the cases shown in Figure 3.48. Arrows have been added intentionally to show the direction of the maximum current

same modal significance. Consequently, at the corner of the ground plane, both eigen currents can be excited with the same intensity (Figure 3.49).

This current distribution has an impact on the polarisation of the radiation pattern since the polarisation is aligned with the maximum current intensity on the ground plane. In this case, the polarisation is not linear along the long axis, as was the case for the rectangular ground plane at the low-frequency region, but aligned along the diagonal since the radiation field has two orthogonal in-phase components, resulting in a linearly polarised electromagnetic field. This is an important consideration as far as point-to-point communication is concerned since both the transmitter and receiver should be aligned with the same polarisation to minimise polarisation mismatch and, therefore, to maximise the received power.

As observed in the bandwidth estimation (Figure 3.48), a first maximum is detected at approximately 960 MHz for the corner case and at 930 MHz for the booster at the centre, corresponding to approximately 0.37λ.

3.11 The ground plane size matters

Once the position of an antenna booster has been evaluated to obtain the best bandwidth and efficiency performance, the next step shows the impact of the ground plane size both on bandwidth and efficiency. First, a comparison with three different ground plane widths is discussed, and second, a wide set of ground planes having different lengths and widths is evaluated. In all cases, the antenna booster is 12 mm × 3 mm × 2.4 mm (height) and is located at the corner of the ground plane, considering a full ground clearance. Furthermore, all cases consider a substrate the same size as the minimum rectangle comprising the ground plane and the antenna booster of 1 mm thick, $\varepsilon_r = 4.15$, and $\tan\delta = 0.017$ to model a more realistic scenario than those PCBs existing in wireless devices.

3.11.1 Ground plane width analysis

A practical design rule in classic antenna theory states, "the thicker, the better". Such a helpful tip is practical when designing linear dipole or monopole antennas, where increasing the thickness helps to increase bandwidth [52]. Following the same rationale, the question addressed in this chapter is: does the increase in the width of the ground plane help to improve bandwidth?

To address the question, a rectangular ground plane with the same length of 120 mm is compared as the width changes from 60 mm, 30 mm, and 17 mm. The size of 17 mm is the 12 mm length of the antenna booster plus the 5 mm of the short horizontal strip feeding the antenna booster.

Bandwidths are computed according to Equations (3.2) and (3.3), showing different results at both low and high frequencies (Figure 3.51). For the low-frequency

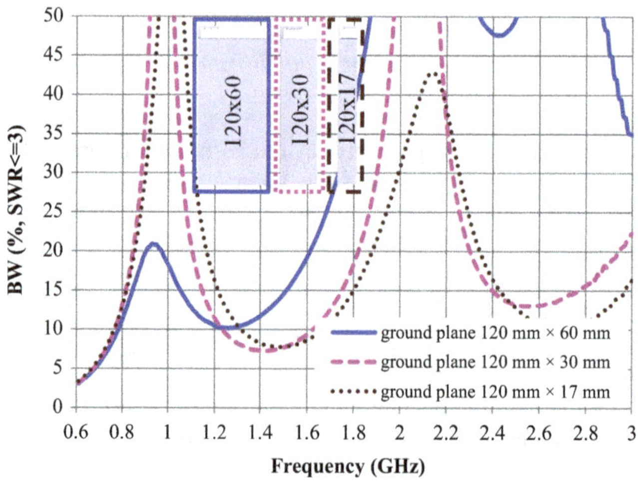

Figure 3.51 Computed bandwidth for three different ground plane widths: 60 mm, 30 mm, and 17 mm with an antenna booster of 12 mm × 3 mm × 2.4 mm (height)

region (frequency less than 1.2 GHz), the best bandwidth occurs for thinner widths of 30 mm and 17 mm cases, providing a 37.0% bandwidth at 0.9 GHz compared to the 20.0% for the 60 mm width. Therefore, for single-band operation, a thinner device of 30 mm provides better bandwidth than the 60 mm width.

As concluded from these results, there are better options than thick. However, for a high-frequency region (frequency larger than 1.7 GHz), the 60 mm wide-width ground plane provides more bandwidth across a wide frequency region than the thinner 30 mm and 17 mm wide-width ground planes. The reason why bandwidth decreases at high frequencies when reducing the width is that horizontal currents aligned with the short edge travel a shorter path. However, for the 60 mm width, such currents travel 0.4λ at 2 GHz. Note that this result is consistent with the current distribution in Figure 3.41. In effect, in that case, the antenna booster at the corner excites horizontal currents aligned with the short edge at 2.1 GHz, whereas the antenna booster at the centre excites shorter currents (the horizontal ones out-of-phase represented by the black arrows) and, as a consequence, less bandwidth is achieved (Figure 3.42). Consequently, for single-band operation at 900 MHz, a 120 mm ground plane with a thin width ($\sim$20 mm) provides a large bandwidth. However, a trade-off exists for multiband operation at 900 MHz and 1.8–3 GHz since the width helps obtain a broad bandwidth across a large frequency region. Depending on the frequency requirements, the size of the ground plane can be adjusted. As a good starting point, 0.4λ ground plane length at the central frequency of the lowest frequency region of operation and 0.4λ ground plane width at the central frequency of the highest frequency region are good references. Reducing such width will increase bandwidth at the lowest frequency region and decrease it at the high-frequency region.

In terms of radiation efficiency, the three cases present values above 90% without any matching network.

To validate the theoretical bandwidth estimation, the 30 mm and 60 mm are matched with a simple matching network comprising a series inductor to provide operation at frequencies around 0.95 GHz (Figure 3.52). Results confirm the theoretical bandwidth calculations showing 32.4% (estimation was 37.0%) and 20.4% (estimation was 20.0%) for 30 mm and 60 mm, respectively. It should be noted that bandwidth estimation based on Equations (3.2) and (3.3) provides a good agreement with the bandwidth that can be physically obtained when adding matching networks, except in cases where the input impedance has loops that tend to provide large bandwidths. Nevertheless, as shown again here, it is a fast tool for comparing several scenarios.

Once the impact of the width has been investigated, a more general analysis is carried out in the next Section 3.11.2, considering several ground planes featuring different lengths and widths.

From a physical standpoint, currents at 900 MHz are computed for the 60 mm and 30 mm widths (Figure 3.53). For the 60 mm width, besides a longitudinal current along the left long edge of 120 mm stronger in intensity than the one at the right edge, a significant current travels along the short 60 mm edge. However, for the 30 mm width, the current is mainly aligned along both long

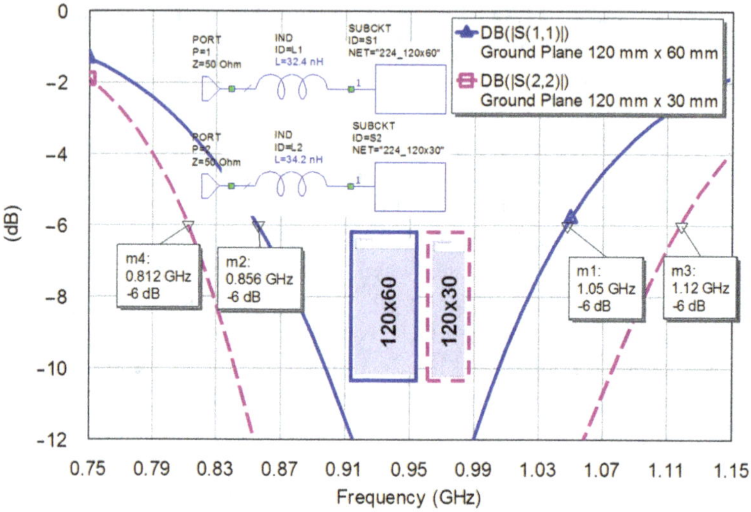

Figure 3.52 Simulated S_{11} for two ground planes having 60 mm and 30 mm widths using an antenna booster of 12 mm × 3 mm × 2.4 mm (height)

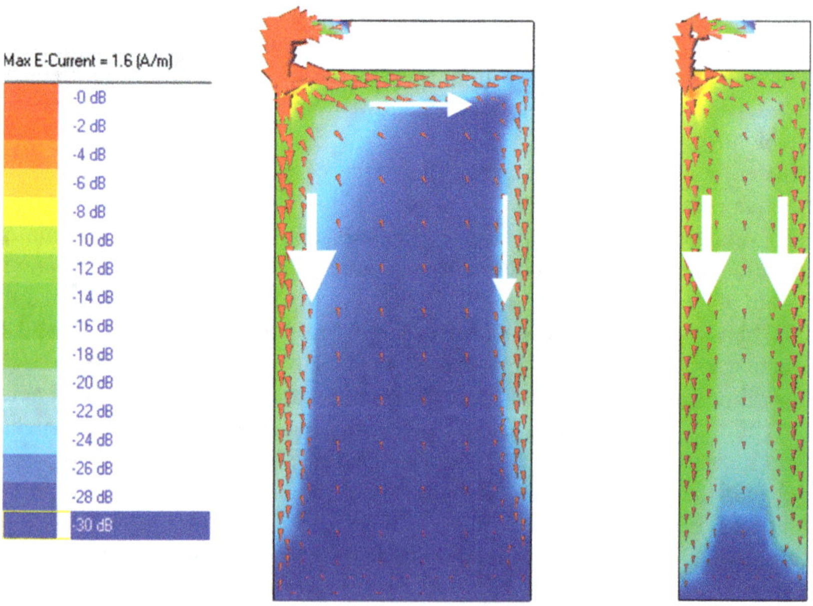

Figure 3.53 Simulated current distribution at 900 MHz for the 120 mm × 60 mm and 120 mm and 30 mm ground planes with a 12 mm × 3 mm × 2.4 mm (height) antenna booster. The same maximum value and dynamic range for both cases. White arrows have been intentionally added to represent the main current distribution.

120 mm edges, making the transversal current almost insignificant. Therefore, the 30 mm wide-width case has a purer mode supported by a longitudinal current than the one of 60 mm wide-width, which combines both a longitudinal and transversal mode. Consequently, the single-mode case achieves much more bandwidth (Figure 3.52).

At this point, it seems that "thicker" ground planes are not always better because, as shown in this case, once it is wide enough, a mode along the width direction appears, decreasing the bandwidth. Can we maintain a longitudinal current for a thick ground plane to obtain better bandwidth? The answer is yes, using two in-phase boosters at each corner of the ground plane [53–56]. This topic is addressed in Chapter 4.

3.11.2 Ground plane of different lengths and widths

As stated in this chapter, the ground plane size matters. The objective of this section is to determine the impact of the ground plane size on the bandwidth and efficiency as well as on the radiation patterns.

Several ground plane sizes are examined in terms of bandwidth, from 20 mm × 20 mm to 200 mm × 200 mm, with steps both in length and width of 10 mm.

The analysis is carried out at a reference frequency of 900 MHz. However, the results can be scaled to other frequencies by applying the scalability principle explained in Chapter 1. All cases use a 12 mm × 3.2 mm × 2.4 mm (height) antenna booster at the corner of the ground plane printed on a thin substrate (1 mm thick, $\varepsilon_r = 4.15$, $\tan\delta = 0.017$). The antenna booster is placed 11 mm from the edge and 5 mm from the corner. A feeding line is then feeding the antenna booster (see Figure 3.25, top case). Bandwidth is computed at 900 MHz when including a matching network of up to two components.

For 0.9 GHz, which is representative of those wireless devices operating at the sub-1 GHz frequency region, it reveals the following conclusions (Figure 3.54):

- Two optimum areas are reaching almost 37.7% bandwidth: for length between 120 mm and 140 mm and width less than 40 mm, and the symmetric situation interchanging L by W. In an in-depth analysis of such areas, the bandwidth is maximum at $L = 140$ mm and $W = 20$ mm, which, in terms of wavelength, represent $L_{opt} = 0.42\lambda$ and $W_{opt} = 0.06\lambda$. This is an essential conclusion since the size of the device can be designed to maximise bandwidth using this optimum size. For example, for the 140 mm × 20 mm ground plane, a BW = 37.1% is achieved ($f_1 = 733$ MHz, $f_2 = 1067$ MHz) with a matching network of only one series inductor. The estimation using Figure 3.54 is 37.7%, which is in excellent agreement.
- Bandwidth decreases if the size *L* x *W* of the device decreases. For example, a ground plane of 40 mm x 40 mm with the 12 mm x 3 mm x 2.4 mm antenna booster can achieve a BW = 2%. If more bandwidth is needed, a broadband matching network, as discussed in Chapter 4, may be designed, which can enlarge the bandwidth by approximately 2.4 times.

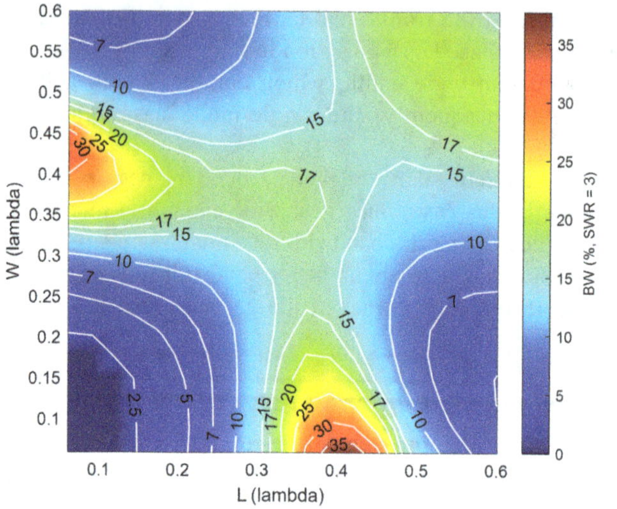

Figure 3.54 Bandwidth evolution at 900 MHz for different ground plane sizes using an antenna booster of 12 mm × 3 mm × 2.4 mm (height) – λ/27.7 length

- It is interesting that for ground planes having dimensions much below L_{opt} and much above L_{opt}, the optimum width is not 0.06λ, but $W = L$. For example, for a ground plane having L = 180 mm (much above L_{opt}), bandwidth is wider if $W = L$ = 180 mm (BW = 19.5%) and not L = 180 mm and W = 20 mm (BW = 8.6%). For instance, if the ground plane is small, with dimensions below L_{opt}, such as L = 60 mm, increasing W to match L can enhance the bandwidth. In this case, when L = 60 mm and W = 60 mm, BW = 4.4%, but if L = 60 mm and W = 20 mm, BW reduces to 3.5%.

Besides bandwidth, efficiency as a function of the ground plane is even more critical since bandwidth does not make sense if efficiency is poor. A map of radiation efficiency for the same set of ground planes has been computed at 900 MHz (Figure 3.55). As observed, the radiation efficiency is very high except for small ground planes (L, W < 60 mm approximately), where radiation efficiency decreases when L and W decrease.

It is essential to analyse the radiation efficiency map carefully, particularly for the small ground plane (L, W < 60 mm at 900 MHz, L, W < 0.18λ at the central frequency of operation). For instance, to enable operation within the frequency range of 863–928 MHz, a ground plane measuring 120 mm × 60 mm (as depicted in Figure 3.25 top) requires just a matching network consisting of two components: a series inductor of 33 nH connected to the antenna booster port and a shunt inductor of 20 nH, both SMD components with a quality factor (Q_m) of approximately 60. The average total efficiency across the band is 87.0%. However, a 60 mm × 60 mm ground plane (as in Figure 3.25 top but with a 60 mm ground

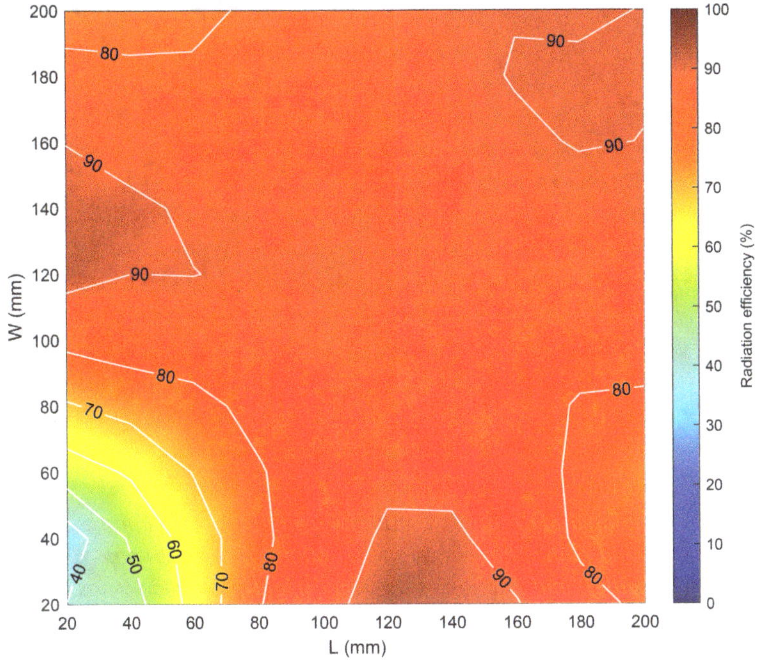

Figure 3.55 Simulated radiation efficiency at 900 MHz for the 12 mm × 3 mm × 2.4 mm antenna booster for different ground planes

plane) needs the same topology but with different values (series inductor of 33 nH from the antenna booster port and a shunt inductor of 8.2 nH, both SMD components with a Q_m ~60) but in this case, the average total efficiency is 51.2%. The significant 2.3 dB efficiency difference does not stem from the radiation efficiency of the 60 mm × 60 mm ground plane, which exceeds 90% at 863–928 MHz. Instead, it results from the high Q_a of the ground plane and antenna booster system ($Q_a = 32$), necessitating a larger Q_m for the matching network components to avoid introducing losses, as indicated by Equation (3.4) and Table 3.1. It is important to note that losses attributed to the matching networks can be approximately calculated using Equation (3.4) (with $Q_a = 32$ and $Q_m = 60$), given that all components share the same nature and Q, resulting in 1.8 dB losses. Therefore, this analysis is revisited from a more practical perspective in Chapter 4, devoted to single-band operation and in Chapter 5, dedicated to multiband operation.

Quality factor of the components

The quality factor (Q_m) of matching network components is crucial; the higher it is, the better, as it helps minimise losses. This importance becomes particularly pronounced for ground planes with small dimensions relative to

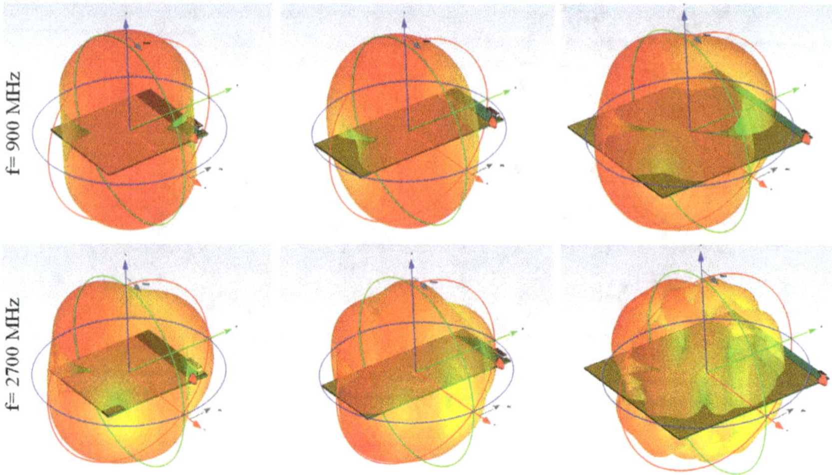

Figure 3.56 Simulated 3D radiation patterns using 30 dB dynamic margin for ground planes having 60 mm × 60 mm (left), 120 mm × 60 mm (middle), and 200 mm × 200 mm (right). The top row is 900 MHz, and the bottom row is 2700 MHz.

the wavelength (approximately $<0.2\lambda$), where the Q_m of the components must be even higher due to the significant Q_a of the ground plane and antenna booster system.

Not only the ground plane size impacts bandwidth and efficiency but also radiation patterns (Figure 3.56). In the end, the radiation pattern mainly depends on the excited currents on the ground plane, not on the currents on the antenna booster. Since ground planes for wireless devices are typically less than a wavelength, radiation patterns are quite isotropic, featuring low directivities, which is convenient for the reception of electromagnetic waves regardless of the incoming angle of arrival.

Radiation patterns for three different ground planes show quasi-isotropic performance (Figure 3.56). Directivity increases as a function of the ground plane size in terms of the wavelength due to the apparition of more modes rather than the fundamental model. For example, directivity is $D = 2.1$ dBi for the 60 mm × 60 mm ground plane at 900 MHz and $D = 5.7$ dBi for the 200 mm × 200 mm at 2.7 GHz.

3.12 On the ground clearance for multiband performance

In the previous sections, antenna boosters were placed at the corner of the ground plane while leaving a free area of grounding or ground clearance. Such ground clearance, however, can be reused to allocate other electronic components, such as sensors. Therefore, the objective of this section is to study the effect of reducing ground clearance and the impact on bandwidth and efficiency.

To obtain a practical design rule, a 12 mm × 3 mm × 2.4 mm antenna booster on a 120 mm × 60 mm ground plane is analysed for several clearance areas of 11 mm × *F* (Figure 3.57). A multiband matching network for operation at 824–960 MHz and 1710–2690 MHz has been designed according to the teachings of Chapter 5. For this section, there is no need to go into detail about the design of such a multiband matching network but to focus on the impact on ground clearance. After such a matching network, a micro-coaxial 50 Ω transmission line is used for testing the S_{11} and total efficiency.

Several experiments of ground reduction have been explored. In particular, the variable *F* has been reduced from 60 mm (full ground clearance) to 50 mm, 40 mm, 30 mm, 25 mm, and 20 mm (Figure 3.57). For each case, the matching network has been tuned, maintaining the same topology. The new values for the matching network are shown in Table 3.5. For each case, total efficiency is measured using 3D pattern integration (Figure 3.58).

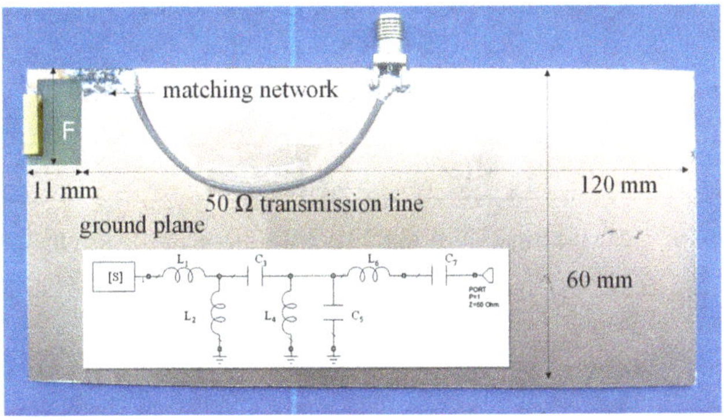

Figure 3.57 A 12 mm × 3 mm × 2.4 mm (height) antenna booster within a ground clearance of 11 mm × F. The multiband matching network comprises seven lumped components (inductors and capacitors). [S] represents the S_{11} of the antenna booster without the matching network.

Table 3.5 Values of the matching network for each ground clearance experiment

F (mm)	*L*1 (nH)	*L*2 (nH)	*C*3 (pF)	*C*4 (pF)	*L*5 (nH)	*C*6 (pF)	*L*7 (nH)
60	4.3	18	0.9	1	13	2	4.5
50	6.5	18	0.8	1	11	1.8	4.5
40	6.5	18	0.7	1	10	1.6	5.1
30	6.5	18	0.7	0.9	11	1.6	4.9
25	6.1	17	0.8	1	11	2	4.5
20	4.9	18	0.8	1	11	1.8	4.5

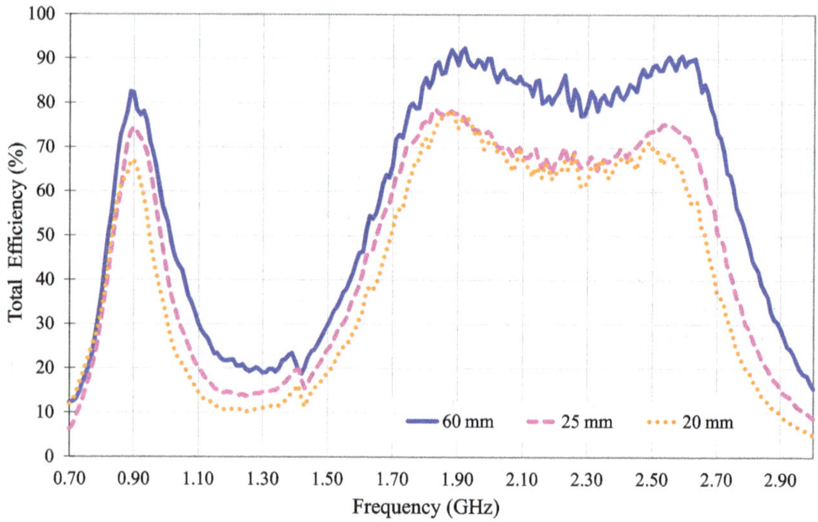

Figure 3.58 Measured total efficiency for three different ground clearance areas: 11 mm × F, where F = 60 mm, 25 mm, and 20 mm

Table 3.6 Measured total efficiency for different ground clearances

	824 MHz–960 MHz			**1710 MHz–2690 MHz**		
***F* (mm)**	**Minimum (%)**	**Maximum (%)**	**Average η_a (%)**	**Minimum (%)**	**Maximum (%)**	**Average η_a (%)**
60	54.3	84.6	73.5	71.1	94.7	84.5
50	54.7	80.3	71.8	57.6	95.7	82.0
40	58.8	78.2	71.1	53.7	92.1	79.6
30	53.6	75.5	67.6	59.9	89.1	76.3
25	46.3	74.6	64.5	47.5	79.4	70.8
20	43.5	68.2	58.7	39.1	78.6	66.4

It is observed from the measured total efficiency that reducing the ground clearance decreases total efficiency. However, such impact is not severe since for the most reduced case ($F = 20$ mm), average efficiency decreases 1 dB both in low and high bands (Table 3.6). This makes the solution compact not only due to the small antenna size of 12 mm × 3 mm × 2.4 mm (h) but also because of the reduced ground clearance of 20 mm × 11 mm.

3.13 Ground plane and ground clearance trade-off

In the previous section, the impact of the ground clearance was analysed while the ground plane length remained the same. In this section, the effect of the ground clearance is again examined, but in contrast to the previous section, the ground clearance is modified. Hence, it increases the distance between the antenna booster and the edge of the ground plane. The objective of the analysis is to determine what is better in terms of bandwidth and efficiency: is it better to increase the ground clearance length while reducing the length of the ground plane or the other way around? Does the ratio between ground clearance length and ground plane length help optimise bandwidth and efficiency? Does such a ratio have a frequency dependence?

All these questions are addressed in this section.

For a given device size, the ratio between ground clearance and ground plane length is studied for an antenna booster feed at the corner of the ground plane. For each scenario, the device size is kept the same and equal to 100 mm × 60 mm, the minimum imaginary rectangle comprising the ground plane and the antenna booster (Figure 3.59). In all cases, the antenna booster is 12 mm × 3 mm × 2.4 mm (height).

At the low-frequency region (frequencies below 1 GHz), the ratio clearance over ground plane length close to one provides the best bandwidth. However, this situation may become impractical since little space is reserved for hosting electronic components on the ground plane. Therefore, as far as only single-band operation at the low frequency is concerned, the ratio of ground clearance over ground plane length can be increased to achieve enough bandwidth as long as the ground plane is enough for the electronics of the wireless device.

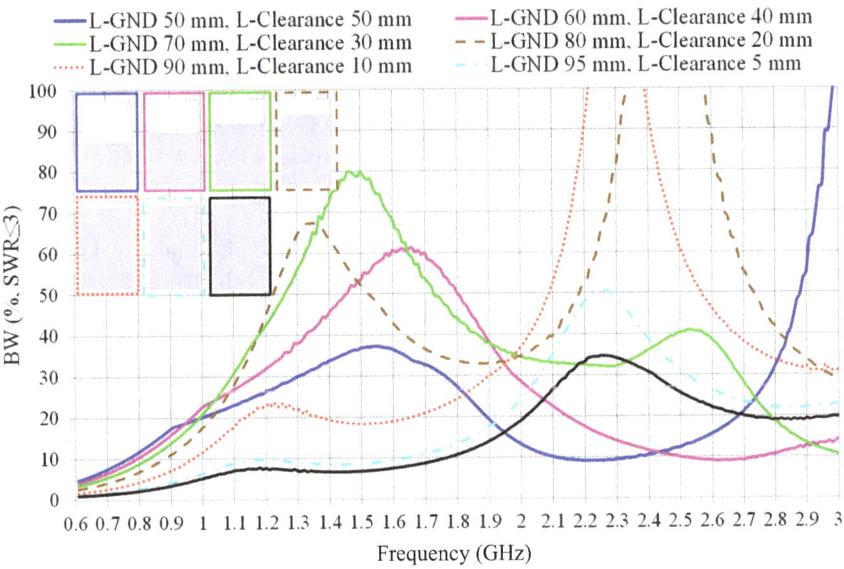

Figure 3.59 Bandwidth for several ratios between the ground clearance length (L-Clearance) to ground plane length (L-GND). All cases feature a total device size of 100 mm × 60 mm.

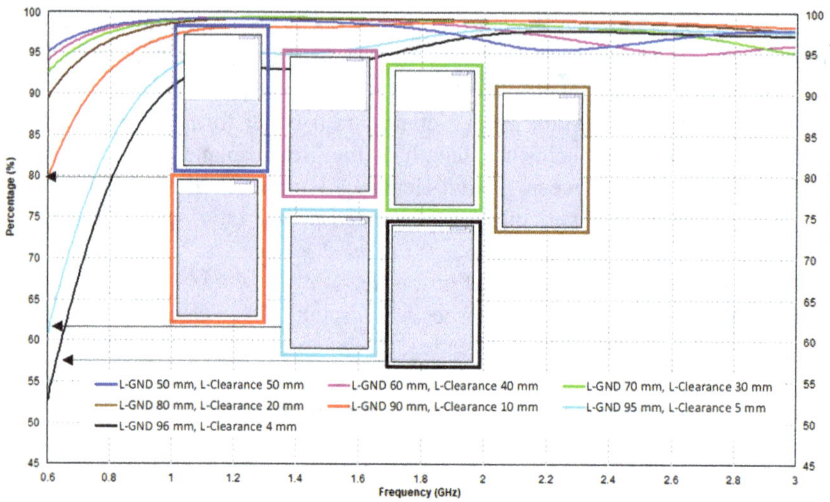

Figure 3.60 Simulated radiation efficiency for several ratios between the ground clearance length (L-Clearance) and ground plane length (L-GND). All cases feature a total device size of 100 mm × 60 mm.

At high frequencies (frequencies over 1 GHz), the trend does not follow the same rule. For example, in the frequency region comprising 1.71–2.69 GHz, the case having a ratio of 10/90 provides a large ground plane size and the best bandwidth. Also, the case 20/80 provides similar bandwidth performance but with less ground plane available for electronics. Now, if a multiband performance is sought, that is, performance at frequencies at the sub-1 GHz and at a higher frequency region, such as the 1.71–2.69 GHz, the ratio 10/90 or 20/80 is a good trade-off.

It is noteworthy that while the ratio of 1 is optimal for the low-frequency range ($f < 1$ GHz), it becomes the worst-case scenario for frequencies above 1.7 GHz. Equally remarkable is the frequency range between 1 GHz and 1.7 GHz, where an optimal ratio between ground clearance length and ground plane length exists, with a ratio of 30/70 = 0.42 (green line in Figure 3.59). This valuable design tip maximises bandwidth within this frequency region, benefiting applications like GNSS, which necessitates about 32% bandwidth (1.164 GHz and 1.605 GHz).

Finally, the radiation efficiency is computed as a function of the same aspect ratios (Figure 3.60). It is observed that in the low-frequency region ($f < 1$ GHz), the larger the ratio of ground clearance length to ground plane length, the better the radiation efficiency. At higher frequencies ($f > 1$ GHz), radiation efficiencies are almost identical for all cases.

Ground clearance to ground plane ratio

- At frequencies where the ground plane is smaller than 0.4λ, increasing the ground clearance to ground plane ratio, improves bandwidth and efficiency.

However, it should be taken into account that such a ground plane reduction is enough for allocating the electronics of the wireless device.

- For multiband applications at two frequency regions, where the ground plane is smaller than 0.4λ at the first frequency region and larger at the second frequency region, a tradeoff exists since increasing the ratio of ground clearance to ground plane enhances bandwidth at the low-frequency region but decreases it at the high-frequency region.
- For single-band applications, the ratio 3/7 maximises bandwidth (the green line in Figure 3.55). Note that these conclusions are valid for the antenna booster configuration shown in Figure 3.55. Using different antenna booster sizes or feeding lines changes the values.

3.14 Other ground plane geometries

The previous section investigated ground planes having rectangular shapes as a good approximation for most practical situations. However, other shapes with more than four corners are studied. In this section, one example of a ground plane featuring a trapezoidal section is discussed (Figure 3.61).

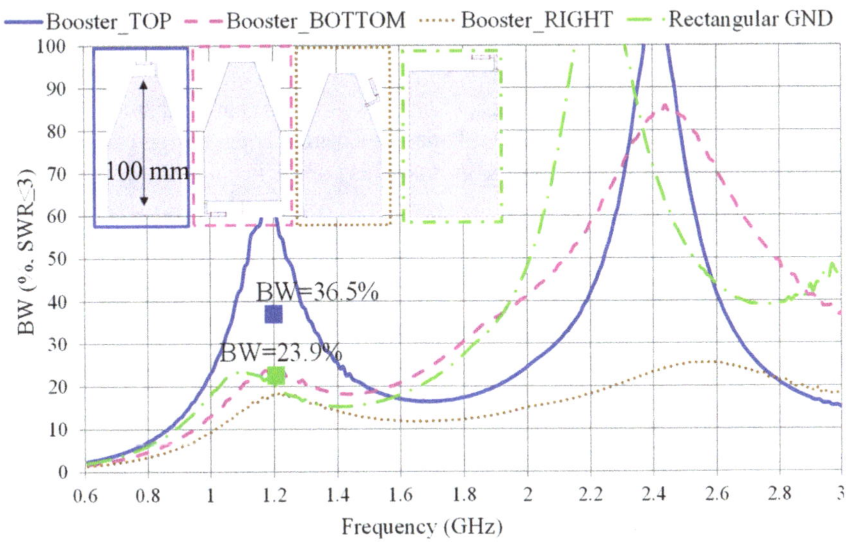

Figure 3.61 12 mm × 3 mm × 2.4 mm (height) antenna booster on a ground plane having a trapezoidal shape compared to a rectangular ground plane of the same length (100 mm). The lines represent the bandwidth estimation with Equations (3.2) and (3.3), whereas the marker shows the exact bandwidth obtained with a two-component matching network, which is a more realistic prediction.

The results indicate that placing the antenna booster is not straightforward. For example, two corner positions (top and bottom) feature a very different performance in terms of bandwidth for frequencies below 1.5 GHz. Therefore, when designing wireless devices with antenna boosters, the first step must be to optimise the placement to reach the best bandwidth and efficiency.

A position away from the corner (booster right) presents the lowest bandwidth among the three tested positions. This is consistent with the results shown in Figures 3.8 and 3.21. However, this should not be considered a drawback but an advantage for some applications where a small bandwidth is required, as long as the antenna booster in such a position provides acceptable efficiency. Having a narrow bandwidth but sufficient help filter harmonics and, therefore, minimising interferences. In effect, consider a transceiver transmitting at f_1; due to a non-linear effect, harmonics appear at $m{\cdot}f_1$, with m an integer. If the bandwidth of the antenna is narrow enough, those harmonics ($2{\cdot}f_1$, $3{\cdot}f_1$, ...) are filtered and mitigated when they reach free space. Consequently, interference is reduced.

It should be noted that, in a low-frequency region, some off-corner positions, like certain centre-edge positions, can offer a similar bandwidth as those located at a corner (Figure 3.38). Again, the position of an antenna booster on the ground plane of a wireless device is key.

The benefits of shaping the ground plane whenever possible are clear in Figure 3.61. Comparing the booster top case with the rectangular ground plane at 1.2 GHz, the bandwidth is larger for the first one: 36.5% compared to 23.9% when matching both cases at 1.2 GHz with a two-component matching network. Applying the scalability principle explained in Chapter 1, the design can be scaled to other frequencies by changing the ground plane while keeping the same antenna booster.

As a side note, the bandwidth prediction diverges at 1.2 GHz and around 2.2–2.4 GHz due to the limitation of Equations (3.2) and (3.3). The correct reading of these plots in those situations should be qualitative rather than quantitative. Even so, the results predict that a better bandwidth at 1.2 GHz is obtained for the booster top with the corners of the ground plane removed than the bandwidth obtained for the rectangular ground plane, which is confirmed when designing a matching network for both situations.

It is interesting to revisit the optimum ground plane size for these scenarios. For the trapezoidal ground plane, a maximum appears at 1.2 GHz, resulting in an optimum length of the ground plane at 0.4λ. However, for the rectangular ground plane, such a maximum is shifted downwards to 1.11 GHz, making the optimum length 0.36λ consistent with the results in Figure 3.51, where the 120 mm × 60 mm ground plane has a maximum bandwidth at 940 MHz, resulting in an optimum length of 0.37λ.

As for the radiation efficiencies of the three positions in Figure 3.61, all feature values over 70% beyond 600 MHz with similar values.

Antenna booster position goes first when designing a wireless device
When designing wireless devices, the first step must be to optimise the placement of the antenna booster to reach the best bandwidth and efficiency. Electromagnetic simulation without a matching network is vital since it allows us to determine the radiation efficiency and potential bandwidth.

Besides shaping the ground plane, slots can be included to enlarge the current path [45–51]. As studied in the chapter, if the current path can be increased to be close to an optimum length, bandwidth can be maximised. In the next experiment, a slot is introduced in a small ground plane with a length of 59 mm. At 900 MHz, this size corresponds to only 0.017λ, far from $\sim 0.4\lambda$. The purpose of introducing a slot in the middle is to enlarge the current path (Figure 3.62). Potential bandwidth is computed with Equations (3.2) and (3.3), where an optimum length is observed at 51 mm, where bandwidth is more than double compared to the case without a slot. All cases embed a 12 mm × 3 mm × 2.4 mm antenna booster at the corner of the ground plane. The slot length controls the frequency corresponding to the maximum bandwidth remarkably; the larger the slot, the lower the frequency. This confirms that the slot length is responsible for enlarging the current path.

The slot length can be reduced by incorporating reactive loading, such as a capacitor positioned at the open edge of the slot. This approach provides additional flexibility in adjusting the optimal length without requiring any modifications to the PCB.

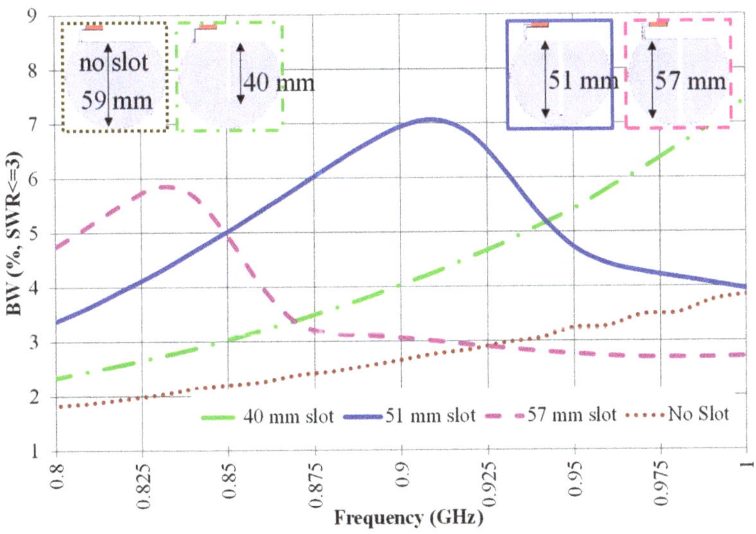

Figure 3.62 Potential bandwidth for different ground planes: without slot and 40 mm, 51 mm, and 57 mm slot. All ground planes are excited by a 12 mm × 3 mm × 2.4 mm antenna booster.

The current distribution is computed at 900 MHz for the ground plane without and with the 51 mm slot, where it is observed that the effect of the slot is to enlarge the current path (Figure 3.63). Therefore, the virtual size of the ground plane becomes larger when the slot is embedded in the ground plane.

A clear advantage is obtained when both cases are matched at the central frequency of 900 MHz (Figure 3.64), where the bandwidth corresponding to the slot

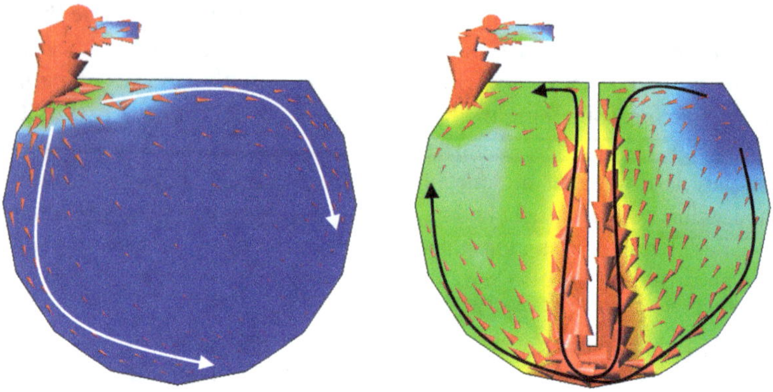

Figure 3.63 Current distribution at 900 MHz for the ground plane without slot (left) and for the 51 mm slot (right). The large arrows have been intentionally added to represent the main current path.

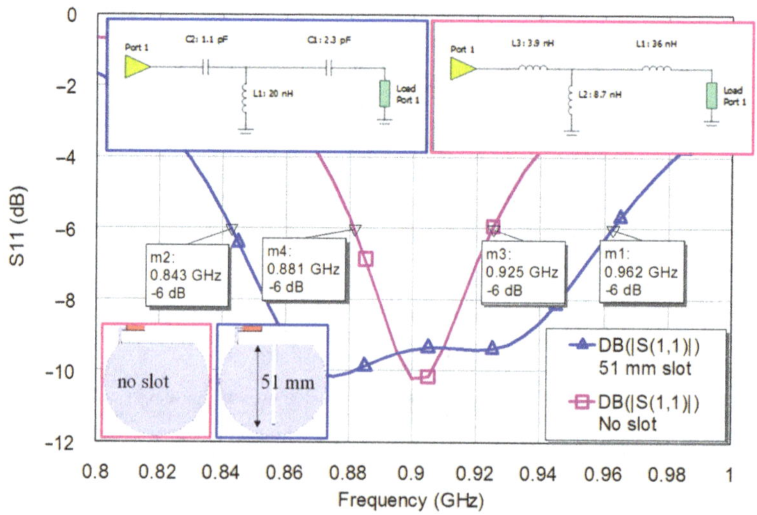

Figure 3.64 Bandwidth for the ground plane without slot and the ground plane with 51 mm slot. The matching networks consider real components (finite Q).

situation is larger (BW_{noslot} = 4.8%, BW_{slot} = 13.2%). Since the bandwidth for the slot case is larger, the Q_a is less than the Q_a for the no-slot case. The Q_a for the no-slot case is 43 at 900 MHz, whereas it is Q_a=17 at the same frequency for the 51 mm slot.

By revisiting Equation (3.4) and considering Q_m = 90, the losses of the matching network for the no-slot case is η_m = 67% and for the slot is η_m = 84%. Since radiation efficiency is almost the same for both cases (can be computed with electromagnetic simulation), total efficiency is ~1 dB (84/67) larger for the case with the slot. In effect, the computed total efficiency at 900 MHz for the no-slot case is 56.5%, whereas it is 79.4%, which is 1.4 dB larger when considering the matching networks shown in Figure 3.64. Note once again the utility of Equation (3.4) to evaluate the impact of the losses of a matching network since although Equation (3.4) considers the same Q_m for all components of the matching network, which is not the real situation, the estimation of the matching network losses is accurate. Once again, the losses of the matching network do not depend on the Q_m but on the relation of Q_m compared to Q_a.

3.15 Conclusions

Table 3.7 answers the objectives given at the beginning of the chapter. It also points to relevant figures and equations to consolidate the concepts learnt. It is also advisable to refresh all essentials included in the grey boxes distributed along the chapter.

Table 3.7 Key aspects to take away

Topic	Review and take away
Antenna booster	It is a small size element (<λ/20) with a non-resonant impedance nature able to excite the ground plane of a wireless device.
Design with antenna boosters	The design of wireless devices based on Antenna Booster Technology relies on the use of antenna boosters, where the frequency of operation is determined by the proper location of the antenna booster on the ground plane of the device and the design of a matching network. Unlike self-resonant antennas, which derive their operating frequency from the design of complex geometries, antenna boosters feature simple structures. They may adopt the form of an off-the-shelf chip. The chip nature allows the integration of an antenna booster on a PCB through a pick-and-place assembly process.
Antenna booster versus resonant antenna elements	Antenna boosters provide more bandwidth and efficiency than self-resonant antennas such as spiral and fractal-shaped (Figures 3.27 and 3.28). Therefore, antenna boosters make better use of the ground plane to attain bandwidth and efficiency.
Architecture	The architecture of an antenna booster system is simple: (a) an antenna booster, (b) a matching network, and (c) a ground plane of the device.

(Continues)

Table 3.7 *(Continued)*

Topic	Review and take away
Design steps	The design of a wireless device embedding antenna boosters implies simple basic steps: 1. Selection of an appropriate antenna booster and location on the ground plane. 2. Design of the matching network. 3. Measure the main parameters, such as the reflection coefficient and total efficiency.
Antenna booster position	The preferred location is typically at the corner of the ground plane (Figure 3.21). However, the recommended position for magnetic boosters is at the centre of the edges, where the magnetic field is at its maximum (Figure 3.23). In this regard, the recommended location of electric and magnetic antenna boosters is different. While electric antenna boosters should be placed at a maximum electrical field, magnetic boosters should be placed at a maximum magnetic field.
Ground plane size	Ground plane size matters since it is a vital player in determining bandwidth, efficiency, and radiation patterns (both shape and polarisation). To maximise bandwidth for single-band operation centred at f_0, a rectangular shape of L_{opt} ~0.42λ and W_{opt} ~0.06λ is a preferred reference. For example, at 900 MHz, this results in L ~140 mm and W ~20 mm. If the maximum dimension of the ground plane is lower than L_{opt}, then to extract more bandwidth and efficiency, the width should be $W = L$. For example, if the centre frequency of operation is at 900 MHz, and the maximum dimension of the ground plane $L = 0.24\lambda$ (80 mm), then 80 mm x 80 mm provides better bandwidth than 80 mm x 20 mm, that is, in this case, W is different from the previous W_{opt} but the same as L. For ground planes having a dimension over L_{opt}, the width $W = L$ again provides better bandwidth (Figure 3.54). For multiband performance at a first frequency region with a centred frequency f_1 and a second frequency region centred at $2{\cdot}f_1$, a good starting size is $L = 0.36\lambda$ and the width $W = 0.18\lambda$ at f_1.
Antenna booster length	For single-band operation, the larger the volume of the antenna booster, the more bandwidth can be obtained. However, for a small antenna booster (<λ/20), bandwidth can be as large as a ten times larger self-resonant antenna. Besides, for single-band operation, bandwidth can be easily enlarged around two times by adding broadband matching networks, as will be explained in detail in Chapter 4. For multiband operation, however, the statement is no longer valid (Table 3.4). It is valid for a certain range (L ~<λ/15) where both bandwidths at the low- and high-frequency regions are optimised.

(Continues)

Table 3.7 (*Continued*)

Topic	Review and take away
	However, the length should be increased if it is not enough to achieve bandwidth at the low-frequency region. Beyond a certain length, the bandwidth at the second-frequency region decreases because it operates in a higher-order mode. Therefore, the length of the antenna booster may be adjusted to satisfy the bandwidth and both frequency regions (Figures 3.33–3.35). A recommended length election is $L = \lambda/20$, $L = \lambda/30$.
Antenna booster width	Width is not as relevant as length. In fact, its impact is meaningless (Figure 3.37).
Antenna booster height	The height of an antenna booster helps modulate the bandwidth (Figure 3.37). Adding volume to the antenna booster helps in achieving bandwidth. However, if the bandwidth is not a limiting factor, a printed antenna booster can also work.
Clearance	Although the larger the clearance area, the better, an antenna booster can operate even with small clearance areas (Figures 3.57 and 3.58).
Ratio clearance/ground plane length	The ratio of clearance to ground plane length matters for a fixed size of the PCB. Its ratio helps to increase the bandwidth at some frequencies (typically the lower frequencies), whereas it decreases at others (Figure 3.59). Its ratio should be optimised considering not only electromagnetic performance but also the size of the ground plane to leave enough area to allocate the electronics of the wireless device.
Ground planes of different shapes	Not only does the size of the ground plane matter, but also the shape (Figure 3.61). Although rectangular and circular ground planes are the most common, other shapes may perform better. Not only can the shape alter the bandwidth, but also, when including slots in the ground plane, it can enlarge the current distribution to emulate a larger ground plane that can excite the ground plane better. As a result, more bandwidth is obtained (Figure 3.64).

3.16 Exercises

A 3D electromagnetic software and a circuit simulator, including matching network synthesis, are needed to solve the following exercises.

3.16.1 Exercise 1

Design a 12 mm × 3 mm × 2.4 mm antenna booster with $\sigma = 5.9 \times 10^7$ S/m as in Figure 3.25 case #1. Compute input impedance from 600 MHz up to 6 GHz.

Determine the electrical size of the antenna booster at 800 MHz. Is the antenna booster matched at 600–960 MHz? Compute the radiation efficiency across the whole frequency range and the potential bandwidth with Equations (3.2) and (3.3).

3.16.2 Exercise 2

For the exercise in Section 3.16.1, assume a matching network with the same Q_m for the lumped components. Compute the matching network losses at 800 MHz when $Q_m = 25$, 50, and 100 with Equation (3.4).

3.16.3 Exercise 3

Considering the antenna booster as in the previous exercises, perform a parametric analysis for a ground plane of 20 mm × 20 mm up to 300 × 300 mm using steps of 20 mm to compute the potential bandwidth at 900 MHz.

3.16.4 Exercise 4

For the antenna booster of exercise 3.16.1, compute the current distributions at 900 MHz, 2.2 GHz, and 3.5 GHz, as well as the radiation patterns. Represent the evolution of directivity as a function of frequency.

3.16.5 Exercise 5

For the configurations of Figure 3.38, compute potential bandwidth from 600 MHz to 6 GHz for a 12 mm × 3 mm × 2.4 mm antenna booster when it is located at the corner of the ground plane, at the middle of the short edge and the middle of the long edge of a 90 mm × 60 mm ground plane. Determine which position provides the best bandwidth below 1 GHz. Compute radiation efficiency. Repeat the experiment when the ground plane is 60 mm × 60 mm and compare bandwidth below 1 GHz with the 90 mm × 60 mm ground plane.

3.16.6 Exercise 6

Design a 12 mm × 3 mm × 2.4 mm antenna booster, as shown in Figure 3.25 top scheme, on a 120 mm × 60 mm ground plane. Consider the ground plane printed on a 1 mm thick substrate with $\varepsilon_r = 4.15$. Compute radiation efficiency from 0.6 GHz to 6 GHz considering $\tan\delta = 0.02$, $\tan\delta = 0.01$, and $\tan\delta = 0.001$. Determine how critical is the loss tangent in the radiation efficiency results.

3.16.7 Exercise 7

For a 12 mm × 3 mm × 2.4 mm antenna booster, as shown in Figure 3.25 top scheme on a 120 mm × 60 mm ground plane, consider the ground plane printed on a 1 mm thick substrate with $\varepsilon_r = 4.15$ and $\tan\delta = 0.017$. Simulate radiation efficiency from 600 MHz to 6 GHz considering a conductivity of both the ground plane and antenna booster of $\sigma = 5.9 \times 10^7$ S/m (copper conductor) and $\sigma = 10^6$ S/m (ITO – Indium Tin Oxide transparent conductor). Compare results. Repeat the exercise when ground plane $L \times W$ is 60 mm × 60 mm and 30 mm × 60 mm.

3.16.8 Exercise 8

For a 12 mm × 3 mm × 2.4 mm antenna booster, as shown in Figure 3.25 top scheme, compute radiation efficiency and bandwidth from 600 MHz to 6 GHz for a ground plane having a total area of 72 cm^2. Compare different ground plane geometries such as a square ground plane, rectangular, circular, triangular, pentagonal, and hexagonal.

3.16.9 Exercise 9

For a 12 mm × 3 mm × 2.4 mm antenna booster, as shown in Figure 3.25 top scheme, compute radiation efficiency and bandwidth from 600 MHz to 6 GHz for a square ground plane of 80 mm × 80 mm. Place the antenna booster at the corner of the ground plane. Compute current distribution at 900 MHz.

3.16.10 Exercise 10

For the previous exercise 3.4.9, make a square hole on the ground plane from 10 mm × 10 mm to 70 mm × 70 mm at 10 mm steps. Compare radiation efficiency and bandwidth from 600 MHz to 6 GHz. Compute current distribution at 900 MHz. Compare radiation patterns at 900 MHz for the solid 80 mm × 80 mm ground plane with the annular ground plane with a hole of 70 mm × 70 mm. Where is the maximum of the radiation pattern for each case?

3.16.11 Exercise 11

Consider a 30 mm × 3 mm × 2 mm (height) antenna booster spaced 10 mm from the edge of a ground plane with a vertical feedline 2 mm width and 8 mm length (Figure 3.33). Consider a central frequency of 400 MHz. Simulate the bandwidth and radiation efficiency from $L = 50$ mm, $W = 50$ mm to $L = 300$ mm, $W = 300$ mm at steps of 20 mm. Determine a map as in Figures 3.54 and 3.55. Which ground plane maximises bandwidth?

3.16.12 Exercise 12

For the previous exercise 3.4.11, what is the minimum Q_m for the matching network components to ensure matching network losses of less than 0.5 dB at 400 MHz if the ground plane size is 100 mm × 100 mm? What kind of components can achieve such Q_m?

3.16.13 Exercise 13

For the ground planes shown in Figure 3.61 (antenna booster top and rectangular ground plane), scale the ground plane size to have the maximum bandwidth at 900 MHz. Compute bandwidth with Equations (3.2) and (3.3).

3.16.14 Exercise 14

For the ground planes shown in Figure 3.62, instead of 59 mm in length, increase the length to 65 mm. Optimise the length of the slot to maximise bandwidth at

900 MHz. Consider a 12 mm × 3 mm × 2.4 mm antenna booster at 11 mm from the edge of the ground plane. Put the antenna booster at the corner of the ground plane.

Once the slot is optimised, reduce it to 15 mm. Add a capacitor at the open edge. Tune the capacitor to maximise bandwidth. Compare the bandwidth with the slot without the capacitor.

3.16.15 Exercise 15

Compute the modal significance from 500 MHz to 5 GHz for the following ground planes: 60 mm × 60 mm, 120 mm × 60 mm, 140 mm × 20 mm, 200 mm × 200 mm. Determine the current distribution for the fundamental mode.

3.16.16 Exercise 16

Compute the modal significance from 800 MHz to 1 GHz for the ground planes with and without slot of Figure 3.64. Can you identify the relation between the modal significance and the bandwidth improvement when using the slot in the ground plane?

3.16.17 Exercise 17

For the antenna booster in Figure 3.25, compute the potential bandwidth from 600 MHz to 6 GHz. Repeat the exercise reducing the ground plane from 120 mm to 110 mm but increasing the clearance by 10 mm so the whole PCB is the same (131 mm × 60 mm). Note that you should increase the length of the feeding line by 10 mm. Repeat for a ground plane with 100 mm and clearance of 31 mm. Which one provides better bandwidth?

References

[1] J. Anguera, A. Andújar, C. Puente, and J. Mumbrú, "Antennaless Wireless Device Capable of Operation in Multiple Frequency Regions," *Patent app.* WO2010/015364, Feb. 2010.

[2] J. Anguera, A. Andújar, C. Puente, and J. Mumbrú, "Antennaless Wireless Device," *Patent US* 8,203,492, June 2012.

[3] J. Anguera, A. Andújar, C. Puente, and J. Mumbrú, "Antennaless Wireless Device capable of Operation in Multiple Frequency regions," *Patent US* 8,237,615 B2, August 2012.

[4] A. Andújar, J. Anguera, and C. Puente, "Ground Plane Boosters as a Compact Antenna Technology for Wireless Handheld Devices," *IEEE Transactions on Antennas and Propagation*, vol. 59, no. 5, pp. 1668–1677, 2011.

[5] J. Anguera, N. Toporcer, and A. Andújar, "Slim Bar Booster for Electronics Devices," *Patent US* 9,960,478 B2, May 2018.

[6] J. Anguera, A. Andújar, and C. Puente, "Antenna-Less Wireless: A Marriage Between Antenna and Microwave Engineering," *Microwave Journal*, vol. 60, no. 10, pp. 22–36, 2017.

[7] A. Andújar and J. Anguera, "Integration of a Non-Resonant Antenna in a Smartphone for Multiband Operation," *European Conference on Antennas and Propagation*, EUCAP 2018, London, UK, April 2018.

[8] HardwareMe Electronics S.L.

[9] J. Anguera, A. Andújar, J. L. Leiva, C. Schepens, R. Gaddi, and S. Kahng, "Multiband Antenna Operation with a Non-Resonant Element Using a Reconfigurable Matching Network," *European Conference on Antennas and Propagation*, EUCAP 2018, London, UK, April 2018.

[10] A. Andújar, J. L. Leiva, J. Anguera, C. Schepens, and R. Gaddi, "Chip Antenna-Antenna Tuner Combo Cover LTE Bands," *Microwave Journal*, vol. 62, no. 1, pp. 70–78, 2019.

[11] J. Anguera, A. Andújar, J. L. Leiva, O. Massó, J. Tonnesen, E. Rindalsholt, R. Brandsegg, and R. Gaddi, "Reconfigurable Multiband Operation for Wireless Devices Embedding Antenna Boosters," *Electronics* 10, 808, 2021.

[12] K. L. Wong, *Planar Antennas for Wireless Communications,* Wiley Inter-Science, 2003.

[13] M. Martínez-Vázquez, O. Litschke, M. Geissler, D. Heberling, A. M. Martínez-González, and D. Sánchez-Hernández, "Integrated Planar Multiband Antennas for Personal Communication Handsets," *IEEE Transactions on Antennas and Propagation*, vol. 54, no. 2, pp. 384–391, 2006, doi:10.1109/TAP.2005.863402.

[14] K. L. Wong, G.Y. Lee, and T.W. Chiou, "A Low-Profile Planar Monopole Antenna for Multiband Operation of Mobile Handsets," *IEEE Transactions on Antennas and Propag.*, vol. 51, no. 1, pp. 121–125, 2003.

[15] K. L. Wong and S.C. Chen, "Printed Single-Strip Monopole Using a Chip Inductor for Penta-Band WWAN Operation in the Mobile Phone," *IEEE Transactions on Antennas and Propagation*, vol. 58, no. 3, pp. 1011–1014, 2010.

[16] H. Kanj and S.M. Ali, "Compact Multiband Folded 3-D Monopole Antenna," *IEEE Antennas and Wireless Propagation Letters*, vol. 8, pp. 185–188, 2009.

[17] J. Ma, Y. Z. Yin, J.L. Guo, and Y.H. Huang, "Miniature Printed Octaband Monopole Antenna for Mobile Phones," *IEEE Antennas and Wireless Propagation Letters*, vol. 9, pp. 1033–1036, 2010.

[18] K. L. Wong and C.H. Huang, "Printed Loop Antenna with a Perpendicular Feed for Penta-band Mobile Phone Application," *IEEE Transactions on Antennas and Propagation*, vol. 56, no. 7, pp. 2138–2141, 2008.

[19] C. Lin and K.L. Wong, "Printed Monopole Slot Antenna for Internal Multiband Mobile Phone Antenna," *IEEE Transactions on Antennas and Propagation*, vol. 55, no. 2, pp. 3690–3697, 2007.

[20] C. H. Wu and K. L. Wong "Hexa-band Internal Printed Slot Antenna for Mobile Phone Application," *Microwave and Optical Technology Letters*, vol. 50, pp. 35–38, 2008.

[21] B.S. Collins, S.P. Kingsley, J.M. Ide, S.A. Saario, R.W. Schlub, and S.G. O'Keefe, "A Multi-band Hybrid Balanced Antenna," *IEEE International*

Workshop on Antenna Technology: Small Antennas and Novel Metamaterials, New York, March 6–8, pp. 100–103, 2006.

[22] C. Lin, K. L. Wong, "Printed Monopole Slot Antenna for Internal Multiband Mobile Phone Antenna," *IEEE Transactions on Antennas and Propagation*, vol. 55, no. 12, pp. 3690–3697, 2007.

[23] J. Anguera, I. Sanz, J. Mumbrú, and C. Puente, "Multi-Band Handset Antenna with a Parallel Excitation of PIFA and Slot Radiators," *IEEE Transactions on Antennas and Propagation*, vol. 58, no. 2, pp. 348–356, 2010.

[24] J. Anguera, C. Picher, A. Bujalance, and A. Andújar, "Ground Plane Booster Antenna Technology for Smartphones and Tablets," *Microwave and Optical Technology Letters*, vol. 58, no. 6, pp. 1289–1294, 2016.

[25] J. Anguera, A. Andújar, G. Mestre, J. Rahola, and J. Juntunen, "Design of Multiband Antenna Systems for Wireless Devices Using Antenna Boosters," *IEEE Microwave Magazine*, vol. 20, no. 12, pp. 102–114, 2019, doi:10.1109/MMM.2019.2941662.

[26] J. Anguera, C. Puente, A. Andújar, R. M. Mateos, D. Vye and M. Lien, "Antenna Library for IoT Devices with Antenna Boosters," *2020 50th European Microwave Conference (EuMC)*, 2021, pp. 1147–1150, DOI:10.23919/EuMC48046.2021.9338246.

[27] J. Anguera and A. Andújar, "Antenna Booster Technology: From R&D to Ignion," *IEEE Antennas and Propagation Magazine*, vol. 63, no. 4, pp. 133–137, 2021, doi: 10.1109/MAP.2021.3085163.

[28] D. Vye, "Network Synthesis Wizard Automates Interactive Matching-Circuit Design," *Microwave Journal*, vol. 61, no. 11, pp. 96–102, 2018.

[29] J. Juntunen, J. Järveläinen, and D. Linden, "MIMO Dual-Band WiFi Antenna Using NI AWR Software, Optenni Lab, and Premix PREPERM Materials," *MMee*, pp. 12–14, March–April 2019.

[30] A. Andújar, J. Anguera, and R. Mª Mateo, "Multiband Non-Resonant Antenna System with Reduced Ground Clearance," *European Conference on Antennas and Propagation*, EUCAP 2017, Paris, France, April 2017.

[31] A. Suárez, F. Ramírez and S. Sancho, "Generalized Stability Criteria for Power Amplifiers Under Mismatch Effects," *IEEE Transactions on Microwave Theory and Techniques*, vol. 63, no. 12, pp. 4415–4428, 2015, doi:10.1109/TMTT.2015.2494578.

[32] S. R. Best, "The Inverse Relationship between Quality Factor and Bandwidth in Multiple Resonant Antennas," *IEEE Antennas and Propagation Society International Symposium*, pp. 623–626, 2006.

[33] J. Rahola, "Bandwidth Potential and Electromagnetic Isolation: Tools for Analyzing the Impedance Behavior of Antenna Systems," *Proceedings of the EuCAP 2009 Conference*, Berlin, March 23–27, 2009.

[34] G. Smith, "Efficiency of Electrically Small Antennas Combined with Matching Networks," *IEEE Transaction on Antennas and Propagation*, vol. 25, no. 3, pp. 369–373, 1977.

[35] R. J. Garbacz, "A Generalized Expansion for Radiated and Scattered Fields," *Ph.D. dissertation*, Ohio State University, Columbus, 1968.

[36] R. F. Harrington and J.R. Mautz, "Theory of Characteristic Modes for Conducting Bodies," *IEEE Transaction on Antennas and Propagation*, vol. 19, no. 5, pp. 622–628, 1971, doi: 10.1109/TAP.1971.1139999.

[37] F. Harrington and J.R. Mautz, "Computation of Characteristic Modes for Conducting Bodies," *IEEE Transactions on Antennas and Propagation*, vol. AP-19, no. 5, pp. 629–639, 1971.

[38] M. Cabedo, E. Antonino, A. Valero, and M. Ferrando, "The Theory of Characteristic Modes Revisited: A Contribution to the Design of Antennas for Modern Applications," *IEEE Antennas and Propag. Magazine*, vol. 49, no. 5, pp. 52–68, 2007.

[39] S. N. Makarov, *Antenna and EM Modeling with MATLAB®,* John Wiley & Sons, New York, 2002.

[40] J. Anguera and A. Andújar, "Ground Plane Contribution in Wireless Handheld Devices Using Radar Cross Section Analysis," *Progress in Electromagnetics Research M*, vol. 26, pp. 101–114, 2012.

[41] T. Y. Wu and K. L. Wong, "On the Impedance Bandwidth of a Planar Inverted-F Antenna for Mobile Handsets," *Microwave and Optical Technology Letters*, vol. 32, pp. 249–251, 2002.

[42] B. L. G. Jonsson, F. Ferrero, S. Shi and L. Wang, "Optimal Bandwidth Positions for a Terminal Embedded Antenna: Physical Bounds and Antenna Design," *IEEE Transactions on Antennas and Propagation*, vol. 69, no. 4, pp. 1931–1941, 2021, doi:10.1109/TAP.2020.3026921.

[43] A. Andújar and J. Anguera, "Magnetic Boosters for Multi-band Operation," *Microwave and Optical Technology Letters*, vol. 55, no. 1, pp. 65–75, 2013.

[44] J. Anguera, C. Puente, E. Martínez, and E. Rozan, "The Fractal Hilbert Monopole: A Two-dimensional Wire," *Microwave and Optical Technology Letters*, vol. 36, no. 2, pp. 102–104, 2003.

[45] P. Vainikainen, J. Ollikainen, O. Kivekäs, and I. Kelander, "Resonator-Based Analysis of the Combination of Mobile Handset Antenna and Chassis," *IEEE Transactions on Antennas and Propagation*, vol. 50, no. 10, pp. 1433–1444, 2002.

[46] A. Cabedo, J. Anguera, C. Picher, M. Ribó, and C. Puente, "Multi-Band Handset Antenna Combining a PIFA, Slots, and Ground Plane Modes," *IEEE Trans. Antennas and Propag.*, vol. 57, no. 9, 2526–2533, 2009.

[47] M. Cabedo, E. Antonino, V. Rodrigo, and C. Suárez, "Análisis Modal de un Plano de Masa Radiante Doblado y con una Ranura para Terminales Móviles," *Proceedings of the XXI National Symposium URSI'06*. Oviedo, Spain, Sept. 2006.

[48] M. F. Abedin and M. Ali, "Modifying the Ground Plane and Its Effect on Planar Inverted-F Antennas (PIFAs) for Mobile Phone Handsets," *IEEE Antennas and Wireless Propag. Lett.*, vol. 2, pp. 226–229, 2003.

[49] R. Hossa, A. Byndas, and M. E. Bialkowski, "Improvement of Compact Terminal Antenna Performance by Incorporating Open-End Slots in Ground

Plane," *IEEE Microwave and Wireless Components Letters*, vol. 14, no. 6, 2004.

[50] B. Sanz-Izquierdo, J. Batchelor, and R. Langley, "Multiband Printed PIFA Antenna with Ground Plane Capacitive Resonator," *Electronics Letters* vol. 40, no. 22, pp. 1391–1392, 2004.

[51] J. Anguera, I. Sanz, A. Sanz, A. Condes, D. Gala, C. Puente, and J. Soler, "Enhancing the Performance of Handset Antennas by Means of Groundplane Design". *IEEE International Workshop on Antenna Technology: Small Antennas and Novel Metamaterials* (iWAT 2006). New York, USA, March 2006.

[52] Y. Duroc and A. I. Najam, "UWB Antennas: Design and Modeling," *IntechOpen*, September 17, 2010, DOI: 10.5772/10002.

[53] J. Anguera and C. Puente, "Distributed Antenna System Robust to Human Body Loading Effects," *Patent US* 9,007,275 B2, April 2015.

[54] J. Anguera, A. Camps, A. Andújar, and C. Puente, "Enhancing the Robustness of Handset Antennas to Finger Loading Effects," *Electronics Letters*, vol. 45, no. 15, pp. 770–771, 2009.

[55] A. Andújar, J. Anguera, Y. Cobo, and C. Picher, "Distributed Antenna Systems for Wireless Handheld Devices," *IEEE Transactions on Antennas and Propagation*, vol. 60, no. 10, pp. 4830–4837, 2012.

[56] A. Andújar, J. Anguera, J. L. Leiva, and C. Picher, "In-phase Versus Outphase Distributed Antenna Systems for Wireless Handheld Devices," *IEEE Transactions on Antennas and Propagation*, vol. 61, no. 1, pp. 346–353, 2013.

Chapter 4

Antenna booster technology for single-band operation

This chapter introduces Antenna Booster Technology for wireless devices requiring a single-band operation. The term single-band refers here to a single-frequency region, for example, a continuous frequency region $[f_1, f_2]$. In many situations, wireless devices require operation at a single frequency region. Some others require multiband operation, including another frequency region $[f_3, f_4]$ spaced from the first region. The topic of designing wireless devices embedding antenna boosters for multiband operation is addressed in Chapter 5, where many concepts learnt in this chapter will be helpful. The chapter will be instructive in understanding the design of wireless devices with antenna boosters and matching networks. As discovered in Chapter 3, the matching network is a crucial player in the architecture of antenna booster technology; it maximises the transfer of power from a generator to the antenna booster system and, as a result, to free space, and vice versa. As we will learn, this design methodology drastically differs from the classic antenna design, which relies on shaping the geometry of an antenna. In fact, the design of wireless devices with antenna boosters looks more like a microwave engineering design rather than an antenna engineering one. Thanks to the non-resonant nature of antenna boosters, the design process is streamlined and efficient, allowing for both human and machine engineering approaches. This simplification significantly enhances accessibility for wireless designers, eliminating the requirement for specialised expertise in antenna design.

The chapter has the following objectives:

1. Design simple matching networks for single-band wireless devices embedding antenna boosters
2. Know the impact of component tolerances in S_{11} and total efficiency to design robust matching networks
3. Understand how the quality factor (Q) of the components of the matching network impacts the performance
4. How the bandwidth of a wireless device embedding antenna boosters can be enhanced with matching networks
5. Scale the designs shown in the chapter to any frequency band
6. Learn how a matching network can be reused across a broad set of wireless devices with ground planes of different sizes

4.1 Introduction

This chapter introduces the design of single-band wireless devices embedding antenna boosters. Several architectures are presented in this chapter (Figure 4.1), with some fundamentals of the role of a matching network (Section 4.2). The first includes simple matching networks with only two lumped components (Section 4.3). A key aspect, along with designing matching networks is determining their robustness to the tolerance of the matching network components. Thus, a discussion on the impact of tolerances of the components is addressed (Section 4.4).

The study in Section 4.5 examines the effect of the finite Q of the lumped components, such as inductors and capacitors, on the overall efficiency of the matching network. An application of a single-band for a wearable wireless device is presented (Section 4.6). It shows how the matching network can be adjusted depending on the proximity of the wireless device to the human body.

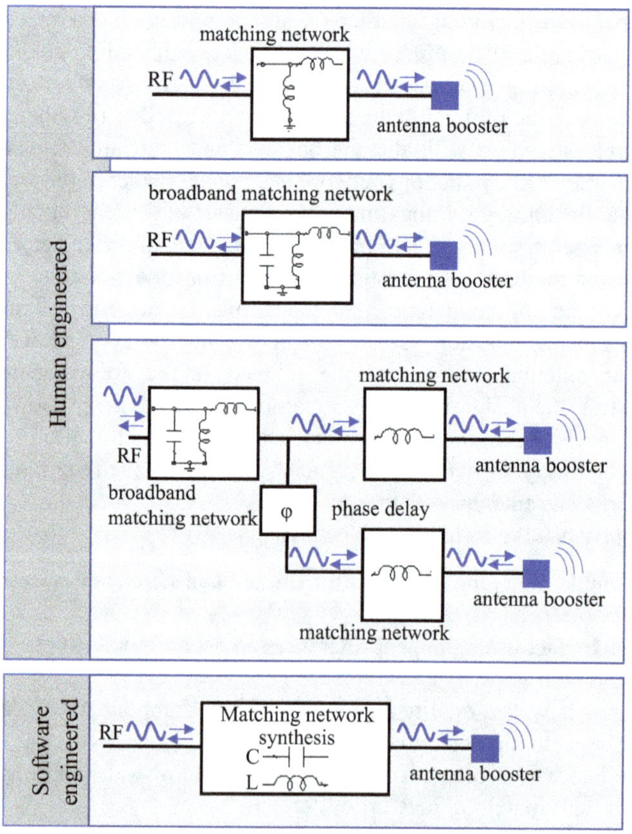

Figure 4.1 From simple matching networks to broadband and matching network synthesis for the design of single-band wireless devices embedding antenna boosters. RF represents a radiofrequency source.

In some situations, a simple matching network with only two components may not be enough to get the required bandwidth. In those cases, a broadband matching network is proposed instead of increasing the antenna element size, as occurs with a geometry-based antenna design. A broadband matching network can not only match the antenna booster system but also increase the inherent bandwidth [1]. Hence, it is interesting to know how much the bandwidth can be enhanced. Can we improve the bandwidth by adding more matching components to the matching network, or is the bandwidth bounded? This topic is addressed in Section 4.7. Once the principles of bandwidth enhancement are understood, broadband matching networks are analysed from two perspectives: one based on human-engineered architectures and the other software-engineered based on matching network synthesis (Section 4.8).

Once a matching network is designed for an antenna booster, including a ground plane, to provide operation at a given frequency range, the question is: can we reuse the same matching network for other ground planes having different sizes? The reuse of matching networks for single-band operation is analysed in Section 4.9.

Other broadband mechanisms are discussed, such as the non-Foster matching networks with negative capacitors (Section 4.10). Then, the combination of two antenna boosters to improve the bandwidth is detailed in Section 4.11.

A loop-based antenna booster scheme is discussed for narrowband applications, particularly suitable for small ground planes (Section 4.12).

The Chu bandwidth limit is revisited in Section 4.13, where a half-wavelength dipole antenna is compared with antenna boosters having different ground plane sizes.

The matching networks discussed throughout the chapter comprise lumped elements (capacitors and inductors). However, the same strategies can be utilised with distributed elements (transmission lines) or a combination of distributed and lumped elements. The reason distributed elements are not used here is that the frequency of operation is not high enough to have small distributed elements. For example, a quarter-wave transformer at 900 MHz, considering a microstrip substrate with $\varepsilon_r = 4$, is 41.6 mm. As a result, since the space available is scarce in a wireless device such as a tracking device, smart-watch, or the like, integrating large and distributed elements that need to share space with other electronic components is challenging.

Conclusions are drawn in Section 4.14. Exercises are presented in Section 4.15. Finally, references for further reading are included in the References section.

4.2 The role of a matching network

The role of a matching network is to transfer the maximum power from the radiofrequency source as the transceiver to the antenna, and vice versa, to transfer the maximum power received by the antenna to the receiver. Since the problem is reciprocal, let us address the situation by considering a radiofrequency source connected to a matching network and an antenna booster (Figure 4.2). Note that

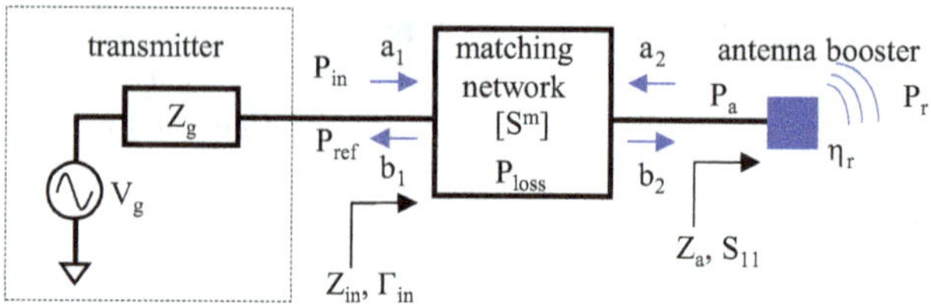

Figure 4.2 A transmitter is connected to a matching network and an antenna booster

$P_{in} = P_{ref} + P_{loss} + P_a$ where P_{in} is the input power, P_{ref} is the reflected power, P_{loss} is the power dissipated by the matching network due to the losses of the matching network components, P_a is the power delivered to the antenna booster, including the ground plane, P_r is the radiated power to the space, and a_1 and a_2 are the incoming power waves to the matching network, whereas b_1 and b_2 are the output power waves. Z_{in} is the input impedance at the matching network port, and Γ_{in} the reflection coefficient associated with such a port. Note that both depend on the matching network and the antenna booster impedance, including the ground plane. Z_a is the impedance of the antenna booster containing the ground plane. Typically, Z_a is reactive, at least at the lowest frequencies of operation. S_{11} is the associated reflection coefficient to the Z_a. Since Z_a is highly reactive, S_{11} is close to 1. However, this mismatch is easily compensated with the matching network, which will provide the flexibility to make the wireless device operative at any frequency band. The mission of the antenna booster is to excite currents on the ground plane, so it becomes the crucial element in the radiation process, determining bandwidth, efficiency, and radiation patterns. Finally, η_r is the radiation efficiency, which is the ratio between P_r and P_a. This parameter depends on the losses of the materials of the antenna booster system, comprising the substrate, the plastic casing of the wireless device, and any other component that absorbs power, excluding the matching network.

The goal of the following steps is to find the role of a matching network to maximise the radiated power P_r, which implies making $P_{loss} = P_{ref} = 0$. First, we define total efficiency η_t as:

$$\eta_t = \frac{P_r}{P_{in}} \quad P_{in} = |a_1|^2 \tag{4.1}$$

where the power wave amplitude a_1 is:

$$a_1 = V_g^2 \cdot \frac{\mathbb{R}\{Z_{in}\}}{|Z_{in} + Z_g|} \tag{4.2}$$

To maximise P_r, P_{in} should be maximised, and thus a_1 should be maximum, which occurs when $Z_{in} = Z_g^*$. As a result, the mission of the matching network is to transform the impedance of the antenna booster Z_a to Z_g^*. Since the reflection coefficient Γ_{in} at the input port of the matching network is:

$$\Gamma_{in} = \frac{Z_{in} - Z_g^*}{Z_{in} + Z_g} \tag{4.3}$$

Then, by conjugate matching, $\Gamma_{in} = 0$.

We define radiation efficiency as the power radiated to the space P_r as the power delivered to the antenna booster system P_a:

$$\eta_r = \frac{P_r}{P_a} \tag{4.4}$$

$$P_a = |b_2|^2 - |a_2|^2 \tag{4.5}$$

Since

$$S_{11} = \frac{a_2}{b_2} \tag{4.6}$$

Then

$$P_a = |b_2|^2 \cdot \left(1 - |S_{11}|^2\right) \tag{4.7}$$

From the S-matrix for the matching network, we have the following:

$$\left.\begin{aligned} b_1 &= S_{11}^m \cdot a_1 + S_{12}^m \cdot a_2 \\ b_2 &= S_{21}^m \cdot a_1 + S_{22}^m \cdot a_2 \end{aligned}\right\} \tag{4.8}$$

By an algebraic manipulation of the right-hand side of the second equation:

$$b_2 = b_2 \cdot \left(S_{21}^m \cdot \frac{a_1}{b_2} + S_{22}^m \cdot \frac{a_2}{b_2}\right) \rightarrow \left(S_{21}^m \cdot \frac{a_1}{b_2} + S_{22}^m \cdot \frac{a_2}{b_2}\right) = 1 \tag{4.9}$$

Resulting in:

$$b_2 = a_1 \cdot \frac{S_{21}^m}{1 - S_{22}^m \cdot S_{11}} \tag{4.10}$$

Substituting Equation (4.10) in (4.7) gives:

$$P_a = |a_1|^2 \cdot \left|\frac{S_{21}^m}{1 - S_{22}^m \cdot S_{11}}\right|^2 \cdot \left(1 - |S_{11}|^2\right) \tag{4.11}$$

With Equations (4.1) and (4.11):

$$P_a = P_{in} \cdot \left|\frac{S_{21}^m}{1 - S_{22}^m \cdot S_{11}}\right|^2 \cdot \left(1 - |S_{11}|^2\right) \tag{4.12}$$

Substituting P_a with P_r, we obtain:

$$P_r = P_{in} \cdot \left| \frac{S_{21}^m}{1 - S_{22}^m \cdot S_{11}} \right|^2 \cdot \left(1 - |S_{11}|^2\right) \cdot \eta_r \tag{4.13}$$

where we finally find the crucial equation for total efficiency η_t:

$$\eta_t = \left| \frac{S_{21}^m}{1 - S_{22}^m \cdot S_{11}} \right|^2 \cdot \left(1 - |S_{11}|^2\right) \cdot \eta_r \tag{4.14}$$

Therefore, the role of the matching network is to maximise total efficiency, hence $P_r = P_{in}$, meaning that total efficiency is 100%. In the following sections, matching networks have precisely this objective. Furthermore, the matching network synthesis techniques explained in this chapter use Equation (4.14). Note that this procedure is different from the route of maximising the captured power from the source and delivering it to the system, that is, $P_{in} - P_{ref}$. The reason for this is that captured power can be maximised by introducing losses to the matching network, which is not desired at all since the power is then not radiated.

Note that in Equation (4.14), the term of $1-|S_{11}|^2$ is typically close to zero for an antenna booster since the impedance is reactive and poorly matched, at least at the lowest frequencies of operation. For example, an antenna booster may feature, at a given frequency, a reactive impedance of $Z_a = 37-j208\Omega$. This results in an $|S_{11}|$ of 0.92. Therefore, it is the mission of the matching network to compensate for such mismatching and, at the same time, not introduce losses in the matching network.

Although Equation (4.14) is crucial, it does not directly show the impact of how relevant it is to provide input impedance matching. To put it differently, so that the mismatching, as well as the efficiency of the matching network, appear explicitly in the equation, we express the total efficiency as:

$$\eta_t = \left(1 - |\Gamma_{in}|^2\right) \cdot \eta_m \cdot \eta_r \tag{4.15}$$

where the first term corresponds to the input power to the matching network and antenna booster system, which directly depends on the matching network. This equation is insightful and clearly states that to maximise the power transferred to the antenna booster, we should:

(a) provide impedance matching by making Γ_{in} as close to zero as possible.
(b) maximise η_m to be close to 1. This can be achieved by using high-quality factor components. However, this efficiency does not only depend on the quality factor of the components Q_m but also on the relationship between the quality factor of the components and the quality factor of the antenna booster system Q_a.
(c) maximise radiation efficiency η_r, which depends on the losses of the radiating system (the dielectric substrate where the ground plane is printed, the antenna booster, plastic covers, as well as any object absorbing power).

Note that conditions a and b can be satisfied with the design of the matching network, whereas condition c depends on the antenna booster and ground plane electrical size, the losses of the substrate where the ground plane is printed, and the losses of nearby components such as plastic casings.

If we define the efficiency of the matching network η_m as:

$$\eta_m = \frac{P_\text{a}}{P_\text{in} - P_\text{ref}} = \frac{P_\text{a}}{P_\text{a} + P_\text{loss}} \tag{4.16}$$

Note that if the matching network is lossless, $\eta_m = 1$ (100%). This efficiency depends on the quality factor of the components of the matching network.

Since the reflected power P_ref equals to:

$$P_\text{ref} = |b_1|^2 \tag{4.17}$$

and

$$\Gamma_\text{in} = \frac{b_1}{a_1} \tag{4.18}$$

We obtain:

$$\eta_m = \frac{P_\text{a}}{P_\text{in} \cdot \left(1 - |\Gamma_\text{in}|^2\right)} = \frac{\left|\frac{S_{21}^m}{1 - S_{22}^m \cdot S_{11}}\right|^2 \cdot \left(1 - |S_{11}|^2\right)}{1 - |\Gamma_\text{in}|^2} \tag{4.19}$$

The input reflection coefficient Γ_in can be expressed as a function of the S-parameters of the matching network and the antenna as:

$$b_1 = S_{11}^m \cdot a_1 + S_{12}^m \cdot a_2 = S_{11}^m \cdot a_1 + S_{12}^m \cdot S_{11} \cdot b_2 \tag{4.20}$$

From Equation (4.10), we can substitute b_2, resulting in the following:

$$\begin{aligned} b_1 &= S_{11}^m \cdot a_1 + S_{12}^m \cdot a_2 = S_{11}^m \cdot a_1 + \frac{S_{12}^m \cdot S_{11} \cdot S_{21}^m \cdot a_1}{1 - S_{22}^m \cdot S_{11}} = \\ &= a_1 \cdot \left(S_{11}^m + \frac{S_{12}^m \cdot S_{11} \cdot S_{21}^m}{1 - S_{22}^m \cdot S_{11}}\right) \end{aligned} \tag{4.21}$$

And finally

$$\Gamma_\text{in} = S_{11}^m + \frac{S_{12}^m \cdot S_{11} \cdot S_{21}^m}{1 - S_{22}^m \cdot S_{11}} \tag{4.22}$$

We should emphasise that Equation (4.15), by taking Equation (4.19), collapse to Equation (4.14) as expected. Although both Equations (4.14) and (4.15) reflect the same reality, Equation (4.15) is more intuitive because it clearly reflects a sequence of how power navigates from the source to the space. In effect, first, the P_in is transformed into $P_\text{in} \cdot (1 - |\Gamma_\text{in}|^2)$, which mainly depends on the impedance of the antenna booster in the ground plane and the matching network. Then, such power is transferred to the space by considering the attenuation of the matching network with the term η_m and with the term η_r that includes the losses of the antenna booster, the dielectric losses of the substrate, and other materials such as the plastic cover of the wireless device. On the other hand, Equation (4.14) is counterintuitive since the term $1-|S_{11}|^2$ is less than 1 for an antenna booster system due to the reactive nature of the impedance of an antenna booster.

It is interesting to note that the efficiency of the matching network η_m also depends on the impedance of the antenna booster. Although not visibly reflected in Equation (4.19), this impact will be clearly exposed by using an alternative formulation in terms of quality factors, as shown in Section 4.5.

Let us put the above equations into practice by considering a simple matching network consisting of a series L_m (Table 4.1) and a transmitter with a harmonic generator V_g with an internal impedance $Z_g = 50\ \Omega$.

Simple circuit theory with Kirchhoff equations can address this simple exercise. The general equations based on the S-parameter discussed above are intended for more complex scenarios, including matching networks with several components combined in different topologies.

The purpose of such a series inductor is to cancel out the reactive impedance of an antenna booster given by a series RC circuit with R_a and C_a. For the matching network, a series R_m resistance is included to account for losses. If $R_m = 0$, the matching network is lossless, and thus, $\eta_m = 100\%$.

Network theory [2] easily finds the values of the S^m parameters, which are needed to obtain the input reflection coefficient and the matching network efficiency.

Table 4.1 A matching network comprising a series L connected to an antenna booster modelled by an RC series circuit

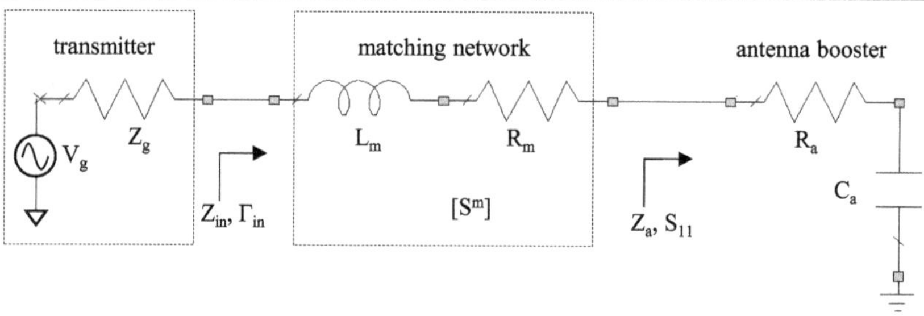

In all cases, $Z_g = 50\ \Omega$, $C_a = 0.88$ pF, $L_a = 35.6$ nH, $f = 900$ MHz, $\eta_r = 100\%$	**$1-\|\Gamma_{in}\|^2$. Eq. (4.22)**	**η_m Eq. (4.19)**	**η_t Eq. (4.14) or (4.15)**
$R_m = 0\ \Omega$, $R_a = 50\ \Omega$	1	100% (0 dB)	100%
$R_m = 25\ \Omega$, $R_a = 25\ \Omega$	1	50% (−3 dB)	50%
$R_m = 10\ \Omega$, $R_a = 40\ \Omega$	1	78.7% (−1 dB)	78.7%
$R_m = 10\ \Omega$, $R_a = 100\ \Omega$	0.858	90.9% (−0.4 dB)	78.1%

For the first case, the input impedance is $Z_{in} = 50\ \Omega$ and thus $\Gamma_{in} = 0$. Since the matching network is lossless, $\eta_m = 100\%$. For the second case, $\Gamma_{in} = 0$ as well. However, in this case, half of the power is lost in the matching network, and, as a result, the $\eta_m = 50\%$. The third case also presents a $Z_{in} = 50\ \Omega$. However, in this case, the loss of the matching network is 1 dB. Finally, the last case combines losses in the matching network and mismatch losses since the input impedance is not 50 Ω but 110 Ω. Note that although $R_m = 10\ \Omega$, the losses of the matching network are less than in the previous case. This clearly reflects that the losses of the matching network do not depend only on the R_m but on the relation between R_m and R_a.

For this simple case, such consideration can be easily derived as follows. By using the matching network definition given in Equation (4.16), we have:

$$\eta_m = \frac{I^2 \cdot R_a}{I^2 \cdot (R_a + R_m)} = \frac{1}{1 + \frac{R_m}{R_a}} \tag{4.23}$$

Equation (4.23) provides insight by illustrating that the efficiency of the matching network relies on how the components of the matching network relate to the real impedance of an antenna booster system. As deeply analysed throughout this chapter, when the electrical size of the ground plane (the main contributor in the radiation process) becomes smaller, R_a becomes smaller as well. Therefore, R_m should also be small.

Let us revisit Equation (4.23) in Section 4.5 in terms of the quality factor of the components of the matching network and their relation to different ground plane sizes. Before, it is key to understand some basics of the role of a matching network component and the impact of the tolerances, as explained next.

4.3 Simple matching networks

The architecture of an antenna booster-based wireless device comprises an antenna booster, a matching network, and a ground plane (Figure 3.7, Chapter 3). The role of the antenna booster position in effectively exciting currents in the ground plane and the impact of the ground plane has been deeply analysed in Chapter 3. This section studies simple matching networks to operate at a given frequency region, comprising f_1 and f_2. Matching networks based on analytic procedures are first presented, and finally, matching network design using network synthesis is also discussed. Understanding these principles will simplify the reading of the following chapters devoted to designing multiband matching networks and the multi-radio architectures presented in Chapters 5 and 6.

First, it is essential to understand the impact on a given impedance $Z_{in}(f)$ of a series/shunt inductor/capacitor (Figure 4.3). The effect is easily derived from knowing the reactance introduced by either a series or shunt inductor or capacitor. For example, let us consider Z_{in} at a given frequency f with a series L_s, which will result in a $Z = j\omega \cdot L_s + Z_{in}$, where $\omega = 2\pi f$. Therefore, the real part of Z does not change, only the imaginary part. Thus, as L_s increase, the new Z will have the same real part as Z_{in}. This represents a clockwise movement along a constant resistance

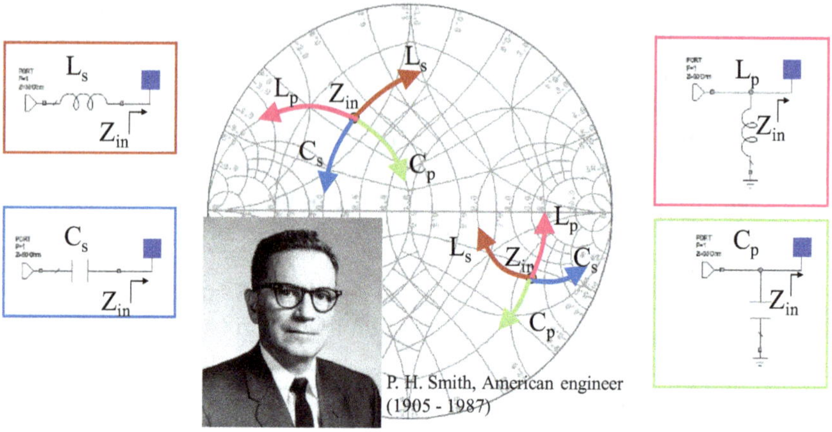

Figure 4.3 Impedance changes as a function of series and shunt L and C. The dot represents the impedance Z_{in} of a booster at a frequency of f.

circle in the Smith chart. Analogously, adding a series C_s produces a counterclockwise movement. Finally, the same principles apply to the shunt cases, but, in this case, an L_p and C_p make counterclockwise and clockwise movements along a constant conductance circle.

With the Smith chart in Figure 4.3 in mind, it is straightforward to match any impedance at a given frequency. To illustrate the simplicity of the matching procedure, let us consider a 12 mm × 3 mm × 2.4 mm (height) antenna booster, including a 120 mm × 60 mm ground plane (Figure 4.4). At 0.9 GHz, the impedance of the antenna booster is highly capacitive, being $Z_{in} = 37.1 - j208.0\ \Omega$. Imagine that the objective is to have a perfect match at 0.9 GHz, that is, 50 Ω; what matching network can provide such impedance? Let us illustrate two possibilities with two components in the matching network, one using a series and shunt L (Figure 4.4) and the other using a shunt L and a series C (Figure 4.5).

As shown for the series and shunt L case (Figure 4.4), the first L series does not change the real part of the impedance. This is true since the L is an ideal and not a real component with losses. Losses change this behaviour slightly due to a resistive part, which can be neglected if components present a high Q. High Q is a must to minimise losses introduced by the matching network, as explained in Section 4.5.

It is worth pointing out how the compact input impedance response of the antenna booster, with any matching from 0.8 GHz to 1 GHz, gets enlarged when it is matched. This is a consequence of the transformation to generate the Smith chart since all values from 0 to infinite are included in the chart. Impedances close to the infinite point are packed in such a small area that one may conclude that the impedance has a broadband behaviour (constant values) when that is not the case.

For the matching network having the shunt L_p and series C_s, the shunt L_p brings Z_{in} to the circle of $R = 50$ and then adding a series C_s does not change that 50 Ω and cancels out the reactance of $j240\Omega$ (Figure 4.5).

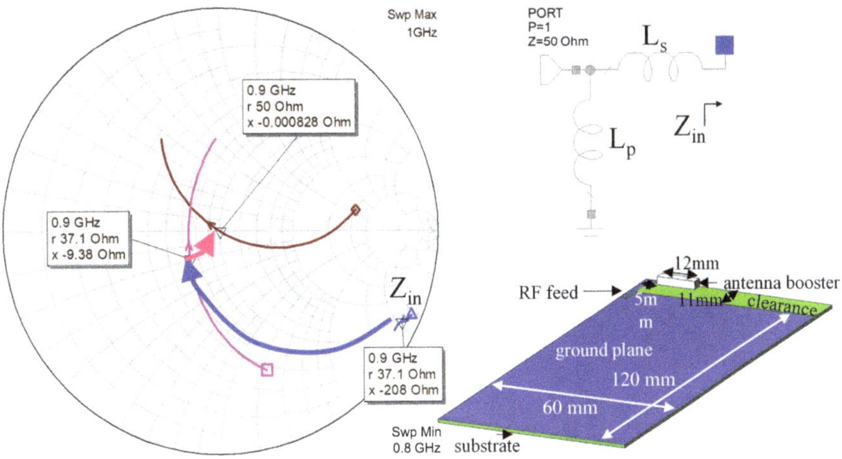

Figure 4.4 *The first series L decreases reactance while maintaining the same real part of the impedance. A shunt L decreases reactance moving along a circle of constant admittance. The ground plane and antenna booster present a conductivity $\sigma = 5.9 \times 10^7$ S/m; the substrate is 1 mm thick, $\varepsilon_r = 4.15$, $\tan\delta = 0.017$.*

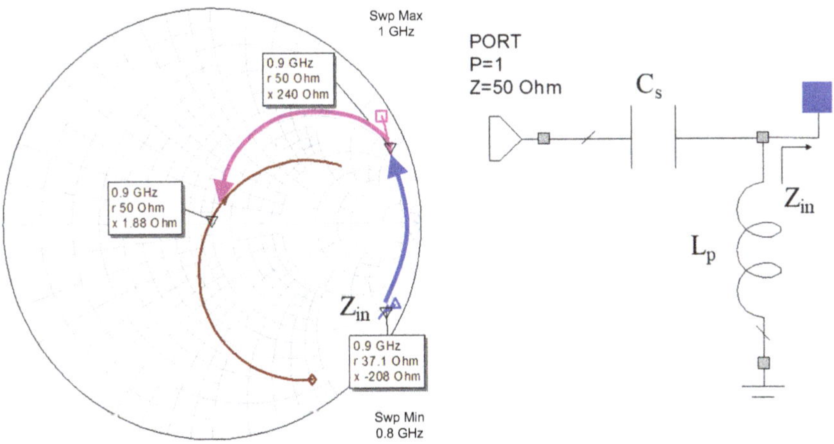

Figure 4.5 *A shunt L moves the impedance Z_{in} to the $R = 50\ \Omega$ circle. Then, the inductive part is cancelled out by adding a series C_s.*

It is worth comparing the responses across the frequencies of both solutions (Figure 4.6). Although both matchings provide a perfect match at 900 MHz, the matching with only inductors provides more bandwidth. This means that matching networks, besides providing matching, can also enlarge bandwidth, as is the case of broadband matching networks explained in Section 4.8.

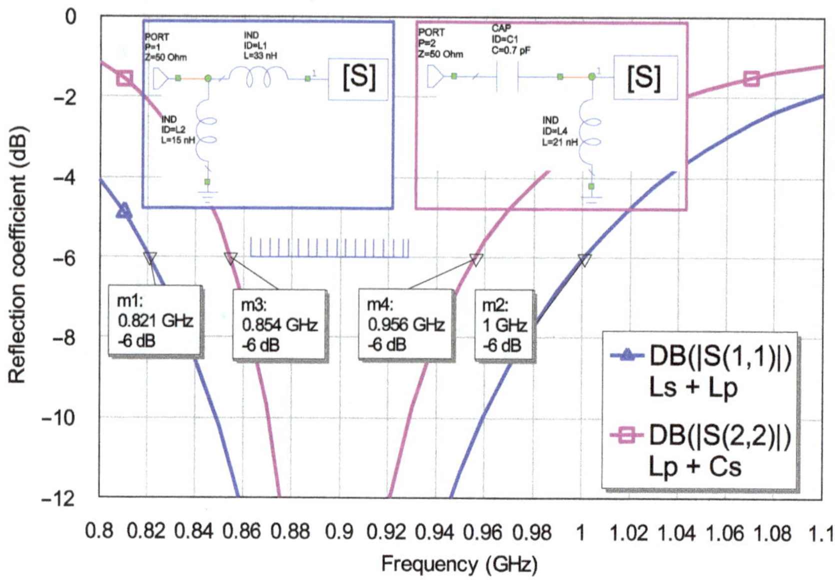

Figure 4.6 Different matching networks can achieve different bandwidths. [S] represents the S_{11} of a 12 mm × 3 mm × 2.4 mm antenna booster, including a 120 mm × 60 mm ground plane, as shown in Figure 4.4.

4.4 About tolerances

Lumped components using SMD present tolerances that change the nominal inductance and capacitance values. Typical tolerances are less than 5%. For the matching networks analysed in the book, tolerances should be less than 2% to guarantee a robust solution. Tolerances are essential as far as mass production is concerned. Thinking of millions of manufactured devices, one must consider the tolerances analysis at several parts of the radio frequency design, the matching network being one of them. Most microwave simulators include applications where the reflection coefficient or total efficiency can be analysed as a function of the tolerance of the lumped components of the matching network. A matching network for an antenna booster should be chosen because of its simplicity (a smaller number of components), size of the SMD components, power handling, impedance bandwidth, and total efficiency. It should also be selected considering its robustness to tolerances. Analysing the impact of tolerances on the reflection coefficient of the simple matching networks designed in Section 4.3 reveals that the one employing only inductors demonstrates greater robustness (Figure 4.6). This may change depending on the topology and the values of the components and, therefore, should be analysed case by case. For the example shown in Figure 4.6, the matching network with only inductors achieves an $S_{11}<-6$ dB in all cases (Figure 4.7). However, the success ratio decreases to 74% for the matching network with the inductor and capacitor.

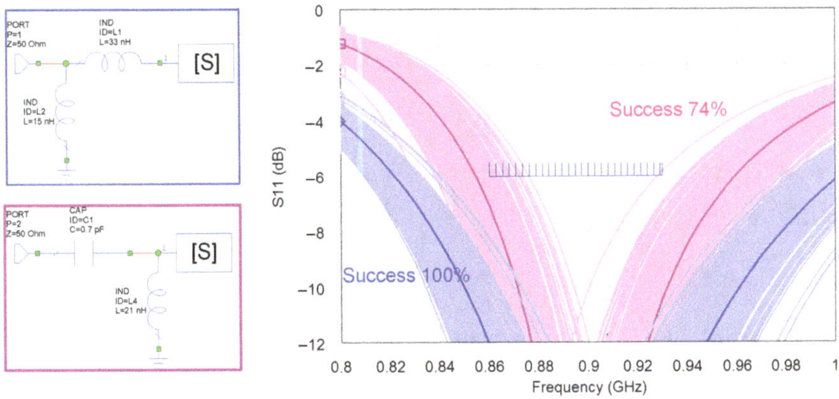

Figure 4.7 Tolerances of the components is 2%, Gaussian distribution, 500 iterations have been tested

Tolerance of the matching network components
Tolerance analysis must be considered to design robust matching networks for antenna boosters. As a practical rule of thumb, component tolerances for inductors and capacitors should be less than 2%.

4.5 On the quality factor of the matching network components

So far, ideal components have been used, but real components (finite Q) must be considered to better estimate bandwidth and, more importantly, total efficiency. Losses can increase bandwidth because reflection is minimised. Therefore, it can lead to the wrong conclusion that the solution is satisfactory because of its bandwidth and a good level of S_{11}. Hence, total efficiency should be analysed as well. Let us consider the same 12 mm × 3 mm × 2.4 mm (height) antenna booster described in Figure 4.4. Let us also consider a single matching network comprising a series and shunt inductors, one having components with the same $Q_m = 23$ and the other with the same $Q_m = 65$ (Figure 4.8). As expected, the total efficiency for the second case is higher: total efficiency averaged in the frequency range of 824–960 MHz is 76.3% and 86.6% for the $Q_m = 23$ and $Q_m = 65$ cases, respectively.

Losses introduced by a matching network (η_m, Equation(4.23)) do not depend only on the lossy part of the components of the matching network but on the relation between said losses and the resistive part of the antenna booster impedance. A general formulation in terms of Q based on Tellegan's theorem [3] shows that the efficiency η_m of the matching network can be stated as a function of the Q of the components of the matching network Q_m and the Q of an antenna Q_a, in this case, Q_a is the Q of the antenna booster, including the ground plane, without any matching

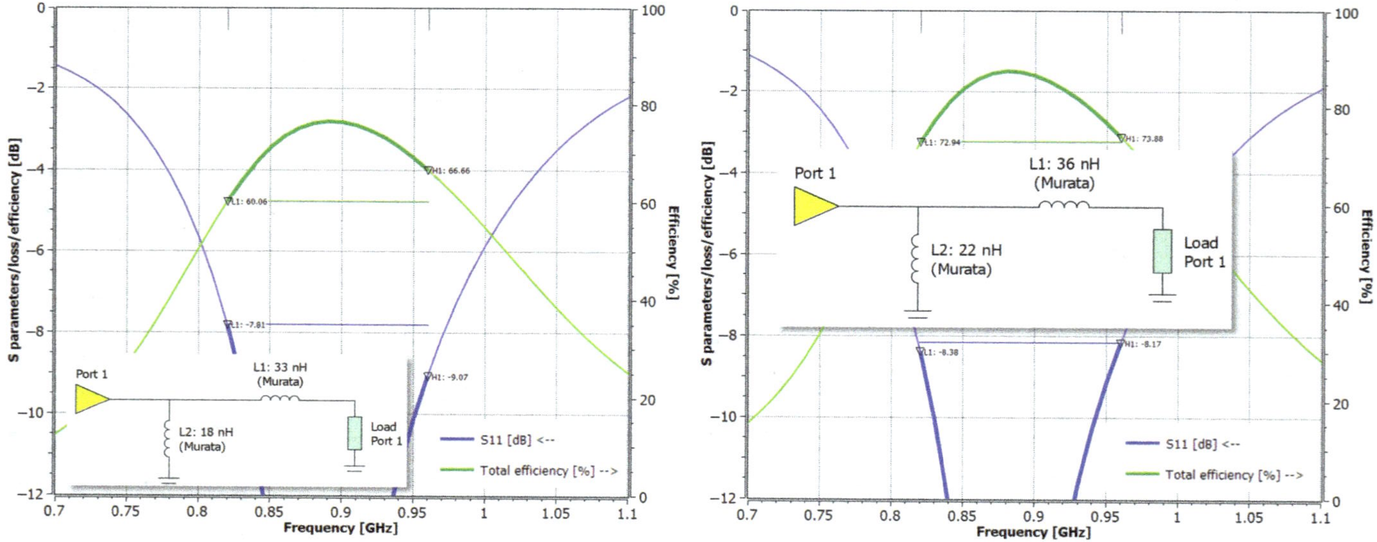

Figure 4.8 *(Left) A matching network with components having a $Q_m = 23$ at 900 MHz. (Right) A matching network having $Q_m = 65$.*

network (Equation (4.24)). As demonstrated, this efficiency relies not on the Q_m of the lumped elements within the matching network but rather on how these Q_m values relate to the Q_a of the antenna booster system in the absence of any matching network [3]. Equation (4.24) has been derived in [3], assuming the same type of lumped element and with the same $Q = Q_m$. Q_a is the quality factor for the antenna booster.

$$\eta_m = \frac{1}{1 + \frac{Q_a}{Q_m}} \tag{4.24}$$

Now, the total efficiency given in Equation (4.15) can be expressed as:

$$\eta_t = \left(1 - |\Gamma_{\text{in}}|^2\right) \cdot \frac{1}{1 + \frac{Q_a}{Q_m}} \cdot \eta_r \tag{4.25}$$

where it is visible the impact of the matching network in the first two terms.

To emphasise the importance of the Q_m of the components, two situations are compared (Figure 4.9), where both use a 12 mm × 3 mm × 2.4 mm antenna booster and the same topology of the matching network comprising a L_{series} and L_{shunt} (Figure 4.10).

Both cases feature approximately the same $Q_m \approx 80$ using SMD components having 0603 sizes (note that for lumped SMD inductors, increasing the size allows a thicker wire to fabricate the inductor, and therefore, less loss/high Q is obtained). The radiation efficiency of the antenna booster, including the ground plane, is computed using MoM (IE3D software) where radiation efficiencies at 900 MHz are $\eta_r = 97\%$ and $\eta_r = 92\%$ for the 120 mm × 60 mm and 60 mm × 60 mm ground planes, respectively. Based on this data, one should expect similar total efficiencies due to a similar radiation efficiency and matching network. However, based on Equation (4.24), Q_m is not what matters, but its relation to Q_a. For small ground plane sizes, Q_a grows, and hence, the Q_m has to be larger than Q_a to minimise losses of the matching network. However, the technology of the SMD components limits Q_m.

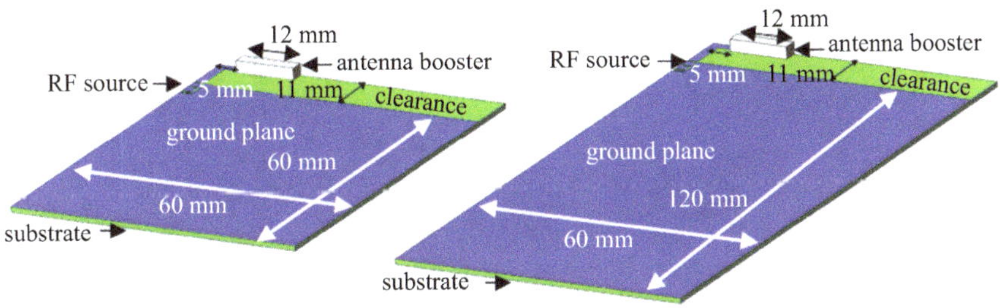

Figure 4.9 (Left) 60 mm × 60 mm ground plane; (Right) 120 mm × 60 mm ground plane. Both cases have a 12 mm × 3 mm × 2.4 mm (height) antenna booster. The antenna booster can be modelled as a conductive 3D parallelepiped with a conductivity $\sigma = 5.9 \times 10^7$ S/m. The substrate is 1 mm thick, $\varepsilon_r = 4.15$, $\tan\delta = 0.017$. RF source consists of a 2 mm × 1 mm gap between the ground plane and the feeding line.

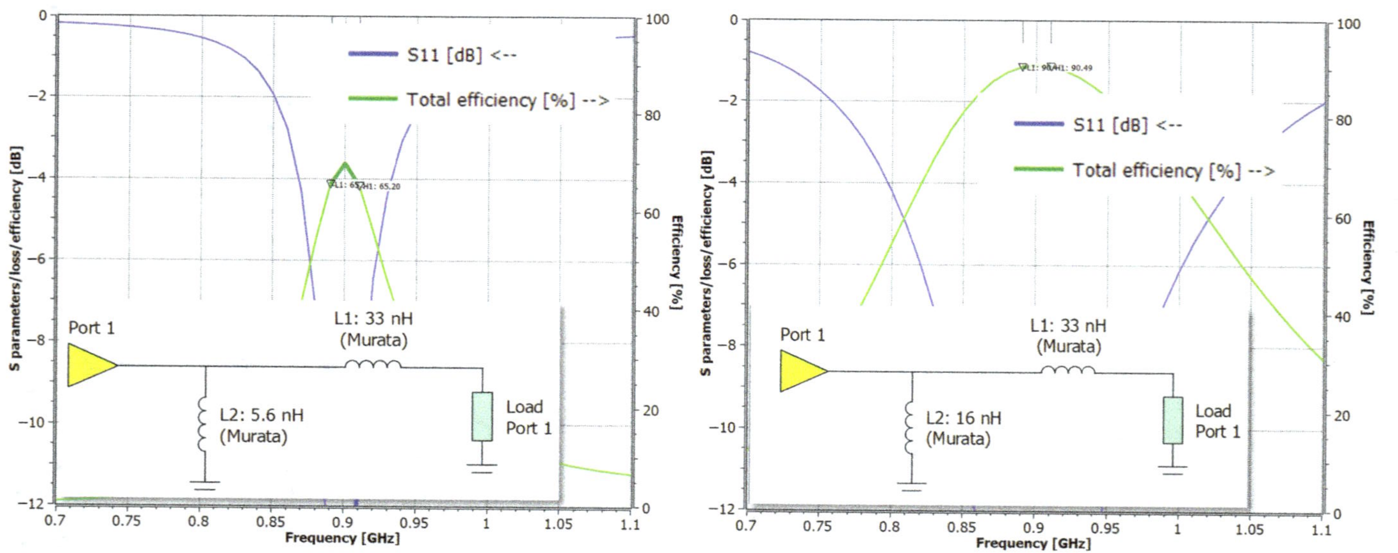

Figure 4.10 (Left) Matching network for the 60 mm × 60 mm ground plane; (right) matching network for the 120 mm × 60 mm ground plane

Q_a is now computed using Equation (4.26) using the input impedance data calculated with MoM of the antenna booster, including the ground plane [4,5]. It results in Q_a (120 mm × 60 mm ground plane) = 5.7 and Q_a (60 mm × 60 mm ground plane) = 31.1.

$$Q_a(\omega) = \frac{\omega}{2R(\omega)}\sqrt{\left[\frac{dR(\omega)}{d\omega}\right]^2 + \left[\frac{dX(\omega)}{d\omega} + \left|\frac{X(\omega)}{\omega}\right|\right]^2} \tag{4.26}$$

With all this data and Equation (4.24), losses η_m due to the matching network are:

$$\eta_m = 10\log\left(\frac{1}{1+\frac{31.1}{80}}\right) = -1.4\text{dB} \qquad \eta_m = \frac{1}{1+\frac{5.7}{80}} = -0.3\text{dB}$$

60 mm × 60 mm ground plane   120 mm × 60 mm ground plane

Due to its higher Q_a, the loss induced by the matching network, with identical topology and Q_m, is 1.1 dB greater for the 60 mm × 60 mm ground plane.

With this simple data, the estimated total efficiencies at 900 MHz are:

$$\eta_t = \frac{0.92}{1+\frac{31.1}{80}} = 66.2\% \qquad \eta_t = \frac{0.97}{1+\frac{5.7}{80}} = 90.5\%$$

60 mm × 60 mm ground plane   120 mm × 60 mm ground plane

This straightforward calculation aligns with the comprehensive circuit simulation conducted in Optenni Lab [6–8]), which accounts for the actual Q_m of individual components and simulates the η_r of the antenna booster, including the ground plane (Figure 4.10). Moreover, as expected after the teachings of Chapter 3, the bandwidth is four times larger for the 120 mm × 60 mm ground plane (20% at SWR = 3) than the bandwidth of the 60 mm × 60 mm (5.1% at SWR = 3).

Matching network losses are dependent on the ratio Q_a/Q_m. To get a general picture, the matching network losses as a function of Q_m and for three different ground plane sizes, from a big one of 120 mm × 60 mm, a medium size of 60 mm × 60 mm, to a small one of 50 mm × 40 mm, are computed (Figure 4.11). As a practical rule, if matching network losses should be less than 1 dB, Q_m is not critical for the large ground plane case of 120 mm × 60 mm but is very relevant for the other two cases. For the 60 mm × 60 mm ground plane, Q_m must be larger than 120 and larger than 200 for the 50 mm × 40 mm to have matching network losses of less than 1 dB. Besides, the radiation efficiency for the smaller ground plane is less than that of the 120 mm × 60 mm ground plane, and thus the product $\eta_r \cdot \eta_m$ is lower.

The quality factor of the components of the matching network

The quality factor of the components (Q_m) in the matching network should be as high as possible to minimise dissipation losses and be larger than the Q_a of the antenna booster including the ground plane. As a practical rule, if $Q_m > 10 \cdot Q_a$, losses introduced by the matching network are less than 0.5 dB. The selection of Q_m is more critical for small ground planes where Q_a is high.

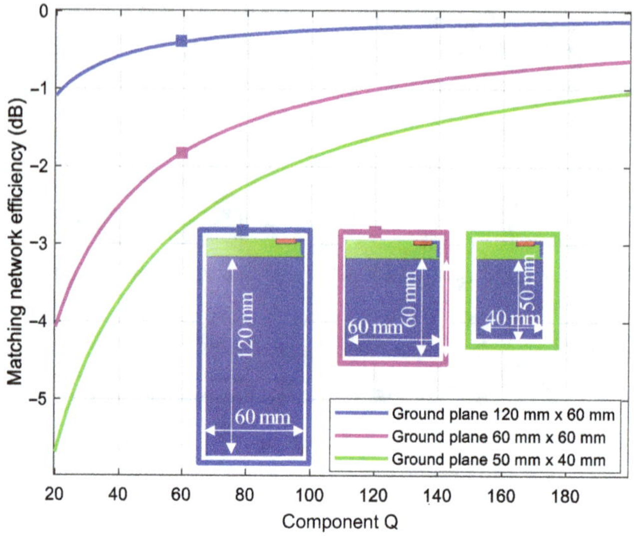

Figure 4.11 Matching network efficiency computed using Equation (4.24) as a function of the matching network Q_m. Q_a = 5.7, 31.1, and 54 for the 120 mm × 60 mm, 60 mm × 60 mm, and 50 mm × 40 mm ground planes, respectively. Q_a is computed at 900 MHz using Equation (4.26).

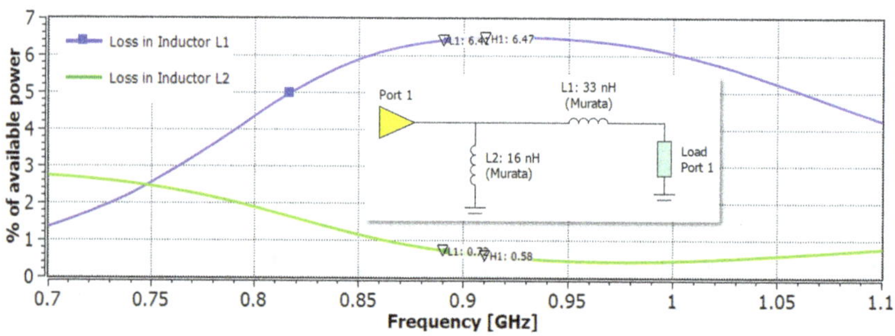

Figure 4.12 Dissipation for the matching network for the 120 mm × 60 mm ground plane

Now that the Q_m impact is clearly understood, the next question is whether all components introduce the same amount of loss. Owing to Ohms law, the ohmic loss for a given component is $I^2 \cdot R_\Omega$, where I is the current through the component, and R_Ω represents the ohmic loss due to the finite Q_m. Therefore, losses should not only depend on Q_m but also on I. Such I rely on the matching circuit topology. For the L_{series} and L_{shunt} case used to match the 12 mm × 3 mm × 2.4 mm with the 120 mm × 60 mm ground plane, the contribution to losses is ten times more for the L_1 = 33 nH (Figure 4.12).

Thus, Q_m for this component is more critical and should be carefully selected to minimise losses. The same analysis will be carried out for other topologies, including more lumped components such as those cases for achieving multiband performance (Chapter 5).

Finally, Q_m for a component such as inductors and capacitors is frequency-dependant (Chapter 1), having an operating frequency window beyond that, the Q_m is low. Besides, the behaviour changes beyond the operating frequency window. For example, an inductor should not be used beyond the so-called self-resonant frequency, where the parasitic capacitances dominate. Thus, inductors behave no more like an inductor than like a capacitor. The same happens with capacitors. Therefore, components must be carefully selected to provide high Q within the frequency range of operation.

4.6 A simple matching for an antenna booster for a body warn device

A design example for a wearable wireless device such as a smartwatch is analysed [9,10]. The size of the antenna booster is a 3D conductive brass-made parallelepiped $L \times W \times H$ (12.0 mm × 3.0 mm × 2.4 mm), representing an electric length of 0.098λ at 2.45 GHz, which is 2.5 less than a quarter-wavelength monopole. Although small, it is not the typical size of antenna boosters discussed in the book, having lengths typically ~λ/30 at the lowest frequency of operation. This size has been chosen to show that the same antenna booster can be utilised across several platforms [5]. The objective here is to use the same geometry, but in this case, for operation in the 2.4–2.5 GHz band, changing only the matching network, not the antenna booster geometry. However, a smaller antenna booster can also be utilised, such as λ/20 or λ/30 in length. For example, the Chapter 5 section shows a 7 mm × 3 mm × 1 mm antenna booster not only for covering 2.4–2.5 GHz but also to cover from 5.9 GHz to 7.2 GHz. In that case, the antenna booster length is ~λ/17.

The ground clearance has been reduced as much as possible (14 mm × 5 mm), so more grounding is left to host the circuitry of the device (e.g., smartwatch). The ground plane is approximately 30 mm × 30 mm, which, although physically small, represents 0.24λ at 2.4 GHz (Figure 4.13). A 50 Ω Coplanar Waveguide (CPW) transmission line is included with an SMA connector to feed the structure. The PCB (Printed Circuit Board) is made of FR4 ($\varepsilon_r = 4.15$ and $\tan\delta = 0.017$), having a ground plane on both the top and bottom layers. The coaxial cable includes ferrites to mitigate the impact of the coaxial transmission line when testing the antenna with the vector network analyser or anechoic chamber. In this manner, the possible current flowing out of the coaxial cable is reduced, minimising the impact on SWR and efficiency.

The antenna booster, with an impedance of $Z_{\text{in}} = 11.3–27.2j\Omega$ at 2.45 GHz, prompts the design of a matching network using lumped components to align the antenna with 50 Ω impedance, as depicted in Figure 4.3. Since the antenna booster is smaller than a quarter wavelength monopole, the impedance is capacitive (Figure 4.14). In this sense, a possible matching network comprises a two-element

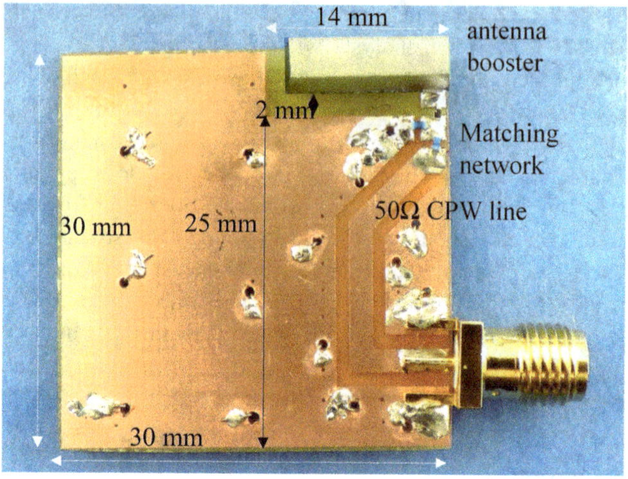

Figure 4.13 The antenna booster element is a 3D brass-made structure of 12 mm × 3 mm × 2.4 mm (height), and it is placed on a grounding clearance area of 14 mm × 5 mm. The matching network uses SMD 0402 lumped components.

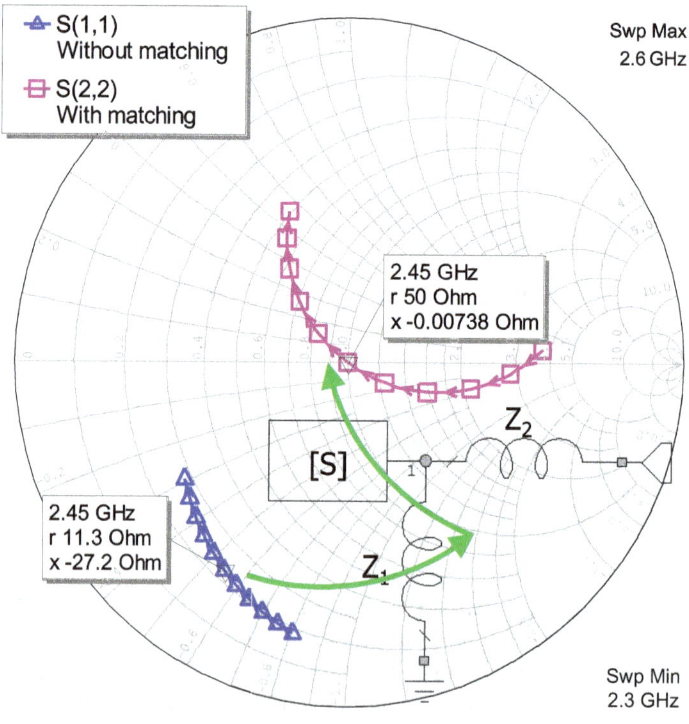

Figure 4.14 Input impedance of the antenna booster, including the ground plane without matching and with a matching network. $Z_1 = 3$ nH, $Z_2 = 2.3$ nH.

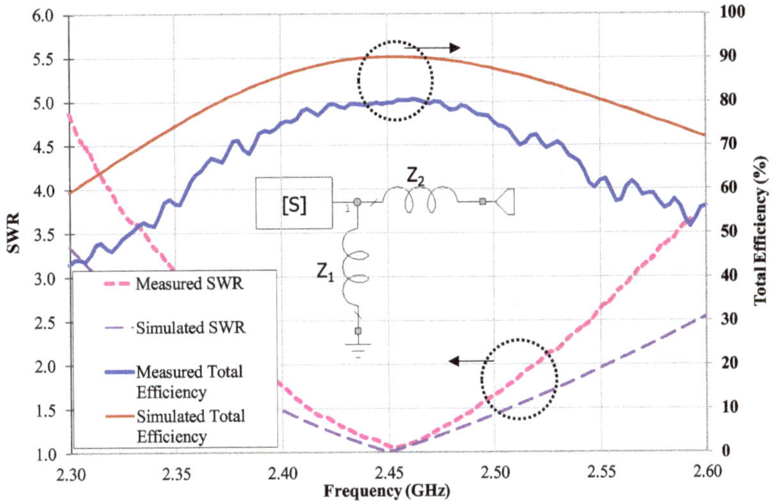

Figure 4.15 Simulated and measured SWR and total efficiency in free space. $Z_1 = 2.3$ nH ($Q_m = 119$) and $Z_2 = 3.3$ nH ($Q_m = 119$). Total efficiency considers both radiation efficiency and mismatching. It is measured using 3D pattern integration with Satimo Stargate-32 anechoic chamber.

matching network, one shunt, and one series inductor. For this case, high-Q ($Q_m > 119$) tight tolerance (2%) SMD 0402 type components are used.

In a free-space scenario, the measured SWR presents values less than 2 for the 2.4–2.5 GHz frequency range and a total efficiency between 70% and 80% (Figure 4.15). The total efficiency (η_t) considers the radiation efficiency (η_r) and mismatching as $\eta_t = \eta_r \cdot (1 - |S_{11}|^2)$. Total efficiency is measured using 3D pattern integration with the Satimo Stargate-32 anechoic chamber.

When comparing total efficiency from 2.4 GHz to 2.5 GHz, the average values are 78.5% and 88.9% for the simulation and measurement, respectively, with an acceptable difference of 0.5 dB.

Since the smartwatch is thought to operate very close to the human hand, its effect must be considered. The following section analyses the impact on efficiency when the distance between the ground plane and a phantom hand varies.

4.6.1 Human hand effect

The effect over total efficiency as a function of the distance between the ground and the phantom hand ($\varepsilon_r = 25.7$, $\sigma = 1.32$ S/m) is analysed for three cases: case *a*, which is the touching position (the bottom ground layer directly touches the phantom hand); case *b* and *c*, where the distance between the bottom ground layer is 3 mm and 6 mm, respectively (Figure 4.16). The objective is to study how critical the distance is to determining design recommendations for this application. Porexpan spacers have been used to separate the PCB from the phantom hand.

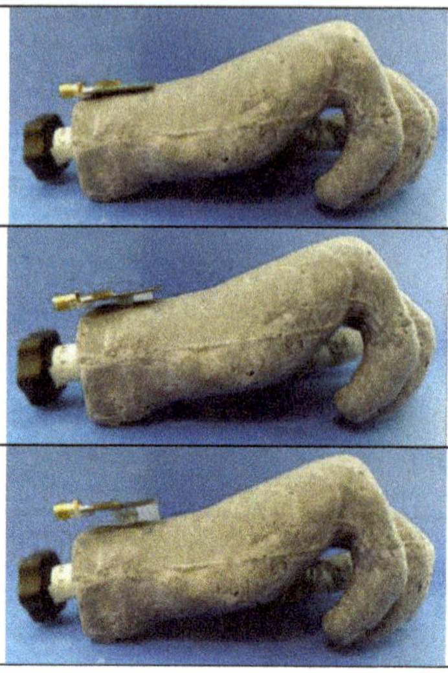

Case *a*
PCB 0 mm from the phantom hand
Z_1=1.5 nH
Z_2=0Ω
(PCB: Printed Circuit Board)

Case *b*
PCB 3 mm from the phantom hand
Z_1=1.8 nH
Z_2=3.3 nH

Case *c*
PCB 6 mm from the phantom hand
Z_1=1.8 nH
Z_2=3.4 nH

Figure 4.16 Different scenarios to evaluate the impact of the phantom hand as a function of the distance between the PCB and the phantom: (a) 0 mm, (b) 3 mm, and (c) 6 mm. Distance is measured between the bottom ground layer of the PCB and the surface of the phantom hand. 0 mm means that the ground is in contact with the phantom hand.

Since the proximity of the human hand detunes the antenna system, a new antenna matching has been designed for each of the three cases under study to minimise mismatch losses (Figure 4.16). At this point, it is interesting that for each case, the geometry of the antenna booster is maintained; only the matching network values change from one case to the other. In this sense, no customised antennas are required. A matching network can also be electronically configured, depending on the distance between the device and the human body. The topic of reconfigurability is deeply analysed in Chapter 7.

The results reflect the critical effect of the hand on the total efficiency (Figure 4.17 and Table 4.2). In particular, the average efficiency across the 2.4–2.5 GHz frequency range is 15.1% in the touching position. It increases to 25.0% and 37.0% for the 3 mm and 6 mm distances, respectively. In summary, the average efficiency increases by about 3 dB from the worst case (touching position) to where the ground plane is spaced apart by 3 mm and 6 mm. Since, in real situations, the ground is not touching the hand, efficiency can be improved if the distance of the ground plane is kept apart from the hand, as long as the aesthetic and mechanical aspects of the smartwatch allow doing that.

It is interesting to observe that the 0 mm case has a very good SWR. This is due to the losses introduced by the human hand. However, this improvement in SWR due to losses degrades efficiency when compared to 3 mm and 6 mm cases.

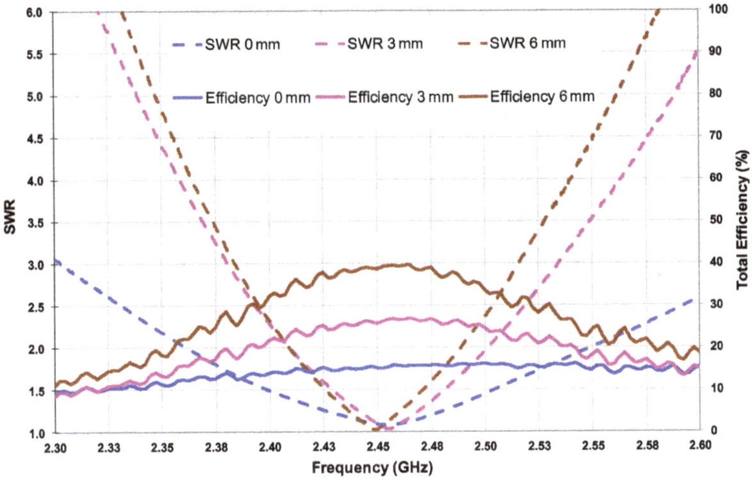

Figure 4.17 Measured SWR and total efficiency for 0 mm, 3 mm, and 6 mm distance cases. Total efficiency is measured using 3D pattern integration with the Satimo Stargate-32 anechoic chamber.

Table 4.2 Summary of the measured results in free space and with the hand effect

	SWR	**Measured total efficiency across 2.4–2.5 GHz (%)**					**Absorption (dB)**
Distance	**2.4–2.5 GHz**	η_t **2.4 GHz**	η_t **2.5 GHz**	η_t **min**	η_t **max**	η_t **average**	
Free-Space	<2:1	76.0	75.5	74.5	81.2	78.6	–
0 mm	<2:1	13.8	16.1	13.8	16.1	15.1	7.1
3 mm	<2.5:1	21.6	24.4	21.6	26.9	25.0	4.9
6 mm	<2.5:1	32.1	33.6	32.1	39.5	37.0	3.2

Absorption due to the phantom hand is computed as the ratio in dB between the average total efficiency in free space and the same considering the phantom hand (Table 4.2). Since the antennas are well-matched (SWR<2.5) for the three cases, the said ratio primarily indicates how much power the hand absorbs. As general numbers: approximately 7 dB are absorbed in the touching position and 3 dB for the 6 mm case.

4.6.2 Improving performance by enlarging the ground layer

To minimise the impact of power absorption due to the hand, this section proposes a simple technique to reduce its influence. Since the space for electronics in a smartwatch is limited, it is reasonable to reuse it using two stacked PCBs. For this

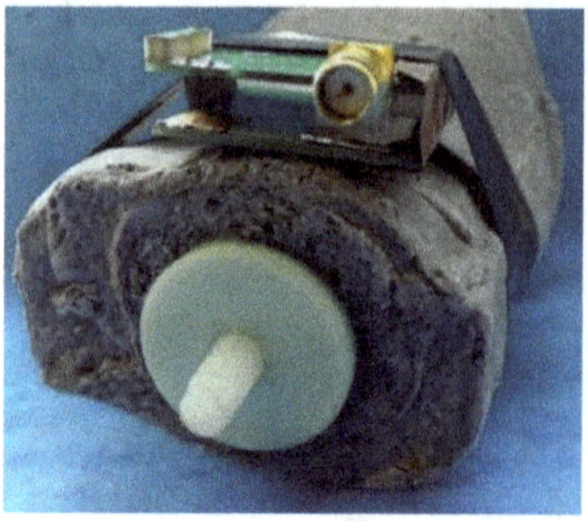

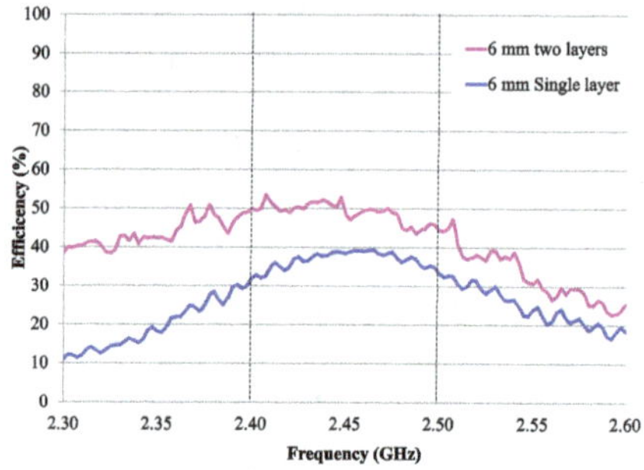

Figure 4.18 (Left) PCB 6 mm above the phantom hand, an extra PCB having a ground layer is connected to the top ground through the edge (the total gap is maintained to 6 mm); (right) total efficiency measured for the PCB 6 mm above the hand and the same case having the configuration shown in the left figure. Porexpan spacers are used.

case, the original case with the PCB 6 mm above the hand is modified, including another PCB with an entire ground plane underneath. To have both PCBs electrically connected, a conductive strip at the edge is proposed (Figure 4.18, left). The matching network is redesigned for this new situation to minimise mismatch losses. The values for the matching network components are $Z_1 = Z_2 = 1.8$ nH, following a different topology from Figure 4.15. In this case, Z_1 is a series component connected to the antenna booster, and Z_2 is a shunt component. This configuration enhances the measured total efficiency from 37.0% (averaged from 2.4 GHz to 2.5 GHz) to 49.1% (Figure 4.18, right), a substantial efficiency increment. This result is close to reported values in [11], where total efficiency with the phantom hand is 65% at the same operating bandwidth but using a total height of 10 mm and a conductive wrist, which is not always possible. Therefore, the results presented here make the solution attractive for smartwatches.

A 12.0 mm × 3.0 mm × 2.4 mm antenna booster with a small ground clearance area of 14 mm × 5 mm operating at the 2.4–2.5 GHz band has been proposed for being integrated into a smartwatch platform. The human hand mainly introduces detuning effects and efficiency decrements. The closer the hand, the lower the total efficiency. A technique to mitigate the hand impact has been proposed, consisting of adding an extra ground plane layer underneath the antenna board. When both layers are connected, measured total efficiency increases from 37.0% (averaged from 2.4 GHz to 2.5 GHz) to 49.1%, a substantial increment of 1.2 dB.

4.7 The Fano limit: theoretical limitations on the matching of arbitrary impedances

As shown in Section 4.3, a matching network can match an antenna booster, resulting in different bandwidths. The question is whether we can improve the bandwidth indefinitely by increasing the number of components or if there is a limit. Fano addressed this issue by showing that a reactive matching network (linear, passive, and lossless) can enhance the bandwidth of a series or shunt RLC circuit by a factor F (Equation (4.27)) [12].

$$F = \frac{\pi\sqrt{S}}{(S-1)\ln\{(S+1)/(S-1)\}} \tag{4.27}$$

where S is a particular SWR level.

In [13], the broad banding enhancement factor as a function of tuning stages is discussed.

An antenna booster, including a ground plane, can be modelled as an RC circuit (electrical booster) for those frequencies where the antenna booster is capacitive. A series L creates an impedance modelled as a series RLC circuit. Thus, the F represents a good approximation of how much the inherent bandwidth can be enlarged for antenna boosters.

For typical $SWR = S = 3$, $F = 3.9$, this is a considerable increment. This has a relevant implication for the design of wireless devices with antenna boosters. Let us imagine that the required bandwidth is BW = 15%. If the antenna booster, including the ground plane, has enough inherent bandwidth (Chapter 3), a simple matching network can be sufficient. However, a BW/F ≈3.9% may be enough if we then enlarge the bandwidth with a matching network. The question is: how can the inherent bandwidth be broadened? This question is addressed in Section 4.8, devoted to the design of broadband matching networks.

One crucial question is whether the simple matching networks with two lumped components (Section 4.2) can increase the inherent bandwidth of the load (the antenna booster, including the ground plane). In the case shown in Figure 4.6, the inherent bandwidth of the 12 mm × 3 mm × 2.4 mm antenna booster, including a 120 mm × 60 mm, can be estimated using Equations (4.2) and (4.28). The resulting inherent bandwidth is BW = 20.2% at $SWR = S = 3$.

$$BW = \frac{S-1}{Q_a\sqrt{S}} \tag{4.28}$$

Calculating the bandwidth for the matching networks in Figure 4.6 results in BW = 19.8% and BW = 11.3% for the $L_{\text{series}} - L_{\text{shunt}}$ and $L_{\text{shunt}} - C_{\text{series}}$ matching networks, respectively. As observed, none of these simple matching networks enhances the inherent bandwidth. Moreover, the series L-shunt L matching network almost maintains the inherent bandwidth, but the shunt L-series L decreases the inherent bandwidth nearly two times.

4.8 Broadband matching networks

For the classic antenna approach, if the bandwidth obtained by the antenna design is not enough, several strategies can be followed, such as adding parasitic elements

(Chapter 2) [14–17]. In Antenna Booster Technology, the design is simpler since there is no need to modify the antenna booster but to change the matching network design. A broadband matching network not only matches the impedance but also enhances the bandwidth achieved by a single matching network consisting of one or two components. This is a considerable advantage since bandwidth can be improved without making a complex design. The benefits of broadbanding are not only helpful in covering the targeted bandwidth but also for being more robust to detuning caused by nearby elements to the device, such as conductive or human/animal bodies, for example.

To understand how a matching network can increase the bandwidth of a given load (load = antenna booster including a ground plane), a simple analysis with an RLC circuit is proposed [18,19]. Assume an impedance of an antenna that can be modelled using a parallel RLC having a resonance at f_0 = 900 MHz with 50 Ω impedance (Figure 4.19).

If we add a series L, the reactance at 1 GHz will increase faster than the frequency at 800 MHz since the added reactance is $X = j{\cdot}2\pi f{\cdot}L$, and therefore X (1 GHz) = $j{\cdot}2\pi{\cdot}1$ GHz$\cdot L$, 1.25 times the X(0.8 GHz) = $j{\cdot}2\pi{\cdot}0.8$ GHz$\cdot L$. Because of this series L, the impedance changes, creating a more compact impedance locus (Figure 4.20). If we add a series C to compensate for the previous series L, the reactance $X = -j/(2\pi f{\cdot}C)$ at 800 MHz is larger than the one at 1 GHz. Therefore, the reactance at 800 MHz decreases faster than the one at 1 GHz, resulting in a compact impedance loop (Figure 4.21).

As seen, series C is selected to cancel the reactance added by the series L at 900 MHz. Thus, the L and C series resonance is 900 MHz. The size of the loop depends on the series L and C values. There is a set of values where the series L and C create an impedance loop with a size of a given SWR. These values can be

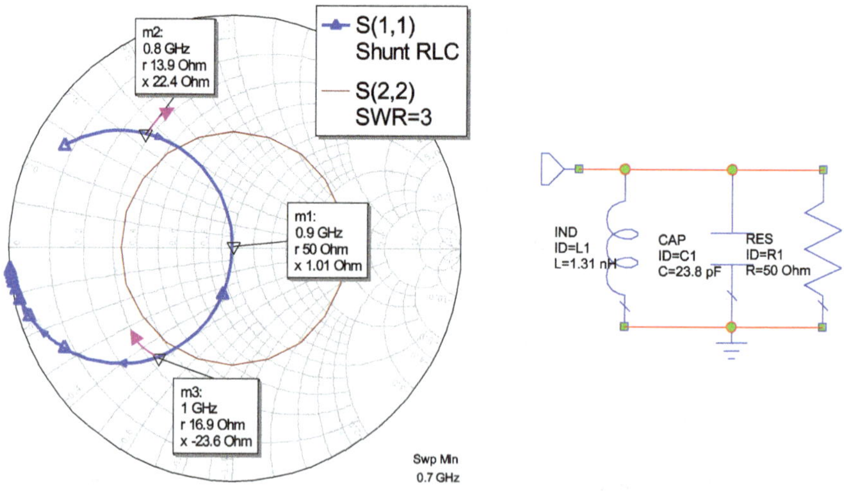

Figure 4.19 Impedance of a shunt RLC circuit resonating at 900 MHz

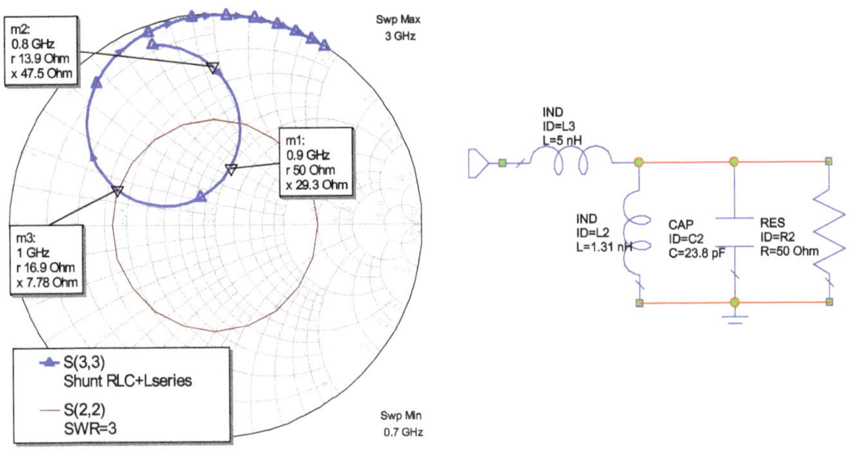

Figure 4.20 A series L added to the RLC circuit creates an input impedance loop

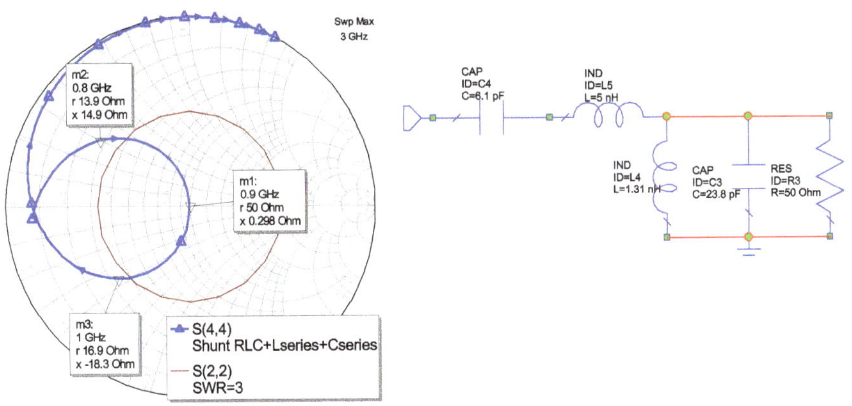

Figure 4.21 The combination of an L and C in series creates an input impedance loop where its size can be controlled by the LC product. The resonance frequency of the series LC is 900 MHz. Two other resonances appear to have the same real impedance.

analytically derived as shown next or by a simple trial and error process, tuning the series L and C till the impedance loop has the size of a given SWR, for example, SWR = 3.

Although an impedance loop has been created, bandwidth has not yet been enhanced because the loop is not centred at the centre of the Smith chart. How then can the impedance loop centred? The straightforward method is to use a matching network to move said loop to the centre of the Smith chart. For example, using Figure 4.3, one possible solution consists of a series L and a shunt C. Note that the series L is then merged with that of the broadband, reducing the number of components compared to the next option. Another possible solution consists of adding a series C and a shunt L. And yet another solution consists of manipulating the impedance at resonance (R) to be R = S·50, with S the SWR target, for example, S = 3. Following this design rule, the impedance loop will be inscribed in a circle of constant SWR = S. However, changing R means modifying the impedance of the antenna, which is not always easy. Therefore, the first method represents the simple approach of having the impedance loop at the centre of the Smith chart.

Having understood how a simple LC circuit can create an impedance loop, let us apply this method to increase the bandwidth of an antenna booster.

The impedance of an electric antenna booster can be modelled as an RC circuit (series R and C) (Figure 4.22). If we add a series L, a series RLC circuit is obtained. The value of such L is selected to locate the resonance at the frequency of interest. A shunt LC network is added following the principles previously explained to close the impedance locus to obtain an impedance loop. Using duality, the LC is not a series LC but a shunt LC. The subindex *m* is used for the values of the matching network. Since the LC is placed in shunt to the series RLC circuit modelling the antenna booster and the series L, it is more convenient to work with admittances.

In this sense, the admittance of a series RLC can be written as:

$$Y_{inRLC}(f) = \frac{R - jRQv}{R^2 + (RQv)^2} \tag{4.29}$$

where

$$v = \frac{f}{f_0} - \frac{f_0}{f} \tag{4.30}$$

and

$$f_0 = \frac{1}{2\pi\sqrt{LC}} \tag{4.31}$$

The bandwidth at a given $SWR = S$ of this series RLC circuit is given by Equation (4.28), where antenna Q_a is given by:

$$Q_a = \frac{2\pi f \cdot L}{R} \tag{4.32}$$

The potential BW of an antenna, with the characteristics described above, can be increased if the correct matching network is chosen. To correctly match the RLC series circuit that models the impedance behaviour of the antenna, a parallel capacitor and a parallel inductor (L_m, C_m) are used, which is a dual situation compared to the previous example using a series RLC case. An accurate mathematical analysis applied to the electrical model in Figure 4.22 leads to the optimum values of these reactances.

The admittance associated with the electrical model of Figure 4.22 is:

$$Y_{in}(f) = Y_{inRLC}(f) + j \cdot 2\pi f \cdot C_m + \frac{1}{j \cdot 2\pi f \cdot L_m} \tag{4.33}$$

Hence, the imaginary and real parts of the input admittance are:

$$\text{Im}\{Y_{in}(f)\} = \frac{-RQ_a v}{R^2 + (RQ_a v)^2} + \omega \cdot C_m - \frac{1}{\omega \cdot L_m} \tag{4.34}$$

$$\text{Re}\{Y_{in}(f)\} = \frac{1}{R \cdot \left(1 + (Q_a v)^2\right)} \tag{4.35}$$

It is important to note that if the imaginary part of the antenna admittance is equated to 0, three resonance frequencies appear (f_0, f_1 and f_2). To maximise the bandwidth, the input impedance locus is forced to fulfil the following requirements that will condition the optimum L_m and C_m values [18] (Figure 4.22).

$$\text{Im}\{Y_{in}(f_0)\} = \text{Im}\{Y_{in}(f_1)\} = \text{Im}\{Y_{in}(f_2)\} = 0 \tag{4.36}$$

$$\text{Re}\{Y_{in}(f_1)\} = \text{Re}\{Y_{in}(f_2)\} = \frac{1}{Z_0 \cdot S} \tag{4.37}$$

$$\text{Re}\{Y_{in}(f_0)\} = \frac{S}{Z_0} \quad \text{Re}\{Z_{in}(f_0)\} = \frac{Z_0}{S} \tag{4.38}$$

The first solution to Equation (4.34) gives the relationship required between L_m and C_m. At the same time, the value of L_m can be easily obtained by substituting Equation (4.39) into Equation (4.34).

$$C_m = \frac{1}{{\omega_0}^2 \cdot L_m} \tag{4.39}$$

$$L_m = \frac{({f_1}^2 - {f_r}^2) \cdot Z_0 \cdot (1 + (Q_a \cdot v_1)^2)}{S \cdot Q_a \cdot ({f_1}^2 - {f_r}^2) \cdot 2 \cdot \pi \cdot f_r} \tag{4.40}$$

where f_1 and f_2 are computed according to (4.40), resulting in the following:

$$f_1 = \frac{f_0\left(\sqrt{S_2 - 1 + 4 \cdot Q_a^2} + \sqrt{S_2 - 1}\right)}{2 \cdot Q_a}$$

$$f_2 = \frac{f_0\left(\sqrt{S_2 - 1 + 4 \cdot Q_a^2} - \sqrt{S_2 - 1}\right)}{2 \cdot Q_a} \tag{4.41}$$

$$BW = \frac{f_1 - f_2}{f_0} = \frac{\sqrt{S^2 - 1}}{Q_a} \tag{4.42}$$

which is the same bandwidth obtained with a parallel RLC resonator using a series LC broadband matching network [18].

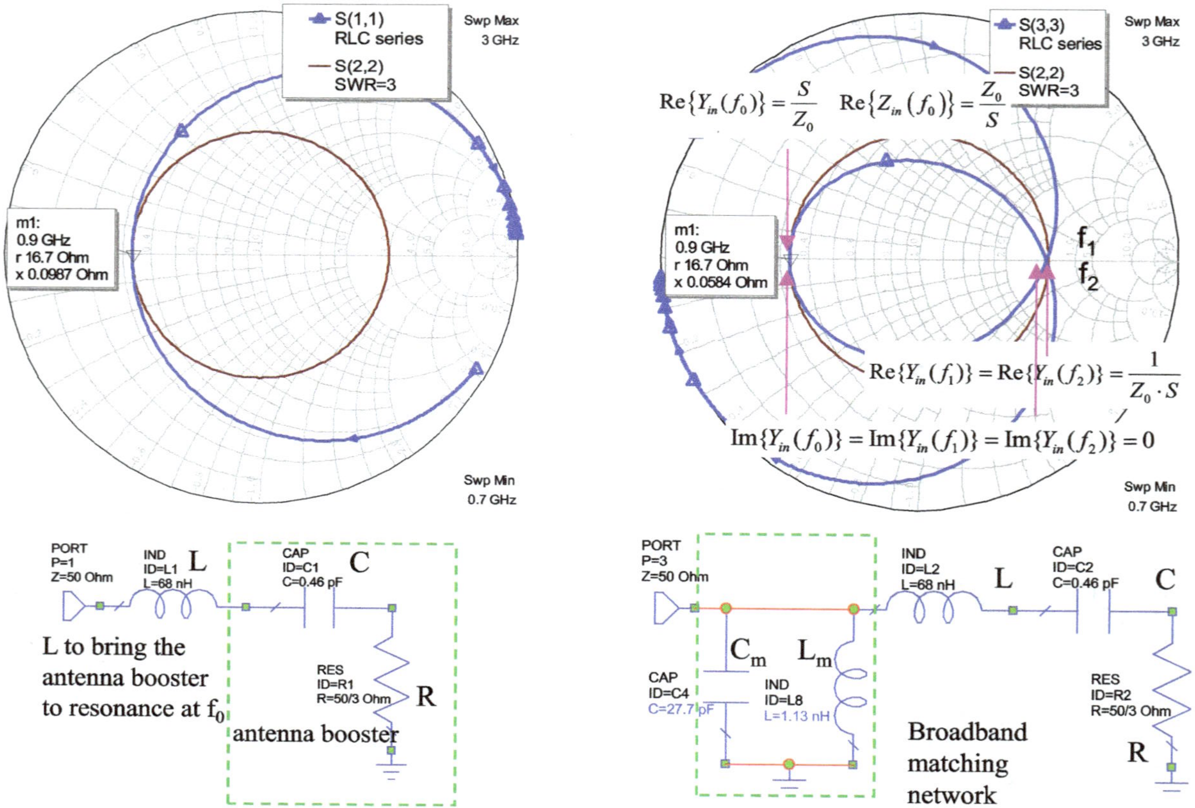

Figure 4.22 *(Left) An RC circuit is modelling an antenna booster, including a ground plane. A series L is added to make the antenna booster system resonate at 900 MHz; (right) a shunt L_m, C_m circuit is added to create an impedance loop inscribed in a circle of $SWR = S = 3$.*

Two important conclusions can be made:

1. The input impedance at f_0 should be Z_0/S. For typical values of $S = 3$, this represents an impedance of 16.6 Ω at the resonance. If this value is not achieved, the impedance loop will be out of the $SWR = S$ circle. However, as demonstrated by the following examples, a simple L-C network can easily move it inside.
2. To obtain the bandwidth enhancement factor F, divide (4.42) by (4.28), yielding Equation (4.43):

$$F = \frac{BW_f}{BW_o} = \frac{\frac{\sqrt{S^2-1}}{Q_a}}{\frac{S-1}{Q_a\sqrt{S}}} = \frac{\sqrt{S^3 - S}}{S - 1} \tag{4.43}$$

For an $SWR = S = 3$, $F = 2.45$, this results in an attractive bandwidth enhancement factor only using two components. It is worth comparing the F given by Equation (4.43) with the one obtained by Fano for an infinite number of components (Figure 4.23). Observations show that the F obtained in Equation (4.43) follows the same trend as F, albeit with a lower magnitude due to the reduced number of components. For S close to 1, F goes to a considerable value. This fact does not mean bandwidth is large because bandwidth, for a low value of S, goes to zero. Therefore, although F increases, the enhanced bandwidth is bounded.

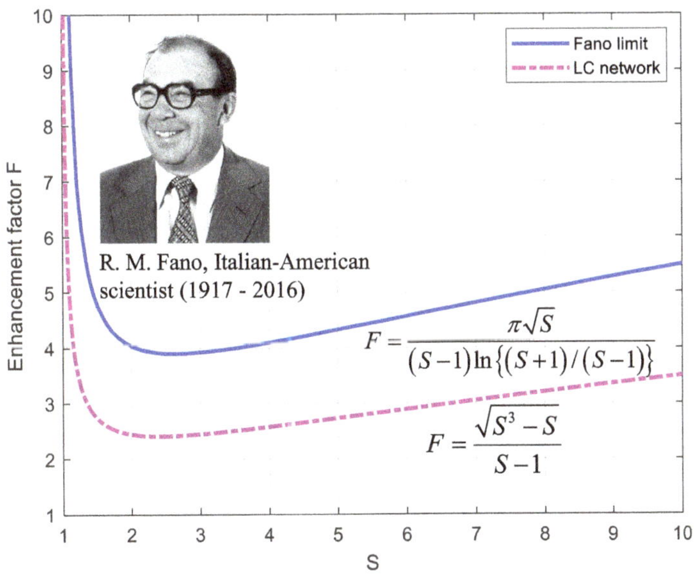

Figure 4.23 The maximum enhancement factor F for a lossless network compared to the F of the LC network

It is interesting to compare the inherent bandwidth obtained with Equations (4.26) and (4.28), and that obtained by modelling the impedance of the antenna as a series RLC circuit, and also that obtained using a generator having a conjugate impedance to the antenna booster, including the ground plane ($Z_g = 37.1+j208\ \Omega$) (Table 4.3). For the RLC circuit, the C value and L can be calculated as follows: $Z_{in} = 37.1-j208\Omega$ at 900 MHz (Figure 4.4), therefore, $R = 37.1\ \Omega$ and the reactance associated with C is $X_c = -208\ \Omega = -1/(2\pi \cdot f \cdot C)$, resulting in $C = 0.85$ pF. To obtain L, we add a series L to make the impedance resonate at 900 MHz. Such L is calculated as $X_L = X_c = 208\ \Omega = 2\pi \cdot f \cdot L$, resulting in $L = 36.7$ nH. Now, bandwidth can be obtained with Equations (4.28) and (4.32).

Inherent bandwidth using the general Equation (4.26) agrees with the inherent bandwidth using the RLC circuit modelling and the bandwidth obtained with a series L and shunt L matching network. However, if the generator has an internal impedance as the complex conjugate of the antenna booster, including the ground plane, the bandwidth could be more significant.

Other analytical broad-banding methods are also available in the literature [20,21].

Note that the broad-banding mechanism explained here is useful for dual-band frequency operation. This can be achieved by increasing the loop size [22].

4.8.1 Broadband design example

The previous design rule was applied to increase the bandwidth of a 5 mm × 5 mm × 5 mm antenna booster on a 120 mm × 60 mm ground plane (Figure 4.24).

Table 4.3 Bandwidth calculations comparisons

Using Equations (4.26) and (4.28)	**Using Equations (4.28) and (4.32)**	**With a L_{series}-L_{shunt} matching (Figure 4.6)**	**With a generator having $Z_g = Z_a^*$**
20.2%	20.6%	19.8%	30.7%

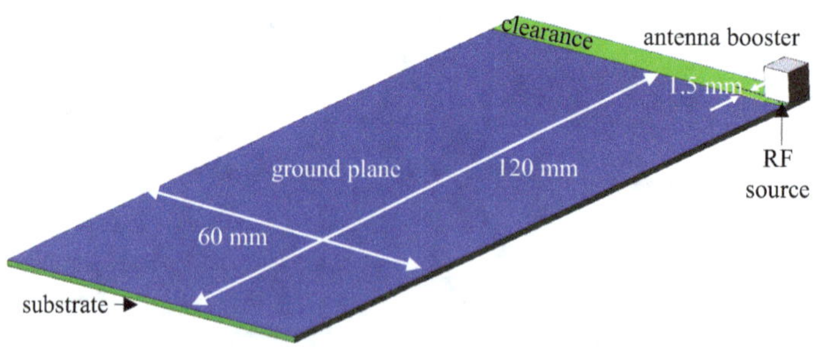

Figure 4.24 A 5 mm × 5 mm × 5 mm antenna booster spaced only 1.5 mm from a 120 mm × 60 mm ground plane

In this case, a different antenna booster is used, with a much smaller gap between the antenna booster and the ground plane. As deeply analysed in Chapter 3, Section 3.13, if the gap between the antenna booster and the ground plane is small, less bandwidth can be achieved. In the present example, we illustrate the benefit of a broadband matching network in a scenario where the gap is small. Therefore, we achieve the bandwidth not by increasing the gap but by including a broadband matching network. In this regard, broadband matching networks are helpful to make the antenna booster solution more compact since the ground clearance is small, allowing a reduction of the wireless device size or gaining ground plane to the clearance area to leave more space to allocate other components in the wireless device.

First, a design using ideal components is illustrated, and then, a real prototype is built to validate the results and show the performance that can be obtained in terms of bandwidth, total efficiency, and radiation patterns.

The input impedance of the 5 mm × 5 mm × 5 mm antenna booster, including the 120 mm × 60 mm ground plane, presents a non-resonance impedance along 0.7–1.2 GHz (Figure 4.25). At 900 MHz, the impedance is 24.4-*j*385Ω. Following the previous teaching, a series $L = 68$ nH is added to create resonance at 900 MHz.

The broadband matching network is designed to create an input impedance loop with a size that can be inscribed in the $SWR = 3$ circle (Figure 4.26). Since the input impedance loop is shifted to high impedances, the LC circuit is adjusted to move the loop to the inductive region while keeping the same size. As a practical rule, the input impedance loop is displaced to have the centre of the loop at the $R = 50\ \Omega$ circle of the Smith chart. Therefore, a series C moves the loop down to be inscribed inside the $SWR = 3$ circle (Figure 4.27).

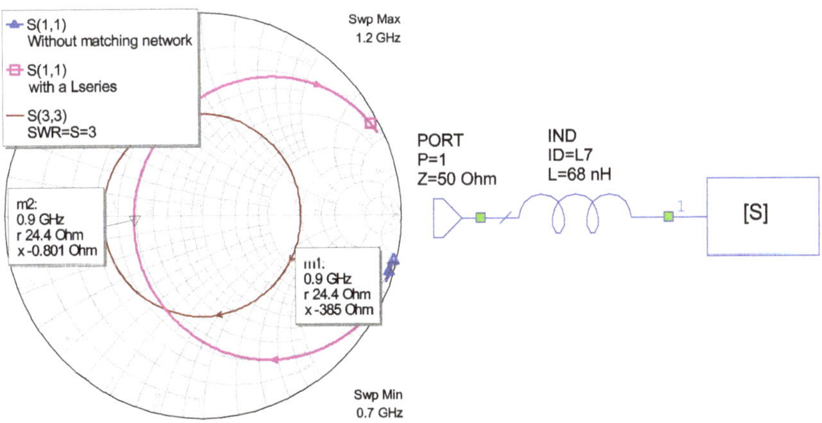

Figure 4.25 Input impedance of the antenna booster, including a 120 mm × 60 mm ground plane represented by [S] when a series L = 68 nH is added in series. The resonant frequency is adjusted to 900 MHz

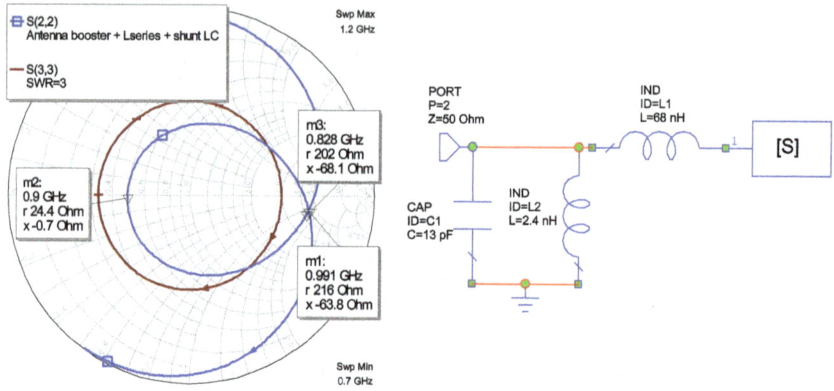

Figure 4.26 An LC shunt network is added to create an impedance loop with a size that can be inscribed in a circle of SWR = 3. The input impedance loop is shifted to high impedances, needing an extra step, as shown in Figure 4.27.

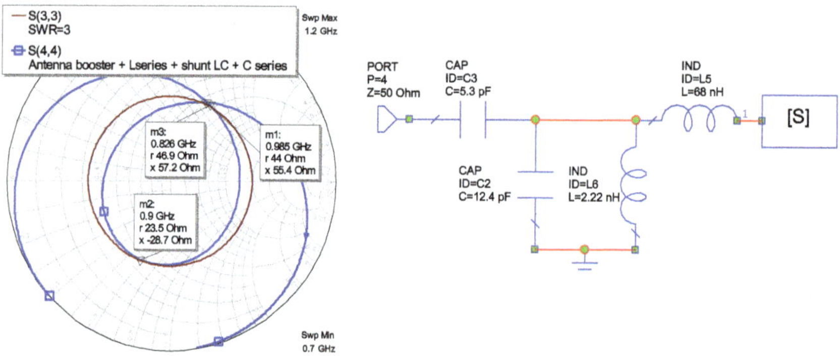

Figure 4.27 A modification of the shunt LC and a series C = 5.3 pF helps to place the impedance loop inside the circle of SWR = 3

The bandwidth with a series L, with the loop at the stage shown in Figure 4.26, and the one inscribed inside the *SWR* = 3 circles are compared, where it is clearly illustrated that the bandwidth at *SWR* = 3 is enhanced 2.1 times, from 8.2% to 17.4%, close to the theoretical value of $F = 2.45$ (Figure 4.28).

So far, ideal components have been used, but, as stated in Section 4.5, losses must be considered to get a better prediction not only on bandwidth but on total efficiency. The single L_{series} and broadband matching networks are simulated with real SMD components (Figure 4.29). When using real components, not all values are possible. Using as a reference the frequency range from 824 MHz to 960 MHz, the case with a L_{series} achieves a minimum efficiency of 36.4%. In contrast, the minimum value for the broadband matching network case is 47.1%, representing

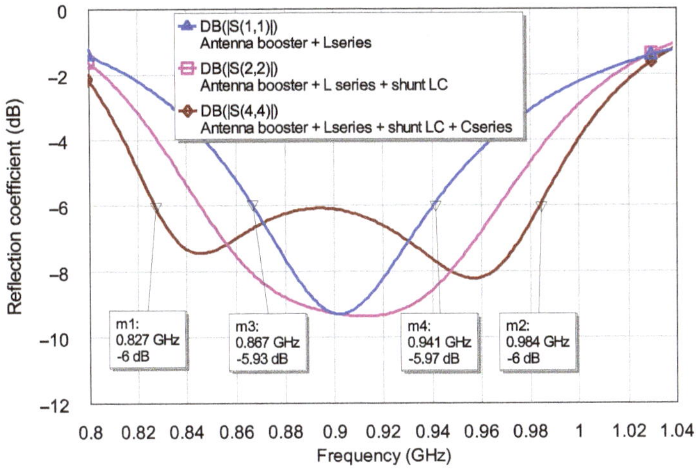

Figure 4.28 Bandwidth enhancement for the shunt LC broadband matching network

1.1 dB more efficiency. The average total efficiency across 824–960 MHz is 52.4%.

Tolerance analysis has been performed based on the real tolerance (2%) where the broadband solution features a stable performance (Figure 4.29).

A prototype is built to validate the numerical data, including the 5 mm × 5 mm × 5 mm antenna booster (a 3D brass-made cuboid) and the 120 mm × 60 mm ground plane printed on an FR4 substrate 1 mm thick (Figure 4.30).

The matching network topology is the same as the simulated one, but in this case, with different values. The reason for this is that the previous simulation did not take into account the effect of the pads for allocating the lumped components. This topic is reserved for Chapter 9, where the design of matching networks is addressed, taking into consideration the effect of the pads. As a result, the matching network in the simulation can be directly built, and a good agreement between simulation and measurement is obtained, hence, no lab adjustment is required. As it will be explained, the design principles are the same and have not been considered here to make the explanation easier. Those effects due to the pads become more relevant as the frequency increases since they introduce delay as well as some impedance, such as inductive effects (a pad is a short transmission line that can be modelled as an inductor).

The SWR of the prototype is measured with a vector network analyser, and the total efficiency is measured using 3D pattern integration inside an anechoic chamber (Figure 4.31). The total efficiency across 824–960 MHz bands is 55.6%, compared to 52.4% in the simulation (Figure 4.29), indicating a minimal difference of only 0.25 dB, which validates the utility of electromagnetic simulation.

Radiation patterns are measured at 890 MHz, showing similar performance as a half-wavelength dipole, for $\varphi = 0°$, the radiation pattern is almost

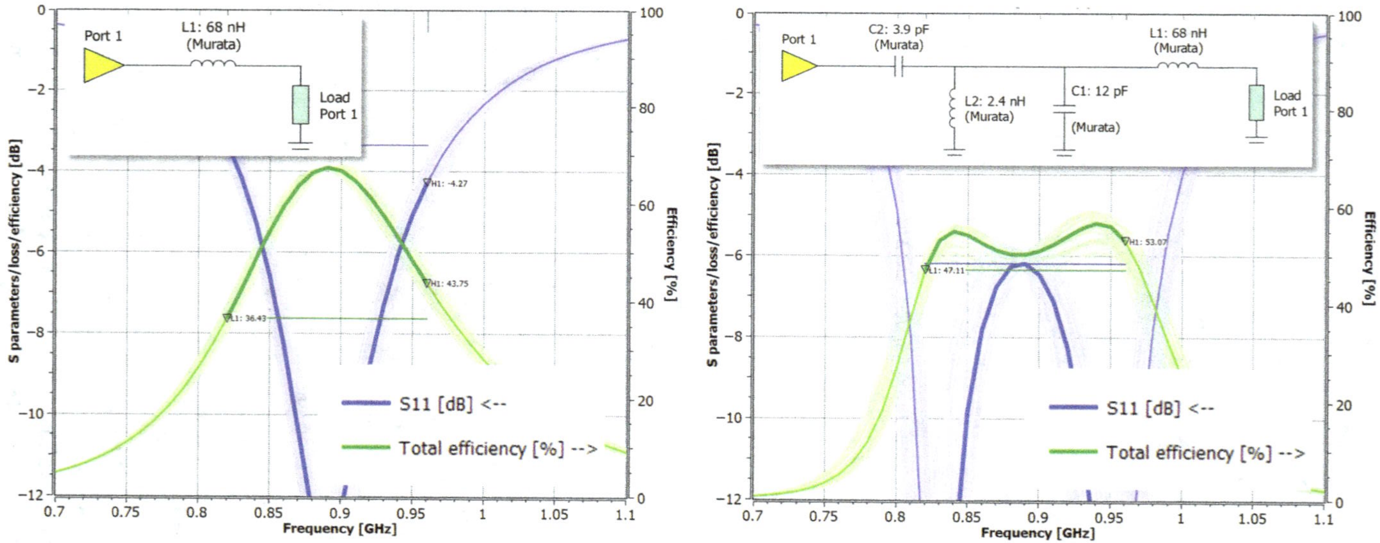

Figure 4.29 Comparison of the performance in terms of S_{11} and total efficiency for a simple matching and a broadband matching network. Load represents the S_{11} of the 5 mm × 5 mm × 5 mm antenna booster, including a 120 mm × 60 mm ground plane.

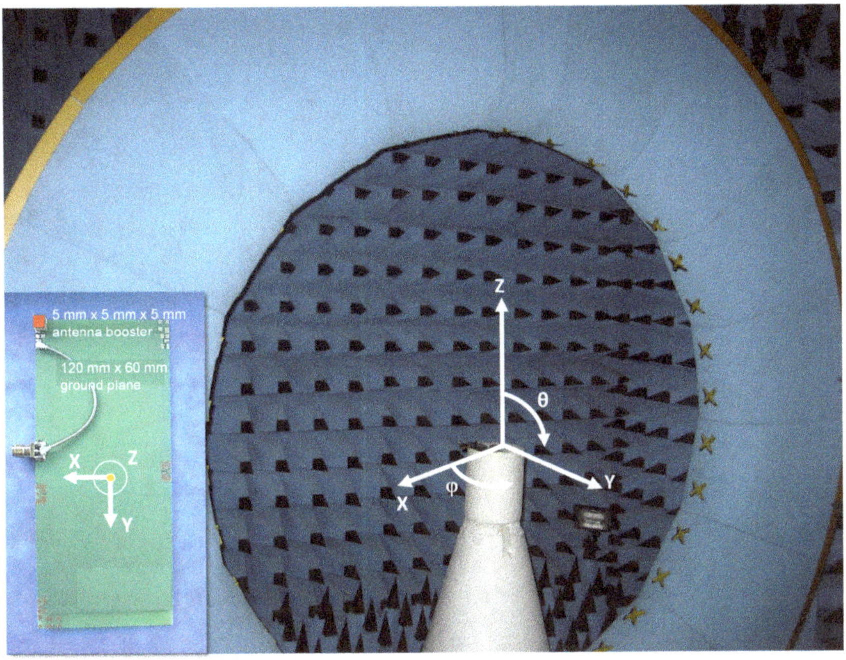

Figure 4.30 *Fabricated prototype with the 5 mm × 5 mm × 5 mm antenna booster (shown in red), including a 120 mm × 60 mm ground plane printed on an FR4 substrate 1 mm thick. The matching network employs SMD 0402 and 0603 components. A 50 Ω transmission line serves to test S_{11} and efficiency. Satimo Stargate 32 is shown.*

omnidirectional, and for $\varphi = 90°$, it has a maximum for the direction perpendicular to the ground plane and a minimum along the longitudinal axis (Figure 4.32). This is due to the currents excited by the antenna booster on the ground plane due to the excitation of the eigen current J_1 (see Chapter 3). As far as polarisation is concerned, it has linear polarisation along the maximum direction (any direction in the $\varphi = 0°$ plane) and aligned with the long axis of the ground plane.

It has been shown the utility of the circuit analysis conducted here to demonstrate how much the bandwidth of the antenna booster system can be increased. It has been shown that due to the RLC shunt characteristic of the antenna booster plus a series inductor, an impedance loop can be created. When properly designed, the size of such an impedance loop can be inscribed inside a circle of a given SWR, and as a result, the bandwidth is increased. Section 4.8.3 addresses the same objective, which is to increase bandwidth. However, the strategy avoids any analytic method. Instead, it relies on machine engineering using matching network synthesis. By no means does the next route disparage the previous method. On the contrary, the analytic approach presented so far has established the bandwidth limitation of a simple matching network, helping to better guide the matching network synthesis.

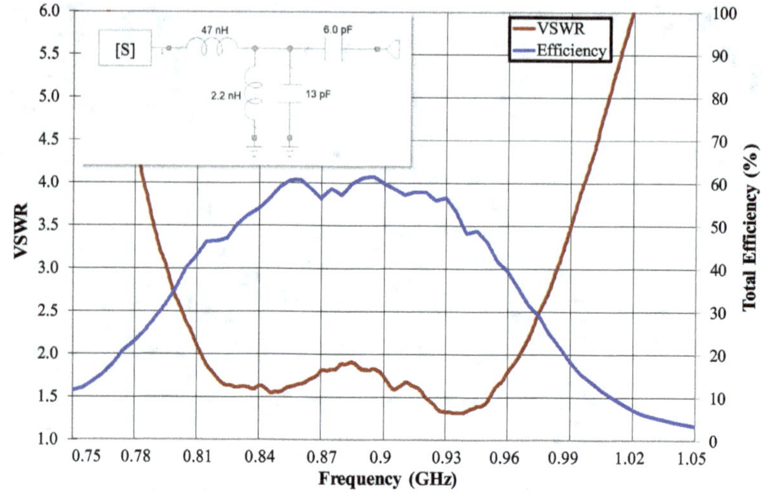

Figure 4.31 *Measured SWR and total efficiency. The matching network includes an SMD 0603 for the 47 nH and SMD 0402 for the rest. [S] represents the S_{11} of the antenna booster, including the ground plane.*

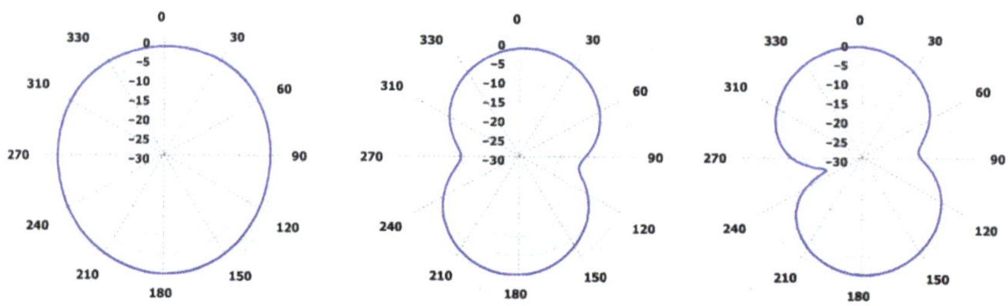

Figure 4.32 *Measured radiation patterns at 890 MHz. The normalised total gain in dBi is represented. The peak realised gain is 1.2 dBi. Axes are shown in Figure 4.30.*

4.8.2 Broadband design example with planar antenna boosters

The next proposed radiating system illustrates not only the usefulness of broadband matching networks but also planar antenna boosters [23] (Figure 4.33).

It comprises one planar antenna booster for each set of communication standards: the first one from 698 MHz to 787 MHz; the second one from 824 MHz to 960 MHz, the third one from 1559 MHz to 1609 MHz, and the fourth one from 1710 MHz to 2690 MHz. They are strategically arranged at each corner of the ground plane. Since the first and second frequency regions are close in frequency,

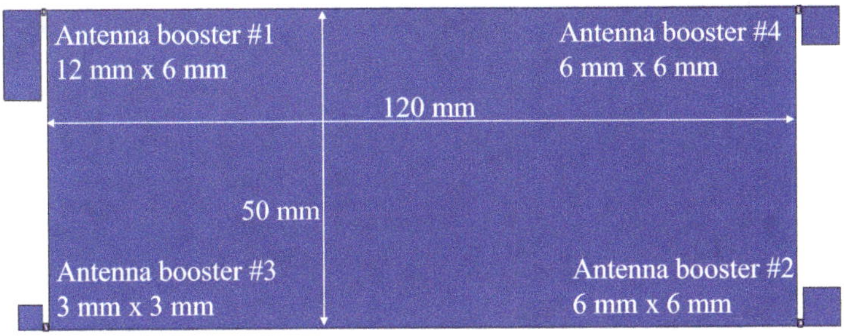

Figure 4.33 Architecture of the proposed small, multiband, and planar radiating structure, comprising four antenna boosters at the corner of a 120 mm × 50 mm ground plane printed on a thin layer of FR4 having 1 mm thick, $\varepsilon_r = 4.15$, tanδ = 0.017. Antenna booster #1 is designed for 698–791 MHz, #2 for 824–960 MHz, #3 for 1561–1606 MHz, and #4 for 1710–2690 MHz. The gap between the antenna boosters and the ground plane edge is 1 mm. The total area occupied by all antenna boosters is only 153 mm^2.

the antenna booster for the first frequency region (antenna booster #1) is located on one short edge of the ground plane, while the antenna booster for the second frequency region (antenna booster # 2) is placed at the opposite edge to obtain high isolation.

After the broadband design rules explained above, all the antenna boosters need a broadband matching network except antenna booster #3 since it only needs to cover a small bandwidth of 3.1% (Figure 4.34). The reflection coefficient shows that the bandwidth satisfies the bandwidth requirements, as shown next in the measurement results.

The current distribution is computed for several representative frequencies across the operating bands at each corresponding port (Figure 4.35). In particular, for the low-frequency region, the fundamental mode in the ground plane is excited, corresponding to an almost linear current aligned across the long edge of the ground plane. For higher frequencies, higher order modes appear not only aligned with the long axis of the ground plane but also with the short axis of the ground plane. In the end, not only bandwidth but also radiation patterns are determined by the current distribution in the ground plane and not in the antenna booster.

A prototype is built for the sake of validating the simulation analysis and the feasibility of the proposal (Figure 4.36). Each antenna booster is matched using the radiofrequency system with the values shown in Figure 4.34.

The measured reflection coefficient shows that each antenna booster operates at the frequency ranges with an $SWR \leq 3$ (Figure 4.37). The isolation between ports shows that the worst case is below 18.5 dB, ensuring low coupling among them (Table 4.4). Total efficiency is measured using 3D pattern integration with the

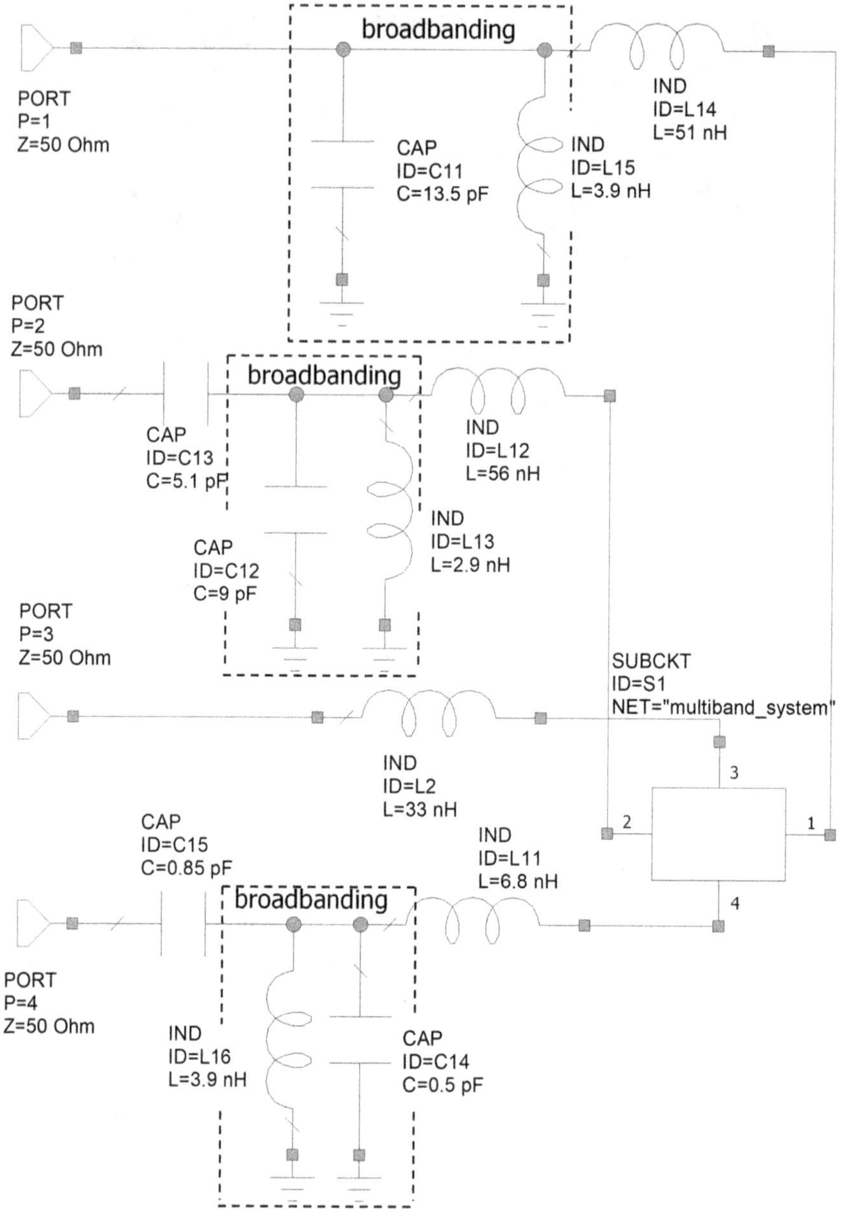

Figure 4.34 Radiofrequency system selected for the radiating structure of Figure 4.33. Port 1, 2, 3, and 4 correspond to 698–791 MHz, 824–960 MHz, 1561–1606 MHz, and 1710–2690 MHz, respectively. Port 1, 2, and 4 include broadband matching networks, whereas port 3 comprises a single series matching network.

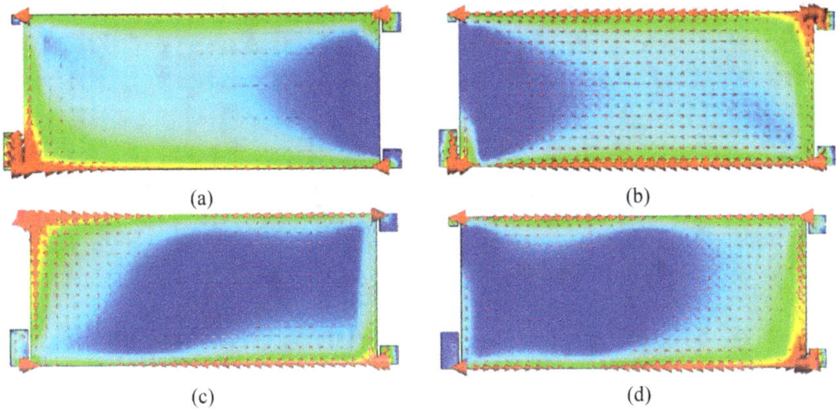

Figure 4.35 *Computed surface current distributions at several frequencies. (a) f = 750 MHz port #1 excited; (b) f = 950 MHz port #2 excited; (c) f = 1575 MHz port #3 excited; (d) f = 2200 MHz port #4 excited. Each antenna booster has the matching network shown in Figure 4.34.*

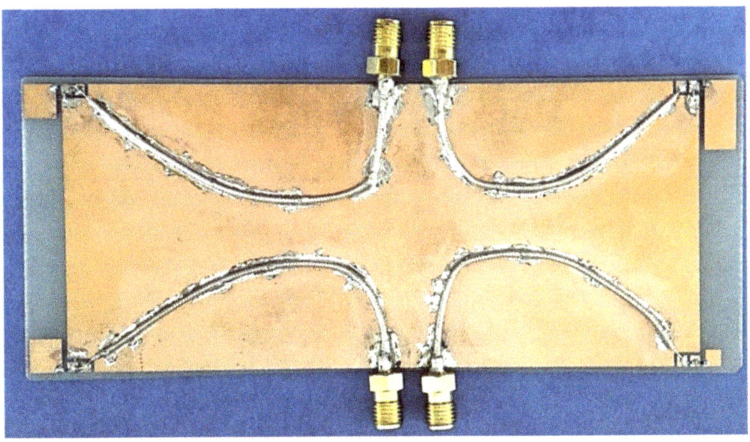

Figure 4.36 *Prototype using the four proposed antenna boosters. Dimensions are disclosed in Figure 4.33.*

Satimo Stargate-32 anechoic chamber (Figure 4.30). For these measurements, 50 Ω loads are connected to each of the not-under-test ports. Each antenna booster can be connected to a different front-end module (FEM) in charge of the operation in each number of bands used in wireless devices.

Total efficiency is moderate for the low-frequency region but still tolerable for mobile communications [24,25]. For antenna booster #3, total efficiency is 58% at

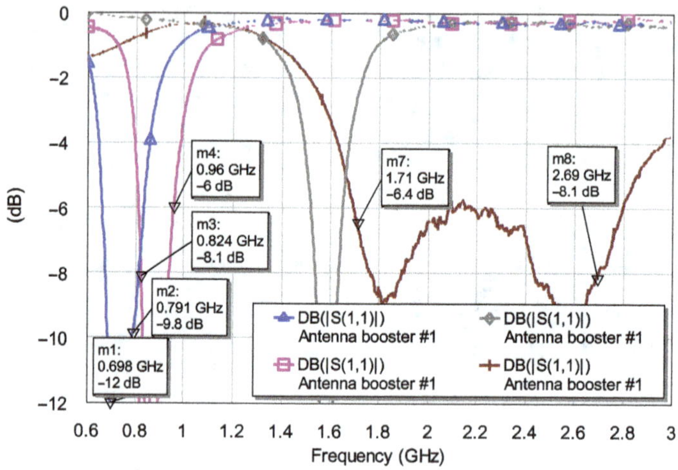

Figure 4.37 Measured reflection coefficient at each port of the prototype presented in Figure 4.36

Table 4.4 Measured isolation between ports

Isolation (dB)	698–791 MHz	824–960 MHz	1561–1606 MHz	1710–2690 MHz
698–791 MHz	–	18.5	31.9	42.6
824–960 MHz	18.5	–	28.4	40.3
1561–1606 MHz	31.9	28.4	–	21.8

1575 MHz with a size of only 3 mm × 3 mm. It becomes the most miniature known GPS antenna with such total efficiency values. For antenna booster #4, the efficiencies are even higher despite the small dimensions of $\lambda/29$ at 1710 MHz, thus emphasising again that antenna boosters effectively excite the ground plane currents.

Measured radiation patterns are omnidirectional in the horizontal plane with similarly measured directivities D suitable for communication environments with unknown direction of the incoming signal: D(750 MHz) = 2.8 dB, D(900 MHz) = 2.7 dB, D(1575 MHz) = 1.9 dB, D(1800 MHz) = 4.5 dB, D(2200 MHz) = 3.8 dB, D (2600 MHz) = 3.8 dB (Figure 4.38).

A small-multiband antenna booster system employing four planar antenna boosters has been proposed to operate at several frequency bands, able to cover cellular and satellite communication systems where the total area considering all antenna boosters is only 153 mm^3, becoming a compact solution. The planar antenna booster element has been chosen because it has acceptable electromagnetic performance and is geometrically simple.

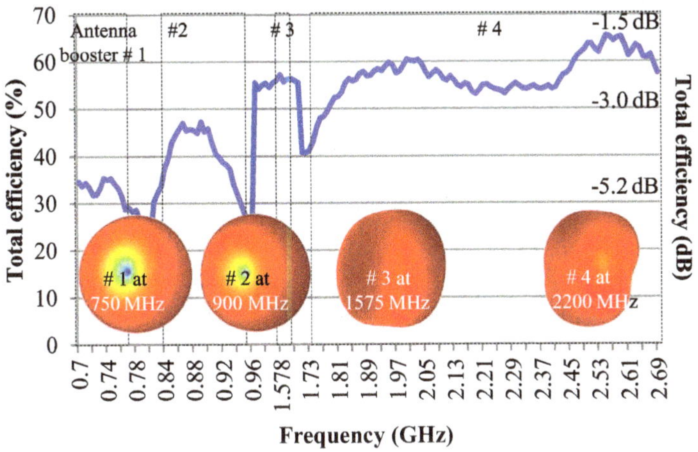

Figure 4.38 *Measured total efficiency η_t ($\eta_t = \eta_r \cdot (1-|S_{11}|^2)$) and radiation patterns for the prototype of Figure 4.36*

Broadband matching networks with an LC resonator

A broadband matching network is a simple way to obtain bandwidth without the need to increase the antenna booster size or adjust the ground plane size/form.

A broadband matching network can increase the bandwidth by two times the bandwidth obtained with a matching network of two lumped components.

4.8.3 Broadband matching networks with automatic network synthesis

Besides broadband matching networks designed using the analytical method presented here, other alternatives based on matching network synthesis are discussed in this section. Matching network synthesis relies on numerical methods, such as genetic algorithms, to synthesise a matching network to achieve a specific goal. For example, total efficiency across a certain frequency range. Using the same bandwidth as the previous case, 824 MHz to 960 MHz, a matching network is synthesised using four components as a maximum, the same as the broadband matching networks discussed before (Figure 4.39). As shown, better total efficiency is obtained in this example with the matching network based on the ladder topology. Nevertheless, it is worth noting that the matching network utilising the broadband LC network exhibits improved S_{11} performance across the 824–960 MHz band. This can be advantageous as it reduces mismatching for an amplifier connected to the matching network, thereby preserving its nominal gain [26].

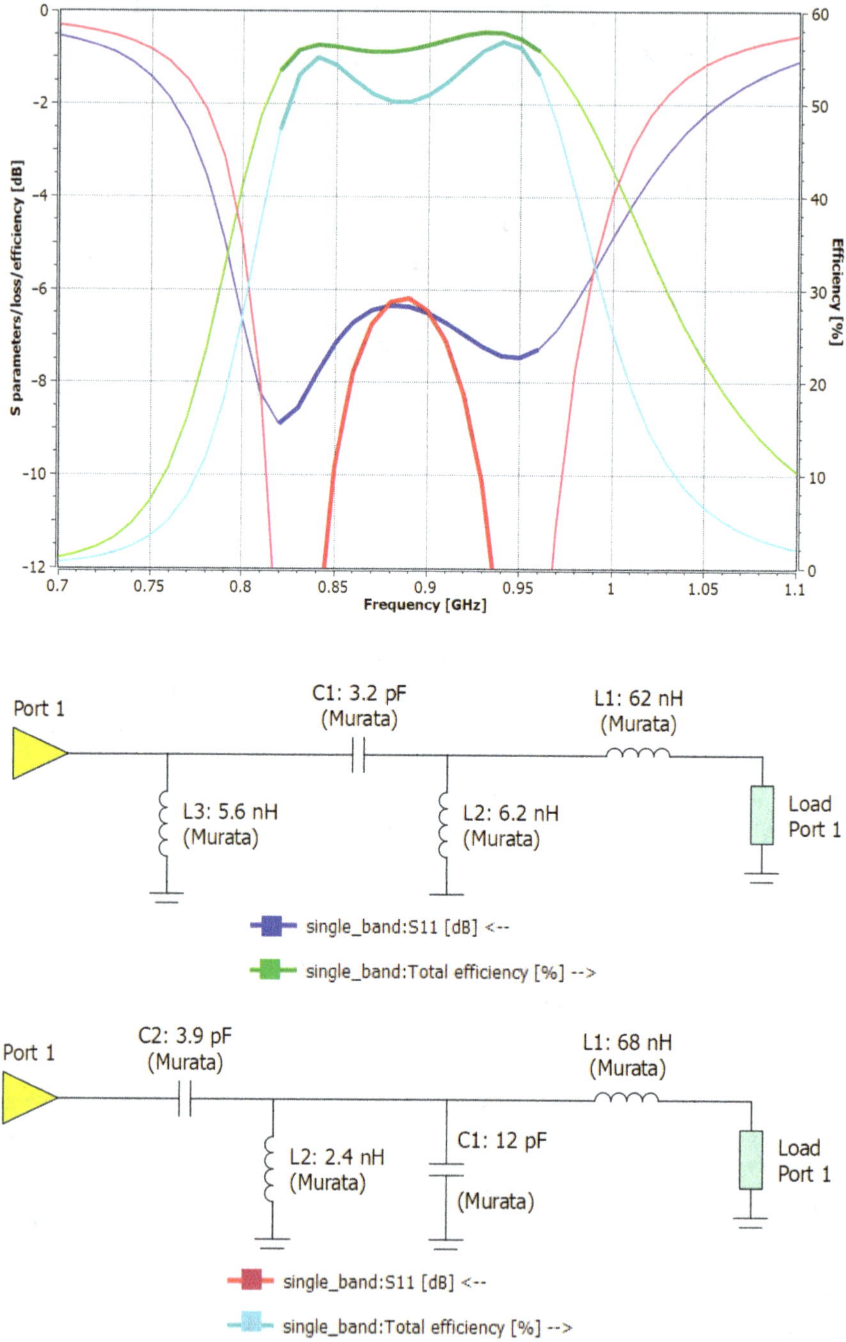

Figure 4.39 Comparison of a broadband matching network using a ladder topology and a shunt LC

4.9 On the reuse of a matching network

A single-band matching network has been designed according to a specific antenna booster and ground plane size (Figure 4.40).

Given a matching network for single-band operation, the question is: can we use the exact matching for other ground plane sizes? This analysis aims to determine how reusable a matching network designed for a given device can be for other devices with different sizes. Hence, no extra design efforts need to be made [27]. To answer this question, 361 ground plane sizes have been simulated, changing their length and width from 20 mm to 200 mm with steps of 10 mm (Figure 4.41).

The reference case is the 120 mm × 60 mm ground plane with a 12 mm × 3 mm × 2.4 mm antenna booster. This reference entails the design of a

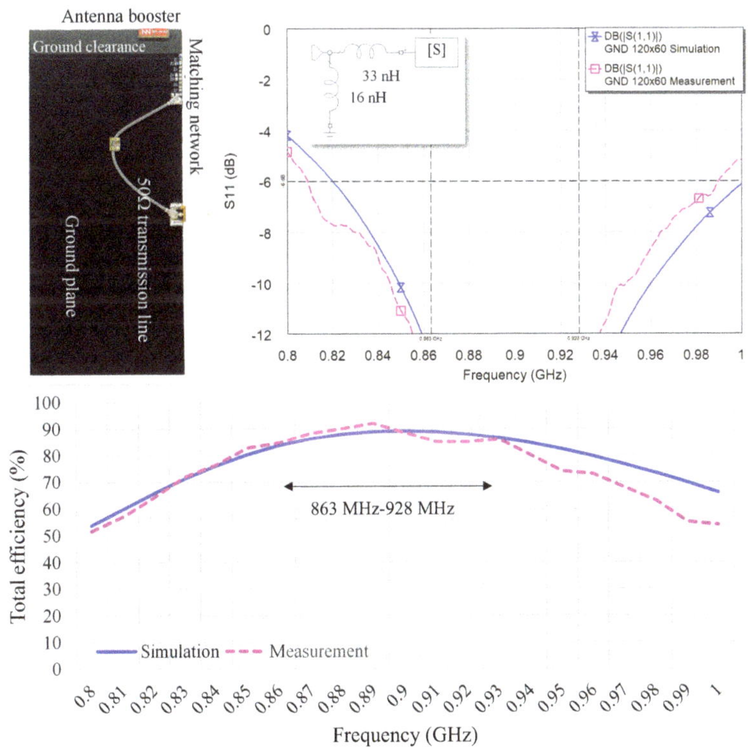

Figure 4.40 (Top left) A 131 mm × 60 mm PCB comprising a 120 mm × 60 mm ground plane, a 12 mm × 3 mm × 2.4 mm (height) antenna booster, a matching network with 0603 SMD inductors with a $Q > 85$ at the 863–928 MHz frequency range. The PCB comprises a substrate 1 mm thick, $\varepsilon_r = 4.15$ and $\tan\delta = 0.017$. (Top right) Simulation and measurement of the S_{11}. (Bottom) Total efficiency simulated and measured.

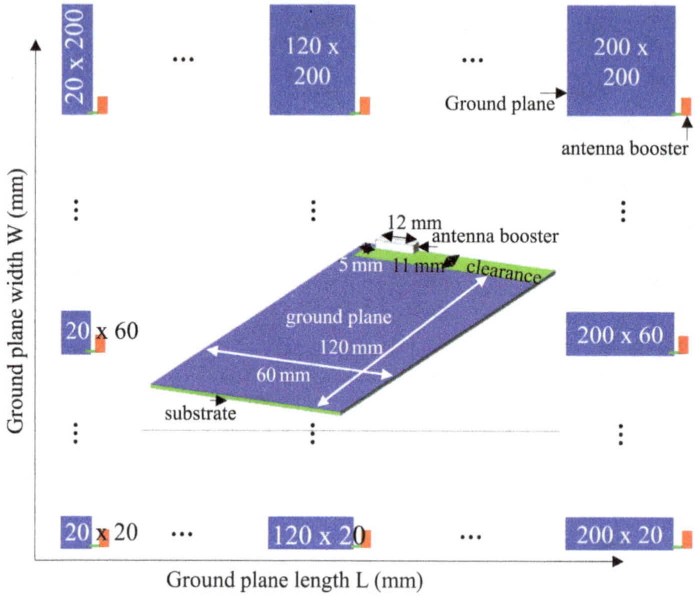

Figure 4.41 *Ground plane sizes (not to scale) from the smallest one, 20 mm × 20 mm, to the biggest one, 200 mm × 200 mm. The 120 mm × 60 mm ground plane with a 12 mm × 3 mm × 2.4 mm antenna booster is the reference case where a matching network has been designed to provide operation at 863–928 MHz. Ground plane and antenna booster present $\sigma = 5.9 \times 10^7$ S/m, the substrate is 1 mm thick, $\varepsilon_r = 4.15$, $\tan\delta = 0.017$. In all cases, the feeding line to feed the antenna booster is the same, as well as the position of the antenna booster.*

matching network, incorporating a 33 nH L_{series} and 16 nH L_{shunt}, while considering actual values derived from the component manufacturer's S parameters. The objective is to enable operation within the frequency range of 863–928 MHz. The Q_m for the components is 85 for the 33 nH and 16 nH at the frequency range of operation.

Once the matching network is designed for the 120 mm × 60 mm ground plane, the same matching network is added to all ground plane sizes. For each case, the S_{11} is computed, and the worst S_{11} in the 863–928 MHz is the reference to evaluate the performance of the matching network. For example, if the matching network for the 80 mm × 40 mm ground plane gives the worst S_{11} of −5.5 dB within the frequency band, the −5.5 dB is stored. This procedure is repeated for all ground plane sizes to obtain, as a result, a map that reflects the performance of the matching network (Figure 4.42).

4.9.1 The cluster

Results reveal a cluster of ground planes where the matching network designed for the reference case can be reused for many other cases. For example, if we fix as a

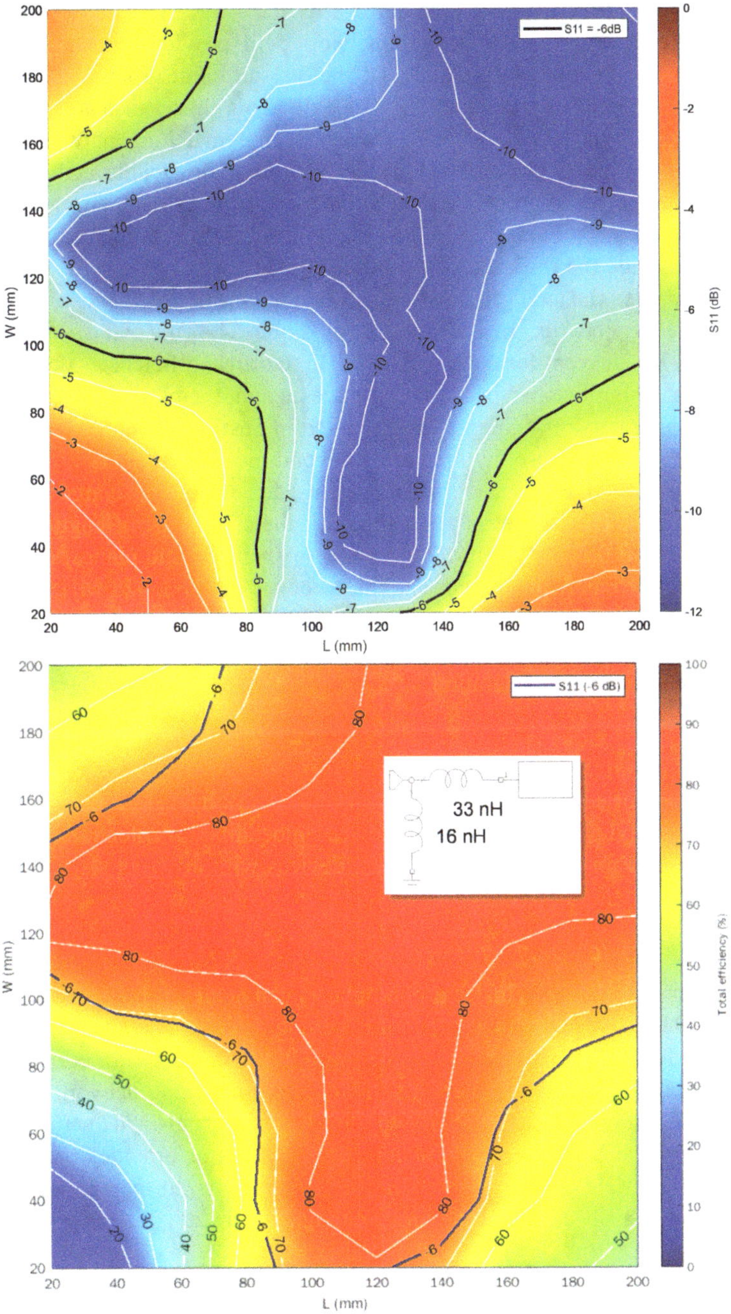

Figure 4.42 *(Top) Map of the worst S_{11} within the 863–928 MHz frequency range for the 33 nH L_{series} and 16 nH L_{shunt} matching network. 66.8% of the cases satisfy an $S_{11} \leq -6$ dB within the frequency range; (bottom) total efficiency considering all the loss mechanisms (substrate of the ground plane), including the losses of the matching network.*

target $S_{11} \leq -6$ dB, the whole area defined by the -6 dB line represents ground planes where the reference matching network will provide an $S_{11} \leq -6$, constituting a large set of ground planes. Of the 361 different ground plane possibilities, 241 ground planes feature an $S_{11} \leq -6$ dB. This cluster means that 66.8% of the cases feature an $S_{11} \leq -6$ dB with the same matching network. This is attractive from a practical point of view since the same matching network can be reused for a large set of ground planes for wireless devices. Naturally, one should respect the same layout and matching network position to avoid introducing phase delays that change the response of the matching network.

We computed the total efficiency for all cases, considering the losses presented by the matching network, in addition to S_{11} (Figure 4.42). Total efficiencies are greater than 70%, except for the critical area. In this area, efficiencies drop not only due to mismatching but also because Q_a increases and, therefore, losses introduced by the matching network increase, as discussed in Section 4.5.

It is interesting to observe three areas beyond the cluster (areas with $S_{11} > -6$ dB). The question is: can a new matching be used to match those areas to achieve an $S_{11} < -6$ dB? The answer is yes, except for the critical area (length and width less than 60 mm, 0.02λ at 900 MHz). For these small ground planes, Q_a is large, and the inherent bandwidth is not enough to cover the 863–928 MHz frequency band, representing 7.2% with an $S_{11} < -6$ dB. Nevertheless, the operation can still be guaranteed as long as the antenna system provides competitive efficiencies for the application under consideration.

Reuse of a matching network
For the sake of simplicity, once a matching network has been designed for a specific ground plane of a wireless device embedding an antenna booster, it can be reused for a wide set of wireless devices with different sizes.

4.9.2 *Off-cluster performance*

It is worth noting that off-cluster ground planes can achieve enough bandwidth to cover 863–928 MHz but with a different matching network than the reference one. This is true for the off-cluster area at the top left and bottom right corners (either large length or large width), where a similar L-type matching network can match the antenna system. For example, for the top left corner, a 47 nH series inductor (connected to the booster) with a shunt 2.7 pF capacitor provides bandwidth to cover 863–928 MHz. A 48 nH series inductor (connected to the antenna booster) with a shunt 1.8 pF capacitor provides enough bandwidth for the right bottom corner.

However, the bottom left corner (small ground plane) requires special treatment since the achievable bandwidth decreases as the area of the ground plane decreases. For this critical zone, some cases can be matched, but more than two components are needed. In this case, the matching network will be a broadband

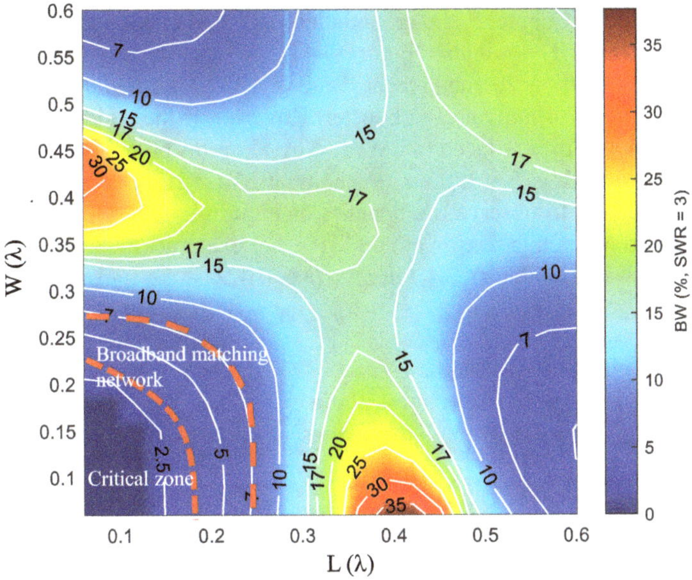

Figure 4.43 Potential bandwidth for a 12 mm × 3 mm × 2.4 mm antenna booster is given in Figure 4.41. The critical area is defined as those cases having a potential bandwidth of less than 3%. The transition area is defined as having a potential bandwidth larger than 3% and less than 7.2%. In this area, a broadband matching network (three or more lumped components) can achieve 7.2% to cover the 863–928 MHz frequency band.

matching network that can enhance the inherent bandwidth of the antenna system by a factor of ~2.4 [18,19].

The potential bandwidth for each off-cluster ground plane can be computed according to the input impedance information [5] (Figure 4.43). It is interesting to observe a first area, called the *broadband matching network*. This area is defined as those cases having a potential bandwidth larger than 3% and less than 7.2%. 7.2% is the required bandwidth to cover the 863–928 MHz frequency range. Therefore, if the potential bandwidth, which is the achievable bandwidth that can be obtained with a simple matching network with one or two lumped components, is less than 7.2%, a more sophisticated matching network is needed. In this case, a broadband matching network comprising three components or more may increase the potential bandwidth by a factor of 2.4 for $S_{11} = -6$ dB. Since broadband matching typically enhances the bandwidth by a factor of 2.4, a broadband matching circuit can achieve a 7.2% bandwidth for a 3% antenna system. Finally, the critical area is defined as those cases having a potential bandwidth of less than 3%. In this area, a passive matching network is not recommended, and a reconfigurable matching network (Chapter 7) can be the solution to cover the 863–928 MHz frequency range.

As evaluated in Chapter 3, the ground plane size that maximises the bandwidth is $L = \sim 0.42\lambda$ and $W = \sim 0.06\lambda$ (Figure 4.43). For example, a ground plane of $L = 140$ mm and $W = 20$ mm reaches a BW = 37.7% at 900 MHz with the 12 mm × 3 mm × 2.4 mm antenna booster. However, if the width is increased to 60 mm, the bandwidth is reduced to BW = 16.6%. For this 60 mm width, the length optimising bandwidth is $L = 120$ mm, reaching a BW = 20.0%.

It is interesting to observe the benefits of Figure 4.43 beyond the frequency analysis carried out in this section at 900 MHz. For example, the ground plane size that maximises bandwidth at 2.4 GHz is $L = 0.42\lambda = 52$ mm and $W = 0.06\lambda = 20$ mm.

Optimising the size of the ground plane
A $L \times W$ rectangular ground plane presents an optimum bandwidth for $L \sim 0.42\lambda$ and $W \sim 0.06\lambda$.

4.9.3 About radiation patterns

Besides impacting impedance, the ground plane affects the radiation properties (Figure 4.44). The current distribution is computed at 900 MHz for the reference 120 mm × 60 mm ground plane and for the four ground planes at each corner of the domain: 20 mm × 20 mm, 20 mm × 200 mm, 200 mm × 20 mm, and 200 mm × 200 mm.

Since all cases feature small sizes in terms of the wavelength, @radiation patterns are low directive, and, like a half-wavelength dipole: directivity is 1.7 dBi for the 20 mm × 20 mm ground plane case, 2.6 dBi for the reference case, and 3.3 dBi for the largest ground plane. The polarisation in the direction of maximum radiation is linear, with an axial ratio (AR) larger than 13 dB. The direction of the polarisation vector changes according to the current distribution on the ground plane. The polarisation is aligned with the horizontal axis for the 20 mm × 20 mm and 200 mm × 20 mm ground planes. For the reference ground plane (120 mm × 60 mm), the polarisation is slanted 20° with respect to the horizontal axis. This is due to currents along the short edge of the ground plane. For the 20 mm × 200 mm, the polarisation is almost vertical. Finally, the polarisation for the 200 mm × 200 mm ground planes is slanted at 45°. This is because the intensity of the current distribution is symmetrical along the horizontal and vertical axes since the antenna booster is located at the corner of the ground plane.

While a matching network and the positioning of the antenna booster (typically favouring corners for increased bandwidth and efficiency [5]) can regulate impedance matching, the properties of the radiation pattern are dictated by the size of the ground plane relative to the operating wavelength. From a practical perspective, since the sizes of the ground planes for wireless devices are small in terms of wavelength (e.g., the largest ground plane analysed here is 200 mm × 200 mm ground plane is 0.6λ at 900 MHz), radiation patterns are quasi-isotropic, convenient for scenarios where the angle of arrival of the incoming electromagnetic signal is

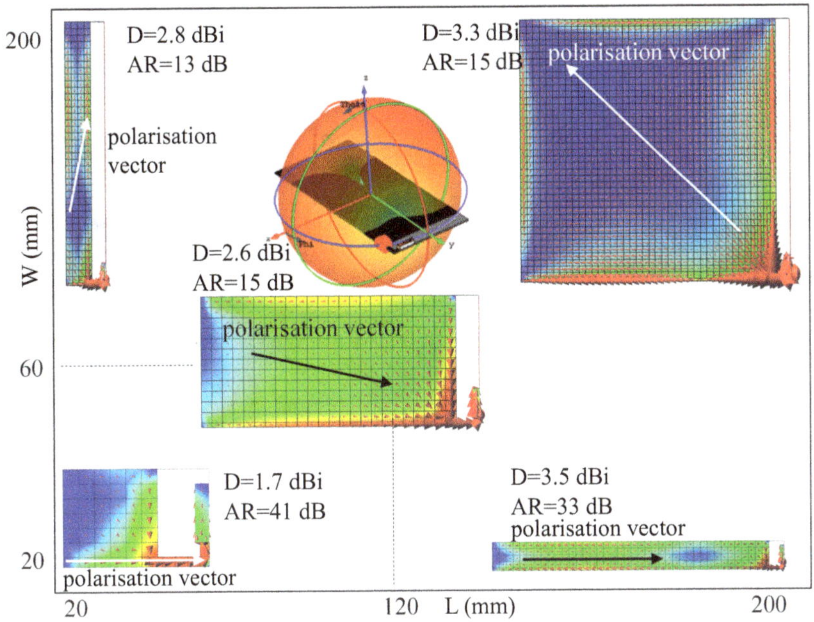

Figure 4.44 *Simulated current distribution, directivity, and the axial ratio at 900 MHz for the 20 mm × 20 mm, 20 mm × 200 mm, 200 mm × 20 mm, 120 mm × 60 mm, and 200 mm × 200 mm ground planes. AR is the axial ratio for $\theta = 0°$ (direction perpendicular to the ground plane). The arrows indicate the direction of the polarisation vector. The 3D radiation pattern is represented in dB with a dynamic range of 30 dB at 900 MHz for the 120 mm × 60 mm ground plane.*

not known. On the contrary, a highly directional antenna is preferred for static point-to-point links.

4.9.4 Discussion about the reuse

In a bid to further highlight the benefit of reusing the same matching network (Figure 4.40), four different cases are studied: the reference case having a 120 mm × 60 mm ground plane, a thinner case with a 120 mm × 20 mm ground plane such as a remote control, a small device having a 90 mm × 40 mm ground plane such as a tracker, and a 200 mm × 200 mm ground plane such as a smart meter (Figure 4.45).

All four cases are within the cluster (Figure 4.42), ensuring an $S_{11} < -6$ dB across 863–928 MHz. The averaged total efficiencies are 87.3%, 79.6%, 72.2%, and 67.1% for the 120 mm × 60 mm, 120 mm × 20 mm, 200 mm × 200 mm, and 90 mm × 40 mm, respectively. As demonstrated in the previous section, total efficiency is competitive if the device has a ground plane belonging to the cluster. Logically, further tuning the matching network can increase total efficiency to have

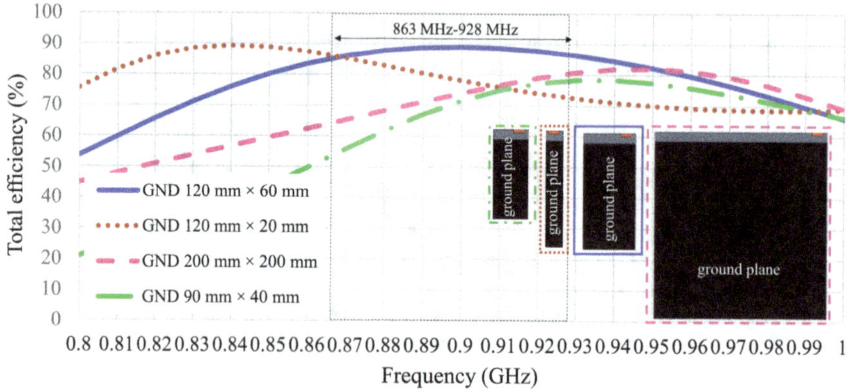

Figure 4.45 Simulated total efficiency for different cases embedding a 12 mm × 3 mm × 2.4 mm antenna booster with the same matching network as in Figure 4.40

a symmetrical response, as in the 120 mm × 60 mm reference case. However, reusing the same matching network avoids any adjustment of the components, and hence, the antenna booster solution follows the concept of plug, play, and forget. As a result, the solution is easy to use and embed in a wide set of wireless devices featuring different form factors since no customisation of the antenna system is needed.

4.10 Non-Foster matching networks

The matching network components utilised so far obey the Foster's theorem [28,29]. The theorem essentially declares that in a lossless passive two-terminal device, the frequency-dependent slope of its reactance X (and susceptance B) must strictly ascend:

$$\frac{\partial X(f)}{\partial f} > 0 \quad \frac{\partial B(f)}{\partial f} > 0 \tag{4.44}$$

An inductor, a capacitor, and combinations thereof are examples of components within that set. A component such as a negative inductor or capacitor is an example of a non-Foster component. Although such components do not exist, they can be emulated with active components such as negative impedance converters.

Although the book focuses on Foster matching networks, an example of a non-Foster matching network is included for illustrative purposes. In the following simple example, a negative capacitor is used in series with a 12 mm × 3 mm × 2.4 mm antenna booster featuring a capacitive impedance below 1.5 GHz similar to an RC circuit. Therefore, a negative capacitor compensates for the positive capacitive behaviour of the antenna booster. A shunt capacitor is then used to match the impedance (Figure 4.46). The bandwidth is 60.5% compared to the

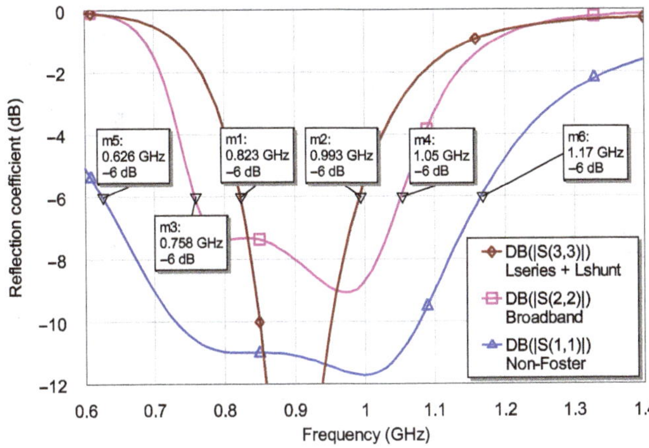

Figure 4.46 *Comparison of a Non-Foster matching network comprising a negative capacitor in series to a broadband (four lumped components) and a simple matching network (two components). The antenna booster system is illustrated in Figure 4.4.*

32.2% of a broadband bandwidth with an LC circuit and 18.7% for a single matching using a L_{series} and L_{shunt}.

Non-Foster elements can achieve impedance transformations that are not possible with conventional passive components alone. This flexibility in impedance transformations opens up new possibilities for designing innovative and efficient RF/microwave circuits.

4.11 Distributed antenna boosters

To increase bandwidth, three techniques have been examined so far. First, increase the size of the antenna booster (Chapter 3). Second, change the size or shape of the ground plane (Chapter 3). And third, adding broadband matching networks. In this section, a new technique called *distributed antenna boosters* is presented. It consists of distributing two or more antenna boosters along an edge of the ground plane and feeding them with a particular phase to increase bandwidth [30–35]. When spreading two or more antenna boosters, the feeding phase between antenna boosters (Figure 4.47) is optimised to cancel the reflection from each antenna booster. When the reflected waves are out of phase (180° in the case of two waves, 120° in the case of three waves, and so on [32]), the S_{11} is minimised for a broad frequency range, and therefore, the bandwidth compared to a single antenna booster solution is enhanced [30,34]. The excitation phase is achieved by a transmission length where the length L is adjusted according to $\varphi = -\beta \cdot L \pm 2\pi$ being $\beta = \frac{2\pi f \sqrt{\varepsilon_{eff}}}{c}$, where ε_{eff} is the effective relative permittivity of the transmission line, and c is the speed of light in vacuum. For example, to achieve an out-of-phase effect, $\varphi = \pm 90°$ since the reflected wave travels from the source to the antenna booster and back,

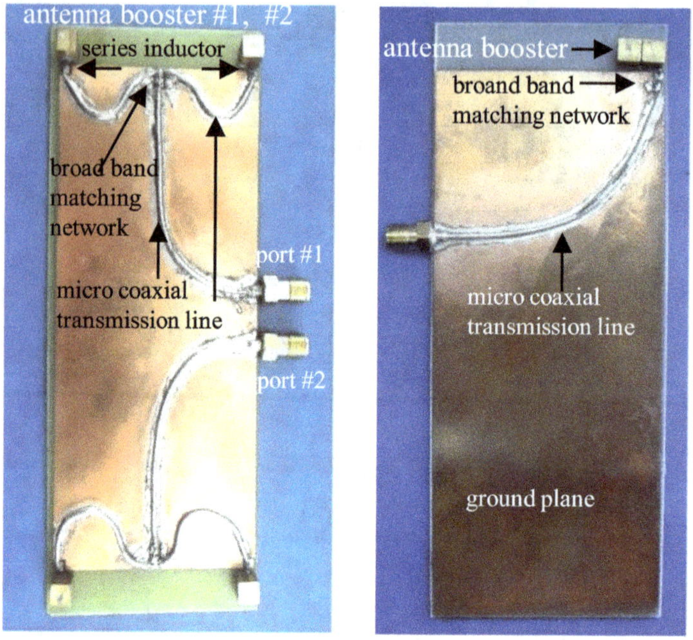

Figure 4.47 *Distributed antenna booster in-phase scheme (left) compared to a single antenna booster with a broadband matching network (right). Port #1 provides an in-phase scheme centred at 900 MHz and port #2 at 2.2 GHz [34]. Ground planes are 120 mm × 50 mm. The antenna boosters for the in-phase schemes are 5 mm × 5 mm × 5mm, whereas it is 10 mm × 5 mm × 5 mm for the single antenna booster to make a fair comparison with the distributed scheme having two boosters.*

gaining a phase of 2φ. Considering a microcoaxial transmission line filled with a substrate having $\varepsilon_r = \varepsilon_{eff} = 2.1$, it results in $L = 51.7/f(\text{MHz})$. For example, at 900 MHz, it results in $L = 57$ mm, which should be the additional transmission line length to achieve the phase delay.

Another technique consists of an in-phase scheme where the antenna boosters are fed with the same phase. Concerning Figure 4.1, an out-of-phase scheme presents a $\varphi = 180°/n$ where n is the number of antenna boosters, and $\varphi = 0°$ for an in-phase feeding scheme. The rationale behind the in-phase scheme is to excite a longitudinal mode along the ground plane to increase bandwidth. In effect, as studied in Chapter 3 as well as in Section 4.9.2, a narrow ground plane provided more bandwidth than a broader ground plane since the fundamental longitudinal mode was better excited for the 20 mm wide ground plane than the 60 mm wide ground plane (Figure 4.43). Therefore, in-phase antenna boosters located at the two corners of the short edge may excite the longitudinal mode better. Simply put, it is like waving a piece of paper by two corners instead of one. Holding the sheet of paper

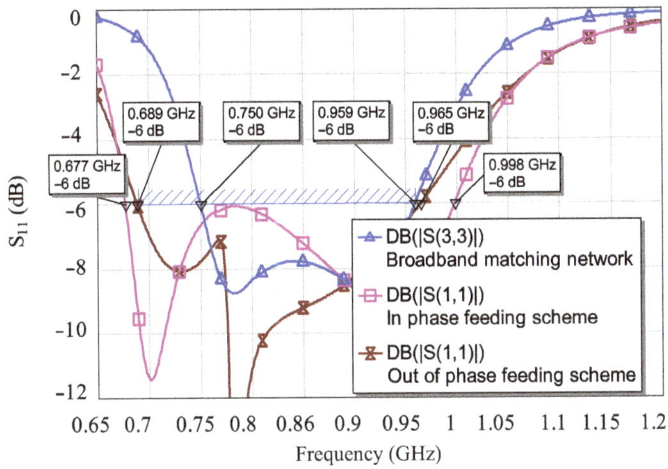

Figure 4.48 Comparison between a single antenna booster, an in-phase and out-of-phase feeding architectures

by the two corners facilitates easy vibration along its longest side. Therefore, from a practical point of view, the in-phase scheme does not require streamlining the transmission line length.

Either in the out-of-phase or the in-phase scheme, the radiofrequency architecture may be the same: a first-series inductor with the antenna booster and a transmission line connecting each antenna booster to a common point. From the common point, a broadband matching network is included to improve the bandwidth further (Figures 4.1 and 4.47, [34]).

The bandwidth of each in-phase and out-of-phase solution is contrasted with that of a single antenna booster equipped with a broadband matching network, as outlined in Section 4.8 (Figure 4.48) at 900 MHz. It is observed that both in-phase and out-of-phase architectures provide more bandwidth than the single antenna booster: 24.4%, 33.3%, and 38.3% for the single antenna booster, for the out-of-phase and in-phase schemes, respectively.

4.12 Loop antenna boosters for narrow band applications in small ground planes

As studied in Section 4.5, when the Q_a of the ground plane, including the antenna booster, is high, the Q_m of the matching network components is critical (Equation (4.24)). This situation occurs for electrically small ground planes (e.g., ground plane length less than λ/π, Chapter 1, Section 1.4.10). In some cases where the required bandwidth is small, the following approach proposes to change the antenna booster's capacitive nature to an inductive behaviour by including a short connection (Figure 4.49). As a result, the matching network does not require inductors but capacitors. Since Q_m for capacitors is typically higher (~3 times), the matching network losses are minimised since $Q_m >> Q_a$ [36].

To demonstrate the benefits of a loop-based antenna booster system, we design a small ground plane of approximately $\lambda/13$ at 450 MHz (measuring 145 mm × 130 mm) and compare it to an antenna booster featuring capacitive impedance (Figure 4.49). In the loop-based antenna booster case, the impedance at 450 MHz is Z_a = 5+291$j\Omega$ with a Q_a = 83. With this impedance, a matching network comprising a C_{series} = 1.3 pF and a C_{shunt} = 15 pF is needed with Q_m = 275. Without the loop, the Z_a = 10−392$j\Omega$ with a Q_a = 45. In this case, the matching network is a L_{series} = 150 nH and a C_{shunt} = 8.2 pF. For this situation, Q_m is considered to be the inductor with Q_m = 60 (Table 4.5).

As shown, although Q_a for the loop is worse (less bandwidth) than the non-loop situation, the Q_m for the matching network is much higher, resulting in a better matching network efficiency. Since radiation efficiency (η_r) is almost the same for both cases, the total efficiency for the loop-based case is better.

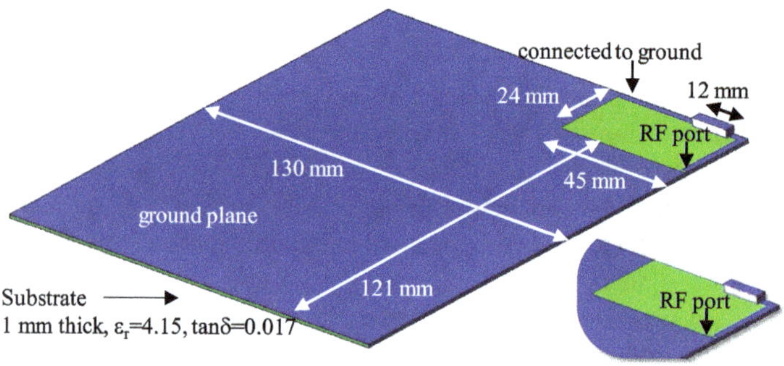

Figure 4.49 A loop-based antenna booster system for narrow band applications in small ground planes. The antenna booster without the loop scheme is shown in the bottom right corner.

Table 4.5 Comparison between an antenna booster with and without a loop scheme 450 MHz

Parameter	**Loop-antenna booster**	**Without loop**
Q_a	83	45
Potential BW (*SWR* = 3)	1.4%	2.6%
η_r	88.7%	89.4%
Q_m considering components of the same size (SMD 0402)	275	60
η_m with Equation (4.24)	77.6%	57.5%
Computed η_m	81.2%	54.9%
η_t	72.4%	50%

Consequently, when dealing with narrow-band applications and small ground plane sizes, the antenna booster utilising a loop architecture becomes a compelling choice.

4.13 Revisiting the Chu limit

The Chu limit was introduced in Chapter 1, showing the bandwidth dependence on the device size: the smaller the electrical size of the device, the less bandwidth can be obtained [37]. In this section, we revisit the Chu limit for some of the representative designs using antenna boosters (Equations (4.45) and (4.46)).

$$Q_{\min} = \eta_r \cdot \left(\frac{1}{(ka)^3} + \frac{1}{ka} \right) \tag{4.45}$$

$$\begin{aligned} BW_{\max} &= \frac{1}{Q_{\min}} \cdot \frac{SWR - 1}{\sqrt{SWR}} \\ BW_{\max} &= \frac{1}{\eta_r} \cdot \left(\frac{1}{(ka)^3} + \frac{1}{ka} \right)^{-1} \cdot \frac{SWR - 1}{\sqrt{SWR}} \end{aligned} \tag{4.46}$$

A $\lambda/2$ dipole antenna (L = 333 mm, W = 2 mm, resonant frequency 900 MHz) is designed as a reference. This dipole can be inscribed inside a sphere of radius a = 166.5 mm, equivalent to $\lambda/4$ at 900 MHz, and therefore, $k{\cdot}a$ = 1.57 since $k = 2\pi/\lambda$. Its simulated bandwidth is BW = 17.6%, which is far away from the theoretical limit of 129% (Figure 4.50). The reason is that the thin dipole does not fill the radian-sphere effectively. Other antenna configurations make better use of the radian-sphere, thus obtaining larger bandwidths than the basic thin half-wavelength dipole [37–39]. As a reminder, a device is small if the total size ($k{\cdot}2a$) including the ground plane and antenna booster satisfies $k{\cdot}a < 1$ (Figure 1.29, Chapter 1).

In this case, however, it is interesting to analyse how antenna boosters are located with respect to the Chu limit. To answer the question, four ground plane sizes have been studied: 120 mm × 60 mm, 120 mm × 17 mm, 60 mm × 60 mm, and an L-shaped ground plane inscribed in a 60 mm × 60 mm square. For each case, the bandwidth is computed at 900 MHz at SWR = 3 with a single matching network consisting of two lumped components and a broadband matching network with a maximum of seven components.

As shown in Chapter 3, the potential bandwidth of a 120-mm-long ground plane was larger for 17 mm than for a 60-mm-wide ground plane (Figure 3.51) since the fundamental mode of the ground plane was better excited. As a result, when designing a matching network for both cases, the 17-mm-wide ground plane achieves a better bandwidth than the 60-mm-wide. In particular, the BW = 20.0% and BW = 35.5% for the 60 mm and 17-mm-wide ground planes, respectively, resulted in a substantial bandwidth increase. When broadband matching networks are designed for both situations, both bandwidths increase to 38.8% and 50.0% for the 60 mm and 17 mm widths, respectively. This results in a bandwidth

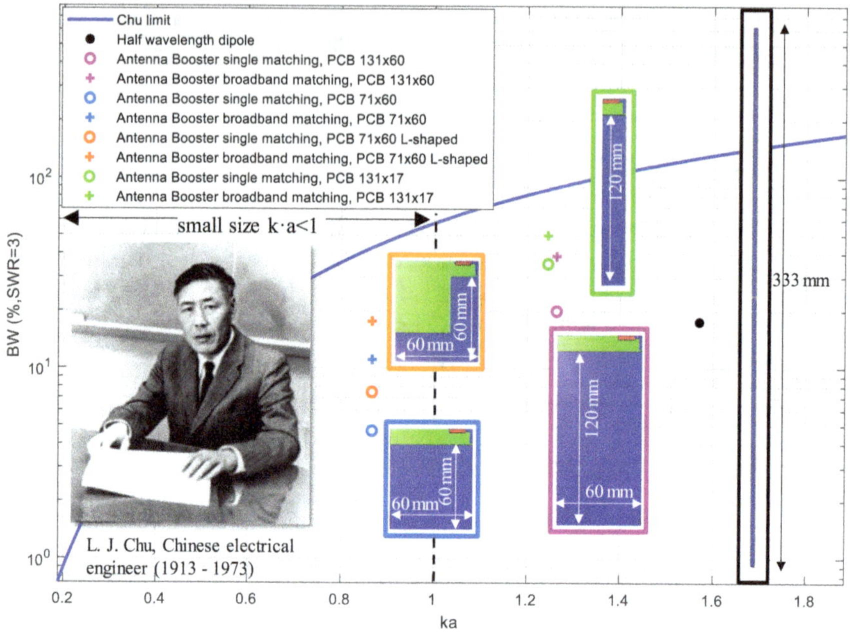

Figure 4.50 Chu bandwidth limit for SWR = 3. The k·a product is determined using a as the radius of the smallest sphere that contains the ground plane and the antenna booster. Figures at scale. The antenna booster is 12 mm × 3 mm × 2.4 mm at 900 MHz.

enhancement of 38.8/20.0 = 1.94 and 50.0/35.5 = 1.4 for the 60 mm and 17 mm width ground planes, respectively.

When the length of the ground plane decreases from 120 mm to 60 mm, bandwidth decreases. For the 60 mm × 60 mm ground plane, bandwidth is BW = 4.7% and 11.1% for the single and broadband matching network cases, respectively. If we compare with the 120 mm × 60 mm ground plane, it is concluded that reducing the area by reducing the length of the ground plane results in a drastic reduction from 20.0% to 4.7% for the single matching network, which is around four times less bandwidth. This reduction is much more than what was expected with the Chu limit. In effect, if the Chu bandwidth limit is computed for the $k{\cdot}a$ corresponding to the 120 mm × 60 mm ground plane and the 60 mm × 60 mm ground plane at 900 MHz, the bandwidth reduction is 2.1.

To enlarge the current path, an L-shaped ground plane is compared. The L-shaped ground plane can be inscribed in a square of 60 mm × 60 mm. As observed, the bandwidth for the L-shaped ground plane with a single matching network is BW = 7.5% instead of 4.7% for the 60 mm × 60 mm ground plane, which is an appropriate increment. Finally, bandwidth is increased to 17.7% for the L-shaped ground plane, with a broadband matching network having a bandwidth similar to

120 mm × 60 mm but with half the area. Once again, the size and shape of a ground plane play an essential role in antenna booster technology.

4.14 Conclusions

Table 4.6 answers the objectives given at the beginning of the chapter. It also points to relevant figures and equations to consolidate the concepts learnt. It is also advisable to refresh all essentials included in the grey boxes distributed along the chapter.

Table 4.6 Key aspects to take away

Topic	**Review and take away**
Matching networks for antenna boosters	To impedance match and also to increase bandwidth (Figure 4.1). Not all matchings provide the same bandwidth (Figure 4.6) and robustness to tolerances (Figure 4.7).
Tolerances	Tolerance of components (inductors and capacitors) should be tight to provide a robust response in terms of matching and total efficiency. A 2% tolerance or less is recommended (Figure 4.7).
Quality factor (Q_m) of the components of the matching network	The higher the Q_m, the better. Losses of the matching network depend on the relation of Q_a/Q_m (Equation (4.24)), the lower, the better. If $Q_m = Q_a$, the matching network will introduce about 3 dB losses. This is critical for ground plane lengths less than ~0.2λ (Figure 4.11).
Bandwidth enhancement with matching networks	A broad matching network not only matches the impedance but also increases the inherent bandwidth of the antenna booster radiating system. From a practical perspective, the bandwidth can be increased by a factor of $F = 2.45$ for an $SWR = 3$ or $S_{11} = -6$ dB (Equation (4.43), Figures 4.23 and 4.28).
Human body	The human body, particularly any close object, typically detunes the impedance and absorbs power. While the power absorption can be minimised by spacing the antenna booster radiating system from the body, the impedance mismatch, if any, can be recovered by the design of the matching network considering the real environment where the device will operate.
Matching network synthesis	Both simple and broadband matching networks for antenna boosters can be designed with software tools based on smart algorithms (e.g., genetic). This is a simple and fast method for matching at the desired frequency band.

(Continues)

Table 4.6 (Continued)

Topic	Review and take away
Reuse of a matching network	A matching network designed for a specific PCB can be reused for other sizes (Figures 4.42 and 4.45). Naturally, an *ad-hoc* matching network for a given PCB optimises performance. However, the reuse of a matching network enables the concept of easy to utilise without any further adjustment depending on the form factor.
Optimum ground plane size	Based on the results of this chapter, the size of a rectangular ground plane ($L \times W$) that maximises bandwidth at a given frequency f is $L = 0.42\lambda$ and $W = 0.06\lambda$ (Figure 4.43).
Loop-antenna boosters	When dealing with narrow bandwidth applications and small ground planes, the loop antenna booster allows to match with capacitors only instead of inductors. Since capacitors typically feature larger Q_m than inductors, matching network losses are minimised and thus, better total efficiency is obtained.
Chu bandwidth limit	The size of the ground plane matters, but also the shape to increase bandwidth (Figure 4.50). The smaller the ground plane size compared to the wavelength, the smaller the bandwidth.
Bandwidth enhancement	After the teachings of this chapter and Chapter 3, there are six routes to increase the bandwidth: 1. Antenna booster volume (Figure 3.37, Chapter 3) 2. Size and shape of the ground plane (Figure 3.54, 3.61, Chapter 3) 3. Including slots in the ground plane (Figure 3.63) 4. Broadband matching networks (Figure 4.28) 5. Non-Foster matching networks (Figure 4.46) 6. Distributed antenna boosters (Figure 4.47) From a practical perspective, route four is simple and effective. It has been concluded that a simple broadband matching network can enhance the bandwidth about 2.45 times (Equation (4.43), Figure 4.23, Figure 4.28).
From broadband to dual-band	The broadband mechanism consisting of an LC resonator can be expanded to design dual-band antenna systems by enlarging the impedance loop.

4.15 Exercises

A 3D electromagnetic software and a circuit simulator, including matching network synthesis, are needed to solve the following exercises.

4.15.1 Exercise 1

Determine total efficiency η_t with Equation (4.14) considering η_r = 100% for an antenna booster as in Figure 4.51 where the impedance at 900 MHz is Z_a = 37.1−208$j\Omega$ and the matching network in Figure 4.10 using components with finite Q taken from a datasheet or consider a Q_m = 80. You can obtain the [S^m] analytically with a circuit simulator. Repeat the exercise when the impedance corresponds to an antenna booster with a 60 mm × 60 mm ground plane given by Z_a = 8.0−206$j\Omega$. Use the matching as in Figure 4.10, keeping the same Q_m = 80. Compare the results to those obtained in Section 4.5. Why the total efficiency is lower in the second case if the η_r = 100% for both cases and with perfect impedance matching?

4.15.2 Exercise 2

Using the schematic in Figure 4.51 comprising a 12 mm × 3 mm × 2.4 mm antenna booster and a 120 mm × 60 mm ground plane, design a matching network for operating at 863 MHz to 928 MHz. Compute S_{11}, total efficiency, and radiation patterns. Repeat the exercise for working at 1.2–1.3 GHz with the same antenna booster and only changing the matching network. Note that a dielectric substrate is included below the ground plane. Consider this fact for your electromagnetic simulations for this and the following exercises.

4.15.3 Exercise 3

For the previous exercise 4.15.2, run tolerance analysis considering matching network components with 5% tolerance. Repeat with 2%.

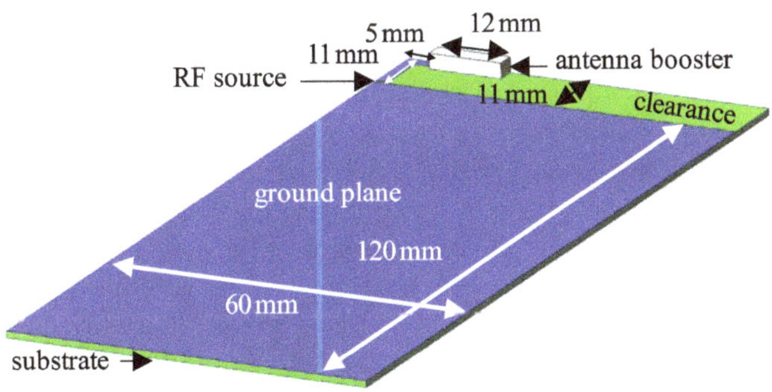

Figure 4.51 A 12 mm × 3 mm × 2.4 mm (height) conductive (copper) parallelepiped modelling an antenna booster on a 120 mm × 60 mm ground plane. The feeding line is 2 mm in width and 10.5 mm in length in the vertical direction. A final 5 mm strip feeds the antenna booster. A 0.5 mm gap between the feeding line and the ground plane allocates an RF source. The substrate is 1 mm thick with ε_r = 4.15 and tanδ = 0.017.

4.15.4 Exercise 4

For the exercise in Section 4.15.2, design a matching network with lumped components having Q_m less than 20 and another matching with $Q_m > 60$. Analyse the impact on total efficiency. Why does the second provide better total efficiency?

4.15.5 Exercise 5

Consider Figure 4.4 with a 120 mm × 60 mm ground plane and two other situations with a 60 mm × 60 mm ground plane and a 40 mm × 60 mm ground plane. Compute Q_a across 0.6–1 GHz using Equation (4.26). Analyse how Q_a increases as the ground plane size decreases. Compute bandwidth for each case according to Equation (4.28) for an $SWR = 3$.

4.15.6 Exercise 6

Given the Q_a obtained in exercise 4.15.5, calculate losses introduced by a matching network (η_m) if Q_m is 10, 20, 50, 100, and 200. Refer to Equation (4.24) to compute η_m. What is the minimum Q_m for each ground plane to have losses η_m no greater than 0.5 dB? And 1 dB?

4.15.7 Exercise 7

Consider Figure 4.51 but with a 50 mm × 60 mm ground plane. Compute potential bandwidth from 0.6 GHz to 1 GHz. Can we cover 30% centred at 850 MHz with $S_{11} \leq -6$ dB? If not, how many matching networks with only two components to obtain $S_{11} \leq -6$ dB are needed? How many matching networks do you need to cover from 698 MHz to 960 MHz? How do you think the problem can be addressed?

4.15.8 Exercise 8

Given a series RC circuit with $R_a = 50/3$ and $C_a = 0.26$ pF representing the impedance of an antenna booster at $f_0 = 1.57$ GHz. Use a series $L_a = 38.9$ nH. Represent the input impedance across $0.5{\cdot}f_0$ to $1.5{\cdot}f_0$. What is the resonant frequency? Although the impedance is not matched, compute the bandwidth at $S = SWR = 3$ with Equations (4.28) and (4.32). Design a broadband matching network with an LC shunt using Equations (4.39)–(4.41). Represent the input impedance and compute the $SWR = 3$ bandwidth. Compare it with the previous case. What is the bandwidth enhancement factor F? Compare it with Equation (4.43).

4.15.9 Exercise 9

Consider Figure 4.51 but with a 60 mm × 60 mm ground plane. Compute Q_a from 0.6 GHz to 1 GHz. What is the potential bandwidth at 900 MHz for $S_{11} \leq -6$ dB? Use Equations (4.26) and (4.28). Based on the broadband concepts explained in Section 4.8, how much bandwidth can be enhanced considering $S_{11} \leq -6$ dB? Design a broadband matching network and compare the obtained bandwidth with the estimated value.

4.15.10 Exercise 10

For the layout shown in Figure 4.13, include a 3 mm × 2 mm × 1 mm (height) antenna booster and a matching network to provide operation at 2.4–2.5 GHz. Compute bandwidth and total efficiency in a free space scenario. Repeat the exercise including a human arm or a dielectric body where the ground plane is spaced 1 mm and 3 mm. How do bandwidth and total efficiency change with distance? Why?

4.15.11 Exercise 11

Consider Figure 4.51 with the same 120 mm × 60 mm ground plane. Design a matching network with only two components to provide operation from 860 MHz to 930 MHz with an $S_{11} < -6$ dB. Change the ground plane to 200 mm × 200 mm, 80 mm × 60 mm, and 60 mm × 60 mm and use the same matching network. Analyse if the exact matching network can provide performance with an $S_{11} < -6$ dB.

4.15.12 Exercise 12

Repeat the exercise in Section 4.15.11 but with a broadband matching network using the teaching of Section 4.8. Use a human-engineered matching network with the LC resonator and a software-engineered matching network. Compare the usability for all ground planes between $L = 20$ mm, $W = 20$ mm and $L = 200$ mm, $W = 200$ mm. Does the broadband matching network improve the reuse compared to a single matching with only two components? Design the reference broadband matching network for the 120 mm × 60 mm ground plane.

4.15.13 Exercise 13

For Figure 4.51, run a matching network synthesiser considering a matching network with 3, 5, 7, 9, and 11 lumped components. Determine the maximum bandwidth that can be obtained centred at 800 MHz. Use ideal components and repeat the exercise with real components considering its measured S data. Compute total efficiency for the real component case for each scenario. Is it worth using many components? Run a tolerance analysis as well.

4.15.14 Exercise 14

Consider a MIMO wireless device embedding a 7 mm × 3 mm × 1 mm antenna booster, including a 200 mm × 200 mm ground plane. The antenna booster is shifted to have a 2 mm gap from the ground plane edge. Place one antenna booster at each corner. Design a matching network to operate from 2.4 GHz to 2.5 GHz. Compute S_{ii} and S_{ij} as well as total efficiencies at each port.

4.15.15 Exercise 15

Design a circular ground plane with the same area as the 120 mm × 60 mm ground plane in Figure 4.51. Compute the bandwidth centred at 900 MHz using a single (one or two components) and a broadband matching network. Compare the results with the rectangular ground plane. Repeat the exercise with a circular ground plane having the same area as the 60 mm × 60 mm ground plane. In all cases, put the antenna booster as shown in Figure 4.51.

4.15.16 Exercise 16

For the 60 mm × 60 mm ground plane in Figure 4.50, add another 12 mm × 3 mm × 2.4 mm antenna booster at the left corner. Utilise a series inductor at each port of the antenna booster to bring them to resonance at 900 MHz. Add two 30 mm-length 50 Ω transmission lines connected to a common port. Simulate the input impedance and then design a broadband matching network. Compute the bandwidth and compare it with the results given in Figure 4.50.

References

[1] J. Anguera, A. Andújar, C. Puente, and J. Mumbrú, "Antennaless Wireless Device," *Patent US 8*,203,492 B2, June 2012.

[2] D. M. Pozar, "Microwave Engineering," Hoboken, NJ: Wiley, 2012.

[3] G. Smith, "Efficiency of Electrically Small Antennas Combined with Matching Networks," *IEEE Transaction on Antennas and Propagation*, vol. 25, no. 3, pp. 369–373, 1977.

[4] S. R. Best, "The Inverse Relationship between Quality Factor and Bandwidth in Multiple Resonant Antennas," *IEEE Antennas and Propagation Society International Symposium*, 623–626, 2006.

[5] J. Anguera, A. Andújar, G. Mestre, J. Rahola, and J. Juntunen, "Design of Multiband Antenna Systems for Wireless Devices Using Antenna Boosters," *IEEE Microwave Magazine*, vol. 20, no. 12, pp. 102–114, 2019, doi:10.1109/MMM.2019.2941662.

[6] J. Rahola, "Optimization of Matching Circuits for Antennas," *European Conference on Antennas and Propagation*, EuCAP 2011, Rome, April 2011.

[7] J. Rahola, "Power Waves and Conjugate Matching," *IEEE Transactions on Circuits and Systems—II: Express Briefs*, vol. 55, pp. 92–96, 2008.

[8] J. Rahola, "Estimating the Performance of Matching Circuits for Antennas," *European Conference on Antennas and Propagation*, EuCAP 2010 Conference, Barcelona, April 12–16, 2010.

[9] J. Anguera and A. Andújar, C. Puente, and J. Mumbrú, "Antennaless Wireless Device," *Patent US 9*,130,259 B2

[10] J. Anguera, A. Andújar, "Ground Plane Booster Antenna Technology for Wearable Devices," *Patent US 10*,734,713 B2, Nov. 2017.

[11] D. Wu, S. W. Cheung, Q. L. Li, and T. I. Yuk, "Slot Antenna for All-metal Smartwatch Applications," *European Conf. Antennas Propagation*, 1144–1147, Davos, Switzerland, 2016.

[12] R. M. Fano, "Theoretical Limitations on the Broadband Matching or Arbitrary Impedances," *Journal of Franklin Institute*, vol. 249, no. 1–2, pp. 57–83 and 139–154, 1950.

[13] A. R. López, "Review of Narrowband Impedance-Matching Limitations," in *IEEE Antennas and Propagation Magazine*, vol. 46, no. 4, pp. 88–90, 2004.

[14] S. Risco, J. Anguera, A. Andújar, A. Pérez, and C. Puente, "Coupled Monopole Antenna Design for Multiband Handset Devices," *Microwave and Optical Technology Letters*, vol. 52, no. 10, pp. 359–364, 2010.

[15] P. Huang and L. Huang, "Design of an Antenna with Ground Plane Slots for LTE/WWAN Mobile Phones," *Progress in Electromagnetics Research Letters*, vol. 83, pp. 15–22, 2019.doi:10.2528/PIERL19012701.

[16] A. Chen, M. Sun, Z. Zhang, and X. Fu, "Planar Monopole Antenna with a Parasitic Shorted Strip for Multistandard Handheld Terminals," *IEEE Access*, vol. 8, pp. 51647–51652, 2020, DOI:10.1109/ACCESS.2020. 2979483.

[17] C. Deng, Y. Li, Z. Zhang, and Z. Feng, "Planar Printed Multi-Resonant Antenna for Octa-Band WWAN/LTE Mobile Handset," *IEEE Antennas and Wireless Propagation Letters*, vol. 14, pp. 1734–1737, 2015, doi:10.1109/ LAWP.2015.2421335.

[18] J. Anguera, C. Puente, C. Borja, G. Font, and J. Soler "A Systematic Method to Design Single-Patch Broadband Microstrip Patch Antennas," *Microwave and Optical Technology Letters*, vol. 31, no. 3, 2001, pp. 185–188.

[19] A. Andújar, J. Anguera, and C. Puente, "A Systematic Method to Design Broadband Matching Networks," *European Conference on Antennas and Propagation*, EuCAP 2010, Barcelona, Spain, pp. 1–5, April 2010.

[20] R. L. Thomas, "*A Practical Introduction to Impedance Matching*," Artech House, 1976.

[21] B. S. Yarman, "*Design of Ultra-Wideband Antenna Matching Networks: Via Simplified Real Frequency Technique*," Springer Science & Business Media, Aug 25, 2008 - Technology & Engineering.

[22] G. Pérez, "Broadbanding Techniques for Antenna Boosters," *Engineering Degree Thesis*, La Salle Engineering-Universitat Ramon Llull. Directed by J. Anguera and A. Andújar, 2022.

[23] J. Anguera, A. Andújar, and C. García, "Multiband and Small Coplanar Antenna System for Wireless Handheld Devices," *IEEE Transactions on Antennas and Propagation*, vol. 61, no. 7, pp. 3782–3789, 2013.

[24] C. W. Yang, Y. B. Jung, and C. W. Jung, "Octaband Internal Antenna for 4G Mobile Handset," *IEEE Antennas and Wireless Propagation Letters*, vol. 10 pp. 817–819, 2011.

[25] C. Rowel, and E. Y. Lam, "Mobile-Phone Antenna Design," *IEEE Antennas and Propagation Magazine*, vol. 54, no. 4, pp. 14–34, 2012.

[26] J. Jeon, Y. Jung, and M. Kang, "Software-Based Adaptive Protection Control against Load Mismatch for a Mobile Power Amplifier Module." *Electronics*, vol. 8, 1226 2019 https://doi.org/10.3390/electronics8111226

[27] J. Gui, A. Andújar, and J. Anguera, "On the Reuse of a Matching Network for IoT Devices Operating at 900MHz Embedding Antenna Boosters," *Electronics*, vol. 11, 1267 2022. https://doi.org/10.3390/electronics11081267.

[28] S. E. Sussman-Fort and R. M. Rudish, "Non-Foster Impedance Matching of Electrically-Small Antennas," *IEEE Transactions on Antennas and Propagation*, vol. 57, no. 8, pp. 2230–2241, 2009, doi:10.1109/TAP.2009.2024494.

[29] J. T. Aberle and R. Loepsinger-Romak, "*Antennas with Non-Foster Matching Networks*," Morgan & Claypool Publishers, 2007.

[30] J. Anguera, A. Camps, A. Andújar, and C. Puente, "Enhancing the Robustness of Handset Antennas to Finger Loading Effects," *Electronics Letters*, vol. 45, no. 15, pp. 770–771, 2009.

[31] J. Anguera, A. Andújar, Y. Cobo, C. Picher, and C. Puente, "Handset Antenna Array to Mitigate the Finger Loading Effect," *European Conference on Antennas and Propagation*, EuCAP 2011, pp. 611–614, Rome, Italy.

[32] A. Andújar, J. Anguera, Y. Cobo, and C. Picher, "Distributed Antenna Systems for Wireless Handheld Devices," *IEEE Transactions on Antennas and Propagation*, vol. 60, no. 10, pp. 4830–4837, 2012.

[33] A. Andújar, J. Anguera, J. L. Leiva, and C. Picher, "In-phase versus Outphase Distributed Antenna Systems for Wireless Handheld Devices," *IEEE Transactions on Antennas and Propagation*, vol. 61, no. 1, pp. 346–353, 2013.

[34] A. Andújar, J. Anguera, and Y. Cobo, "Distributed Systems Robust to Hand Loading Based on Non-Resonant Elements," *Microwave and Optical Technology Letters*, vol. 55, no. 10, pp. 2307–2317, 2013.

[35] A. Andújar and J. Anguera, "Scattered Virtual Antenna Technology for Wireless Devices," *Patent US 10*,062,973 B2, Aug. 2018.

[36] J. Anguera, A. Andújar, and J. L. Pina, "Loop Booster for Small IoT Devices," *Patent Application* US US63339992

[37] L. J. Chu, "Physical Limitations of Omni-directional Antennas," *Journal of Applied Physics*, vol. 19, no. 12, pp. 1163–1175, 1948.

[38] M. Capek, M. Gustafsson, and K. Schab, "Minimization of Antenna Quality Factor," *IEEE Transactions on Antennas and Propagation*, vol. 65, no. 8, pp. 4115–4123, 2017, doi:10.1109/TAP.2017.2717478.

[39] D. F. Sievenpiper, D. C. Dawson, M. M. Jacob, *et al.*, "Experimental Validation of Performance Limits and Design Guidelines for Small Antennas," *IEEE Transactions on Antennas and Propagation*, vol. 60, no. 1, pp. 8–19, 2012, doi:10.1109/TAP.2011.2167938.

Chapter 5
Antenna booster technology for multiband operation

This chapter introduces Antenna Booster Technology for wireless devices requiring a multiband operation. The term multiband refers here to frequency bands divided into at least two different frequency regions, for example, a first frequency region $[f_1, f_2]$ and a different frequency region $[f_3, f_4]$, with a frequency gap between both regions. For example, a first frequency region $[f_1, f_2]$ = 698–960 MHz and a second frequency region at $[f_3, f_4]$ = 1710–2690 MHz. This is referred to hereafter as multiband. In a similar manner as in Chapter 4, the multiband operation is achieved not by shaping the geometry of a complex antenna but by using antenna booster architectures in combination with matching circuits. Besides matching networks, other circuits, such as resonators and filters, are used for the multiband operation. The concepts and design explained throughout the chapter can be extended to frequency regions other than the ones shown here.

The chapter has the following objectives:

1. To understand the design of multiband wireless devices with antenna boosters using multiband matching networks
2. To get familiar with the design of automatic multiband matching networks
3. To know the impact of the ground plane on achieving multiband performance
4. To realise that the quality factor of components of the matching network and tolerance should be considered to achieve efficient and robust designs
5. To be aware of the frequency range of operation of lumped inductors and capacitors
6. To be able to design multiband matching networks with distributed elements
7. To be able to scale designs to any frequency range

5.1 Introduction

Besides single-band applications, wireless devices also need to operate at different bands at different frequency regions. Radiofrequency transceivers can support multiband application features, typically in a single input/output port, and therefore, the antenna system should feature a single input port. But, if the transceiver can support a first frequency region in one port and the other in a second port, the architectures explained in Chapter 4 are applied. In effect, one antenna booster and a matching network can provide operation at the first frequency region. In the same manner, a second antenna booster and another matching network are designed to provide operation at the second frequency region.

The first situation is addressed here, which needs to provide multiband operation with a single port solution and, therefore, be compatible with those transceivers supporting multiband operation in a single input/output port.

This chapter shows several strategies (Figure 5.1). The first one includes one antenna booster for each frequency region and a broadband matching network. Then, both ports are combined into a single port using notch filters with SMD components (Section 5.2). The advantage of this architecture is that behaviour can

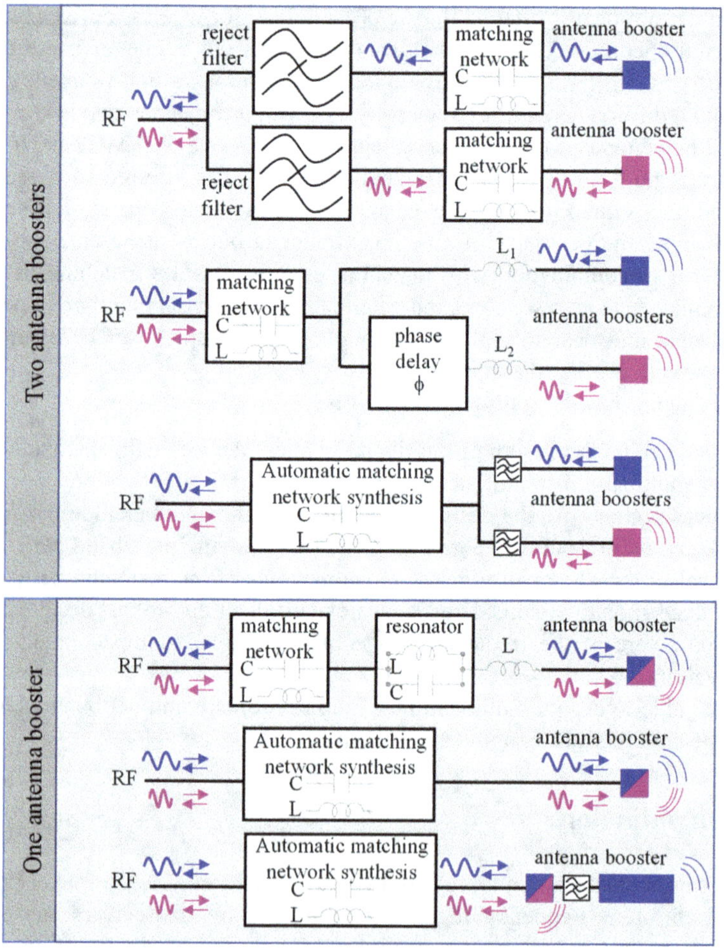

Figure 5.1 Multiband architectures with antenna boosters. RF represents the radiofrequency path to connect a multiband source such as a transceiver.

be independently optimised for each frequency region. However, it needs two antenna boosters and a combined circuit.

To simplify the number of lumped components, another architecture is proposed with two antenna boosters and a simplified matching network architecture (Sections 5.3 and 5.4).

A new set of architectures based on a single antenna booster is presented. A single antenna booster solution with a multiband matching network is shown to reduce the number of antenna boosters from two to one. This simplifies the matching network architecture and reduces the number of antenna boosters from two to one (Section 5.5). These first two architectures (Figure 5.1) employ human engineered matching networks, whereas the last one uses software synthesis. In this case, a network synthesis procedure (Section 5.6) automatically obtains the multiband matching network, following the same idea of a single booster and one matching network as the second one. Automatic matching network synthesis presents more flexibility since the software will end up with a solution. If there is no solution, the software approach will provide the matching network with the closest performance to the target specifications. Multiband matching networks with lumped elements as well as distributed elements are discussed. One more architecture comprising an antenna booster module divided into two parts is presented in Section 5.7.

Multiband 4 × 4 MIMO (Multiple Input Multiple Output) is addressed in this chapter using the architecture presented in Section 5.6 (Section 5.8).

Note that this section covers multiband architectures and not multi-radio, that is, frequency bands belonging to different communications protocols. For example, suppose performance is sought in global navigation systems and short-range communication. In that case, a different strategy is analysed in Chapter 6, where instead of using one antenna booster for each radio protocol, a single antenna booster is employed and split into two ports to be connected to each input/output port of the different transceivers. The topic of multi-radio is studied in Chapter 6.

The examples illustrated throughout the chapter can be scaled to other frequency regions following the well-known rules of frequency scaling of Maxwell equations.

The conclusion part summarises the main findings (Section 5.9). Besides the contents, a list of exercises is proposed at the end of the chapter (Section 5.10), together with a references section at the end.

5.2 Multiband performance with two antenna boosters

The following architecture takes the broadband solutions analysed in Chapter 4 to have one antenna booster operating at one frequency region and another at a second frequency region [1–4]. For illustration purposes, the first frequency region (low-frequency region) comprises 824–960 MHz, and the second frequency region comprises (high-frequency region) 1710–2170 MHz (Figure 5.2). It will be easy for the readers to expand this present architecture to other frequency regions. Each

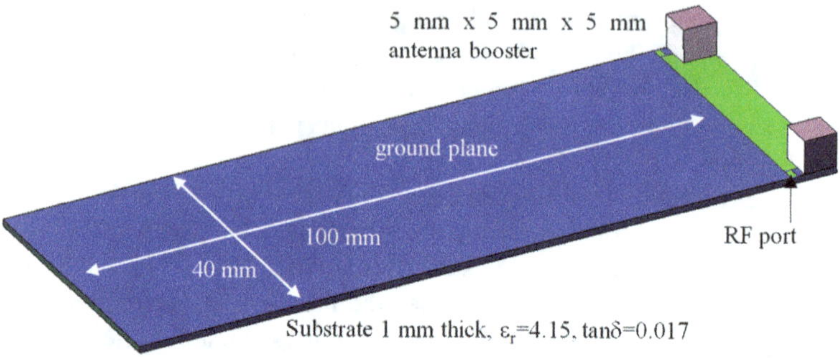

Figure 5.2 Detailed view of the radiating structure obtained from the simulated layout comprising two non-resonant 5 mm × 5 mm × 5 mm ground plane boosters (conductive cubes with σ = 5.9 × 10^7 S/m) and a 100 mm × 40 mm ground plane. The ground plane boosters are located at a 2 mm distance from the edge of the ground plane.

5 mm × 5 mm × 5 mm antenna booster is placed at the corner of a 100 × 40 mm ground plane since the corner of the ground plane typically provides the best bandwidth and efficiency (Section 3.6, Chapter 3), and both boosters are spaced only 2 mm from the edge of the ground plane. Antenna boosters are modelled with a 3D conductive cube ($\sigma = 5.9 \times 10^7$ S/m). The ground clearance is 40 mm × 7 mm.

The quality factor (Q_a) of the antenna booster alone in a monopole configuration (infinite ground plane), as explained in Chapter 3, is computed, resulting in $Q_a = 2{,}250$ and $Q_a = 265$ for 900 MHz and 2000 MHz, respectively. This shows that an antenna booster alone is a very high-Q element. Once coupled to a ground plane, it can excite radiating currents, where the Q of the ground plane mainly determines the final Q. Therefore, an antenna booster is not considered an antenna, as its high Q confirms, but an exciter of the ground plane.

The impedance of the antenna booster is non-resonant across the whole frequency region (Figure 5.3). In both frequency ranges, an L-series generates a response akin to that of a series RLC circuit, while a broadband matching network forms a condensed impedance loop (Figure 5.3).

Once the bandwidth has been optimised at each frequency region, two ports are connected to a single port using notch filters (Figure 5.4). The top part provides a high impedance at the high-frequency region; similarly, the bottom part provides a high impedance at the low-frequency region. With this architecture, both parts can be connected with a minimum loading impact on each other. It is important to note that the notch-filter scheme comprises two stages: a first series LC in shunt and a second stage with a shunt LC. Although a single stage may be enough, two stages provide more selectivity, and the impedance loading of one frequency region over the other is minimised. Furthermore, the second stage is a shunt LC circuit, which is the same as the broadband matching network, and therefore, both can be merged into a single LC circuit. In this manner, the shunt LC works to broaden the

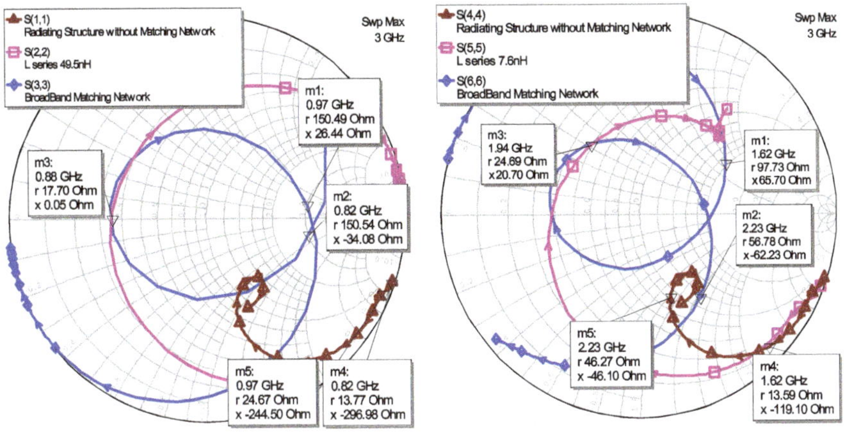

Figure 5.3 (Left) The input impedance of the antenna booster, including the ground plane without a matching network, with a series L = 49.5 nH, and with the broadband matching network using a shunt LC for providing operation at the first frequency region; (right) same as left but to provide operation at the second frequency region

bandwidth and minimise the loading of the other frequency region. In the end, ten SMD components are needed to provide the multiband performance with a single port. Note that an off-the-shelf diplexer could have been used too.

A prototype is built to validate the proposal (Figure 5.5). S_{11} and total efficiency are measured with a vector network analyser and anechoic chamber, respectively (Figure 5.6). As for the S_{11}, it is below −6 dB for both frequency regions. Total efficiency averaged in the low- and high-frequency regions is 35.3% and 61.3%, respectively.

Antenna boosters and broadband matching circuits

The impedance nature of an antenna booster is suitable for adding a broadband matching circuit. By combining two antenna boosters at two different frequency regions, a multiband behaviour is obtained where each frequency region is independently adjusted.

Other variations of this concept make use of the same two antenna boosters but place one close to the other [5–8], resulting in a compact solution since both antenna boosters are located in the same corner of the wireless device (Figure 5.7). The architecture follows the architecture explained above, but in this case, the two boosters are close to one another, including reject filters to improve isolation. Two ports are provided for each frequency range. If needed, both ports can be merged either by a combiner with lumped elements, as shown in Figure 5.4 or using an

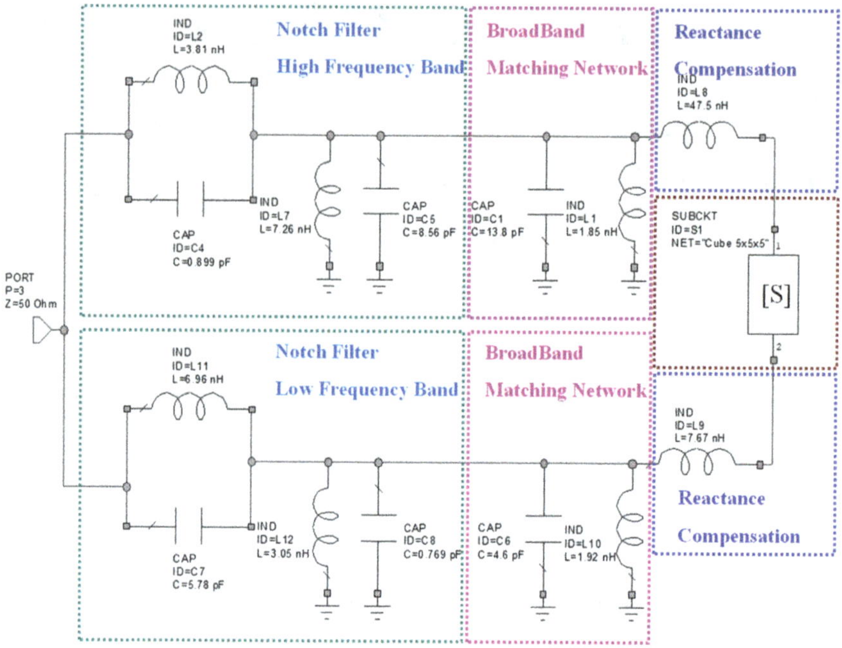

Figure 5.4 Matching network architecture comprising a reactance cancellation element, a broadband matching network, and a notch filter for each frequency region. Note that the first stage of the notch filter and the broadband matching network can be simplified using only two components. The two-port box is the simulated input impedance of the antenna booster system in Figure 5.2.

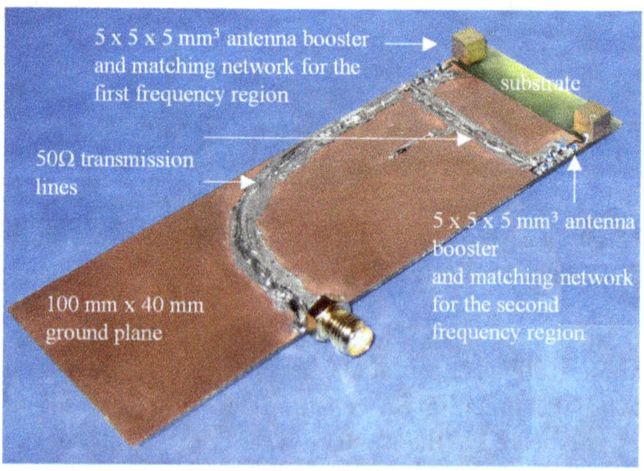

Figure 5.5 Two 5 mm × 5 mm × 5 mm antenna boosters with their respective matching network connected with a diplexer using reject filters

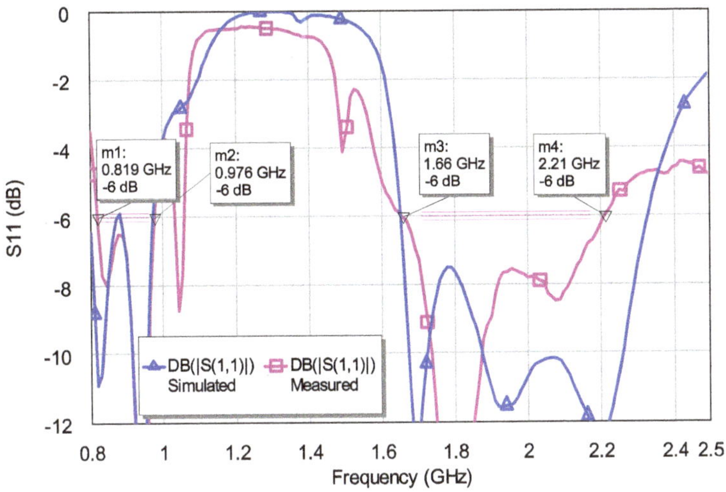

Figure 5.6 Simulated and measured S_{11} of the prototype of Figure 5.5

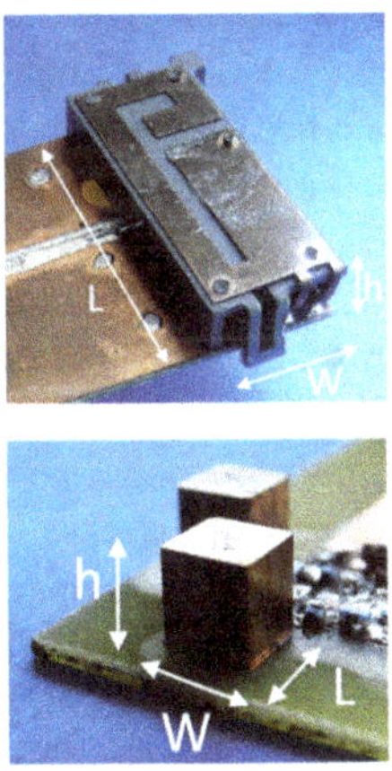

	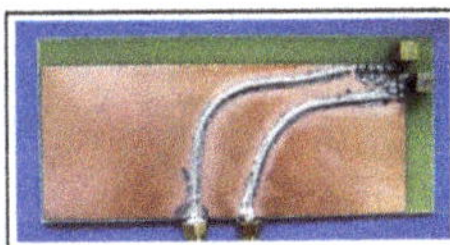	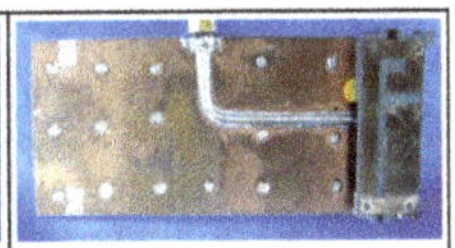
Height (h)	5 mm (1 booster)	6 mm
Length (L)	5 mm (1 booster)	40 mm
Width (W)	5 mm (1 booster)	15 mm
Volume	250 mm³ (2 boosters)	3600 mm³
Ground plane length	100 mm	100 mm
Ground plane width	40 mm	40 mm
BW_{f1-f2} [ROE ≤ 3]	16.4 %	14.6 %
BW_{f3-f4} [ROE ≤ 3]	25 %	8.2 %
$\overline{\eta_t}$ (824 – 960 MHz)	44 %	49.7 %
$\overline{\eta_t}$ (1710 – 2170 MHz)	55 %	44.2 %

Figure 5.7 A comparison between a multiband PIFA (Printed Inverted-F Antenna) and a multiband solution with two antenna boosters

off-the-shelf diplexer. This architecture is also helpful for multi-radio applications where each port operates on a different radio protocol. Both ground planes feature the same 100 mm × 40 mm size. As shown, the total efficiency (η_t) averaged at 824–960 MHz and 1710–2170 MHz is practically the same, with the advantage that the antenna booster solution occupies 14 times less volume. Antenna boosters should be packaged in a small PCB (Printed Circuit Board) space for practical applications.

5.3 Multiband performance with two antenna boosters: a simplified matching network architecture

Ten lumped components were needed to provide multiband operation in the previous section. This section proposes a simplified scheme that reduces the number of lumped components to only four.

The proposed radiating structure comprises two antenna boosters characterised by their reduced dimensions of 5 mm × 5 mm × 5 mm and high-quality factors (Q_a ≈2,250 for the low-frequency region (824–960 MHz) and Q_a ≈265 for the high-frequency region (1710–2690 MHz)) as explained in Section 5.2 [3]. This Q_a is computed considering only the antenna booster in a monopole configuration on an infinite ground plane (Chapter 3). Said antenna boosters are distributed along the shortest edge of a 120 mm × 50 mm ground plane. They are capable of exciting ground plane modes. However, as explained throughout the book, the simple excitation of said ground plane mode is not enough to guarantee operability since a radiofrequency architecture is needed for impedance-matching purposes. The proposed radiating structure is not resonant in any of both frequency regions. Thus, it is characterised by a capacitive behaviour that can be compensated by adding a series inductance (Figure 5.8(a)).

Unlike the radiofrequency system proposed in Section 5.2, in this case, neither broadband matching networks nor notch filters are needed to provide multiband operation. On the contrary, the proposal attains compact impedance loops centred in the circle $SWR \leq 3$ by the combination of adequately shifted input impedances achieved by adding a transmission line [9,10].

More specifically, the proposal relies on providing reactance cancellation stages at the input port of both antenna boosters to make them resonant at the centre frequencies of the low and high-frequency regions (824–960 MHz and 1710–2690 MHz), respectively (Figure 5.8(a) and (b)). Said input impedances are combined into a single input/output port by adding a phase delay, in this case, a transmission line (Figure 5.8(c)). Said phase delay allows the creation of compact impedance loops in both frequency regions, which can be easily centred through the addition of a fine-tuning stage composed of an L-C circuit, thus attaining multiband operation (Figures 5.8(d) and 5.9).

To validate the simulated results, a prototype is built, and main parameters such as the reflection coefficient and total efficiency are measured (Figure 5.10). As presented in the former section, the topology of the radiofrequency system mainly relies on two reactance cancellation elements, a phase delay, and a fine-tuning stage of two elements. In this case, the phase delay primarily consists of a microcoaxial transmission line with a characteristic impedance of 50 Ω. The results demonstrate the effectiveness of the proposal since the radiating system can provide operation in low and high-frequency regions (824–960 MHz and 1710–2690 MHz) with enhanced electromagnetic performance (Figures 5.11–5.13). The average measured total efficiency is 47.7% and 66.1% at the low- and high-frequency regions, respectively.

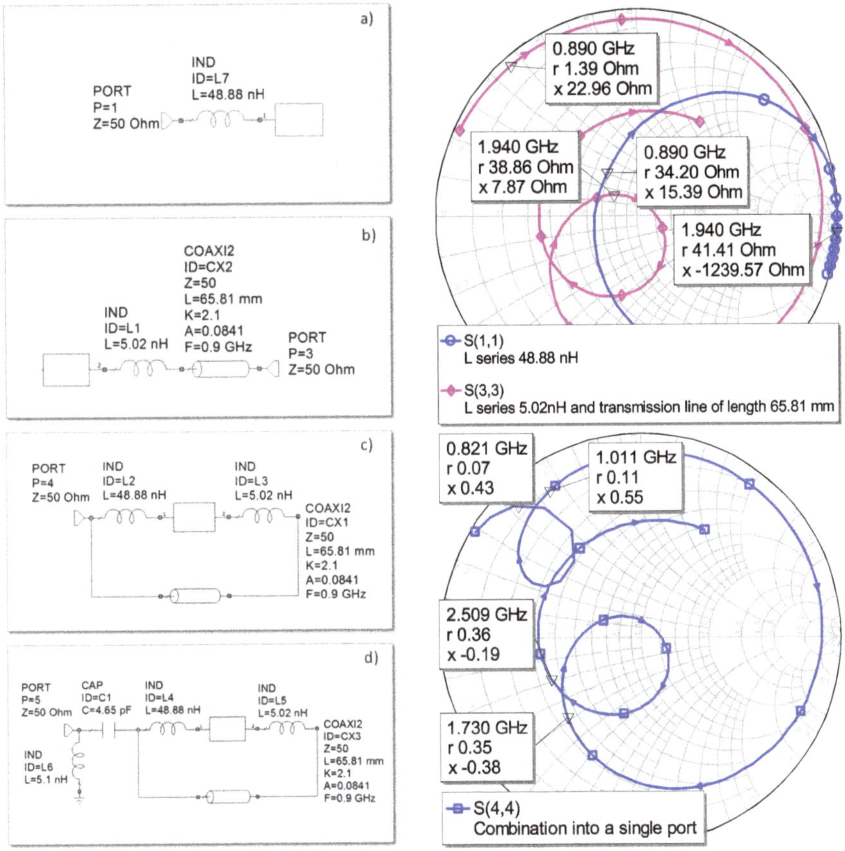

Figure 5.8 Matching network topology and input impedance response associated with the different stages that constitute the proposed radiofrequency system. (a) Reactive cancellation for the low-frequency region; (b) reactive cancellation for the high-frequency region; (c) combination into a single port; (d) fine-tuning stage

Current distributions
The current distribution simulation provides physical insight into the behaviour. It helps to understand how antenna boosters contribute to each frequency region as well as the current modes supported by the ground plane.

The antenna booster elements can effectively stimulate the efficient radiation mode linked with the conductive ground planes of wireless devices. However, said

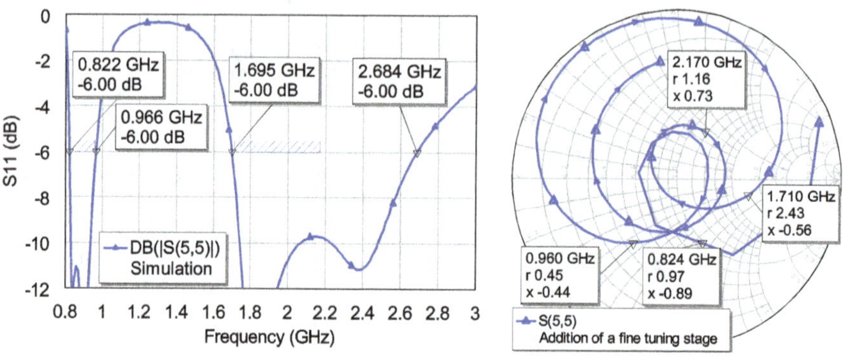

Figure 5.9 Simulated reflection coefficient and input impedance obtained from the proposed radiofrequency system depicted in Figure 5.8(d)

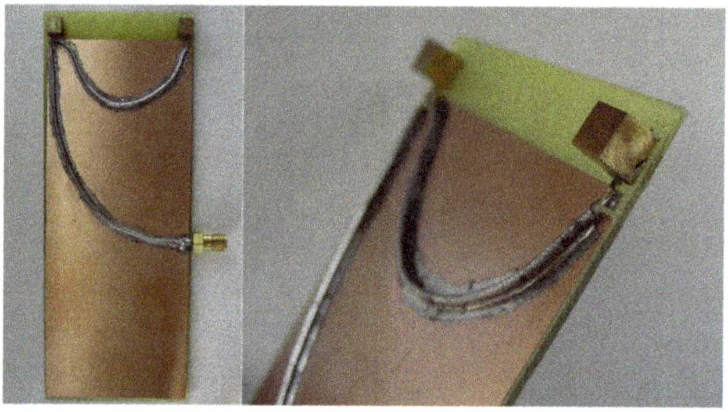

Figure 5.10 Radiating structure comprising two brass-made 5 mm × 5 mm × 5 mm antenna boosters and a 120 mm × 50 mm ground plane etched over a 1 mm thick FR4 substrate. The transmission line connecting the two antenna boosters is a 50 Ω microcoaxial line. Another 50 Ω microcoaxial line is included to feed the antenna booster system ended with an SMA connector.

excitation does not guarantee proper performance since a radiofrequency system capable of matching the resulting input impedance is required. In this sense, the appropriate design of radiofrequency systems becomes an essential part of wireless devices embedding antenna boosters. The radiofrequency system proposed in this section not only allows minimising the number of reactive components by a factor of two concerning previous solutions but also simplifies its integration while enhancing its electromagnetic performance. The proposed technique achieves the multiband operation with two small 5 mm × 5 mm × 5 mm antenna boosters and a radiofrequency system of only five reactive components.

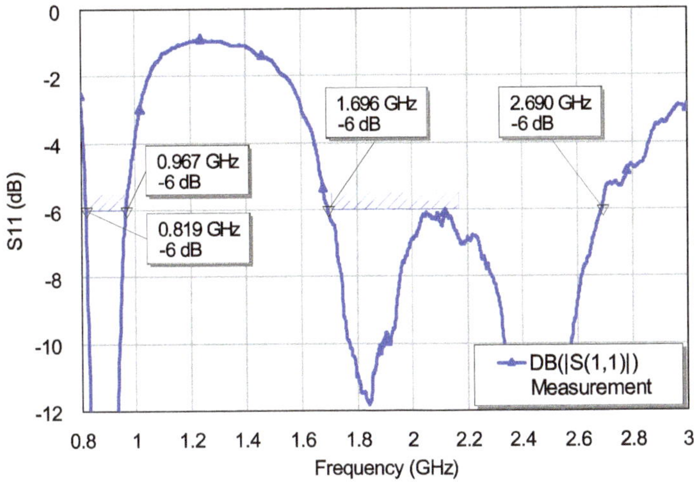

Figure 5.11 *Measured reflection coefficient (S_{11}) associated with the prototype depicted in Figure 5.10*

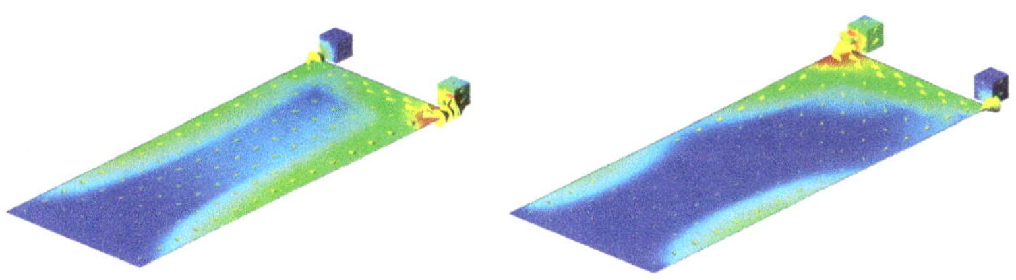

Figure 5.12 *Simulated current distribution for 900 MHz (left) and 2 GHz (right). The right antenna booster is responsible for providing operation at the low-frequency region, whereas the left antenna booster is mainly responsible for the high-frequency region.*

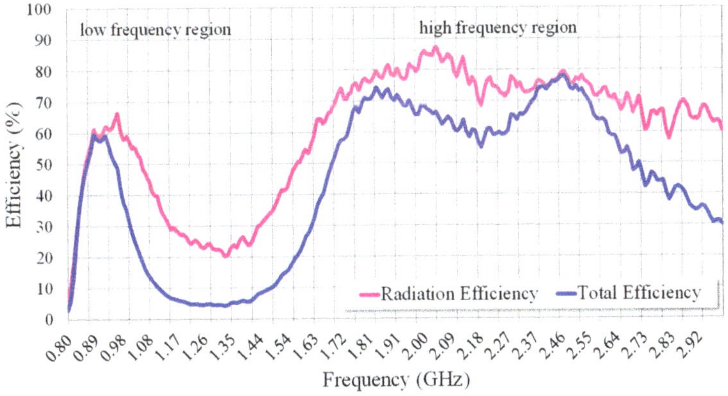

Figure 5.13 *Measured radiation (η_r) and total efficiency ($\eta_t = \eta_r \cdot (1-|S_{11}|^2)$) associated with the prototype depicted in Figure 5.10*

5.4 Multiband performance with a multi-structure antenna booster system

An architecture comprising two 12 mm × 3 mm × 2.4 mm antenna boosters, including a 120 mm × 60 mm ground plane, is studied in this section (Figure 5.14) [11]. The goal is to have one antenna booster operating in a first-frequency region and a second booster for a second-frequency region. A notch filter at each one is

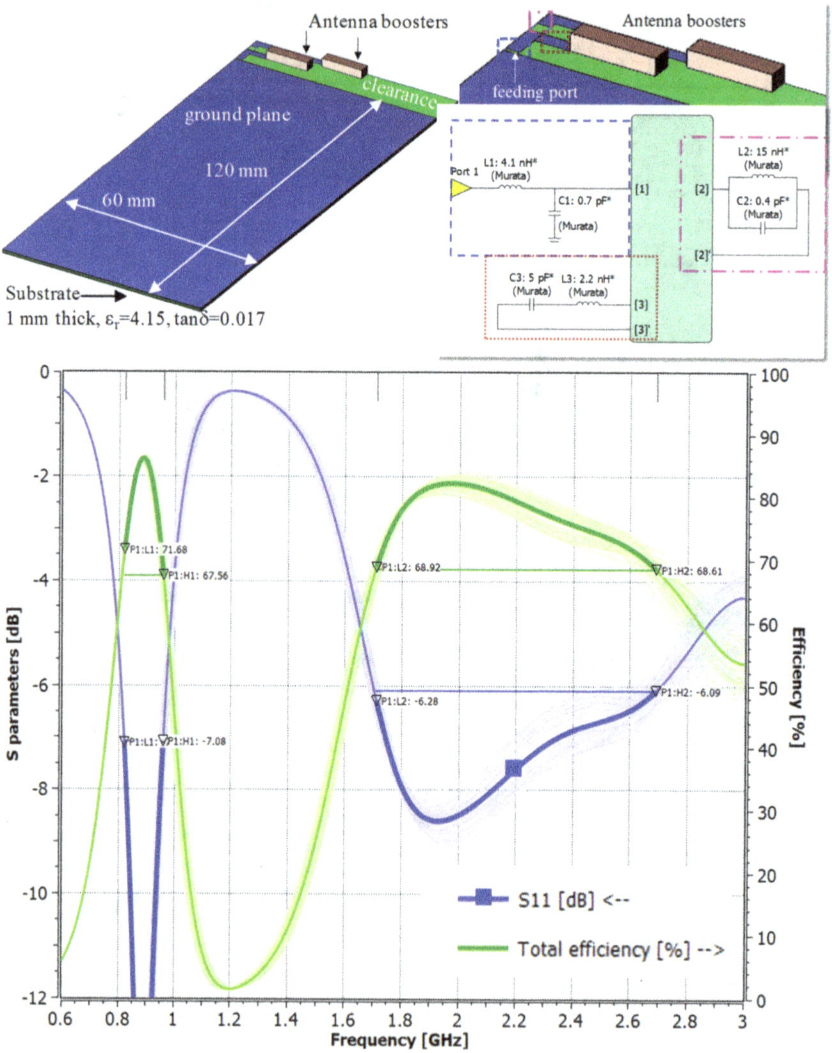

Figure 5.14 (Top) Antenna booster system with two antenna boosters; (bottom) simulated S_{11} and total efficiency

included for one antenna booster, not to load the other. A simple matching network is added for fine-tuning purposes. The architecture is fully optimised using an automatic procedure explained in Section 5.6. To achieve operation at 824–960 MHz and 1710–2690 MHz, six lumped components are needed: four for the filters and two for the fine-tuning network. Considering real component tolerances (∼2%), tolerance analysis shows a very stable performance at the first frequency region and admissible behaviour at the second frequency region (Figure 5.14).

The automatic synthesis procedure ends with two different filters: one series LC, or passband, and a shunt LC, or reject band (Figure 5.14). In this scenario, the S matrix is an $S_{3\times3}$ where ports #2 and #3 are the ones to allocate the filters, whereas port #1 is the one to connect the matching network and the 50 Ω terminated feeding source.

Results in S_{11} and total efficiency are pretty much the same compared with the architectures with four, six, and seven components discussed in Section 5.6.1: it is better at the low-frequency region but worse at the high. Therefore, this is an alternative for applications with more stringent targets at the low-frequency region. Besides, the degree of freedom can help the device designer optimise the structure further to improve performance at the required frequency bands of operation.

This two-antenna booster architecture is compared in Section 5.7 among the architectures comprising one antenna booster as well as with an architecture using a modular antenna booster.

5.5 Multiband performance with a single antenna booster

The techniques presented above integrate two antenna boosters to achieve performance in two frequency regions. This section analyses an architecture that only uses an antenna booster [5,12,13].

A matching network is required to match the capacitive impedance of the booster alone (Figure 5.15). The architecture combines a single antenna booster and

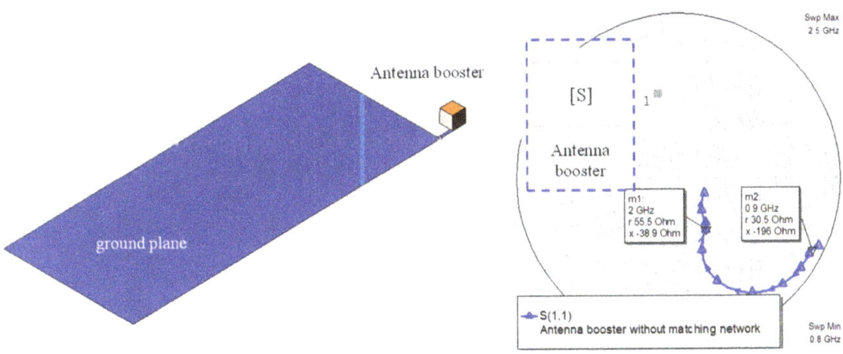

Figure 5.15 A 5 mm × 5 mm × 5 mm antenna booster connected to the 120 mm × 50 mm ground plane with a short strip of 4 mm × 1 mm

a multiband network in charge of matching at two different frequency regions, particularly in a low-frequency region (LFR: 824–960 MHz) and a high-frequency region (HFR: 1710–2170 MHz). As a consequence, this architecture simplifies the number of booster elements from two to one with respect to previous Sections 5.2 and 5.3 [3,5] thanks to the design of the multiband matching network [12].

To illustrate the design process, a 5 mm × 5 mm × 5 mm antenna booster, including a 120 mm × 50 mm ground plane, is utilised. The input impedance features a non-resonant impedance across the low and high-frequency regions (Figure 5.15).

The architecture of the proposed matching network consists of a first stage bringing the capacitive impedance to resonance at a frequency in the middle of the first and second frequency regions by using a series inductor L_1 (Figure 5.16). The purpose is to measure the impedance of the low- and high-frequency regions with similar real parts and opposite signs in their reactive parts. In particular, the impedance at the high-frequency region is inductive, whereas it is capacitive at the low-frequency region. At this point, a series capacitor is needed to decrease the inductive behaviour of the high-frequency region, and a series inductor is required to increase the capacitive behaviour of the low-frequency region (Figure 5.17).

The question is: how to achieve such an objective at the same time? A possible solution proposed here is to use an LC resonator (stage #2) in series with the former inductor L_1 (stage #1) (Figure 5.18).

In this way, at frequencies belonging to the low-frequency region, the LC resonator behaves mainly as a series inductor, making the capacitive effect

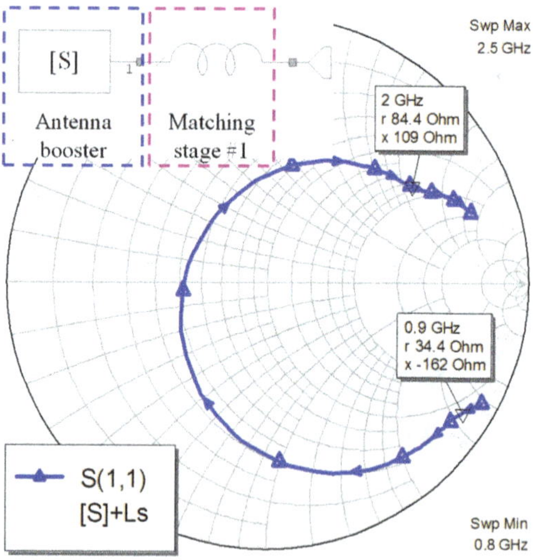

Figure 5.16 Matching stage #1 comprises a series inductance to create a resonance between the first and second frequency regions. [S] is the S_{11} of the antenna booster, including the ground plane

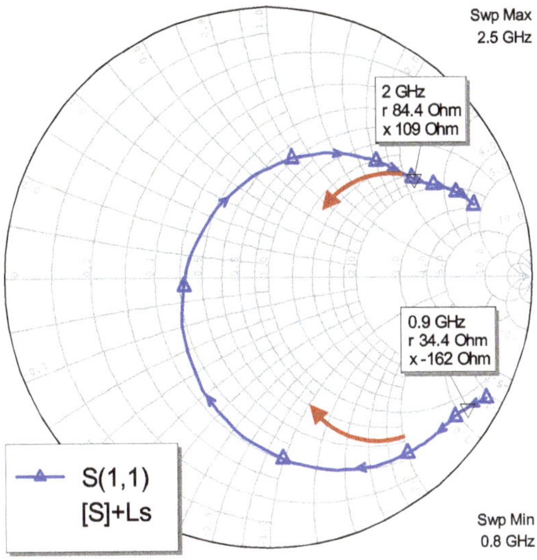

Figure 5.17 *To match the low-frequency region, a series inductance is needed, whereas a series capacitor is necessary to match the high-frequency region*

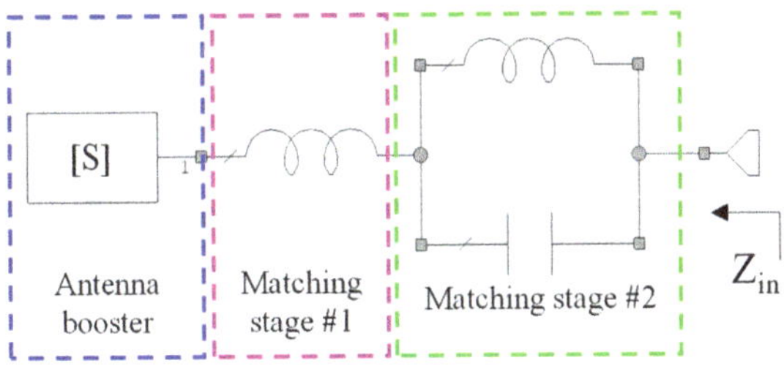

Figure 5.18 *A network for creating a dual-frequency region response. [S] represents the S_{11} of the antenna booster, including the ground plane; stage #1 is a series L to bring the resonance to a frequency in the middle of the first and second frequency regions; stage #2 creates the two frequency regions*

negligible (Figure 5.19). Similarly, the LC resonator acts primarily as a series capacitor at frequencies belonging to the high-frequency region. As a result, the inductive behaviour of the low-frequency region becomes resonant, and in the same way, the capacitive behaviour of the low-frequency region goes to resonance (Figure 5.20). Although this step might achieve dual resonance, if the response is not centred at the centre of the Smith chart, extra stages are added to place the input impedance of both frequency regions inside a circle of $SWR = 3$ (Figure 5.21).

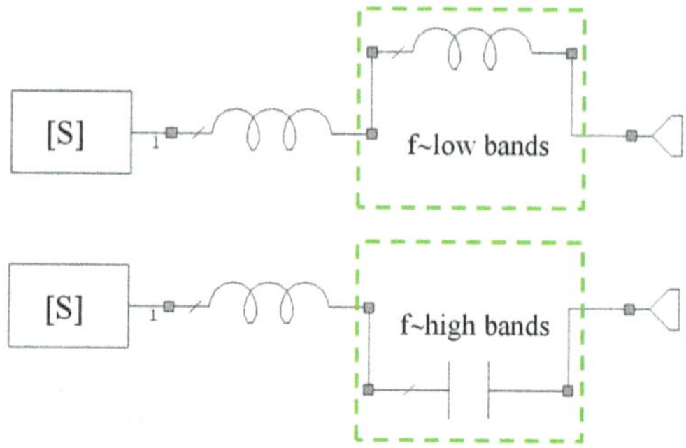

Figure 5.19 *Stage #2 at Figure 5.18 behaves as a series L for the first frequency region and a series C for the second frequency region*

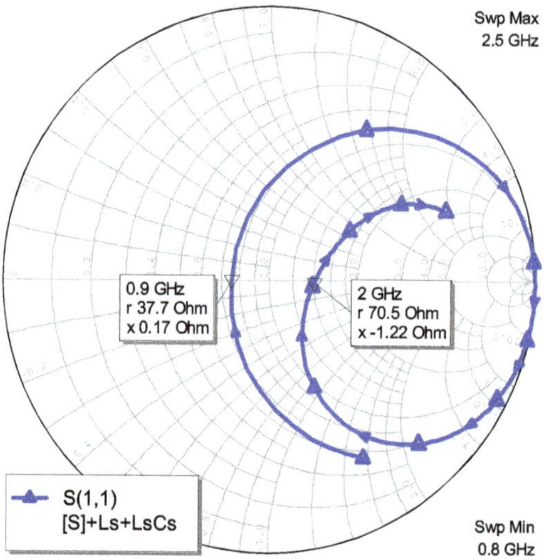

Figure 5.20 *Dual-frequency region response of the matching circuit given in Figure 5.18*

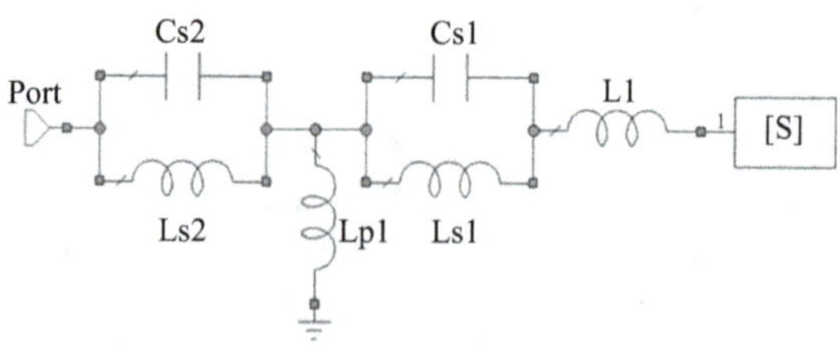

Figure 5.21 *A new stage including L_{p1}, C_{s2}, and L_{s2} to provide an extra degree of freedom for adjusting the two frequency regions*

The impedance of an LC resonator
Consider f_1 and f_2, where $f_1 \ll f_2$, where f_1 and f_2 correspond to the central frequencies of a first and second frequency region, respectively. A shunt LC resonator with a resonant $f_0 \approx (f_1+f_2)/2$ behaves as an inductance for f_1 and as a capacitor for f_2. This feature adds a degree of freedom for designing multiband matching networks for antenna boosters.

Following the teachings of the multiband matching network described in the previous section, three cases have been designed to demonstrate the engineering application of the proposed architecture. Three representative wireless devices are considered with the following ground plane sizes: 120 mm × 50 mm, 200 mm × 135 mm, and 240 mm × 170 mm (Figure 5.22). The ground plane portion is printed on a thin substrate of FR4 ($\varepsilon_r = 4.15$, $\tan\delta = 0.017$) 1 mm thick.

Since bandwidth not only depends on the platform and matching network but also on the antenna booster volume and its distance to the ground plane, a parametric

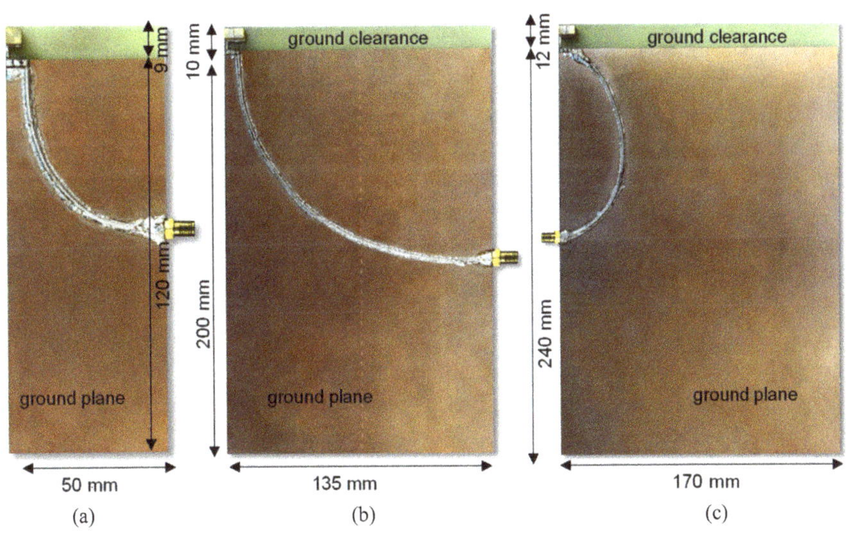

Figure 5.22 Demonstrators including antenna boosters (brass made). Not to scale. (a) 120 mm × 50 mm ground plane, $L_1 = 8.7$ nH, $C_{s1} = 4.3$ pF, $L_{s1} = 5.6$ nH, $L_{p1} = 16$ nH; $C_{s2} = 3$ pF, $L_{s2} = 24$ nH. One booster of 5 mm × 5 mm × 5mm, 4 mm from the edge of the ground plane. (b) 200 mm × 135 mm ground plane, $L_1 = 8.2$ nH, $C_{s1} = 3.3$ pF, $L_{s1} = 4.1$ nH, $L_{p1} = 12$ nH; $C_{s2} = 1.1$ pF, L_{s2} = none. One booster of 5 mm × 10 mm × 5 mm, 5 mm from the edge of the ground plane. (c) 240 mm × 170 mm ground plane, $L_1 = 6.3$ nH, $C_{s1} = 8.6$ pF, $L_{s1} = 15$ nH, $L_{p1} = 15$ nH; $C_{s2} = 1.1$ pF, L_{s2} = none. One booster of 5 mm × 10 mm × 5 mm, 7 mm from the edge of the ground plane.

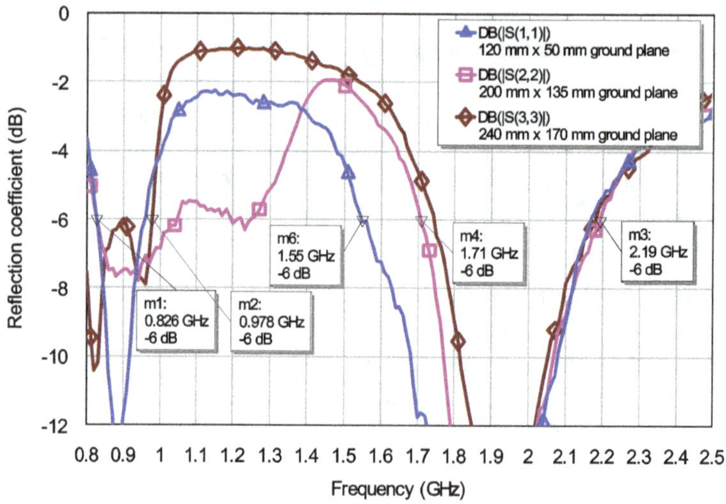

Figure 5.23 Measured S_{11} for the demonstrators shown in Figure 5.22

analysis has been done for each case to obtain the minimum volume and minimum distance to guarantee the following criteria: bandwidth at $SWR \leq 3$ from 824–960 MHz and 1710–2170 MHz (Figures 5.22 and 5.23). Lumped elements are 0402 SMD components of high-Q tight tolerance to minimise ohmic losses. Such lumped components are close to the edge of the ground plane (Figure 5.22) and placed on 2 mm × 2 mm pads to easily solder the components. The effect of those pads is addressed in Chapter 9 since they change the values of lumped components of the matching network slightly when no pads are considered. In real devices, the size of these pads can be further reduced since the precision of the mechanical soldering process is naturally higher than the manual process followed here. Note that the antenna booster is brass-made. However, it can be easily built as an SMD component, simplifying the integration in wireless platforms because it can be soldered on the PCB as other components, such as inductors or capacitors [14].

To better understand the proposed multiband matching network, a step-by-step procedure is shown to analyse the effect of each section (Figure 5.24). As observed, the impedance of the antenna booster is capacitive across the frequency region of interest. First, a series L_1 is added, creating a resonance between both frequency regions. After that, the L_{s1}_C_{s1} creates the dual resonance at the frequency regions of interest (Figure 5.24(a)). However, extra stages are required since the impedance loop is not correctly centred at the centre of the Smith chart. In particular, a shunt L_{p1} moves the impedance loop of the low-frequency region to the circle of $r = 1$ (constant 50 Ω resistance circle). In contrast, the impedance loop at the high-frequency region remains almost in its original place since the shunt L_{p1} moves the impedances belonging to the low-frequency region more than those belonging to the high-frequency region (Figure 5.24(b)). The L_{s1}_C_{s2} stage primarily moves the impedance loop of the low-frequency region. In some cases, only C_{s2} may be needed. This systematic procedure allows a highly capacitive impedance to be matched at two different frequency regions, as shown in this case.

To assess the radiation performance, the total efficiency (η_t) is measured inside an anechoic chamber (Satimo StarGate-32) using 3D pattern integration (Table 5.1). η_t considers radiation efficiency (η_r) and mismatch losses as $\eta_t = \eta_r \cdot (1-|S_{11}|^2)$. The average total efficiency values are in the range of 50% and 75% in the low and high-frequency

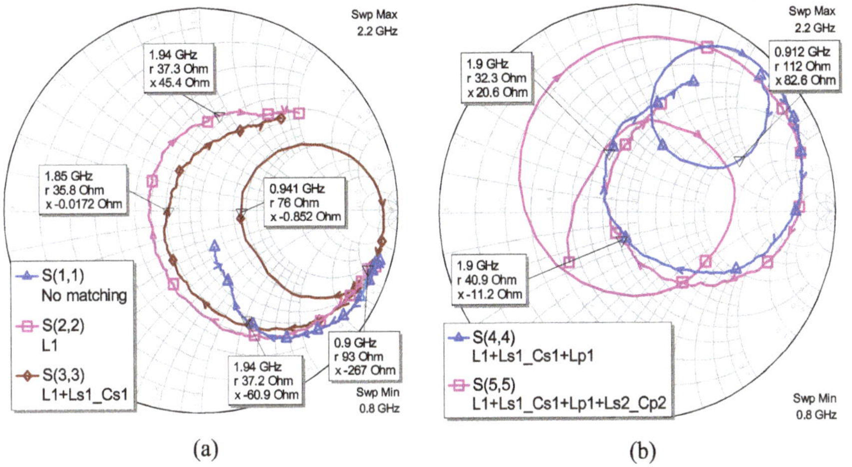

Figure 5.24 Illustration of the multiband matching. (a) S_{11} is the capacitive impedance of the booster, S_{22} is the impedance after the addition of L_1 (Figure 5.21), and S_{33} is the previous impedance after the addition of L_{s1} and C_{s1}; (b) S_{44} is the impedance after the addition of L_{p1} and S_{55} is the impedance having the whole network.

Table 5.1 Summary of the features for each demonstrator, including total efficiency measurements η_t. S_{11} bandwidth is referred to SWR $\leq$ 3. $\bar{\eta}_t$ stands for average value across the frequency range

Ground plane	**120 mm × 50 mm**	**200 mm × 135 mm**	**240 mm × 170 mm**
Matching network	6 elements	5 elements	5 elements
f_1 (GHz)	0.792	0.825	0.823
f_2 (GHz)	0.977	1.046	0.957
$BW_{f1\text{–}f2}$ (%)	20.9	23.6	15.1
f_3 (GHz)	1.744	1.707	1.571
f_4 (GHz)	2.172	2.192	2.219
$BW_{f3\text{–}f4}$ (%)	21.9	24.9	34.2
$\bar{\eta}_t$(LFR) (%)	47.4 (0.791–0.960 GHz)	57.5 (0.824–0.960 GHz)	59.6 (0.824–0.960 GHz)
$\bar{\eta}_t$ (HFR) (%)	68.8 (1.710–2.170 GHz)	67.2 (1.710–2.170 GHz)	75.3 (1.710–2.170 GHz)

regions, respectively, achieved by a small volume of the antenna booster of 125 mm^3 for the 120 mm × 50 mm and 250 mm^3 for the other two ground planes.

Besides efficiency, directivity is also measured across the frequency range for each device. Directivities remain low, particularly for the 120 mm × 50 mm ground plane demonstrator (below 4 dB) and below 6.5 dB for the other two platforms. This higher directivity of the larger devices with respect to the 120 mm × 50 mm ground plane case is due to the larger electrical size of the ground plane that supports higher order modes (Figure 5.25). It is observed that for the 120 mm × 50 mm ground plane

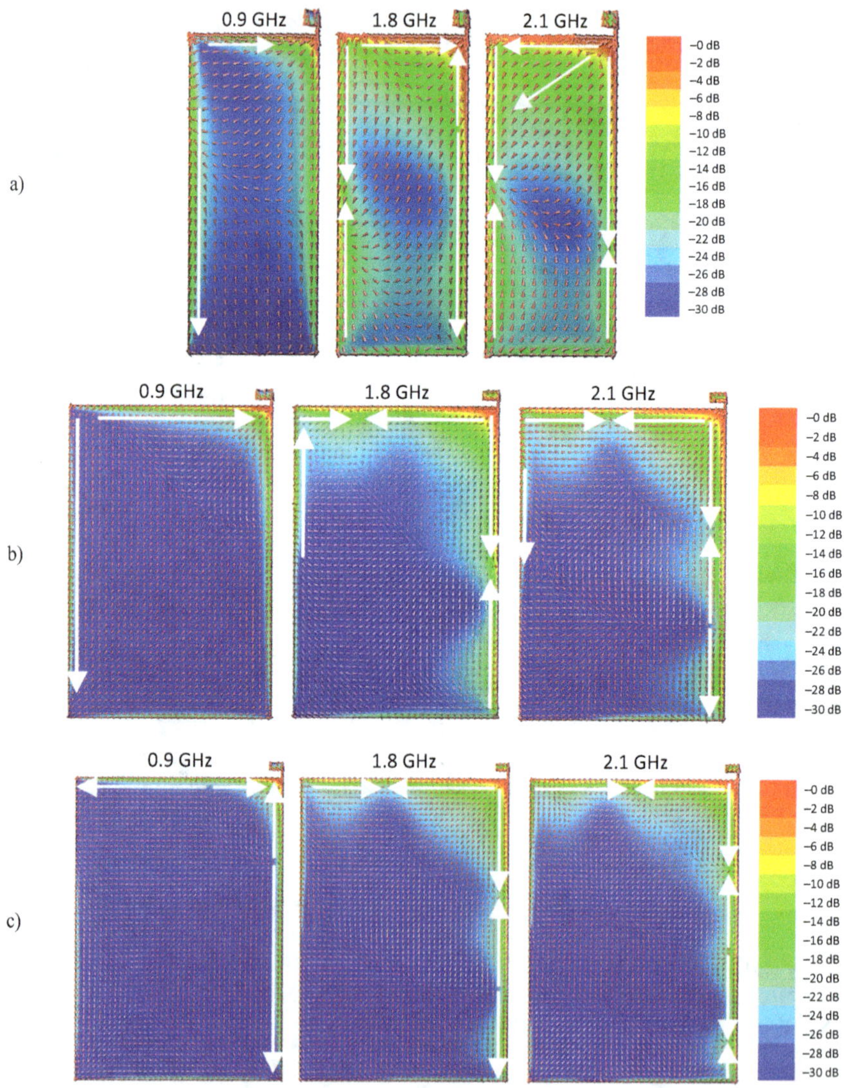

Figure 5.25 Computed current distributions. The intentionally white arrows indicate the main current contribution to radiation. Drawings not to scale.

case at 0.9 GHz, the current is almost vertical, indicating a fundamental mode. This fundamental mode results in a similar dipole pattern directivity of 3 dB. When the frequency increases or the platform size increases (the electrical size of the ground plane increases), current distributions present more variations due to the combination of higher modes both in the long axis and short axis of the ground plane. Ultimately, this current distribution is responsible for determining the radiation pattern. Nevertheless, these directivities are adequate for environments requiring low directive radiation patterns ($D \sim <5$ dBi) where the direction of the incoming electromagnetic wave is not known for the wireless device.

The advantage of this approach is that it embeds a mechanically simple element, the antenna booster, which is easy to fabricate and integrate into a PCB. The multiband behaviour is achieved using a multiband matching network, which can also be applied to other frequency bands.

5.6 Automatic matching network synthesis

A distinct approach hinges on the computational force, leveraging the non-resonant characteristics of antenna booster technology alongside automatic matching network synthesis [15–17]. Optimisation algorithms are helpful in finding the suitable matching network topology as well as the values of the inductors and capacitors used in the matching networks. Besides, for those frequencies where lumped components present limitations due to the self-resonant frequency or low-Q, matching networks are synthesised with distributed elements (transmission lines and stubs), as shown in this section. Optimisation algorithms allow optimising not only S_{11} but also total efficiency. Real components are considered to obtain more realistic results and to avoid solutions that cannot be built into practice. For example, if ideal components are included in the optimisation process, one may end up with a combination of inductors and capacitors that do not exist in practice. Therefore, from a practical perspective, the synthesis of the matching networks should be done using real components. Having this consideration in mind, the goal of the synthesis process should be not only S_{11} but total efficiency. In effect, provided one fixes S_{11} as the unique goal, the resulting matching network can provide a satisfactory S_{11} response across the target frequency bands. However, the solution may achieve poor efficiency because the synthesis procedure was looking for the best $S_{11,}$ not total efficiency, and therefore, a topology with more losses but providing an acceptable S_{11} may be the solution. This issue is, however, easily solved, with total efficiency as the primary goal. Since total efficiency includes both losses of the matching network and S_{11}, the best matching network in terms of total efficiency is ensured.

The automatic matching network synthesis provides other degrees of flexibility as well, such as including in the synthesis the best matching network up to N components. For example, one can start the procedure by fixing $N = 7$ components. If the synthesis procedure finds a suitable solution reaching the target with $N = 4$, it will be selected. This is a very convenient tool from a design perspective since it helps choose matching networks with fewer components.

Another degree of freedom in the synthesis procedure is the multi-goal targets, such as optimising total efficiency for several frequency regions. This section applies the procedure to design a multiband matching network providing operation at two different frequency regions, one low-frequency and another high-frequency. As observed, antenna booster technology is unique since performance will be selected in the design process by automatically synthesising matching networks.

This section illustrates the automatic matching network synthesis tools with the Optenni Lab software [18–20]. However, other synthesis tools, such as those based on genetic algorithms, may be used as well [21].

To illustrate the advantages of the automatic matching network synthesis, no pads for the components are considered. The automated synthesis matching network considering pad effects is analysed in Chapter 9, which essentially follows the same teaching studied here. Also, the synthesis is applied to an impedance obtained from electromagnetic simulation but can also be used to measure the impedance of an antenna booster, including its ground plane. In Chapter 9, this automatic synthesis procedure is employed, considering the pad layout needed to host the designed matching networks. The simulated results are compared to the measured ones, confirming the design method as easy, fast, and automatic.

Several scenarios are considered. The first one with a 12 mm × 3 mm × 2.4 mm antenna booster, including a 120 mm × 60 mm to provide operation at 824–960 MHz and 1710–2690 MHz; the second one with a smaller ground plane of 60 mm × 60 mm; a third one to illustrate triple frequency region operation; fourth, an example of higher frequencies 2.4–2.5 GHz and 5.9–7.2 GHz on a 200 mm × 200 mm ground plane. The reason for operating at higher frequencies is to illustrate how the antenna booster should be scaled from 12 mm × 3 mm × 2. 4 mm of the two previous examples to a physically smaller one. The reason for increasing the ground plane to 200 mm × 200 mm shows the potential of antenna boosters for other wireless platforms.

All simulations consider real lumped inductors and capacitors with measured data provided by the manufacturer. Therefore, losses across the whole frequency band are considered to give realistic S_{11} and total efficiency results. Finally, an example with transmission lines is discussed.

5.6.1 Synthesis to operate at 824–960 MHz and 1710–2690 MHz for a 120 mm × 60 mm ground plane

Instead of using human-engineered architectures like those studied in Sections 5.2, 5.3, and 5.4, machine-engineered architectures using an automatic synthesis procedure are illustrated. This gives designers more flexibility to address the ever-increasing number of applications.

To show the advantage of antenna booster technology combined with automatic synthesis, an example to provide operation across 824–960 MHz low-frequency region) and 1710–2690 MHz (high-frequency region) with a 12 mm × 3 mm × 2. 4 mm antenna booster including a 120 mm × 60 mm ground plane is discussed (Figure 5.26) [22]. In the first situation, a matching network with

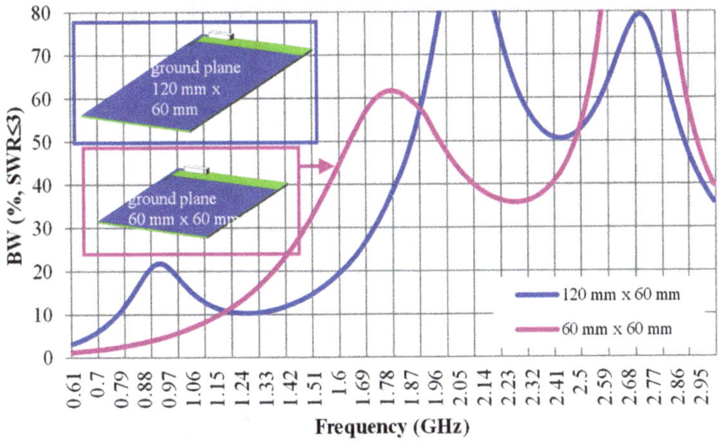

Figure 5.26 Potential bandwidth for a 12 mm × 3 mm × 2.4 mm antenna booster, including 120 mm × 60 mm and 60 mm × 60 mm ground planes

four components is obtained where real components are considered. The goal is to obtain the maximum total efficiency across the bands, which have been fixed at 80% at both low- and high-frequency regions. The total efficiency includes the radiation efficiency of the antenna booster and the ground plane, the mismatching, and the losses of the matching network. In this case, the radiation efficiency is high since the ground plane is large in terms of the wavelength (120 mm is 0.36λ at 900 MHz) being over 97% from 824 MHz and beyond.

Before carrying out any synthesis, the question is: is it physically possible to reach enough bandwidth at $SWR = 3$ to simultaneously cover 824–960 MHz and 1710–2690 MHz? This represents a BW = 15.2% and 44.5% at the low- and high-frequency regions, respectively. This question is easily addressed by calculating the potential bandwidth using Equations (3.2) and (3.3) (Chapter 3) with the input impedance of the antenna booster, including the ground plane (Figure 5.26). Based on the potential bandwidth calculation, the bandwidth that can be achieved at 900 MHz is 19.7% and more than 50% at 2.4 GHz. Therefore, one should expect that both bandwidths can be achieved simultaneously. However, when the ground plane is reduced to 60 mm × 60 mm, the potential bandwidth at 900 MHz decreases drastically to 3.7% but is more than 40% at the high-frequency region. Hence, one should expect that the automatic synthesis process will not find a matching network to cover the whole frequency region at 824–960 MHz but will succeed in achieving bandwidth at the high-frequency region. At this point, one may wonder if a matching network can enhance the bandwidth at the low-frequency region, as per the teachings of Chapter 4. Following the Fano limit, $F = 3.9$ at $SWR = 3$. Thus, $3.7\% \cdot F = 14.4\%$, close to the 15.2% needed to cover 824–960 MHz. However, the matching network also needs to cover the high-frequency region. Consequently, it can be predicted that it will be challenging to cover the whole 824–960 MHz bandwidth for the 60 mm × 60 mm ground plane. The automatic synthesis process in the next section addresses this. In summary, the potential bandwidth is an excellent tool to analyse whether it is possible to obtain the required bandwidth before diving into the automatic synthesis procedure.

After the automatic synthesis procedure, the resulting matching network with four lumped components provides an average total efficiency across the low- and high-frequency regions of 67.6% and 85.1%, respectively (Figure 5.27). As

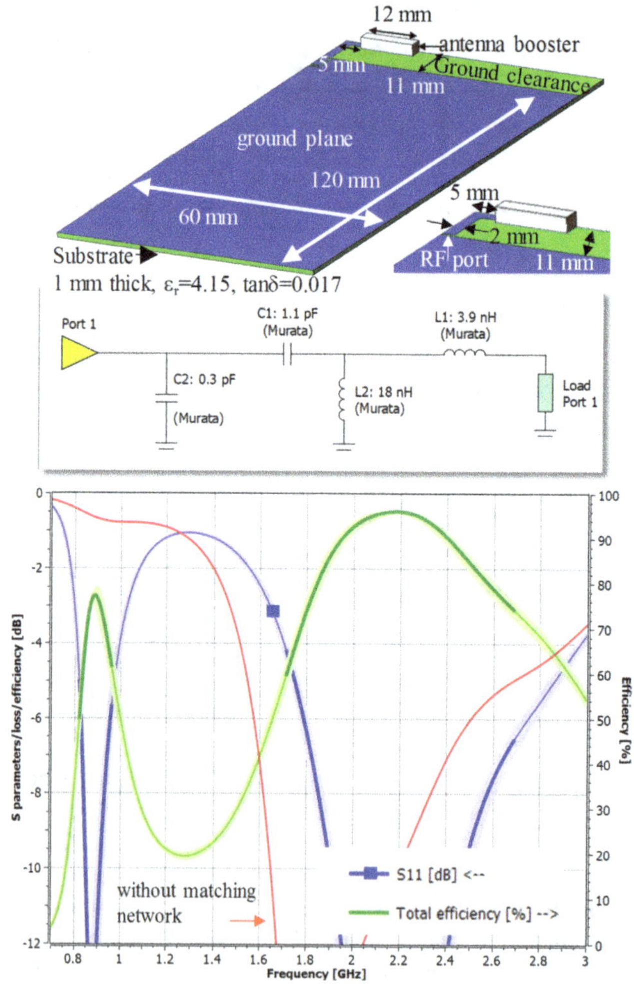

Figure 5.27 S_{11} and total efficiency achieved with the automatic matching network synthesis. Tolerance analysis considers real component tolerance (~2%). Load corresponds to the simulated S_{11} of the antenna booster.

observed, the S_{11} without a matching network presents very poor matching at the low-frequency region due to the non-resonant and highly capacitive nature of the antenna booster (Z = 31.4–j233.4Ω and Z = 37.5–j198 Ω at 824 MHz and 960 MHz, respectively). The high-frequency region provides an S_{11} below −6 dB at almost the entire frequency region.

However, the S_{11} is not below −6 dB (−4.9 dB and −4.1 dB are the worst S_{11} values for the low- and high-frequency regions). At first sight, as long as total efficiency is high, having S_{11} above −6 dB should not be an issue. However, it may be that some radiofrequency amplifiers reduce their gain depending on the mismatching degree, and therefore, performance is degraded [23]. Thus, the question is: can we find a matching network with similar total efficiency but with better S_{11}? Consequently, the S_{11} is also included as an optimisation target (Figure 5.28). With this new goal, several matching networks with six and seven components are found (only two are illustrated). A first matching includes seven components with a much better S_{11} at both frequency regions than the four-component matching network. Besides, the average total efficiency in this case for the low- and high-frequency regions is 69.1% and 87.0%, which is slightly better.

To illustrate that the automatic process may result in several solutions, a matching network with six lumped components is also obtained (Figure 5.28 bottom). Utilising a matching network comprising seven components results in improved S_{11} compared to configurations with four components, while maintaining a very similar total efficiency of 66.0% and 89.1% in the low- and high-frequency regions, respectively. Both outcomes confirm that sufficient bandwidth with S_{11} = −6 dB is achieved in both frequency regions for the respective matching networks, validating the bandwidth estimations depicted in Figure 5.26.

S_{11} and total efficiency

A multiband matching network for an antenna booster, besides providing competitive total efficiency, should also provide a good S_{11} (typically $S_{11} < -6$ dB). Having a solution with a poor S_{11} for transmission ($\ll -6$ dB), may reduce performance since the transceiver connected to the matching network and antenna booster may decrease its output power.

In every scenario, tolerance analysis considering actual component tolerances (with a tight tolerance of approximately 2%) demonstrates that all solutions, including those with seven components, remain robust to component variations (Figure 5.28).

As discussed in Chapter 4, it is interesting to check if there is a critical component responsible for heat dissipation and, thus, loss. This loss is produced when a current, either static or dynamic, circulates across an inductor or capacitor due to Joule's effect (Figure 5.29). Losses are proportional to the square of the current through the component (L or C) multiplied by its resistance. The higher the Q_m of a component, the lesser its associated resistance, and thus, fewer losses are introduced in the matching network.

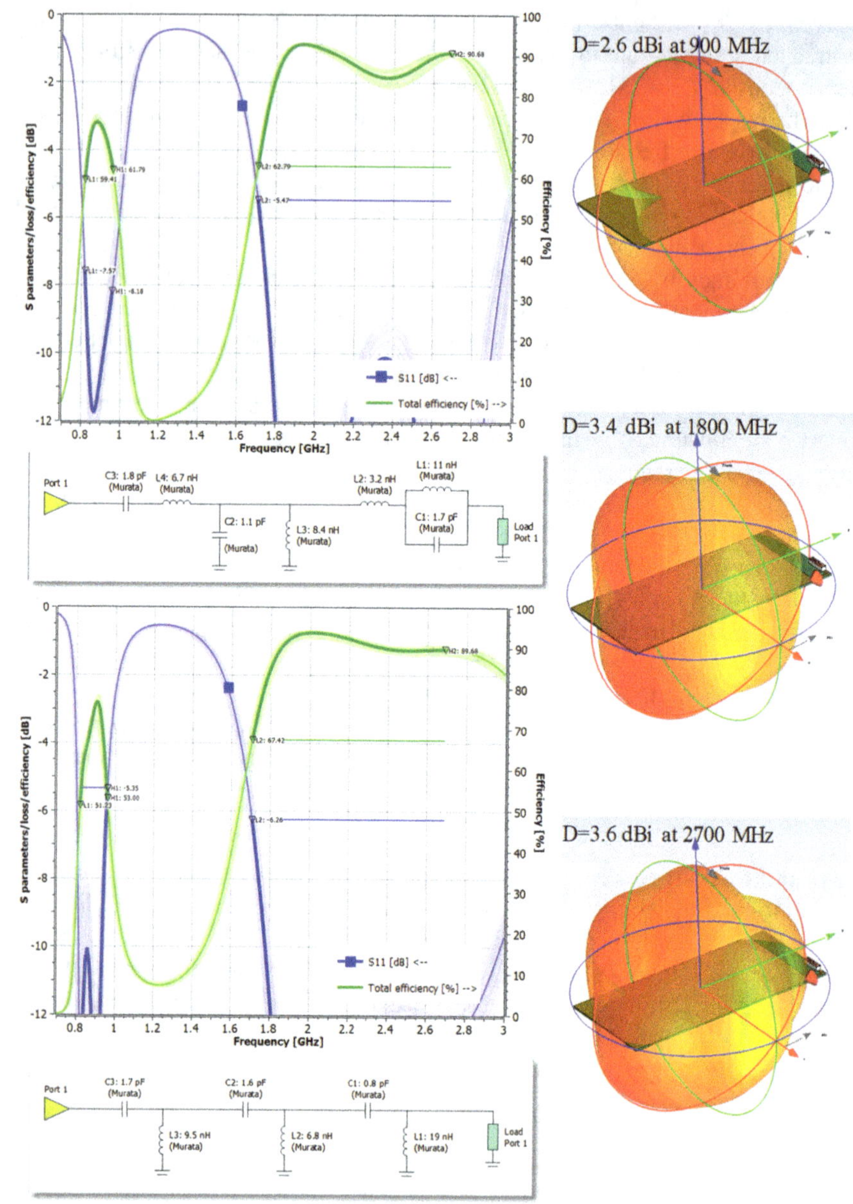

Figure 5.28 Automatic synthesis of multiband matching networks. Load is the simulated S_{11} for the 12 mm × 3 mm × 2.4 mm antenna booster, including the 120 mm × 60 mm ground plane. The matching network considers real lumped inductors and capacitors. Radiation patterns feature quasi-isotropic behaviour (dB scale, 30 dB dynamic range).

Figure 5.29 (Left) James Prescott Joule, an English physicist (1818–1889), discovered Joule's first law in 1841, that "the heat is proportional to the square of the intensity of that current, multiplied by the resistance to conduction which it experiences"; (right) Gustav Kirchhoff, a German physicist (1824–1887), proposed the fundamental behaviour of voltages and currents in a circuit, known as Kirchhoff laws

By analysing the three matching networks (Figure 5.30), it is observed that for the four-component matching network, L_2 is the critical component, L_1 for the matching network with seven, and L_1 and L_2 for the matching network with six components. For this last case, L_2 is not only relevant at the low frequency but also at the high-frequency region. Therefore, there is not a critical component in the same location but changes according to the topology of the matching network, which makes sense according to Kirchhoff laws (Figure 5.29). Depending on the matching network topology, the currents for each component change. Therefore, the component achieving the larger $I^2 \cdot R_\Omega$ is relevant to losses (Figure 5.31). This is confirmed in Figure 5.31, where currents are computed for each of the four components of the matching network.

Currents along L_2 and C_1 are more intense at the low-frequency region. Since the Q_m (quality factor of the component) is lower for L_2 ($Q_m = 80.8$ at 900 MHz) than C_1 ($Q_m = 680$ at 900 MHz), R_Ω is greater for L_2 than C_1, incurring more losses in L_2, as revealed in Figure 5.30. This is not intuitive because one should expect the current across L_1 to be more intense since it is the one reaching the antenna booster. Thus, we cannot conclude that the component close to the antenna booster is the most critical to losses because, in the end, it depends on the matching network topology.

A general trend is that matching losses are less relevant at the high-frequency region ($f > 1.5$ GHz) than the losses at the low-frequency region ($f < 1$ GHz). The reason is that Q_m is much larger than Q_a (quality factor of the antenna booster and the ground plane) in the high-frequency region compared to the low-frequency region (Equation 3.4). In effect, with Figure 3.54 (Chapter 3), considering the worst bandwidth value at 824–960 MHz results in BW = 12.9%, and thus in $Q_a = 8.9$.

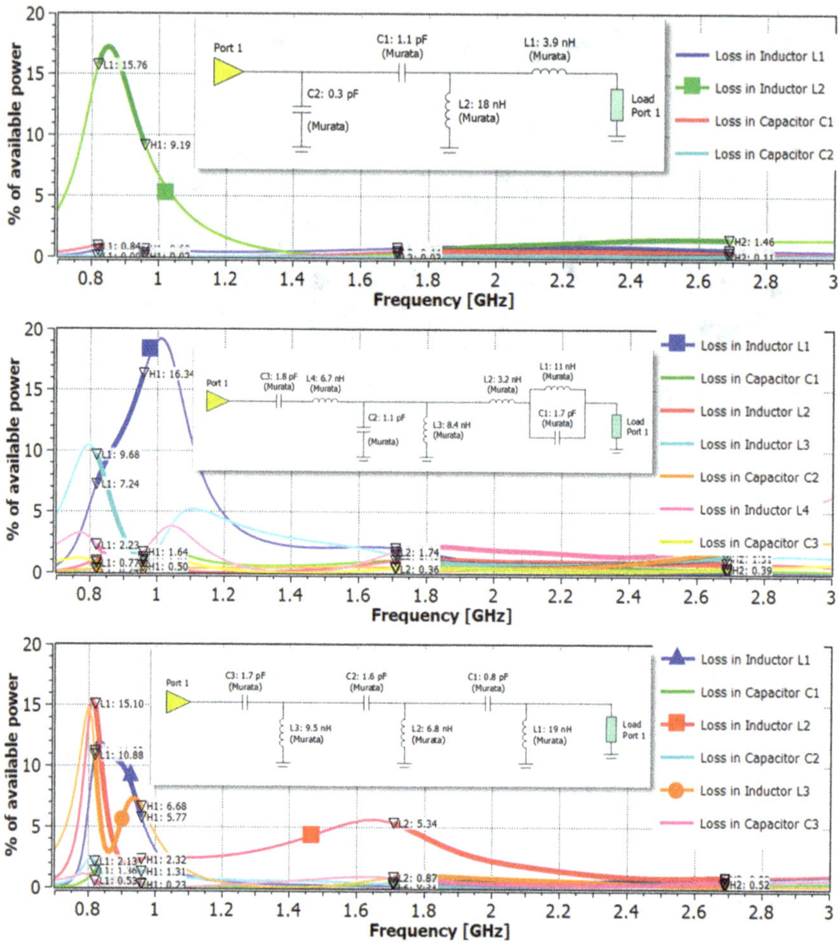

Figure 5.30 Contribution to losses of each of the lumped components of the synthesised matching networks

Analogously, the worst bandwidth value for the high frequency is BW = 27.7%, resulting in $Q_a = 4.3$, two times less than Q_a at the low-frequency region.

5.6.2 Revisiting the ground clearance to ground plane ratio

As studied in Chapter 3, the ground clearance to ground plane ratio can be optimised to obtain more bandwidth. This section shows how this ratio can be useful to improve the efficiency of a multiband solution at 824–960 MHz and 1710–2690 MHz first and, in a second example, to expand the bandwidth at the second frequency region to 698–960 MHz. The same antenna booster scheme is shown in Figure 5.27. For that case, the clearance is 11 mm, and the ground plane is

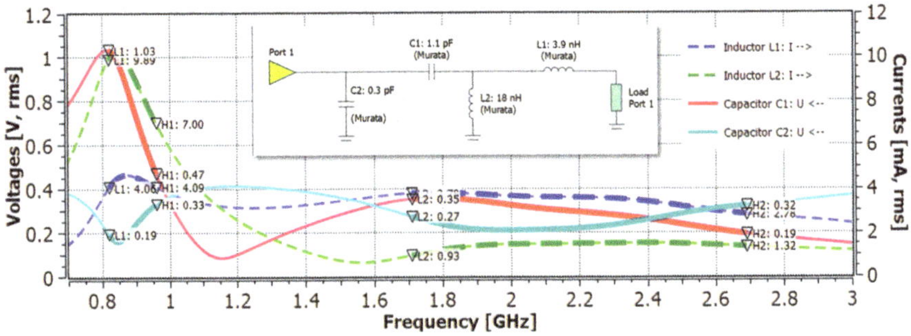

Figure 5.31 *Currents across each component of the synthesised multiband matching network of Figure 5.27. As observed, the current at component L_2 is more intense at the low-frequency region, resulting in more contribution to losses than the rest, as Figure 5.30 confirms.*

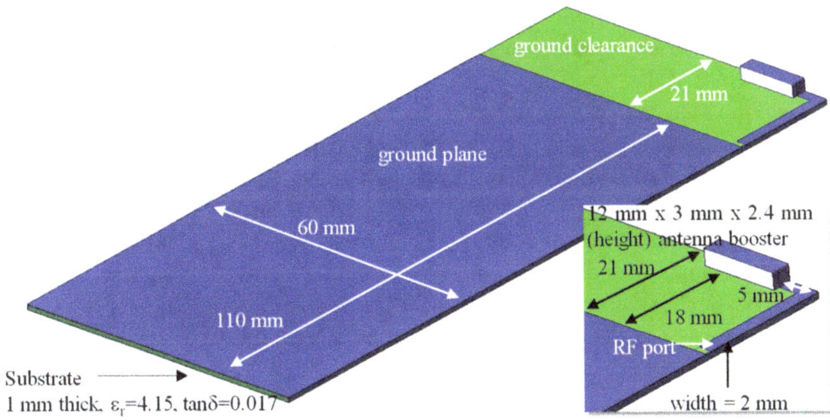

Figure 5.32 *Ground clearance to ground plane ratio equals 0.19*

120 mm. Now, the clearance is increased to 21 mm, and the ground plane is reduced to 110 mm (the ratio is 21/110 = 0.19), so the total length of the PCB is the same, that is, 131 mm (Figure 5.32).

The potential bandwidth at 900 MHz for r = 0.09 is 19.7% at 900 MHz, whereas it is 28.3% for r = 0.19. For the high-frequency region, both bandwidth potentials are practically the same. This result ends up with better total efficiency for r = 0.19 (Table 5.2). Therefore, the ground clearance to ground plane ratio should be taken into consideration in the design for optimising bandwidth and efficiency. This is another example of how an antenna booster is very flexible since it fits small and larger ground clearances due to the small size of the antenna booster.

Table 5.2 Simulated total efficiency. Matching networks consider real lumped components (finite Q)

Frequency (MHz)	Ratio 0.9 (Figure 5.27)	Ratio 0.19 (Figure 5.32)
824–960	69.9%	85.1%
1710–2690	77.6%	87.0%
698–960	47.8%	64.5%
1710–2690	72.4%	85.1%

Matching 1 operates from 824–960 and 1710–2690 and matching 2 from 698–960 and 1710–2690.

5.6.3 Synthesis to operate at 824–960 MHz and 1710–2690 MHz for a 60 mm × 60 mm ground plane

In order to highlight the importance of the ratio between Q_a and Q_m and its impact on the ground plane's bandwidth, the subsequent exercise demonstrates the same dimensions (12 mm × 3 mm × 2.4 mm) as the previous section, but with a reduced ground plane size of 60 mm × 60 mm (Figure 5.33). The same objective as in the earlier cases is set: to operate with maximum total efficiency across 824–960 MHz and 1710–2690 MHz.

Two important conclusions are obtained: the first is that it is impossible to obtain bandwidth across 824–960 MHz as predicted using the potential bandwidth (Figure 5.26), at least with a simple matching network of one or two lumped components. The target has been relaxed to cover 880–960 MHz, but still, it is not possible to achieve 880–960 MHz as S_{11} reaches only −3 dB. The S_{11} = −6 dB bandwidth obtained with the five-component matching network at the low-frequency region is 4.4% (894 MHz to 935 MHz). The potential bandwidth calculation estimated that it was possible to obtain 3.7% at 900 MHz. As observed, the matching network enhances that 3.7% to 4.4% (1.18 times), which is far away from the Fano limit because the matching network also needs to achieve bandwidth at the high-frequency region. The question arises: how can the entire bandwidth be attained? This is thoroughly examined in Chapter 7, where a reconfigurable matching network is extensively analysed. Essentially, given the limitation that not all bandwidths can be simultaneously covered, proposed architectures for reconfigurable matching networks utilise digitally tunable capacitors and radiofrequency switches to cover narrower bandwidths that can be tuned to encompass the entire required frequency range.

Finally, the peak total efficiency at the low-frequency region is 53% at 900 MHz, which is lower than the one obtained with the 120 mm × 60 mm ground plane, even for a larger frequency range of 824–960 MHz. Why is it so low? First, radiation efficiency has been numerically computed, being over 88% beyond 824 MHz (η_r = 92% at 900 MHz). Therefore, the losses come from the matching network. Now, it is clear what the impact of Q_a/Q_m is. At 900 MHz, Q_a = 31.1 whereas it was 5.7 for the 120 mm × 60 mm ground plane. Assuming Q_m = 63 at 900 MHz for the L_3 inductor, which has the highest power dissipation, makes the matching network efficiency at 900 MHz (Equation 3.4, Chapter 3) η_m = 66.9%.

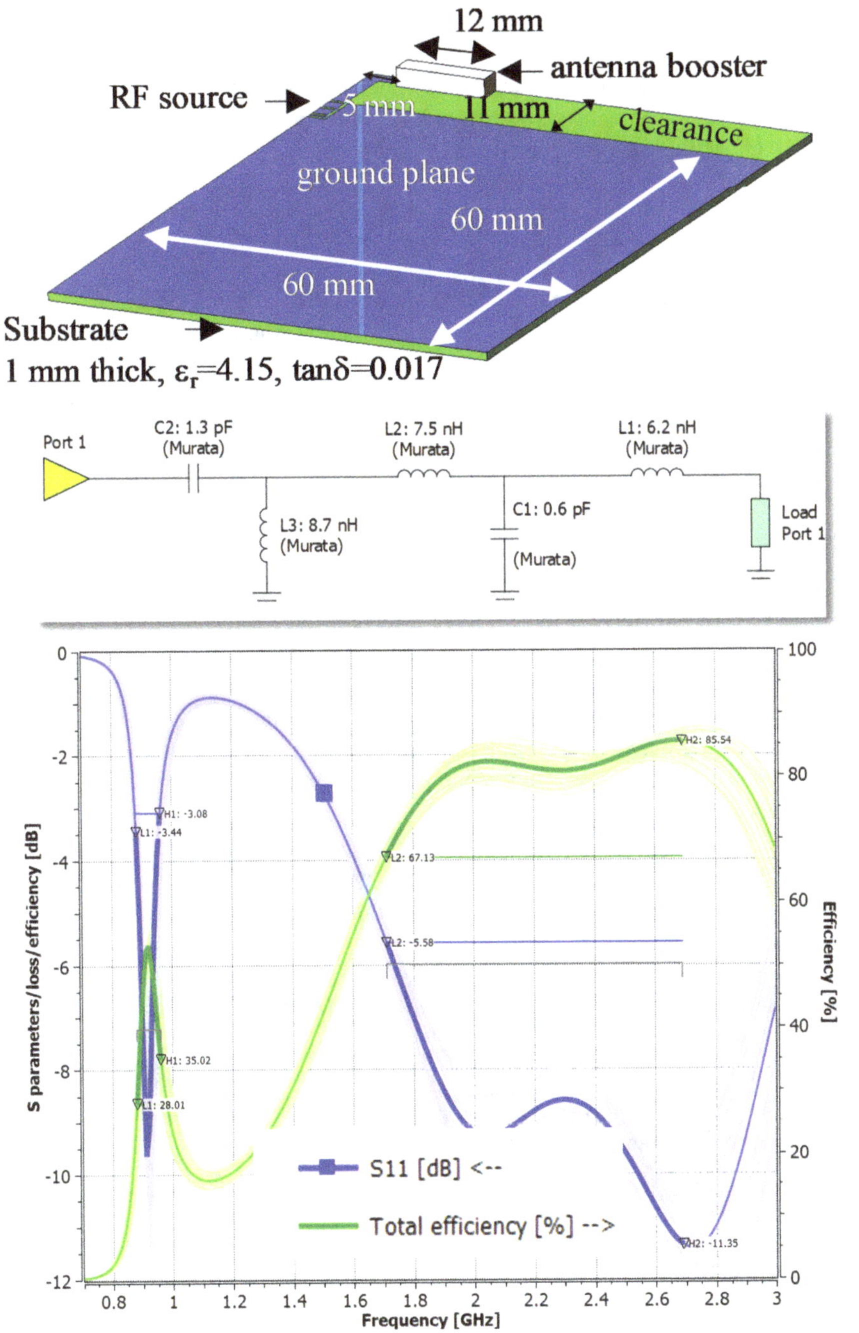

Figure 5.33 S_{11} and total efficiency for the 12 mm × 3 mm × 2.4 mm antenna booster, including a 60 mm × 60 mm ground plane considering a matching network with five lumped components. Load represents the simulated S_{11} of the antenna booster, including the ground plane without the matching network. Tolerance analysis considers real component tolerance of about 2%.

Note that if this operation is repeated considering the 120 mm × 60 mm ground plane (assuming $Q_m = 63$ and $Q_a = 5.7$), it results in $\eta_m = 91.7\%$, which corresponds to 1.3 dB fewer losses. Consequently, as concluded in Chapters 3 and 4 and confirmed here, losses of a matching network depend not on Q_m (the Q of the lumped components of the matching network) but more on the Q_a/Q_m relation; the lower, the better.

Tolerances and Q
When designing multiband matching networks for antenna boosters, tolerance analysis must be carried out to select solutions, that besides S_{11} and total efficiency, are robust to the tolerances of the lumped components of the matching network. Tight tolerances of no higher than 2% are recommended to result in a robust multiband matching network.

Q_m for the components should be as high as possible to minimise losses introduced by a matching network, and thus, maximise total efficiency.

5.6.4 *Matching network synthesis for triple frequency region operation*

An example of triple region operation is included to further illustrate the flexibility of the matching network synthesis (Figure 5.34). In this case, one frequency region is centred at 915 MHz, the second at 1.4 GHz, and the third one at 2.45 GHz. Note that the matching network synthesis can also force mismatching out of the bands if needed. This approach may avoid including a filter to minimise harmonics, as long as the matching provides acceptable attenuation out of the band.

Some of the multi-radio techniques explained in Chapter 6 deal with a combination of a multiband matching network, as shown in this chapter, and the addition of a multiplexer to connect each of the outputs to the respective transceiver.

5.6.5 *Synthesis to operate at 2.4–2.5 GHz and 5.9–7.2 GHz for a 200 mm × 200 mm ground plane*

In the previous examples, antenna booster technology was applied for operating below 3 GHz. In this new scenario, the purpose is to design a multiband device embedding an antenna booster at a higher frequency range and, thus, how the antenna booster can be scaled.

For illustrative purposes, the frequency region comprises a first frequency region from 2.4 GHz to 2.5 GHz and a second frequency region from 5.9 GHz to 7.2 GHz, including a 200 mm × 200 mm ground plane, representing a wireless device for short-range wireless communications.

In the previous section, the antenna booster was 12 mm × 3 mm × 2.4 mm, representing an electrical size at the lowest frequency of operation of $\sim\lambda/30$ at 824 MHz. The same antenna booster may also be used, but to illustrate that antenna boosters can be scaled according to the frequency region of operation, a new

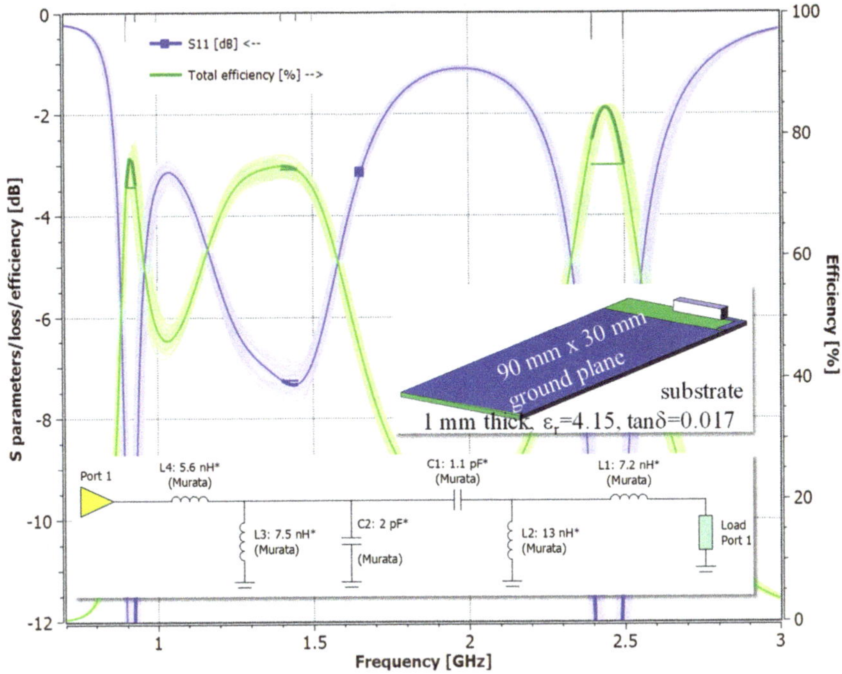

Figure 5.34 *S_{11} and total efficiency for the 12 mm × 3 mm × 2.4 mm antenna booster, including a 90 mm × 30 mm ground plane considering a matching network with six lumped components. Load represents the simulated S_{11} of the antenna booster, including the ground plane without the matching network. Tolerance analysis considers real component tolerance of about 2%.*

antenna booster is designed. For the lowest frequency of operation, 2.4 GHz, $\lambda/30$ represents 4.1 mm. In this case, the antenna booster length is fixed to 7 mm × 3 mm × 1 mm, which, in terms of the wavelength, is larger than the 12 mm × 3 mm × 2.4 mm for providing operation at 2.4 GHz. The reason is to make a small antenna booster that is still usable at frequencies below, even at 0.824 GHz. This new antenna booster has a size of: $\sim\lambda/18$ at 2.4 GHz and $\lambda/50$ at 824 MHz. With this new size, the antenna booster is non-resonant at the first frequency region, having an impedance of $Z = 21.5 - j124.7\ \Omega$ and $Z = 21.1 - j114.1\ \Omega$ at 2.4 GHz and 2.5 GHz, respectively, and it has a first resonant frequency at 4.76 GHz. Without a matching network, it is poorly matched at the low-frequency region (2.4–2.5 GHz) but well matched at the high-frequency region (5.9–7.2 GHz) (Figure 5.35).

The same automatic synthesis procedure explained before is applied here to achieve the multiband operation (Figure 5.35). A matching network comprising four lumped components achieves an S_{11} below −11 dB in both frequency ranges, exhibiting average total efficiencies of 85.1% and 93.3% in the low- and

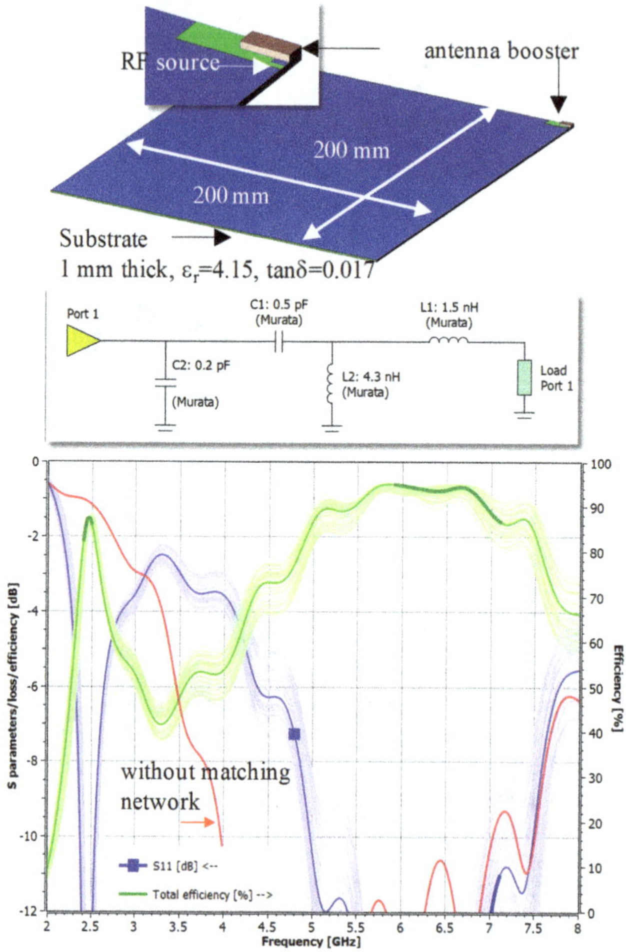

Figure 5.35 Synthesised multiband matching network considering real components compared to the original situation without matching network where it is clearly shown that the matching network allows matching at both low- and high-frequency regions with a simple matching network of four components. Load represents the simulated S_{11} *of the 7 mm × 3 mm × 1 mm antenna booster, including the 200 mm × 200 mm ground plane. Tolerance analysis considers real component tolerances (~2%).*

high-frequency bands, respectively. This reaffirms the high efficiency of the system when paired with an antenna booster element of approximately $\lambda/18$.

The synthesis procedure is applicable for mitigating spurious emissions by crafting a matching network that aligns with the desired frequency bands while simultaneously suppressing unwanted emissions that could potentially disrupt other devices, thereby offering a simpler alternative to complex antenna configurations found in current state-of-the-art solutions [24].

5.6.6 Matching network synthesis with distributed elements

When the frequency of operation goes beyond ~10 GHz, the quality and behaviour of some inductors and capacitors can deviate from their standards. For example, a 5 pF 0402 type SMD presents a self-resonant frequency at 3.8 GHz. Beyond the self-resonant frequency, capacitors no longer behave as capacitors but as inductors. Besides, even if the frequency of operation is below the self-resonant frequency, Q should be analysed since it becomes small because the frequency is close to the self-resonance. For example, the Q of the 5 pF capacitor at 1 GHz is $Q = 283$, whereas it is only $Q = 5.4$ at 3.6 GHz. Therefore, a matching network with distributed elements is an alternative in those scenarios where the lumped components may introduce losses or unexpected behaviour (Figure 5.36). A matching network with distributed elements comprises transmission lines and stubs either open or shorted to the ground plane. Microstrip is one of the simplest forms of distributed matching networks. As explained in Chapter 1, a microstrip line is a conductive line over a substrate layer with a ground plane at the bottom layer. Once the height of the substrate is chosen (typically, it is the height of the substrate of the PCB), the length and width of the microstrip determine the impedance of the matching network in a similar way inductors or capacitors do in a lumped matching network.

A simple matching network example for the 7 mm × 3 mm × 1 mm antenna booster in Figure 5.35 is illustrated (Figure 5.36). In this case, the matching network synthesis procedure proposes a series strip of 8.89 mm in length and 0.967 mm in width with a stub to the ground of 5.9 mm in length. The performance in terms of total efficiency is, in this case (Figure 5.37), similar to the matching network using four components (Figure 5.35).

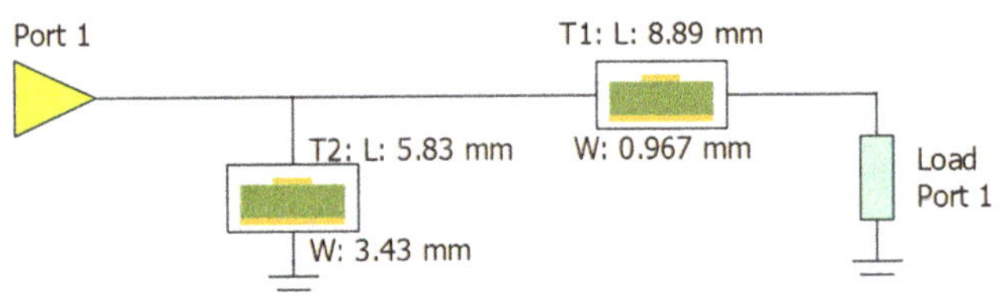

Figure 5.36 Matching network comprising a transmission line and a stub shorted to ground. L and W represent the length and width of the conductive strips on a 1 mm thick substrate with $\varepsilon_r = 4.15$, $\tan\delta = 0.02$. The characteristic impedance of T1 and T2 are $Z_0 = 73.5\ \Omega$ and $Z_0 = 35.0\ \Omega$, respectively. Load represents the S_{11} of the 7 mm × 3 mm × 1 mm antenna booster on the 200 mm × 200 mm ground plane (Figure 5.35).

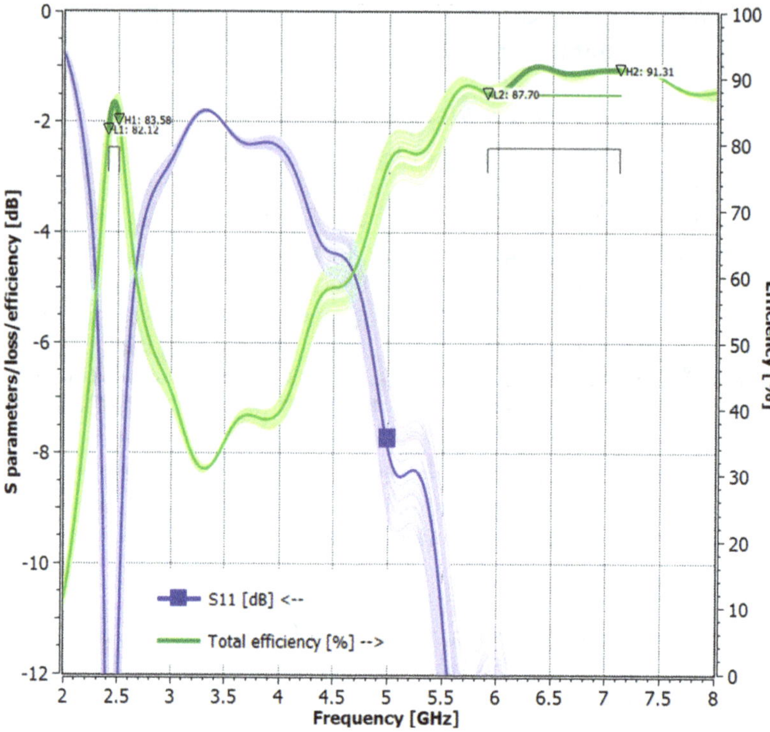

Figure 5.37 Simulated S_{11} and total efficiency of the antenna booster system (Figure 5.35), including the distributed matching network shown in Figure 5.36

5.7 Multiband matching network with a modular structure

This section introduces a novel architecture aimed at delivering multiband performance, relying on a modular antenna booster element integrated with a reject filter based on lumped components [25]. For illustrative purposes, as in Sections 5.4, 5.6.1, and 5.6.2, the same low- and high-frequency regions, 824–960 MHz and 1710–2690 MHz, respectively, are used here.

Why do we need a reject filter? To increase bandwidth at the low-frequency region, the conductive element size is increased to 30 mm. However, this increase is not beneficial for the high-frequency region; on the contrary, it worsens the potential bandwidth (Chapter 3, Figure 3.33). Therefore, how do we have a long, non-resonant element (30 mm) for the low-frequency region and a short one for the high-frequency region? With a reject filter at the high-frequency region at a location where the 30 mm long element is split into two parts (Figure 5.38). Therefore, optimising the reject filter, the high-frequency region currents find a high impedance path when trying to reach the second part of the element. In this manner, the

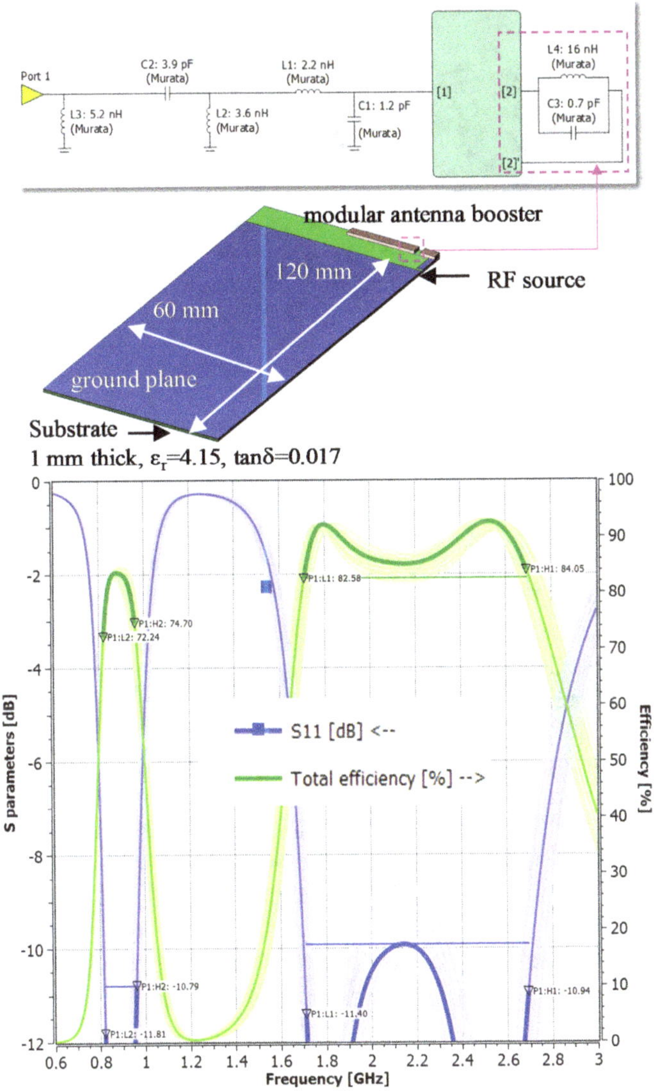

Figure 5.38 Simulated S_{11} and total efficiency for a 30 mm modular antenna booster architecture

effective length for the high-frequency region is the short first portion, which can be considered an antenna booster due to its small electrical size.

Proceeding with the automatic synthesis method, a matching network is synthesised. The reject filter comprises a shunt LC, whose values are optimised at the same time the matching network is synthesised. In this architecture, the S-matrix comprises two ports: port #2 to connect the reject filter and port #1 to connect the matching network.

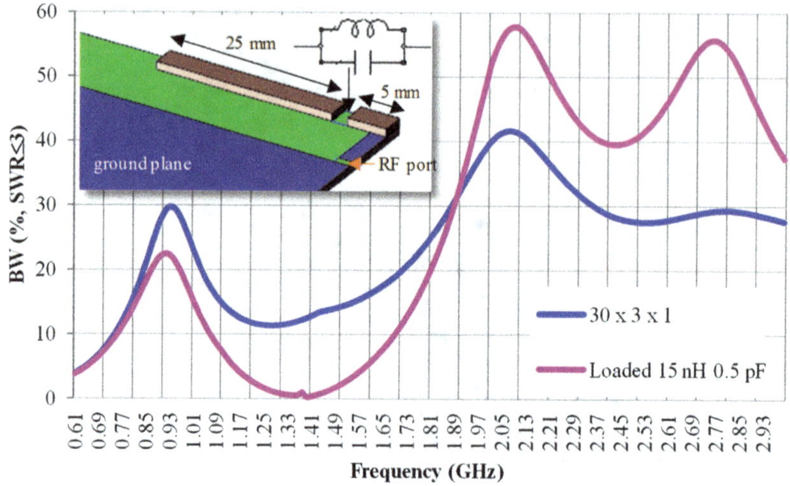

Figure 5.39 Simulated potential bandwidth for a 30 mm × 3 mm × 1 mm element and a modular antenna booster loaded with a series 15 nH in shunt with 0.5 pF

As a design note, if the reject filter is changed by a short circuit, the synthesis process fails to achieve an acceptable bandwidth at the high-frequency region, confirming the usefulness of the potential bandwidth calculations (Figure 5.39).

The addition of a reject filter allows for streamlining the bandwidth at the high-frequency region since 30 mm is too large, as discussed in Chapter 3, Table 3.4. In fact, revisiting those results, bandwidth at the high-frequency region was maximised for an antenna booster between 5 mm × 3 mm × 1 mm and 15 mm × 3 mm × 1 mm. Therefore, the position of the reject filter is to host such an antenna booster in the 30 mm element. For this example, the reject filter LC breaks the 30 mm length into two portions of 25 mm and 5 mm (Figure 5.39). As a result, at high frequencies, the mission of the reject filter is to block the high frequency to reach the 25 mm part, and as a result, the effective antenna booster length is only 5 mm, which has more bandwidth than the 30 mm length (Figure 5.39).

Sizing an antenna booster for multiband operation

The size of the ground plane is the main factor in determining bandwidth and total efficiency. However, antenna booster size can also help in adding some extra bandwidth. As a rule of thumb, the maximum recommended size of the antenna booster is less than $\lambda/20$ at the lowest frequency of operation, resulting in a compact solution while preserving bandwidth and total efficiency thanks to the proper excitation of currents in the ground plane.

Table 5.3 Comparison of total efficiency for different multiband architectures, including a 120 mm × 60 mm ground plane

		Total efficiency (%)	
	Architecture	824 MHz–960 MHz	1710 MHz–2690 MHz
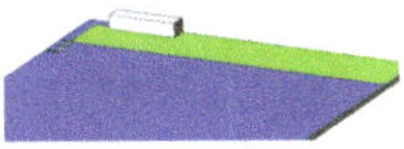	1 antenna booster, 4 components Figure 5.27	67.6	85.1
	1 antenna booster, 6 components Figure 5.28	66.0	89.1
	1 antenna booster, 7 components Figure 5.28	69.1	87.0
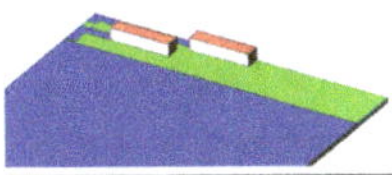	2 antenna boosters, 6 components Figure 5.14	79.3	77.6
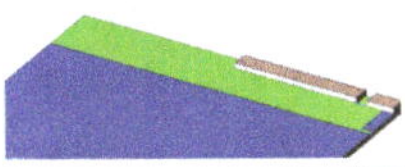	1 modular antenna booster, 8 components Figure 5.38	83.1	89.1

After the calculation, S_{11} presents an $S_{11} < -10$ dB for the whole low- and high-frequency regions. Comparison is made between the overall efficiency of the previous solution and that of a multiband matching network employing either four, six, or seven lumped components, as well as one incorporating two antenna boosters (Table 5.3). Note that all include the same 120 mm × 60 mm ground plane and 60 mm × 11 mm clearance area. The results show that the architectures with the 12 mm × 3 mm × 2.4 mm antenna booster provide similar behaviour. For the architecture with two antenna boosters, performance can be improved at the low-frequency region. The downside is that two antenna boosters are needed. Finally, with the architecture using a 30 mm × 3 mm × 1 mm modular antenna booster, performance is better than the others at both frequency regions, although the size of the element has been increased.

In contrast to the configuration employing 12 mm × 3 mm × 2.4 mm antenna booster with a matching circuit (Figure 5.28), the modular antenna booster architecture offers a broader bandwidth within the low-frequency region. This distinction is evident when comparing the S_{11} data in Figures 5.28 and 5.38, wherein the S_{11} consistently remains below −10.7 dB throughout the entirety of the low-frequency spectrum for the modular antenna booster. Therefore, this architecture is convenient if more bandwidth is required at the low-frequency region. For example, let us now force a bandwidth of 698–960 MHz on both architectures, the one with the 12 mm × 3 mm × 2.4 mm antenna booster and the one with the 30 mm modular antenna booster. The high-frequency region is kept the same at 1710–2690 MHz. Following the automatic synthesis procedure, the 12 mm × 3 mm × 2.4 mm achieves an average total efficiency of 56.2% and 81.2% for the low- and high-frequency regions, respectively. However, the 30 mm modular antenna booster achieves an average total efficiency of 61.6% and 87.9% for the low- and high-frequency regions, respectively. Moreover, the worst efficiency value within 698–960 MHz is 28.3% for the 12 mm × 3 mm × 2.4 mm antenna booster but 35.3% for the modular antenna booster. This 1 dB difference can be crucial in those applications where a wireless device needs to fulfil a TRP certification, where 1 dB may represent a difference between failure or success.

5.8 Multiband antenna boosters for MIMO

Antenna Booster Technology is also attractive for MIMO due to its reduced size, simplifying the embedding of the antenna booster in a wireless device [26–29].

The following example illustrates the architecture of a 4 × 4 MIMO multiband antenna system with four 10 mm × 3.2 mm × 3.2 mm (height) antenna boosters placed at each corner of a 156 mm × 90 mm ground plane (Figure 5.40).

Although the antenna boosters are electrically very small ($\lambda/37$ at 824 MHz), an efficient operation can be achieved thanks to the excitation of the ground plane. Since the ground plane is comparable to the wavelength of operation (0.42λ at 824 MHz), several modes are supported, and therefore the ground plane contributes to radiation. Integrating a 4 × 4 MIMO multiband antenna system in this platform size may be attractive for allowing the transmission and reception of high-speed data rates [30].

Since the input impedance of the antenna booster is not matched across the frequency bands of operation (824–960 MHz and 1710–2400 MHz), a passive multiband matching network following the teachings of Section 5.6 is designed, allowing simultaneous matching into a single port of the low- and high-frequency regions with five lumped components. For a more realistic simulation, the losses of the components have been considered.

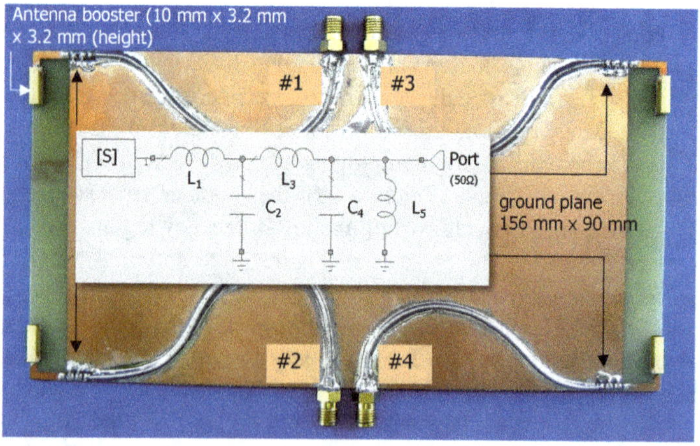

Figure 5.40 Prototype of the 4 × 4 MIMO multiband antenna system with 10 mm × 3.2 mm × 3.2 mm antenna boosters on a 156 mm × 90 mm ground plane. The matching network values for each port are L_1 = 7.1 nH, C_2 = 0.8 pF, L_3 = 9.5 nH, C_4 = 1.6 pF, and L_5 = 5.12 nH. Microcoaxial transmission lines have 50 characteristic impedance.

Automatic synthesis of multiband matching networks for antenna boosters

The automatic synthesis procedure is not only easy, fast, and automatic to design multiband matching networks for antenna boosters but also to optimise a MIMO antenna system, considering the coupling between antenna boosters.

For a MIMO antenna system, not only S_{ii} and S_{ij} ($i \neq j$) must be optimised (low mismatch loss and high isolation), but also a low correlation is required. To combine all these parameters into a single figure of merit, the multiplexing efficiency (η_{MUX}) has been proposed in the literature [31,32]. Multiplexing efficiency is defined as the difference in dB of the SNR (Signal to Noise Ratio) of an ideal MIMO system and the required SNR of the MIMO system under test to achieve the same channel capacity [31]. Multiplexing efficiency considers total efficiencies for each antenna, including radiation efficiency, mismatch loss, and coupling with neighbour antennas. Each total efficiency η_k ($k = 1,\ldots,4$) is included in Equation (5.1), valid for high SNR [32]. To calculate the correlation matrix R, the complex correlation coefficients r_{ij} between antennas i and j are needed, which are computed using full 3D complex patterns as in Equation (5.2), where each pattern contains the complex far-field components $E_\theta(\theta,\varphi)$ and $E\varphi(\theta,\varphi)$ for one antenna while the others are terminated with a 50 Ω load. For Equation (5.2), the scenario is an isotropic and non-polarized field.

$$\tilde{\eta}_{\mathrm{MUX}} = \left(\prod_{k=1}^{M} \eta_k\right)^{\frac{1}{M}} \cdot \det(R)^{\frac{1}{M}} \tag{5.1}$$

$$r_{ij} = \frac{\iint_{4\pi} \vec{E}_i(\theta,\varphi) \cdot \vec{E}_j^{\,*}(\theta,\varphi) d\Omega}{\sqrt{\iint_{4\pi} |\vec{E}_i(\theta,\varphi)| d\Omega \cdot \iint_{4\pi} |\vec{E}_j(\theta,\varphi)| d\Omega \cdot}} \tag{5.2}$$

As for Equation (5.1), it can be simplified as follows [29]. First, considering $r_{ij} = r_{ji}$

$$R = \begin{pmatrix} 1 & r_{12} & r_{13} & r_{14} \\ r_{21} & 1 & r_{23} & r_{24} \\ r_{31} & r_{32} & 1 & r_{34} \\ r_{41} & r_{42} & r_{43} & 1 \end{pmatrix} = \begin{pmatrix} 1 & r_{12} & r_{13} & r_{14} \\ r_{12} & 1 & r_{14} & r_{13} \\ r_{13} & r_{14} & 1 & r_{12} \\ r_{14} & r_{13} & r_{12} & 1 \end{pmatrix} \tag{5.3}$$

Then

$$\det(R) \approx \begin{vmatrix} 1 & r_{14} & r_{13} \\ r_{14} & 1 & r_{12} \\ r_{13} & r_{12} & 1 \end{vmatrix} = 1 - r_{12}^2 - r_{14}(r_{14} - r_{12} \cdot r_{13}) + r_{13}(r_{14}r_{12} - r_{13}) \approx 1 - r_{12}^2 - r_{13}^2 - r_{14}^2 \tag{5.4}$$

where the other minors have been neglected assuming $r_{ij} \ll 1$. Thus,

$$\tilde{\eta}_{\text{MUX}} = \sqrt[4]{\eta_1 \cdot \eta_2 \cdot \eta_3 \cdot \eta_4 \cdot \left(1 - r_{12}^2 - r_{13}^2 - r_{14}^2\right)} \tag{5.5}$$

Based on S_{ii} measurements, the prototype can operate from 824 MHz to 960 MHz and 1710–2400 MHz (Figure 5.41).

Regarding the coupling, the most critical situation is the coupling between ports #1 and #2, reaching −8 dB at the low-frequency region and less than −10 dB at the high-frequency region. For the other port combinations (S_{13}, S_{14}), the coupling is less than −15 dB, approximately, ensuring a low coupling between ports. It is worth pointing out that coupling does not always follow the rule that the larger the separation between the antennas, the less the coupling becomes. The coupling

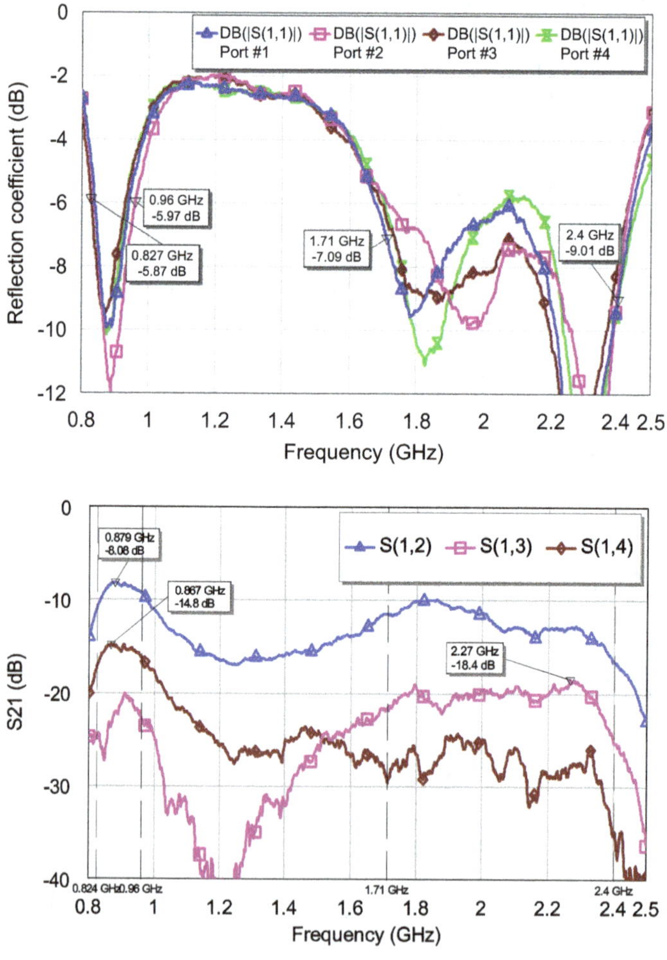

Figure 5.41 Measured S_{ii} and S_{ij} for each port of the prototype of Figure 5.40

not only depends on the distance between antenna ports but also depends on the modes excited in the ground plane [33,34]. For example, the coupling between ports #1 and #3 at the low-frequency region is −20 dB, but it is −14.8 dB between ports #1 and #4, which are further apart (Figure 5.41). In this regard, decoupling techniques are available to improve isolation between ports [35–40].

Results reveal average efficiencies of 36% and 54% for the low- and high-frequency regions, respectively. Since efficiency is not enough to characterise the MIMO antenna system, envelope correlation has been measured with the measured 3D far-field data. It is shown that correlations are below 0.3 across both low- and high-frequency regions [29].

Taking into account the measured total efficiencies at 900 MHz and 2000 MHz for each antenna booster and the correlation matrix R, the multiplexing efficiencies are $\eta_{MUX}(900\ \text{MHz}) = 36.2\%$ (−4.4 dB) and $\eta_{MUX}(2000\ \text{MHz}) = 53.0\%$ (−2.7 dB). As a reference, to achieve 15 bits/s/Hz, an ideal 4 × 4 MIMO requires an S/N ratio of 14 dB. If $\eta_{MUX} = 50\%$, it requires an additional 3 dB, whereas a SISO system would require an S/N ratio of 44 dB. This clearly demonstrates the advantage of a MIMO antenna system in increasing capacity while remaining compact, thanks to the small volume of antenna boosters.

5.9 Conclusions

The following Table 5.4 answers the objectives given at the beginning of the chapter. It also points to relevant figures and equations to consolidate the concepts learnt. It is also advisable to refresh all essentials included in the grey boxes distributed throughout the chapter.

Table 5.4 Key aspects to take away

Topic	**Review and take away**
Multiband performance with N boosters	Multiband performance can be achieved by using as many antenna boosters as frequency regions are needed (Figure 5.1). A first route requires a matching network for each antenna booster and a combiner to merge both ports of each antenna booster into a single port (Figure 5.1). The advantage of this path is that thc bandwidth at each band can be optimised and even enhanced with a broadband matching network (Figures 5.4 and 5.5). Antenna boosters can share the same corner of the ground plane to save space (Figure 5.7). However, coupling between antenna boosters must be taken into account with reject filters. Another strategy relies on integrating two antenna boosters connected through a transmission line (Figures 5.8 and 5.10). With this scheme, no filtering stages are needed. Hence, the number of lumped components of the system is reduced

(Continues)

Table 5.4 (Continued)

Topic	Review and take away
	compared to the previous designs. Although a transmission line is necessary, it can be easily printed on the PCB. A different architecture also comprises two antenna boosters, but in the first stage, a filtering network is included, and the second stage uses a matching network. The purpose of the filtering stage is that when both antenna booster feeding lines are connected at a common point, one branch does not load the other (Figure 5.14). Note that this architecture is different from the one in Figure 5.4 since, in this case, the first stage (seen from the antenna booster) is filtering, and the second stage includes a matching network that is the opposite of the previous scheme.
Multiband performance with a single antenna booster	To simplify the number of antenna boosters for multiband operation to only one antenna booster, multiband matching networks are proposed (Figure 5.1). The design of multiband matching networks can be addressed with human-based steps (Figures 5.15–5.20) and software-engineered processes on matching network synthesis procedures (Figures 5.27 and 5.28). Besides optimising bandwidth at each frequency region, modular antenna boosters may also optimise the bandwidth at two different frequency regions (Figure 5.38). From a practical perspective, the single antenna booster solution is simple to implement. However, the two-antenna booster solution provides flexibility to optimise bandwidth independently at each frequency region.
About component losses-Matching network with distributed elements	The quality factor of the matching network components (Q_m) should be as high as possible (high-Q). This is relevant for low frequencies of operation where Q_m may be closer to Q_a, and even more relevant when the electrical size of the ground plane is small because Q_a is high. Please refer to the conclusions outlined in Chapter 4 regarding this topic. A multiband matching network can also be implemented with distributed elements (Figure 5.36). The frequency region of operation determined its usage, typically beyond 5 GHz, where the size of transmission lines is compact enough to fit in the PCB of a wireless device.
Multiband MIMO	Due to the small size of antenna boosters, many of them can be distributed along the ground plane for MIMO purposes. In this regard, it is preferable to use a single booster rather than distributing multiple boosters in the PCB (Figure 5.40).

5.10 Exercises

A 3D electromagnetic software and a circuit simulator, including matching network synthesis, are needed to solve the following exercises.

5.10.1 Exercise 1

Consider a matching network with inductors and capacitors. Each inductor and capacitor may vary from a minimum to a maximum value, adopting 50 different values for each one. Calculate how many different matching networks with up to seven components are possible considering series-shunt combinations. Does it make sense to test the whole space?

5.10.2 Exercise 2

For a 12 mm × 3 mm × 2.4 mm antenna booster with a 120 mm × 60 mm ground plane, as shown in Figure 5.27, design a multiband matching network to provide operation at 824–960 MHz and 1710–2690 MHz. Do the first design with components having a low Q ($Q < 30$) and the second design with components having a better Q ($Q > 60$). Compute S_{11} and total efficiency for both solutions and analyse the impact of Q on total efficiency.

5.10.3 Exercise 3

Repeat the exercise in Section 5.10.2 with the matching network with better Q using component tolerances of 5% and 2%. Determine the impact of tolerance on the S_{11} and total efficiency.

5.10.4 Exercise 4

For a 12 mm × 3 mm × 2.4 mm antenna booster with a 120 mm × 60 mm ground plane shown in Figure 5.27, reduce the ground plane to 80 mm × 60 mm and 40 mm × 60 mm. Design a matching network for each case, providing operation at 824–960 MHz and 1710–2170 MHz. What are the drawbacks of reducing the ground plane?

5.10.5 Exercise 5

Consider a 12 mm × 3 mm × 2.4 mm antenna booster with a 120 mm × 60 mm ground plane, as shown in Figure 5.27. Increase the ground plane to 130 mm × 60 mm, 140 mm × 60 mm, and 200 mm × 200 mm. Design a matching network for each case for operation at 698–960 MHz and 1710–2170 MHz. What ground plane offers better performance? Why?

5.10.6 Exercise 6

Consider a 12 mm × 3 mm × 2.4 mm antenna booster as in Figure 5.27, but with an 80 mm × 70 mm ground plane instead of 120 mm × 60 mm. Design a multiband matching network to cover 863–928 MHz and 2400–2500 MHz. Use the same ground clearance as shown in Figure 5.27. Repeat the exercise, including a new frequency band from 1560 MHz to 1610 MHz. With the same set-up, design a matching network covering 2400–2500 MHz and 4900–5900 MHz. Then, try to expand the bandwidth at the second frequency region to 7000 MHz. Observe that in all cases, the antenna booster can be used, and only the matching network changes.

5.10.7 Exercise 7

For a 12 mm × 3 mm × 2.4 mm antenna booster with a 120 mm × 60 mm ground plane as the one shown in Figure 5.27, design a matching network for providing operation at 824–960 MHz and 1710–2170 MHz. Reduce the ground clearance from full clearance to 30 mm × 11 mm (Figure 5.42). Analyse the impact on S_{11} and total efficiency.

5.10.8 Exercise 8

For the scheme shown in Figure 5.43, design a multiband matching network for 824–960 MHz and 1710–2170 MHz. Compare your results with those in [16]. Expand the bandwidth to 698–960 MHz and 1710–2170 MHz. Finally, enhance the bandwidth at the second frequency region to include 1710–2690 MHz while keeping the 698–960 MHz. Note that the feeding line width is 2 mm and that the horizontal part is 5 mm.

5.10.9 Exercise 9

For exercise 5.10.7, analyse the impact of adding a plastic cover parallel to the ground plane at 2.5 mm above the ground plane, that is, 0.1 mm above the top face of the antenna booster. Model the plastic cover as a thin substrate 1 mm thick, $\varepsilon_r = 3$

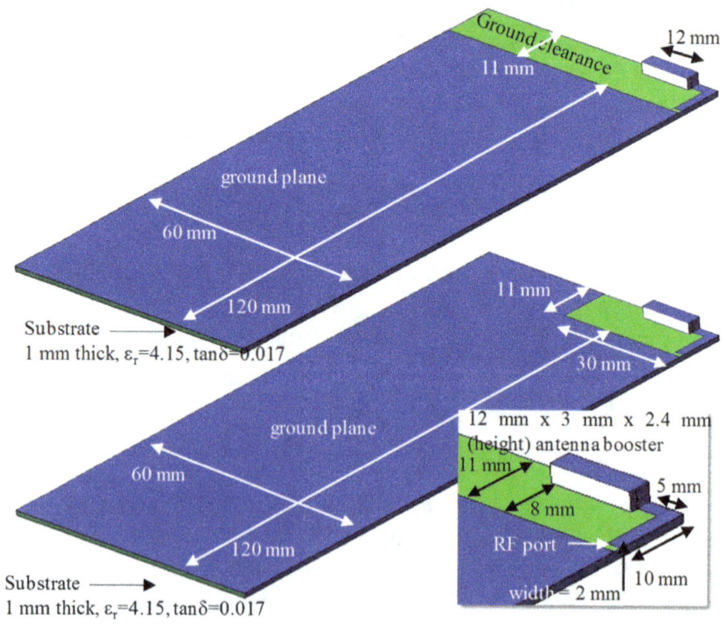

Figure 5.42 A 12 mm × 3 mm × 2.4 mm (height) antenna booster on a 120 mm × 60 mm ground plane with full and reduced ground clearance. Note that PCB size is 131 mm × 60 mm for both cases.

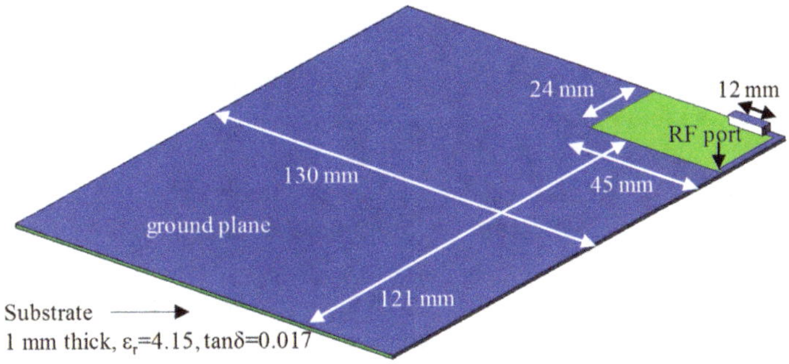

Figure 5.43 A 12 mm × 3 mm × 2.4 mm antenna booster on 145 mm × 130 mm PCB with a 45 mm × 24 mm ground clearance

and with a tanδ = 0.02. Analyse the impact on the S_{11} response as well as radiation efficiency. Tune the matching network if you observe any frequency shifting. Repeat the exercise by increasing the plastic cover to 3 mm above the ground plane. Compare the results concerning the 2.5 mm height.

5.10.10 Exercise 10

Calculate the potential bandwidth achievable with the modular antenna booster depicted in Figure 5.38, considering the scenario where the reject filter's placement divides the 30 mm element into two identical sections measuring 15 mm × 3 mm × 1 mm. Try several values of LC to find out the best way to obtain a trade-off between bandwidth at 824–960 MHz and 1710–2690 MHz. Once the reject filter is optimised, design a multiband matching network to cover said frequency range.

5.10.11 Exercise 11

Compare the 7 mm × 3 mm × 1 mm and 12 mm × 3 mm × 2.4 mm antenna boosters for achieving operation, including a 120 mm × 60 mm ground plane. Analyse for operating at 824–960 MHz and 1710–2170 MHz. Use potential bandwidth and design the multiband matching networks.

5.10.12 Exercise 12

Using the 7 mm × 3 mm × 1 mm antenna booster on the 200 mm × 200 mm ground plane of Figure 5.35, design a 4 × 4 MIMO system, placing each antenna booster at the corners of the ground plane. Besides S_{ii} and total efficiency at each port, compute isolation between ports 1 and 2, 1 and 3, and 1 and 4. Calculate the multiplex efficiency.

5.10.13 Exercise 13

Repeat the exercise 5.10.12, placing the four antenna boosters on the same edge. Compare the results.

References

[1] J. Anguera, A. Andújar, C. Puente, and J. Mumbrú, "Antennaless Wireless Device Capable of Operation in Multiple Frequency Regions," *Patent app.* WO2010/015364, Feb. 2010.

[2] J. Anguera, A. Andújar, C. Puente, and J. Mumbrú, "Antennaless Wireless Device Capable of Operation in Multiple Frequency Regions," *Patent* US. 8,237,615 B2, Aug. 2012.

[3] A. Andújar, J. Anguera, and C. Puente, "Ground Plane Boosters as a Compact Antenna Technology for Wireless Handheld Devices," *IEEE Transactions on Antennas and Propagation*, vol. 59, no. 5, pp. 1668–1677, 2011.

[4] A. Andújar, "Ground Plane Booster Antenna Technology for Wireless Handheld Devices," *Ph.D. thesis* at Universitat Politècnica of Catalunya, Spain. April 2013. Director: J. Anguera.

[5] J. Anguera, C. Picher, A. Andújar, and C. Puente, "Concentrated Antennaless Wireless Device Providing Operability in Multiple Frequency Regions," *Patent US 9*,379,443 B2, June 2016.

[6] J. Anguera, C. Picher, A. Andújar, and C. Puente, "Concentrated Antennaless Wireless Device Providing Operability in Multiple Frequency Regions," *Patent app.* WO2014/012796 A1, Jan. 2014.

[7] C. Picher, "Concentrated Ground Plane Booster Antenna Technology for Multiband Operation in Wireless Devices," *Ph.D. thesis* at Universitat Politècnica of Catalunya, Spain. Dec. 2015. Director: J. Anguera.

[8] C. Picher, J. Anguera, A. Andújar, and C. Puente, "Concentrated Ground Plane Booster Antenna Technology for Multiband Operation in Handset Devices," *Radioengineering*, vol. 23, no. 4, pp. 1061–1070, 2014.

[9] A. Andújar and J. Anguera, "Wireless Device Using an Array of Ground Plane Boosters for Multiband Operation," *Patent US* 10,224,631 B2, 2019.

[10] A. Andújar and J. Anguera, "On the Radiofrequency System of Ground Plane Booster Antenna Technology," *Electronics Letters*, vol. 48, no. 14, pp. 815–817, 2012.

[11] J. Anguera and A. Andújar, "Multi-structure Antenna for Multiband Operation," *Patent US* 10,347,983 B2, July 2019.

[12] J. Anguera, C. Picher, A. Bujalance, and A. Andújar, "Ground Plane Booster Antenna Technology for Smartphones and Tablets," *Microwave and Optical Technology Letters*, vol. 58, no. 6, pp.1289–1294, 2016.

[13] J. Anguera and A. Andújar, "A Non-Resonant Element for Multiband Operation in Smartphones and Tablets," *European Conference on Antennas and Propagation*, EUCAP 2016, Davos, Switzerland, 2016.

[14] J. Anguera, A. Andújar, and C. Puente, "Wireless Handheld Devices, Radiation Systems and Manufacturing Methods," *Patent US* 9,331,389 B2, May 2016.

[15] J. Anguera, A. Andújar, J. Rahola, and J. Juntunen, "Synthesis of Multiband Matching Networks for Non-Resonant Antennas," *Computing and Electromagnetics CEM'17*, Barcelona, Spain, 2017.

[16] J. Anguera, A. Andújar, G. Mestre, J. Rahola, and J. Juntunen, "Design of Multiband Antenna Systems for Wireless Devices Using Antenna Boosters," *IEEE Microwave Magazine*, pp. 102–114, 2019.

[17] J. Anguera, A. Andújar, J. L. Leiva, and R. M. Mateos, "Antenna Boosters: Easy Design of IoT Devices," *European Conference on Antennas and Propagation*, EUCAP 2021, Düsseldorf, Germany, 2021.

[18] J. Rahola, "Optimization of matching circuits for antennas," *European Conference on Antennas and Propagation*, EuCAP 2011, Rome, 2011.

[19] J. Rahola, "Power Waves and Conjugate Matching," *IEEE Transactions on Circuits and Systems—II: Express Briefs*, vol. 55, pp. 92–96, 2008.

[20] J. Rahola, "Estimating the performance of matching circuits for antennas," *European Conference on Antennas and Propagation*, EuCAP 2010 conference, Barcelona, 2010.

[21] D. Vye, "Network Synthesis Wizard Automates Interactive Matching-Circuit Design," *Microwave Journal*, pp. 96–102, 2018.

[22] J. Anguera, N. Toporcer, and A. Andújar, "Slim Radiating Systems for Electronic Devices," *Patent US* 9,960,478 B2, May 2018.

[23] J. Jeon, Y. Jung, and M. Kang, "Software-Based Adaptive Protection Control against Load Mismatch for a Mobile Power Amplifier Module." *Electronics*, vol. 8, p. 1226, 2019, https://doi.org/10.3390/electronics8111226

[24] T. W. Koo, H. S. Lee, J. G. Yook, K. Yoo, J. Cheon and S. Lee, "Radiated Spurious Emission Reduction Using Parasitic Element for Mobile Applications," *2014 International Symposium on Electromagnetic Compatibility*, Gothenburg, Sweden, pp. 760–764, 2014, doi:10.1109/EMCEurope.2014.6931006

[25] J. Anguera, A. Andújar, C. Puente, and R. M. Mateos, "Modular Multi-Stage Antenna System and Component for Wireless Communications," *Patent US* 11,482,772 B2, Oct. 2022.

[26] A. Andújar, J. Anguera, C. Puente, and C. Picher, "Wireless Device Capable of Multiband MIMO Operation," *Patent US* 8,952,855 B2, Feb. 10, 2015.

[27] A. Andújar and J. Anguera, "MIMO Multiband Antenna System Combining Resonant and Non-Resonant Elements," *Microwave and Optical Technology Letters*, vol. 56, no. 5, pp. 1076–1084, 2014.

[28] A. Andújar and J. Anguera, "MIMO Multiband Antenna System with Non-Resonant Elements," *Microwave and Optical Technology Letters*, vol. 57, no. 1, pp. 183–190, 2015.

[29] J. Anguera, A. Andújar, R. Mª Mateos, and S. Kahng, "A 4 × 4 MIMO Multiband Antenna System with Non-Resonant Elements for Smartphone Platforms," *European Conference on Antennas and Propagation*, EUCAP 2017, Paris, France, 2017.

[30] M. S. Sharawi, *"Printed MIMO Antenna Engineering,"* Artech House, 2014. ISBN 13: 978-1-60807-881-9.

[31] S. Zhang, A.A. Glazunov, Z. Ying, and S. He, "Reduction of the Envelope Correlation Coefficient with Improved Total Efficiency for Mobile LTE

MIMO Antenna Arrays: Mutual Scattering Mode," *IEEE Transactions on Antennas and Propagation*, vol. 61, no. 6, pp. 3280–3291, 2013.

[32] R. Tian, B. K. Lau and Z. Ying, "Multiplexing Efficiency of MIMO Antennas," *IEEE Antennas and Wireless Propagation Letters*, vol. 10, pp. 183–186, 2011.

[33] S. Zhang, K. Zhao, Z. Ying, and S. He, "Investigation of Diagonal Antenna–Chassis Mode in Mobile Terminal LTE MIMO Antennas for Bandwidth Enhancement," *IEEE Antennas and Propagation Magazine*, vol. 57, no. 2, pp. 217–228, 2015.

[34] J. Ilvonen, J. Holopainen, K. Rasilainen, A. Lehtovuori, and V. Viikari, "Suitable Multiantenna Placement in Mobile Handsets Based on Electromagnetic Isolation," *The 10th European Conference on Antennas and Propagation*, Eucap 2016, Davos, Switzerland.

[35] H. Islam, S. Das, T. Ali *et al.*, "Bandstop Filter Decoupling Technique for Miniaturized Reconfigurable MIMO Antenna," *IEEE Access*, vol. 10, pp. 19060–19071, 2022, doi:10.1109/ACCESS.2022.3150348.

[36] S. C. Chen, Y. S. Wang, and S. J. Chung, "A Decoupling Technique for Increasing the Port Isolation Between Two Strongly Coupled Antennas," *IEEE Transactions on Antennas and Propagation*, vol. 56, no. 12, pp. 3650–3658, 2008.

[37] C.-M. Luo, J.-S. Hong, and L.-L. Zhong, "Isolation Enhancement of a Very Compact UWB-MIMO Slot Antenna with Two Defected Ground Structures," *IEEE Antennas Wireless Propag. Lett.*, vol. 14, pp. 1766–1769, 2015.

[38] R. Anitha, P. V. Vinesh, K. C. Prakash, P. Mohanan, and K. Vasudevan, "A Compact Quad Element Slotted Ground Wideband Antenna for MIMO applications," *IEEE Transactions on Antennas and Propagation*, vol. 64, no. 10, pp. 4550–4553, 2016.

[39] S.-W. Su, C.-T. Lee, and F.-S. Chang, "Printed MIMO–Antenna System Using Neutralization-Line Technique for Wireless USB-Dongle Applications," *IEEE Transactions on Antennas and Propagation*, vol. 60, no. 2, pp. 456–463, 2012.

[40] M. M. Bait-Suwailam, M. S. Boybay, and O. M. Ramahi, "Electromagnetic Coupling Reduction in High-Profile Monopole Antennas Using Single Negative Magnetic Metamaterials for MIMO Applications," *IEEE Transactions on Antennas and Propagation*, vol. 58, no. 9, pp. 2894–2902, 2010.

Chapter 6

Antenna booster technology for multi-radio applications

This chapter introduces Antenna Booster Technology for multi-radio applications. Multi-radio refers to an antenna system needing operation at different radio protocols, each of which can be a single band or multiband. For example, a first radio provides operation at two frequency regions $[f_1, f_2]$ and $[f_3, f_4]$ as discussed in Chapter 5, a second radio provides operation at $[f_5, f_6]$, and a third radio at $[f_7, f_8]$. For instance, the first radio may provide operation for cellular communications, the second for geolocalisation, and the third for short-range communication. In this example, the radiofrequency may comprise a multi-radio transceiver or three independent transceivers. One way or the other, with the knowledge we have so far, we need one antenna booster for each frequency region. The question is: can we have a single antenna booster for multi-radio operation? This question is addressed in this chapter.

The chapter has the following objectives:

1. To understand the design of multi-radio architectures for antenna boosters for two radios
2. To scale the concept for designing architectures for any number of radios
3. To become familiar with multiplexers for antenna boosters

6.1 Introduction

Wireless devices encompass a variety of radio protocols, including cellular, geolocation, and short-range communication. Consequently, each wireless device typically necessitates separate antennas for each specific protocol (Figure 6.1 left). However, as the number of radios incorporated within a device increases, utilising a single antenna element with multi-radio capabilities becomes advantageous. This approach simplifies the antenna requirements and allows for more efficient utilisation of space, enabling the integration of additional electronic components (Figure 6.1, right). The objective of this chapter is to explore the design of wireless devices that integrate a single antenna booster with multi-radio functionality. Several architectures, from two to four radios, will be examined (Figures 6.2 and 6.3).

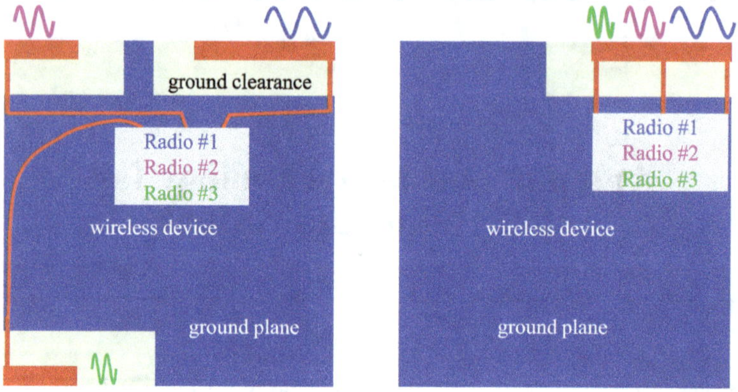

Figure 6.1 (Left) A wireless device for triple radio operation, including three antenna boosters. The waves represent the central frequency of each radio protocol. (Right) Advantages of using a multi-radio antenna booster: space for the antenna system is reduced, and hence, more space is available for integrating other components or making the device smaller (as long as the ground plane size allows competitive performance).

The principle of operation is based on creating as many circuit branches as radios. For example, let us suppose we seek functionality at two radios, one at the frequency region $[f_1, f_2]$ and the other at the frequency region $[f_3, f_4]$. The first branch will present a high impedance at $[f_3, f_4]$ and the other at $[f_1, f_2]$. For example, this high impedance can be achieved with a reject filter, where each filter can be designed to have a notch at the central frequency of each frequency region (Figure 6.2). Suppose we focus now on the branch having the notch filter at $[f_3, f_4]$. In that case, the reactance of the filter at $[f_3, f_4]$ and the reactive impedance behaviour of the antenna booster, including the ground plane, are matched with a matching network to operate at $[f_1, f_2]$. Similarly, the alternate branch features a reject filter positioned at $[f_1, f_2]$ and a matching network designed to match the impedance of the antenna booster at $[f_3, f_4]$. This includes accounting for the ground plane and residual impedance of the reject filter at $[f_1, f_2]$. This strategy allows an independent design of each radio frequency path and ensures high isolation between radios thanks to the notch filter. Other techniques, such as including higher-order filters, can be used. Several examples of two radio antenna boosters are discussed in Sections 6.2 and 6.3.

The question is now: how can we expand the architecture to support more than two radios? The strategy is based on a modular system (Section 6.4). Using Figure 6.2 as a reference, the small wave signals have the following codification: brown, blue, pink, and green, from the lowest to the highest frequency of operation. Each one corresponds to a central frequency of a given radio. In this manner, for a three-radio system, the strategy is to filter the frequency, which is much more

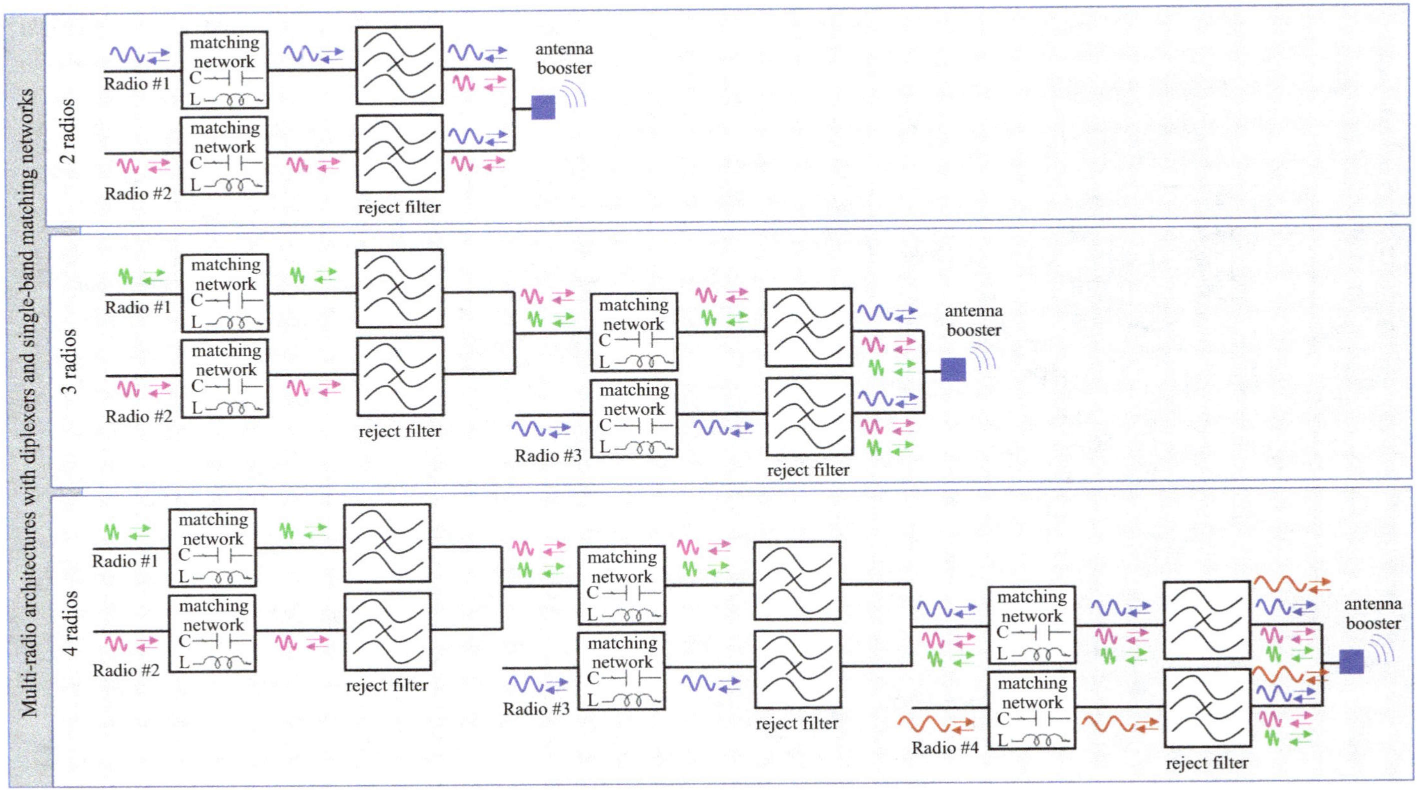

Figure 6.2 Multi-radio antenna booster strategies, from two up to four radios. More radios can be included following the modularity of the design method.

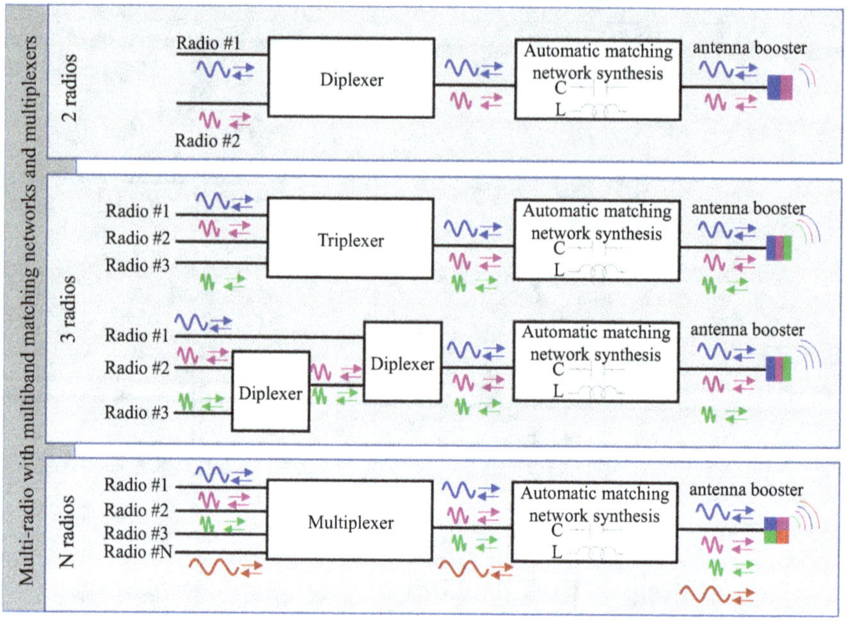

Figure 6.3 Multi-radio architectures based on multiband matching networks and multiplexers. Note that a diplexer, triplexer, or multiplexer can be substituted by a switch, as explained in Chapter 7, Exercise 10.

different than the others. For example, imagine $f_1 = 900$ MHz, $f_2 = 1600$ MHz, and $f_3 = 2500$ MHz. Since $f_1 \ll (f_2, f_3)$, a first stage separates f_1 in one branch and f_2 and f_3 in the other branch. In the following state, we have the same problem as a dual-radio, and thus, a new dual-branch system separates f_2 and f_3.

The process can now be scaled to four radios. Let us consider four radios with a central frequency of $f_1 = 900$ MHz, $f_2 = 1600$ MHz, $f_3 = 2000$ MHz, and $f_4 = 2400$ MHz. Similarly, as before, as $f_1 \ll (f_2, f_3, f_4)$, a first notch branch separates f_1 from the others, f_2, f_3, and f_4. In the second stage, a first filter separates f_2, and another filter divides f_3 and f_4. And last, f_3 and f_4 are again split into two branches with their corresponding filters.

It is important to emphasise that the filter should provide a high impedance to the blocking frequency so that the two branches can be connected. In effect, if one combines two impedances in parallel, Z_1 and Z_2, if $Z_1 \gg Z_2$, Z_1 is not loading Z_2. This is precisely the effect of each reject filter, as explained in more detail in Sections 6.2 and 6.3. These architectures are called, hereafter, self-diplexed.

Besides the self-diplexing architectures, another multi-radio strategy comprises a multiband stage, as proposed in Chapter 5, and a multiplexer (Figure 6.3). The multiplexer can be designed ad hoc or use an off-the-shelf component. With this approach, the first stage comprises a multiband antenna stage, providing matching to as many bands and radios as possible. Therefore, as a second stage, a multiplexer divides the one-path into several branches, as many branches as radios. For a dual-radio system, a diplexer is included; for a triple-radio system, a triplexer, and so forth. A combination of two diplexers may also provide three radios (Figure 6.3). For example, let us imagine a three-radio system for $[f_1, f_2]$, $[f_3, f_4]$, and $[f_5, f_6]$ in such a way that a triplexer is not available to split the three regions. One possibility is to include a diplexer to separate $[f_1, f_2]$ in one branch and $[f_3, f_4]$ and $[f_5, f_6]$ in another branch. After this stage, another diplexer splits $[f_3, f_4]$ and $[f_5, f_6]$. Note that this modular design is used for the self-diplexing structures in Figure 6.2. This approach has the advantage that the multiplexing stage can be ad-hoc as well as based on off-the-shelf multiplexers. The reason is that off-the-shelf multiplexers have 50 Ω antenna ports. In this case, the multiband matching network achieves a good matching, naturally not 50 Ω across the frequency region to match, but at least with an acceptable SWR, usually less than three, enabling the multiplexer to split the bands correctly. However, this is not the case with the self-diplexed strategy since the diplexing state works directly with the antenna booster, which is mainly non-resonant across the frequency bands of operation.

Another advantage of this architecture is that it is software-engineered since the multiband matching network can be automatically designed with a matching network synthesis, as explained in Chapter 5.

A design example using a diplexer and triplexer and a multiband matching network is detailed (Sections 6.5 and 6.6).

Finally, another strategy uses as many closely-spaced antenna boosters as radios (Figure 6.4) [1,2]. This architecture differs from a classic approach, as shown in Figure 6.1 left, since, in this new architecture, the antenna boosters are closely spaced to have a compact solution where the antenna booster can even be packed with the same package. For the scheme in Figure 6.1, each antenna booster is electrically spaced from one to the other, and thus, the mutual coupling is negligible. This means that each antenna booster has a matching network to provide operation at its corresponding radio. However, when antenna boosters are electrically close, like the mechanism proposed in Figure 6.4, a filtering stage is usually needed to ensure proper isolation between radios. For example, for a dual radio system centred at f_1 and f_2, a first reject filter at f_2 is added at the branch operating at f_1. Similarly, a reject filter at f_1 is included at the branch working at f_2. After the filtering stages, a matching network at each frequency region is added (Section 6.7).

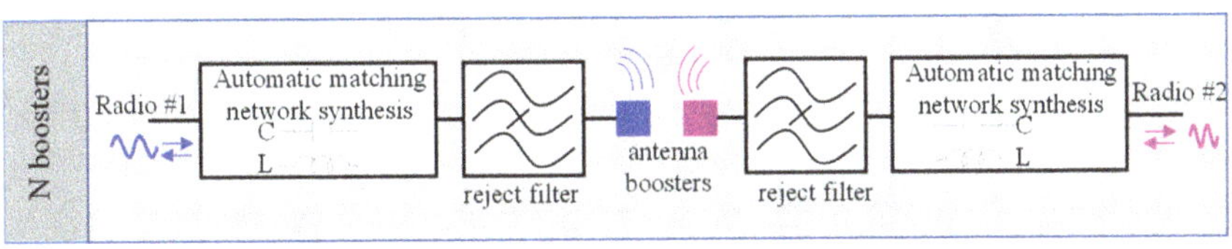

Figure 6.4 Multi-radio architecture comprising two closely spaced antenna boosters

For multi-radio architectures, besides S_{ii} at each port, the mutual coupling (S_{ij}, $i \neq j$) must be considered in the total efficiency for antenna booster i as:

$$\eta_{ti} = \eta_{ri} \cdot \left(1 - |S_{ii}|^2 - \sum_{\forall j \neq i}^{N} |S_{ij}|^2\right) \tag{6.1}$$

where η_{ri} is the radiation efficiency of the antenna booster i. Such radiation efficiency is computed for exciting port i while other ports j are loaded with 50 Ω.

If we reduce (6.1) to $N = 2$, the total efficiency for antenna boosters 1 and 2 is:

$$\eta_{t1} = \eta_{r1} \cdot \left(1 - |S_{11}|^2 - |S_{12}|^2\right) \quad \eta_{t2} = \eta_{r2} \cdot \left(1 - |S_{22}|^2 - |S_{12}|^2\right) \tag{6.2}$$

where S_{11} and S_{22} are the reflection coefficients at the port of antenna boosters 1 and 2, respectively. For example, if $S_{11} = -6$ dB and $S_{12} = -3$ dB, $\eta_{t1} = \eta_{r1} \cdot (1-0.25-0.5) = 0.25 \cdot \eta_{r1}$. However, if the mutual coupling is reduced to $S_{12} = -10$ dB, $\eta_{t1} = \eta_{r1} \cdot (1-0.25-0.1) = 0.65 \cdot \eta_{r1}$, representing 4.1 dB more radiated/received power. Therefore, the mutual coupling should be as low as possible, with −10 dB as a bearable reference, representing 10% of the power lost due to mutual coupling.

Different multi-radio techniques comprise classic antenna pairs such as dipole slots to ensure high isolation [3] and co-design with the front-end module [4]. Other approaches use reconfigurable solutions, which are useful as long as no simultaneous operation is required and the cost of the tunable component is competitive [5,6].

The conclusion part summarises the main findings (Section 6.8). Besides the contents, a list of exercises is proposed at the end of the chapter (Section 6.9), together with references at the end.

6.2 Easy self-diplexed design

The present section proposes a solution comprising only one antenna booster (a small metallic cuboid) to provide multi-radio operation in a wireless device (Figure 6.5). The term multi-radio refers here to the situation where each of the ports of the antenna booster is connected to a different radio module or a different

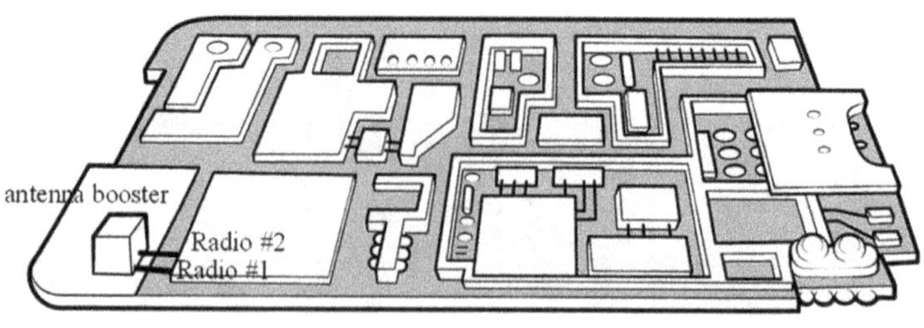

Figure 6.5 A proposed compact multi-radio concept with only one antenna booster to facilitate the integration of other components in a wireless device [7]

port of a multi-radio transceiver embedding several radios in the same chip. For example, imagine a situation that requires operation for geolocalisation (1.57 GHz) and short-range wireless connectivity (2.4 GHz). Instead of using one antenna booster for operating at 1.57 GHz and the other for 2.4 GHz, only one antenna booster is needed where each port is connected to the different ports of the multi-radio transceivers.

The multi-radio operation is achieved by a self-diplexed matching network that splits the radiofrequency signal into two paths, one for a low-frequency region and the other for a high-frequency region [7,8].

The name of the *self-diplexed matching network* comes from the fact that from a single booster element, two branches are created to provide two frequency regions of operation. This way, each band is matched and optimised independently from the others. The main advantage of the present solution is its small volume since only one antenna booster is needed.

A design example with a 12 mm × 3 mm × 2.4 mm (height) antenna booster including a 59 mm × 50 mm ground plane embedding a self-diplexing architecture with two notch filters, one centred at 1.58 GHz and the other at 2.44 GHz. After the filtering stage, a simple matching network with only one capacitor is enough to match each frequency region (Figure 6.6). Another route is explained in Section 6.5, where a diplexer chip is used instead of the one proposed here with lumped elements.

The S_{11} response provides at 1.45 GHz to 1.76 GHz (BW = ~19%, SWR = 3) and 2.21 GHz to 2.62 GHz (BW = ~17%, SWR = 3) for the first and second radios, respectively. The isolation between radios is below 10 dB (worst case), with two deep nulls at each of the resonant frequencies of the notch filters (Figure 6.7). This

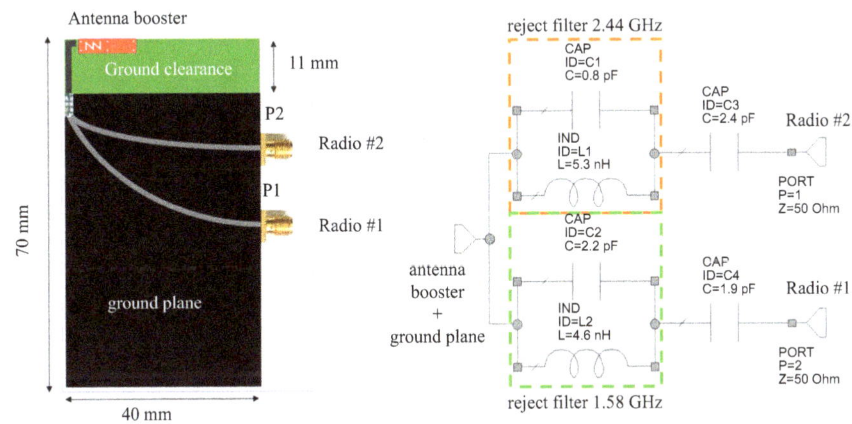

Figure 6.6 (Left) A 12 mm × 3 mm × 2.4 mm (height) antenna booster, including a 59 mm × 40 mm ground plane. (Right) A self-diplexed architecture for providing operation at two radios, one centred at 1.58 GHz and the other at 2.44 GHz.

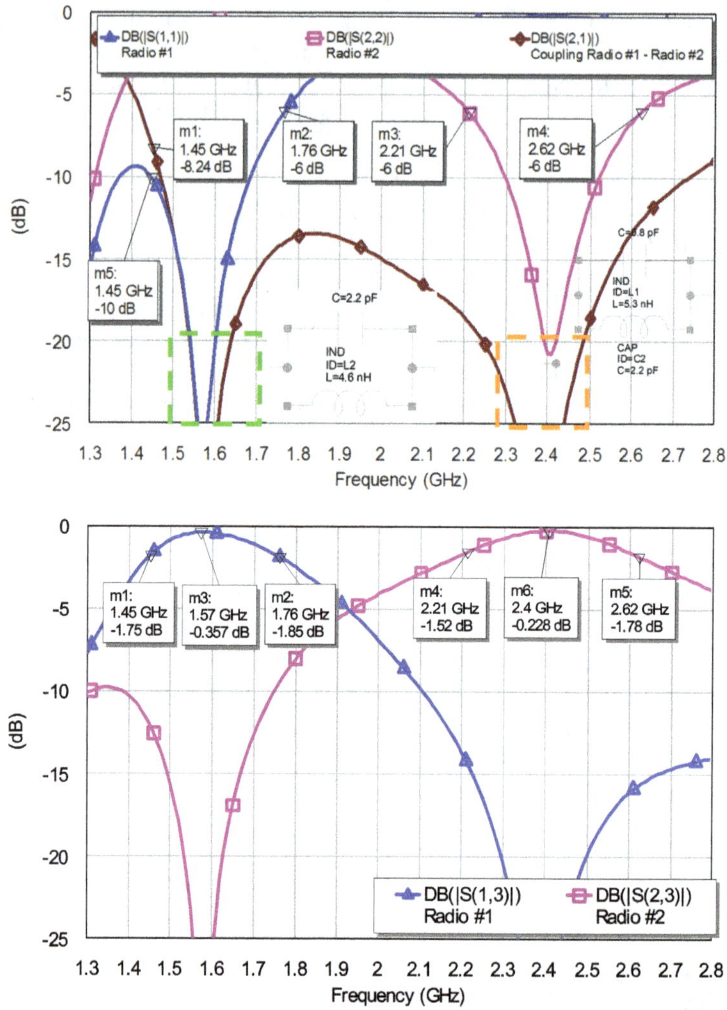

Figure 6.7 (Top) Reflection coefficients for radios #1 and #2 and coupling between ports. The two notches of S_{21} correspond to the notch filters shown in Figure 6.6. (Bottom) Transmission coefficients show the transferred power from radio #1 to the antenna booster and from radio #2 to the antenna booster. Real components with high Q have been considered for the simulation.

isolation can be easily shifted by modifying the resonant frequency of each filter to ensure the best isolation at the frequency of operation of each radio.

Note that for the first frequency region, the edge limit has been defined at 1.45 GHz (Figure 6.7 top), having an $S_{11} = -8.2$ dB and not -6 dB to ensure isolation across the frequency region better than 10 dB.

The insertion loss of the system is also represented by considering a port modelling the antenna booster impedance, including the ground plane. This way, the insertion loss for each radio port to the antenna booster is calculated (Figure 6.7 bottom). Insertion loss considers the S_{11} and S_{22} and the losses introduced by the finite quality factor (Q) of the components. If we take as a reference the frequency of minimum S_{11} and S_{22}, the insertion loss reflects the losses of the self-diplexing system, which are 0.35 dB and 0.22dB for the first and second radio, being a competitive value.

It is worth noting that reject filters can be avoided if a matching network can provide matching at the frequency of interest and, at the same time, a high impedance at the reject band, as shown in Section 6.3.

As deeply discussed in Chapters 4 and 5, tolerance analysis should be carried out to ensure a robust architecture system. This is typically achieved by selecting tight tolerance components with tolerances less than 2%.

Self-diplexed antenna booster architecture
A self-diplexed antenna booster architecture is convenient when an antenna booster needs to provide operation at different frequency regions, each connected to an input/output port of a transceiver. This solution avoids the need to use one antenna booster for each input/output port of the transceiver.

As far as the notch filter design is concerned, it is important to note that there are *infinite* LC pairs that result in a specific notch frequency. The product LC determines the bandwidth of the filter. Thus, the notch filter should be narrow enough not to introduce excessive reactance at the passband, making it challenging to match. For instance, the LC notch filter, depicted at the top of Figure 6.7 and consisting of 2.2 pF and 4.6 nH, exhibits an S_{21} below -10 dB between 1.446 GHz and 1.707 GHz, mirroring the passband's S_{11} characteristics. It is interesting for further comprehension to analyse the S_{21} response of the filters presented in this section and Section 6.4 to gain physical insight into the correct design of the notch filters, which is left as an exercise.

As observed, the design process is simple and can be scaled to any frequency region by tuning the resonant frequency of the filters and designing the matching networks.

6.3 Self-diplexed architecture for multi-radio applications

A design example illustrates a self-diplexed architecture to operate at two frequency regions comprising 0.79–0.972 GHz and 1.72–2.47 GHz. In this case, a broadband matching network consisting of a shunt LC is added to enhance the bandwidth at each frequency region, as explained in Chapter 4. The concept of self-diplexity is expanded to achieve more than two different frequency regions (Section 6.4).

6.3.1 The radiating system

The radiating structure of the proposed design consists of a 120 mm × 50 mm ground plane and a single 10 × 5 × 5 mm^3 antenna booster located in the upper corner of the ground plane. The antenna booster is non-resonant at the frequencies of operation (Figure 6.8) and features a small electrical size. For example, the 10 mm long antenna booster features only $\lambda/33$ at 0.9 GHz. As Chapter 3 explains, its strategic position helps the ground plane be efficiently excited and provides good radiation performance.

Besides the correct position of the antenna booster in the ground plane (at the corners, preferably), a matching network is needed to convert the reactive impedance into two well-matched frequency regions, one at a low-frequency region (frequencies close to 0.9 GHz) and another at a high-frequency region (frequencies close to 1.8 GHz). A self-diplexed matching structure achieves this objective by employing two independent paths for matching at each frequency region.

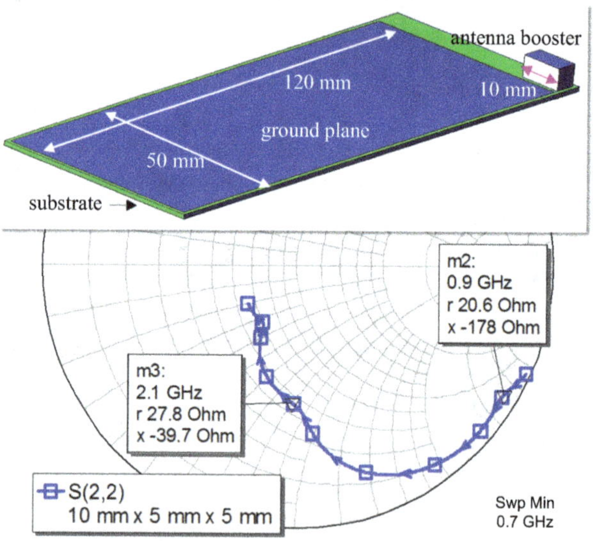

Figure 6.8 (Top) The radiating structures have a 120 mm × 50 mm ground plane with a 10 × 5 × 5 mm^3 antenna booster. The gap between the antenna booster and the ground plane is 1 mm. The ground plane is etched on a thin FR4 substrate 1mm thick, ε_r = 4.15, tanδ = 0.017. (Bottom) The antenna booster does not resonate at the desired frequency bands as it shows capacitive behaviour from 0.8 GHz to 3 GHz. A matching network is required to provide two matched frequency regions.

6.3.2 Self-diplexed architecture

The proposed solution to obtain a multiband performance with only one antenna booster uses a 2-port matching network, one for each frequency region. The lumped elements to match the proposed radiating structure form a self-diplexed matching network (Figure 6.9) with two branches (hereafter, the low- and the high-frequency branch).

The first element of the matching network (Figure 6.9) after the antenna booster is the inductor ($L_{s\ High}$). The value and position of this inductor are set to fulfil two main requirements. First, it must bring the reactive part of the input impedance of the booster at the high-frequency region (1.8 GHz) close to the resonance, and second, it also has to approach the low-frequency region to the resonance. Since the resonance for the low-frequency region is not reached by only this inductor $L_{s\ High}$, another series inductor ($L_{s\ Low}$) is added in the low-frequency branch to attain the resonance at 0.85 GHz.

Once the inductors are added, the element resonates separately at 0.85 GHz and 1.8 GHz. That is the first inductor, $L_{s\ High}$ (Figure 6.9), brings the antenna booster to resonance at one frequency of the high-frequency region, and then, with the $L_{s\ Low}$ together with the previous $L_{s\ High}$, both bring the antenna booster to resonance at one frequency of the low-frequency region. However, the coupling between regions makes the high-frequency region disappear. This is why a notch filter responsible for isolating both frequency regions becomes essential here. The notch filter comprises a single LC resonator stage, with the notch centred at

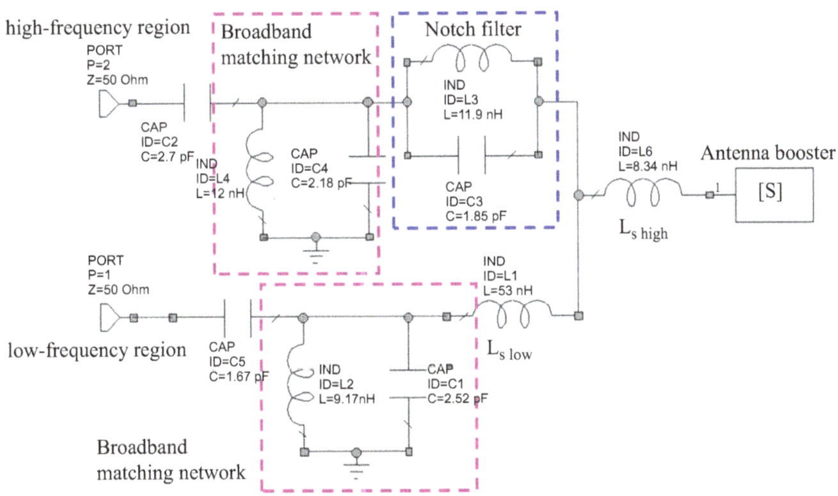

Figure 6.9 Self-diplexed architecture for dual-radio operation. [S] represents the input impedance of the antenna booster shown in Figure 6.8. Note that the low-frequency region branch does not have any notch filter since the series $L_{s\ low}$ (53 nH) to match the low-frequency region is high enough to block the high-frequency

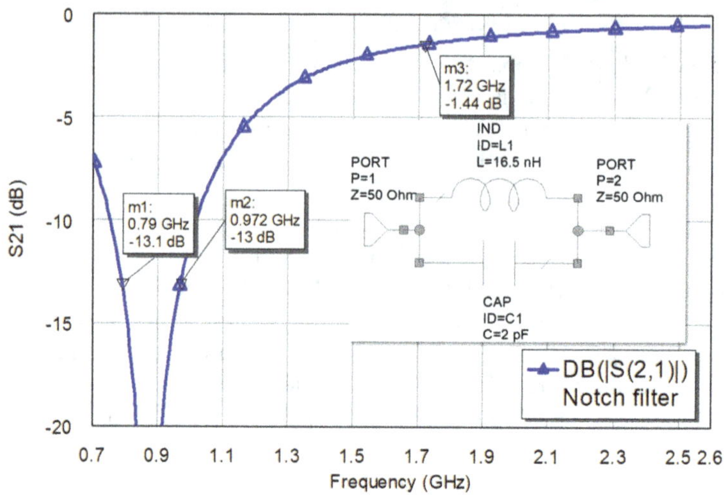

Figure 6.10 S_{21} of the notch filter designed to reject the frequencies of the low-frequency region (S_{21} at 0.9 GHz = −67.3 dB)

0.9 GHz (Figure 6.10). One key aspect to consider is that only one notch filter is needed because the second inductor in the low-frequency branch ($L_{s\ Low}$) presents a very high impedance at the high-frequency region, so it also acts as a notch filter besides bringing the impedance to resonance at the low-frequency region. Note that the nominal filter having $L = 16.5$ nH and $C = 2$ pF (Figure 6.10) is optimised in the whole chain (Figure 6.9) since the nominal filter has been designed considering 50 Ω ports at both ends, which is different from the real situation.

The filter should block the low-frequency region (0.79–0.972 GHz) and be transparent for the high-frequency region (1.72–2.47 GHz). Although the filter attenuates 1.4 dB at 1.72 GHz, the matching network compensates for the mismatching.

After the filtering stage, a broadband matching circuit, as explained in Chapter 4, is required to enhance the bandwidth at each frequency region. As it has been described, for RLC series circuits, the broadband network circuit is composed of a capacitor and an inductor, both connected in parallel. Their primary function is to create an impedance loop around the centre of the Smith Chart. Since the antenna booster (RC response) plus an inductor behave similarly to an RLC series circuit, a shunt LC network modifies the impedance to form a loop to obtain more bandwidth. Finally, as a fine-tuning stage, some other elements can be added (in this case, a series capacitor) to centre the impedance loop created by the broadband matching network at the centre of the Smith chart.

One of the main aspects to consider in a 2-port design is the transmission coefficient (S_{21}), which will be directly related to the isolation between regions. In the current scenario, the S_{21} parameter is lower than −14 dB, guaranteeing good performance in terms of efficiency in all operating bands. Besides, it presents good

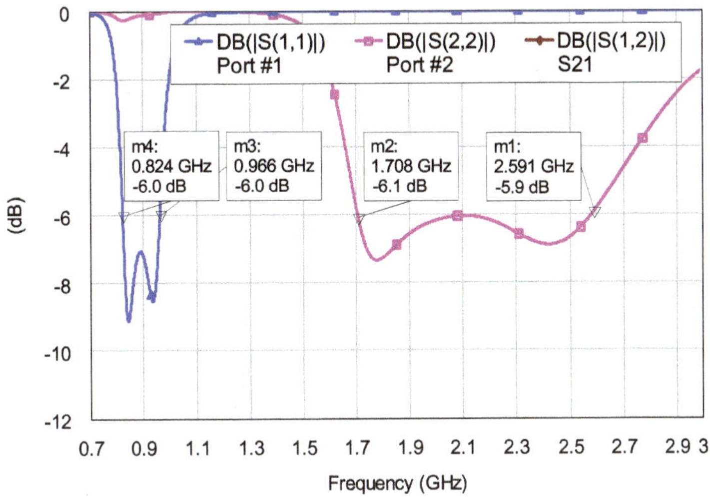

Figure 6.11 Simulated performance for the 10 × 5 × 5 mm³ antenna booster on a 120 × 50 mm² ground plane and the matching network of Figure 6.9

impedance matching (SWR $\leq$ 3) from 0.824 to 0.966 GHz and from 1.708 to 2.591 GHz in the high-frequency region (Figure 6.11). This scheme is selected to build a prototype to validate the simulation analysis (Section 6.3.3).

To give a physical insight, current distributions at both frequency regions on the ground plane have been simulated using MoM (Method of Moments). Figure 6.12(a) shows that the predominant mode (first mode) of the radiating structure at 0.9 GHz presents a current distribution similar to that produced by a half-wavelength dipole (null at the short edges of the ground plane and maximum at the centre on the long edge). At higher frequencies (Figure 6.12(b)–(d)), the current distribution differs from that obtained in the low-frequency region. Although the first mode is still predominant (the main current follows the short edge), other modes are also visible. This impacts the radiation patterns, as shown in the next section: at the low-frequency region, one *E*-field component predominates. However, in the high-frequency region, there is both E_θ and E_φ. Nevertheless, in the end, radiation patterns are low-directivity, making them suitable when the direction of the incoming signal is not known.

6.3.3 Experimental validation

Once the best design parameters of the radiating structure and the matching network are analysed through the simulation experiments, a prototype is made and tested to validate the simulation results. The ground plane is etched on a 1 mm thick FR4 substrate, and the antenna booster is made of brass (Figure 6.13), following the same dimensions as those in the simulation.

The final matching network of the built prototype consists of nine lumped elements, SMD 0402 (high *Q* tight tolerance). The series inductor (L_s high = 6.2 nH),

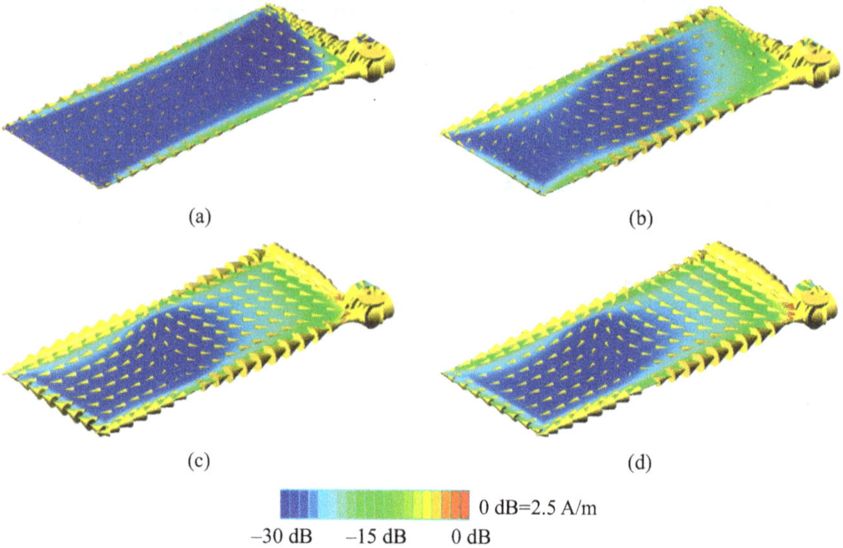

Figure 6.12 Current distributions of the radiating system considering the matching network at 0.9 GHz (a), at 1.8 GHz (b), at 2.16 GHz (c), and 2.4 GHz (d) (current max: 2.5 A/m). The colours and arrows indicate the density and direction of the current distribution, respectively.

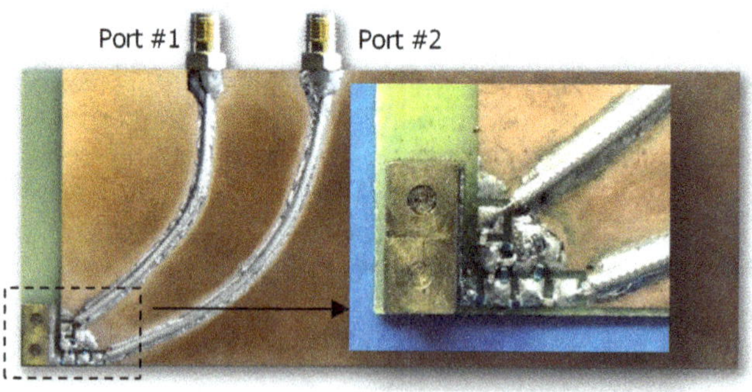

Figure 6.13 The built prototype follows the teachings of the simulation process. The ground plane size is 120 mm × 50 mm, the antenna booster is 10 × 5 × 5 mm^3, and two ports are required for the self-diplexed matching network shown in Figure 6.14. The gap between the booster and the ground plane is 1 mm. The ground plane is etched on a 1 mm thick FR4 substrate (ε_r = 4.15, tanδ = 0.017). Ports #1 and #2 are in charge of the low- and high-frequency regions, respectively.

which is in charge of providing resonance at the central frequency of the high-frequency region, is also responsible, with the $L_{s\ Low}$ = 56 nH, for providing resonance at the central frequency of the low-frequency region. After that, in the low-frequency branch, a parallel inductor (L_{PC} = 40 nH) and series capacitor (C_S = 1 pF) are needed to set the impedance inside the SWR = 3 circle. For the high-frequency branch, a notch filter (L_N = 12 nH and C_N = 1.8 pF) to isolate this branch from the low frequencies, and a broadband matching network (L_{BB}=3.4 nH, C_{BB}=0.75 pF) with a fine-tuning component (C_F = 1.3 pF) is used (Figure 6.14).

The prototype depicted in Figure 6.13, along with the matching network illustrated in Figure 6.14, yields a bandwidth of 20.7% (SWR $\leq$ 3) in the low-frequency region and 35.7% (SWR $\leq$ 3) in the high-frequency region. Hence, the overall radiating system can operate at the following frequency regions: 0.791–0.960 GHz and 1.710–2.484 GHz. Good isolation between ports is also shown in Figure 6.15 ($S_{21} < -11.5$ dB at both frequency regions).

Regarding the total efficiency measured using 3D pattern integration in the Satimo Stargate-32 anechoic chamber, the average total efficiency is 43.4% and 49.2% at 0.791–0.960 GHz and 1.71–2.69 GHz frequency regions, respectively (Figure 6.16).

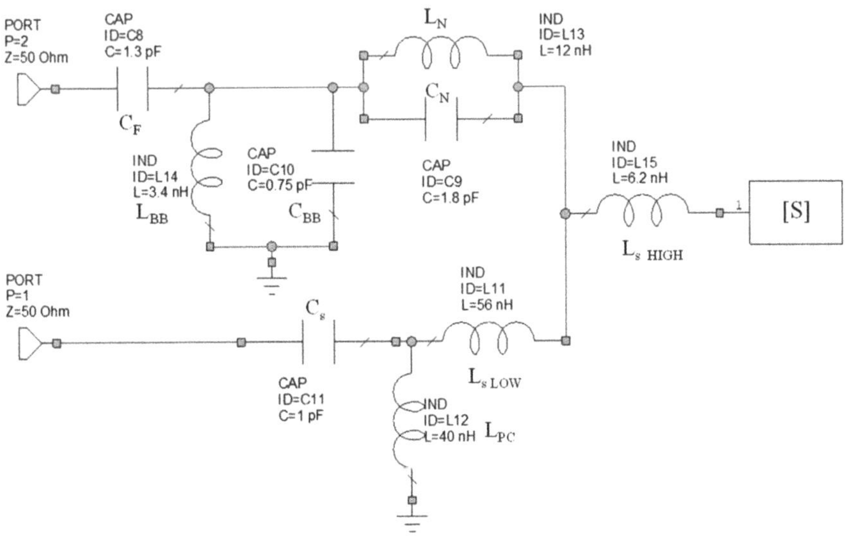

Figure 6.14 The matching network included in the prototype follows the stages described in Figure 6.9. In the low-frequency region, the broadband matching network adjusts slightly to compensate for the impedance difference resulting from the insertion of the pads in the layout. Ports #1 and #2 are in charge of the low- and high-frequency regions, respectively. [S] represents the input impedance of the antenna booster, including the ground plane without any matching network.

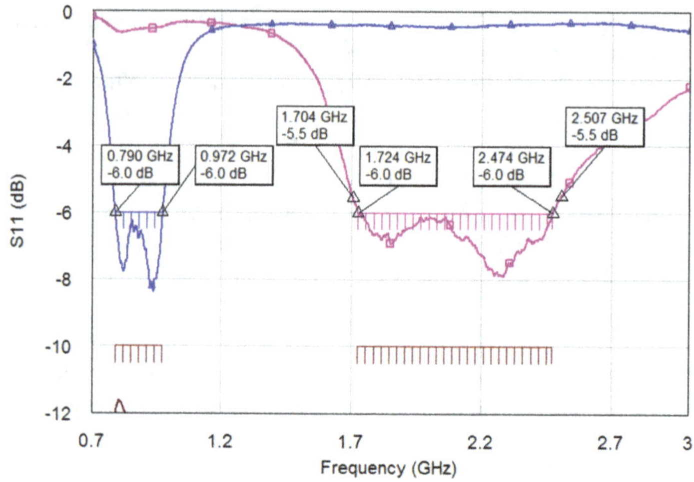

Figure 6.15 Measured impedance results from the radiating structure shown in Figure 6.13 and the matching network shown in Figure 6.14

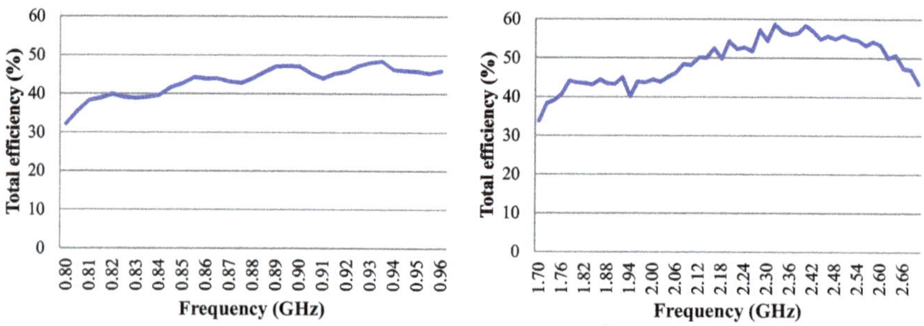

Figure 6.16 Measured total efficiency ($\eta_t = \eta_r \cdot (1 - |S_{11}|^2)$) of the built prototype

The measured directivity shows the low directivity behaviour of the present structure, which is convenient for those applications where the direction of the incoming signal is unknown (Figure 6.17).

Radiation patterns have also been measured in the anechoic chamber (Table 6.1). The main cuts normalised to the maximum gain ($\varphi = 0°$ and $\varphi = 90°$) are shown at 0.9 GHz, 1.8 GHz, and 2.3 GHz. This prototype presents an omni-directional behaviour in $\varphi = 0°$ and a drop in the y-axis at both frequency regions, which is the typical dipole-type pattern for this electrical size. Although a slightly different shape is observed at 1.8 GHz and 2.3 GHz with respect to the one obtained at 0.9 GHz, the radiation pattern presents a low directivity suitable for wireless applications. It is interesting to relate the radiation patterns to the current distribution shown in the previous section. In effect, at the low-frequency region, since the fundamental mode of the ground plane predominates, that is, the current is

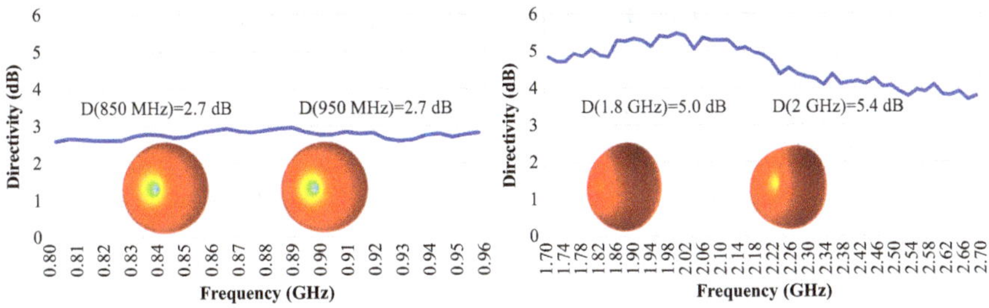

Figure 6.17 The measured directivity of the prototype is shown in Figure 6.13 for the low- and high-frequency regions. Low-frequency directivity values are around 3 dB, whereas in the high-frequency region, although the prototype shows higher values, directivity is always lower than 5.5 dB. Directivity is measured using 3D pattern integration.

Table 6.1 Measured radiation patterns at some of the frequencies of interest. The built prototype is aligned with the y-axis and lies in the x–y plane

	$\varphi = 0°$	$\varphi = 90°$	
0.9 GHz	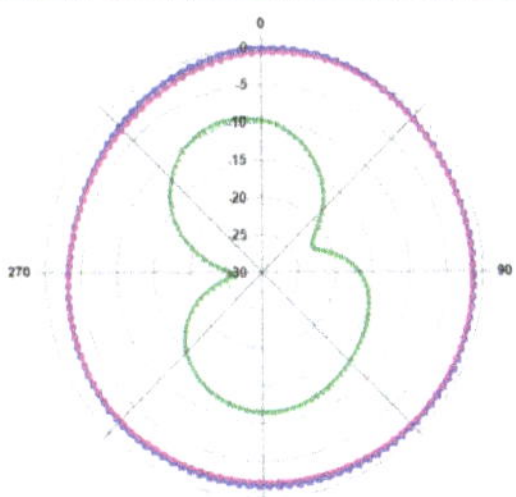	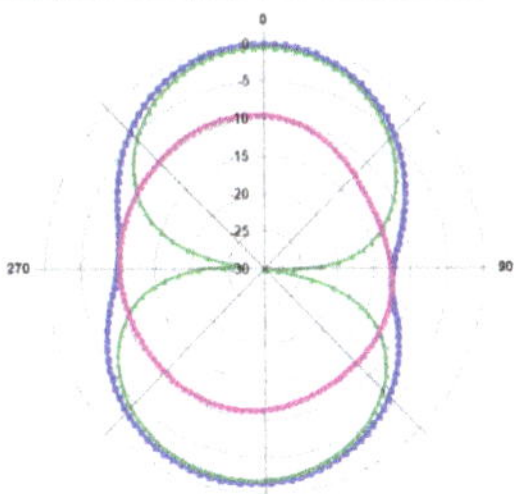	
1.8 GHz	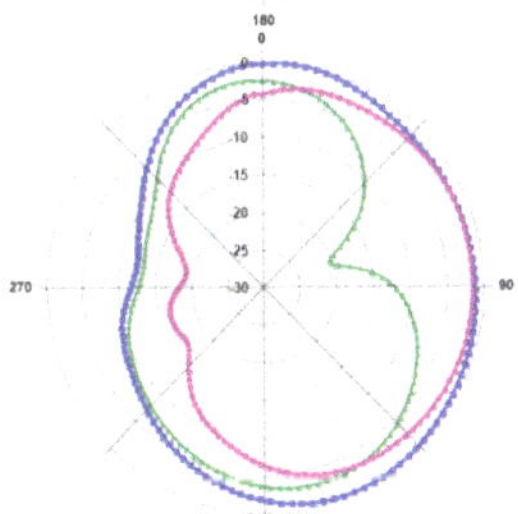	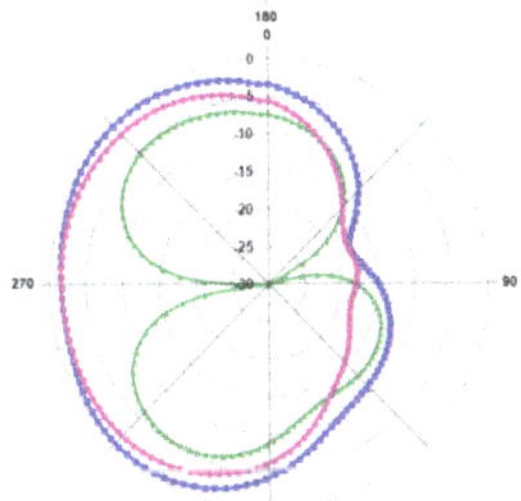	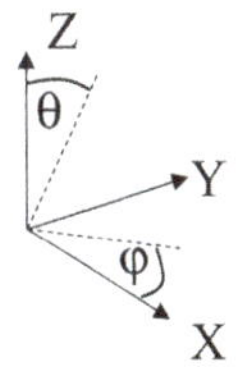
2.3 GHz	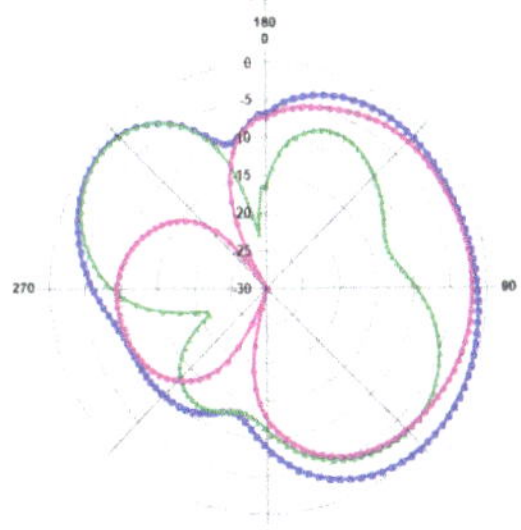	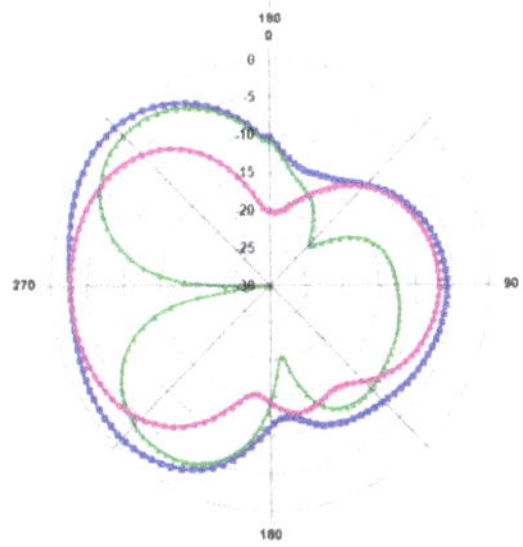	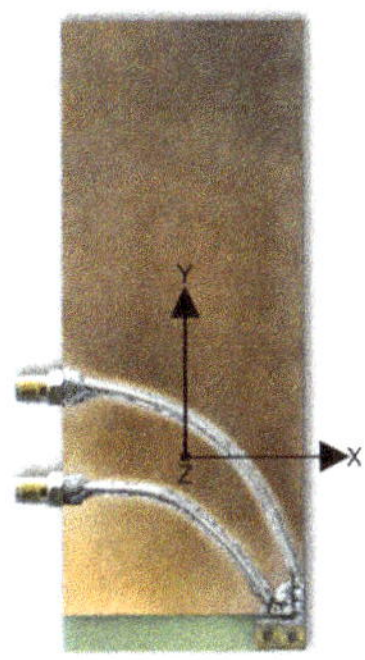

Table 6.2 Summary of the measurements results of the built prototype of Figure 6.13

Separation antenna booster and ground plane	1 mm
Matching network	Two ports: 9 elements
Volume	$10 \times 5 \times 5$ mm^3
f_1 [SWR = 3]	0.790 GHz
f_2 [SWR = 3]	0.972 GHz
BW_{f1-f2}	20.7%
f_3 [SWR = 3]	1.724 GHz
f_4 [SWR = 3]	2.474 GHz
BW_{f3-f4}	35.7%
Isolation between ports	>11.5 dB
$\bar{\eta}_t$ (LFR)	43.4% (0.791–0.960 GHz)
$\bar{\eta}_t$ (HFR)	49.2.% (1.710–2.484 GHz)

aligned with the long axis of the ground plane, the polarisation is linear along the long axis of the ground plane, showing an omnidirectional pattern at $\varphi = 0°$ and null along the long axis of the ground plane.

The results show that the antenna booster of only 250 mm^3 can operate at several frequency bands (Table 6.2). It should be emphasised that 250 mm^3 is a small volume compared to other technologies; in particular, an approximate reduction of a factor of ten is reached, as illustrated in Chapter 10, Table 10.3.

To summarise, a radiating system with a single-element solution and a self-diplexed matching network based on antenna booster technology have been presented. This technology comprises antenna boosters to efficiently excite currents on the ground plane. In particular, in this section, only one antenna booster is required, thanks to the self-diplexed matching network used to achieve two frequency regions of operation.

The measured results demonstrate that good matching and efficiency values can be obtained from 0.729 GHz to 0.972 GHz and from 1.71 GHz to 2.484 GHz using a 10 mm $\times$ 5 mm $\times$ 1 mm antenna booster.

Based on the obtained results, it can be concluded that the proposed technique is a valuable antenna solution because it can operate in two different radios while only requiring a small volume (250 mm^3) of a wireless device compared to other antenna solutions relying on complex geometries.

6.4 Modular design: from two to four radios

Now that the principles of dual-radio are understood, we move forward to scale the architectures to include more than two radios. In this design example, a dual-radio design is scaled to design a three- and four-radio architecture based on a modular design, as the concept shown in Figure 6.2. As a reference, a 12 mm $\times$ 3 mm $\times$ 2.4 mm (height) antenna booster, including a 120 mm $\times$ 60 mm ground plane, is used for the design (Figure 6.18).

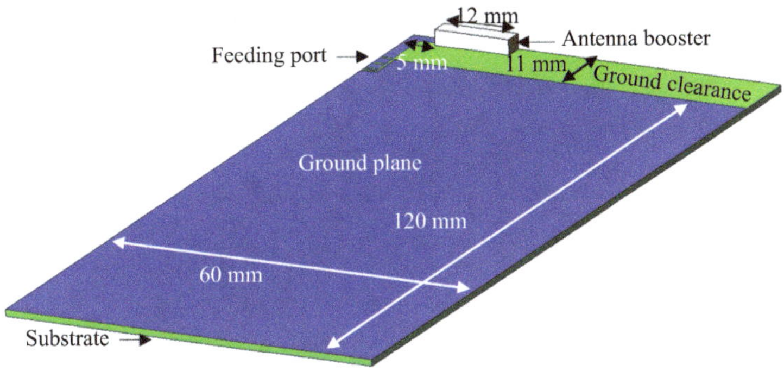

Figure 6.18 12 mm × 3 mm × 2.4 mm (height) antenna booster with a 120 mm × 60 mm ground plane. The ground plane is printed on a 1 mm substrate with $\varepsilon_r = 4.15$ and $\tan\delta = 0.017$.

The goal is to design a quad-radio operating at the central frequencies of 0.9 GHz, 1.57 GHz, 2 GHz, and 2.5 GHz. As observed, 0.9 GHz is far away from the other frequencies, and thus, the first stage is to design a dual-radio architecture where the first radio operates at the central frequency of 0.9 GHz and the second at 2.5 GHz, following the teachings of Section 6.2 (Figure 6.19). The architecture comprises a first filtering stage with two notch filters, one centred at 865 MHz and the other at 2.5 GHz, obtained after tuning the values to optimise performance. The second stage comprises a matching circuit to match each radio. With this scheme, each matching can be adjusted independently from the other.

After adjusting the filters and matching circuits (Figure 6.20), the architecture provides performance at a first-frequency radio from 0.862 GHz to 0.929 GHz ($S_{11} < -6$ dB), representing a bandwidth BW = 7.4%, and at a second-frequency radio from 1.55 GHz to 2.72 GHz, resulting in a bandwidth BW = 54.9%. The filters control isolation between radios, ensuring that isolation is better than 23 dB and 10 dB at the first- and second-frequency radios, respectively. The next step is to divide the second frequency range, as explained next.

The following step consists of reusing the dual-radio architecture to include a new third radio. In fact, the large bandwidth at the second radio of the dual-radio architecture is reused to split the frequency region into two frequency regions, one for the second radio and a third one for the new radio (Figure 6.21).

At this point, it is interesting to outline that either the self-diplexing scheme or an off-the-shelf diplexer can be employed since impedance at radio #2 (Figure 6.19) is well-matched and, thus, can be directly connected to a 50 Ω-based off-the-shelf diplexer. In this case, a lumped-based diplexer is designed (Figure 6.21) to create a second radio at 1.58 GHz and a third radio centred at 2.2 GHz (Figure 6.22).

For this new architecture, three radios are obtained: the first radio from 0.862 GHz to 0.929 GHz, a second radio from 1.56 GHz to 1.61 GHz, and a third radio covering from 1.69 GHz to 2.7 GHz, ensuring isolation better than 20 dB.

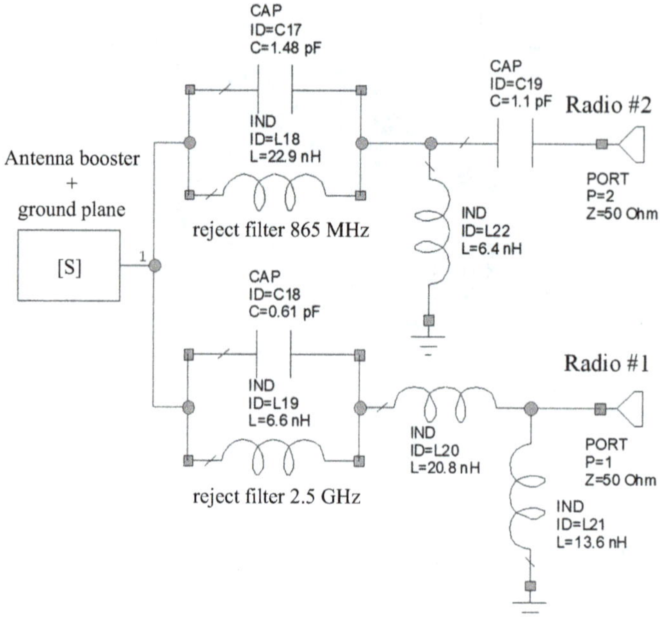

Figure 6.19 A 12 mm × 3 mm × 2.4 mm (height) antenna booster includes a 120 mm × 60 mm ground plane with the self-diplexing architecture to provide operation at two radios, one centred at 0.9 GHz and the other at 2.5 GHz as shown in Figure 6.20. The box [S] represents the simulated S_{11} of the antenna booster, including the ground plane, as shown in Figure 6.18.

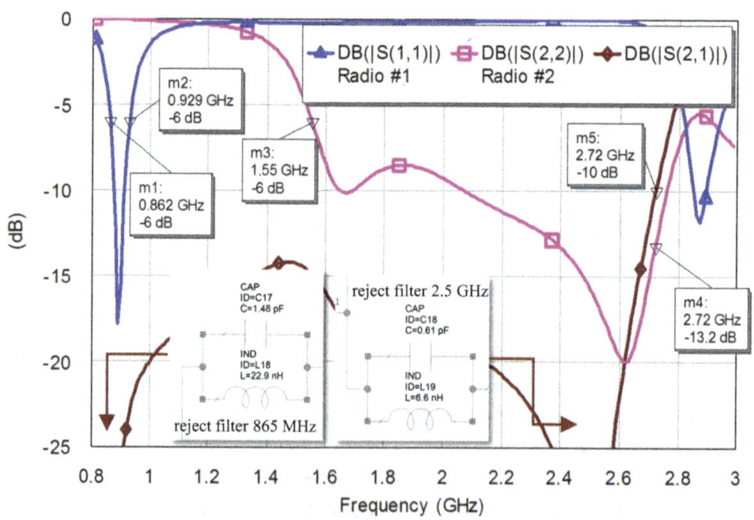

Figure 6.20 Simulated reflection coefficients for radio #1 and #2 as well as coupling between ports. The two notches of S_{21} correspond to the notch filters.

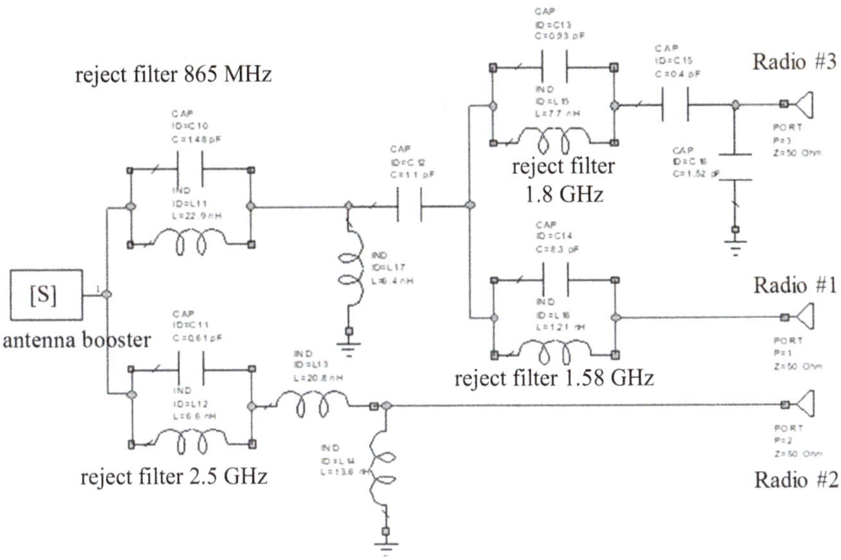

Figure 6.21 *12 mm × 3 mm × 2.4 mm (height) antenna booster including a 120 mm × 60 mm ground plane with the self-diplexing architecture to provide operation at three radios, radio #1 one centred at 0.9 GHz, radio #2 at 1.58 GHz, and radio #3 at 2.2 GHz as shown in Figure 6.22. The [S] represents the simulated S_{11} of the antenna booster, including the ground plane.*

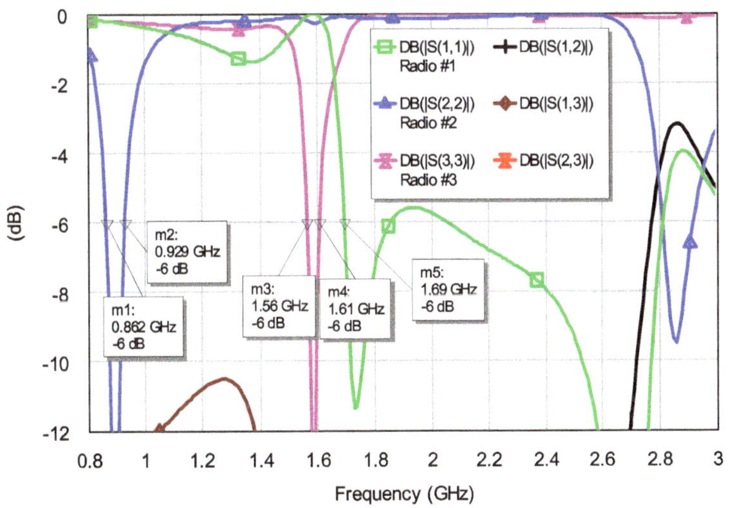

Figure 6.22 *Simulated reflection coefficients for radios #1, #2, and #3 as well as coupling between ports. Coupling is well below −20 dB, as shown in detail in Figure 6.23.*

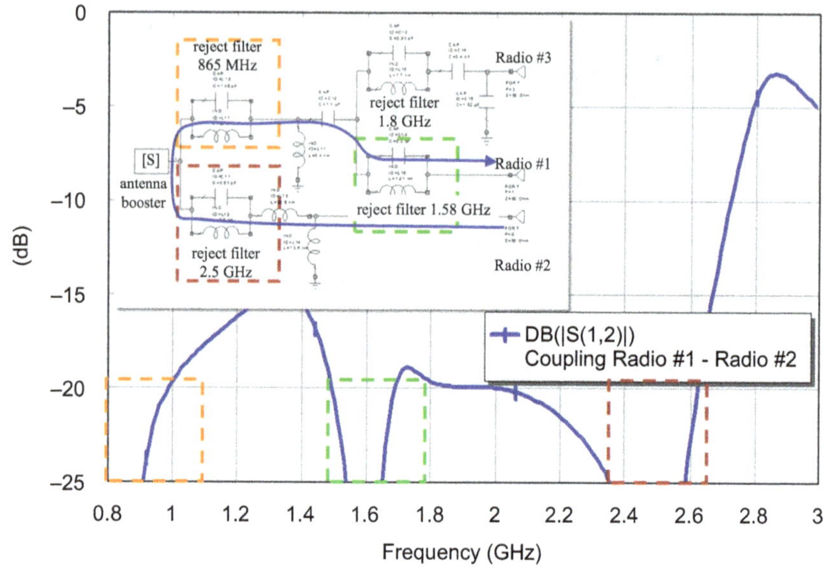

Figure 6.23 Example of coupling from radio #1 and radio #2, where the three notches correspond to each notch filter. Other couplings are in the same order (~−20 dB).

Similar to the dual-radio case, the isolation between each radio port can be adjusted by tuning the resonant frequency of the filters (Figure 6.23). As observed, each reject filter ensures high isolation (>20 dB) between radios (Figure 6.23). For example, the path between radios #1 and #2 includes three reject filters at 865 MHz, 1.58 GHz, and 2.5 GHz, causing notches at those frequencies (Figure 6.23).

Finally, a four-radio architecture is designed by adding a new diplexing stage at the third frequency region of the three-radio scheme (Figure 6.24). In this case, the third frequency region of the three-radio scheme of Figure 6.22, covering from 1.61 GHz to 2.7 GHz, is divided into two frequency regions: one from 1.68 GHz to 2.18 GHz and a second frequency region from 2.36 GHz to 2.54 GHz. This new frequency region is dedicated to the third and fourth radios, respectively (Figure 6.25).

As shown, the modular design first creates two frequency regions $[f_1, f_2]$ and $[f_3, f_4]$. Of these two, $[f_3, f_4]$ is split into two more, and so on. This design process provides flexibility to match a single antenna booster at several frequency radios, provided there is enough bandwidth, depending on the ground plane size, as studied in Chapter 3.

From a practical perspective, the architecture can be first designed, including the notch filters with the notch frequency at the desired frequency. Then, the matching circuit matches each frequency region corresponding to the radio port. Once the general architecture is established, the component values can be further

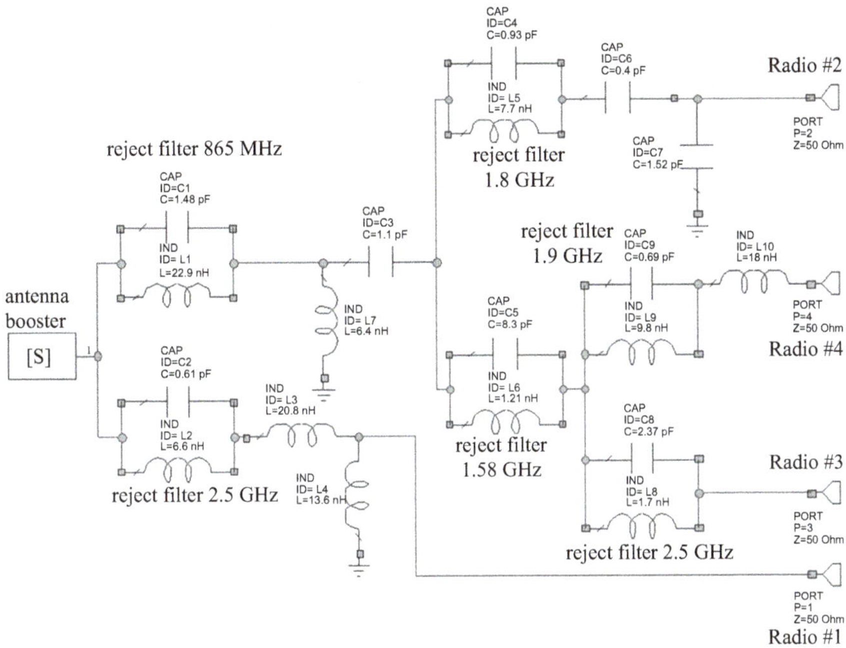

Figure 6.24 A 12 mm × 3 mm × 2.4 mm (height) antenna booster includes a 120 mm × 60 mm ground plane with the self-diplexing architecture to provide operation at four radios, radio #1 centred at 0.9 GHz, radio #2 at 1.58 GHz, radio #3 at 2 GHz, and radio #4 at 2.4 GHz, as shown in Figure 6.25. The [S] represents the simulated S_{11} of the antenna booster, including the ground plane.

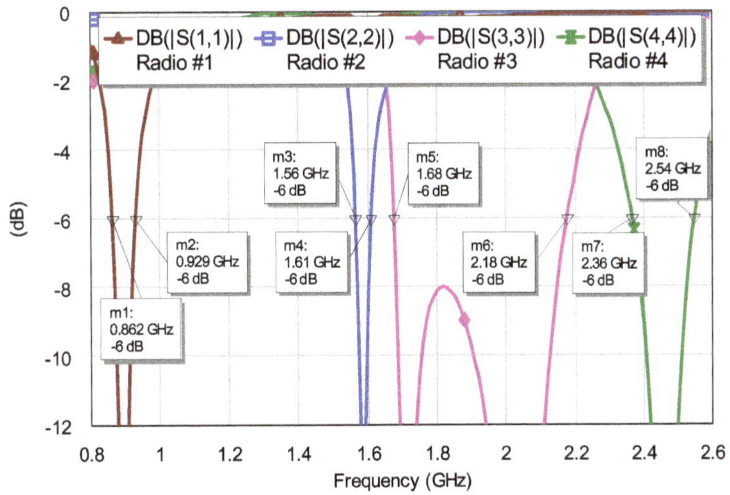

Figure 6.25 Simulated reflection coefficients for radio #1, #2, #3, and #4

optimised to squeeze performance, such as better isolation or bandwidth at a specific frequency band.

In the subsequent sections, Section 6.5 and Section 6.6, alternative design approaches are explored, focusing on multi-radio architectures utilising multiband matching networks and off-the-shelf multiplexers, as depicted in Figure 6.3.

6.5 Multi-radio comprising a multiband matching and a diplexer

This section proposes multi-radio architectures using multiband matching networks and off-the-shelf diplexers. The procedure consists of first matching the antenna booster, including the ground plane, at the frequency regions of the different radio ports. The design of the multiband matching networks is addressed in this section using the automatic synthesis matching network procedure, as deeply analysed in Chapter 5. As observed in this architecture, the multiband matching network must provide operation at the frequency bands of all radios the wireless device needs to operate at. Therefore, covering the whole frequency band with a multiband matching network may be a challenge. In those situations, the self-diplexing scheme is an alternative since it considers matching less bandwidth since it is carried out after the diplexing stage, where at least two frequency ranges have been divided.

Note that this architecture integrates a multiplexer that can be either designed as the previous sections with lumped elements or off-the-shelf components. Off-the-shelf multiplexers are based on technologies such as SAW (Surface Acoustic Wave) and BAW (Bulky Acoustic Wave) [9,10]. In this section, designs with off-the-shelf multiplexers are illustrated, first with a diplexer and then with a triplexer.

A first design example comprises a 12 mm × 3 mm × 2.4 mm (height) antenna booster, including a 120 mm × 60 mm ground plane (Figure 6.18). The input impedance is matched with an automatic synthesis procedure according to the teachings of Chapter 5 (Figure 6.26) [11]. The matching network comprises real components with high Q and tight tolerance (2%). It is designed to provide two frequency regions, one centred at 900 MHz and a second centred at 2.2 GHz. Therefore, an off-the-shelf diplexer is connected to provide two input/output ports.

It is observed that the reflection coefficient after adding the diplexers is almost the same, with some more bandwidth due to some losses introduced by the diplexer (Figure 6.27). If needed, fine-tuning or impedance broadening can be applied at radio ports #1 and #2. Isolation is below 13 dB and 18 dB at the first and second frequency bands of operation.

Finally, the self-diplexing (Figure 6.19) and the multiband/diplexer schemes (Figure 6.26) are compared (Figure 6.28). For the self-diplexing, eight lumped components are needed (four for the filters and four for the matching network). For the multiband/diplexer, there are five lumped elements and an off-the-shelf diplexer. These numbers should be part of the cost equations when choosing a solution for a wireless device design. In Figure 6.28, the performance comparison

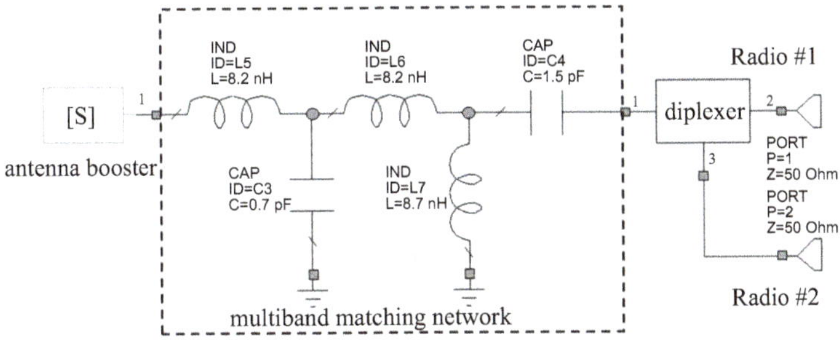

Figure 6.26 A 12 mm × 3 mm × 2.4 mm (height) antenna booster includes a 120 mm × 60 mm ground plane with the multiband+multiplexer architecture to provide operation at two radios. For the matching network, real components (high-Q) have been used. The diplexer features measured S-parameters from an off-the-shelf component. The [S] box represents the simulated S_{11} of the antenna booster, including the ground plane.

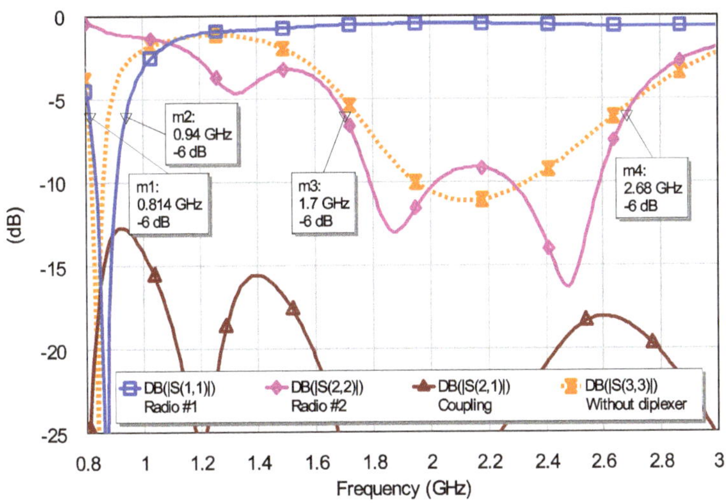

Figure 6.27 Simulated reflection coefficients for radios #1 and #2 as well as coupling between ports. The reflection coefficient before adding the diplexer shows the multiband response.

of insertion loss (technically, the negative of insertion loss) is presented for both schemes depicted in Figures 6.19 and 6.26. Both schemes incorporate real components for the lumped inductors and capacitors, as well as for the off-the-shelf diplexer. As shown, the solution with the self-diplexing scheme is more efficient, about 0.5 dB. In terms of engineering skills, the multiband/diplexer scheme can be addressed by software force; therefore, it is straightforward to design using synthesis procedures.

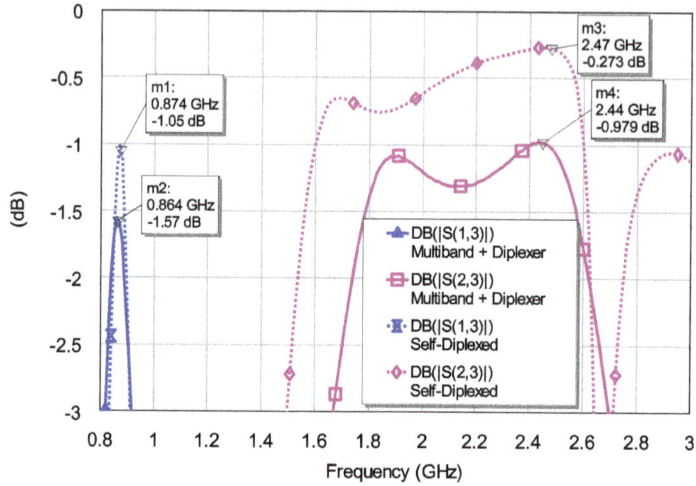

Figure 6.28 Transmission coefficient from each port to the antenna booster for the self-diplexing and multiband/diplexer schemes

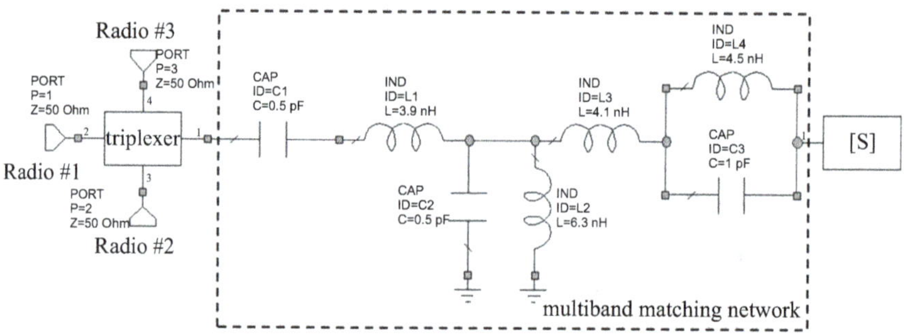

Figure 6.29 A 7 mm × 3 mm × 1 mm (height) antenna booster on a 200 mm × 200 mm ground plane is represented by [S]. A multiband matching network is synthesised and connected to a triplexer for multi-radio operation.

6.6 Multi-radio comprising a multiband matching and a triplexer

Similarly, as in the dual-radio architecture presented in Section 6.5, a triple-radio architecture is designed. A 7 mm × 3 mm × 1 mm antenna booster, including a 200 mm × 200 mm ground plane, is utilised as a reference. First, a multiband matching network is designed with an automatic matching network synthesiser according to the teachings of Chapter 5 (Figure 6.29). In this design example, a first frequency region is centred at 1.58 GHz, a second at 2.5 GHz, and a third at 5 GHz. Note that the antenna booster is $\lambda/27$ at 1.58 GHz. The matching network considers

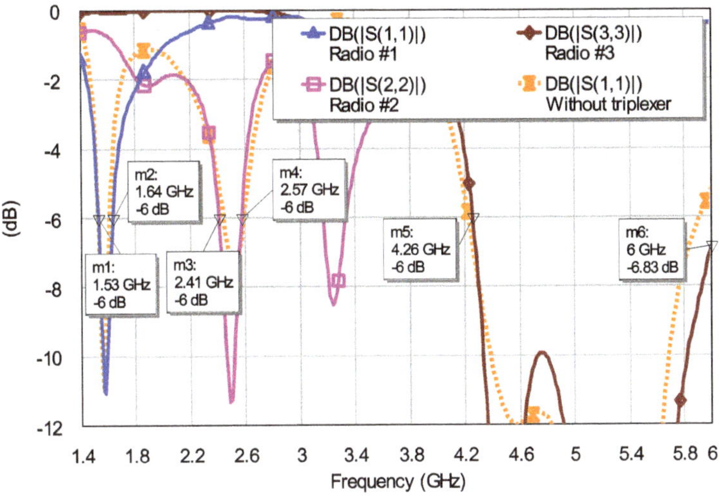

Figure 6.30 Simulated reflection coefficients for radios #1, #2, and #3. Isolation between ports is below 17.5 dB (worst case)

real components, that is, components with measured *S*-data. The S_{11} response provides three bands with 6.9%, 6.4%, and 34.1% SWR = 3 bandwidth at the first, second, and third frequency regions, respectively (Figure 6.30). Then, an off-the-shelf triplexer is connected, where the reflection coefficient response at each port is preserved. This design method is simple and systematic since the multiband matching network is addressed using a matching network synthesiser.

Multi-radio architecture with multiband matching networks and multiplexers
As long as a multiband matching network permits matching the antenna booster at the different radios, the multi-radio architecture may comprise a multiband matching network connected to the antenna booster and, after, a multiplexer.

6.7 Multi-radio with several closely spaced antenna boosters

Another multi-radio architecture comprises several closely spaced ($d < 0.02\lambda$) antenna boosters, as many as radios (Figure 6.4). If the antenna boosters are not closely spaced ($d \gg 0.02\lambda$), the design follows the design routes explained in Chapters 4 and 5 since there is negligible interaction between antenna boosters. Thus, the design of the matching networks for each radio is independent of each other. However, when the antenna boosters are closely spaced to have a compact solution, the interaction between antenna boosters must be taken into account.

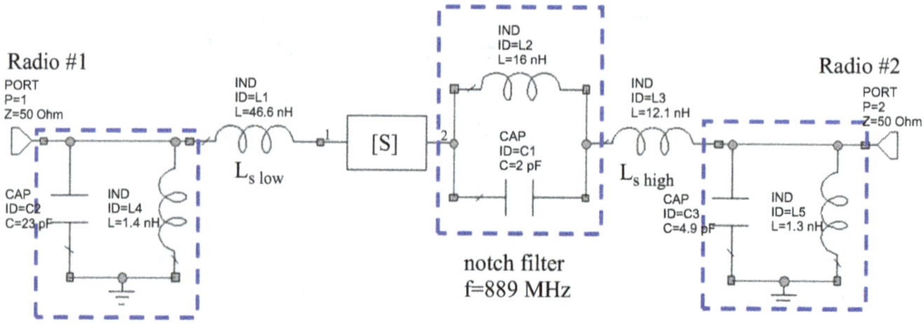

Figure 6.31 Dual radio architecture operating at 900 MHz and 2000 MHz, comprising two 5 mm × 5 mm × 5 mm antenna boosters closely spaced on a 100 mm × 40 mm ground plane. The [S] represents the S-data for the antenna boosters without any components. Broadband matching networks are optional.

One of the critical aspects is controlling the mutual coupling between the different radios (Equations (6.1) and (6.2)). If the mutual coupling is tight, part of the power from radio #1 will flow to radio #2, and thus, less power is radiated to space. To mitigate the mutual coupling, a notch filter is included at each branch [7]. For the present case, a notch filter having a notch at 889 MHz is included in radio #2, operating at a central frequency of 2 GHz (Figure 6.31). Other mitigation techniques to reduce mutual coupling include neutralisation lines (a conductive strip between the antenna pair) [12,13], stubs or slots on the ground plane [14,15], electromagnetic band gaps [16], and dielectric walls [17]. However, these techniques deal with antenna pairs operating at the same frequency where filters do not apply.

Since the 5 mm × 5 mm × 5mm antenna boosters, in this case, are non-resonant at any frequency of operation, a series inductor $L_{s\ high}$ is added after the teaching of Chapter 4. After the series $L_{s\ high}$, a broadband matching with a shunt LC network increases bandwidth. A notch filter can also be included for the low-frequency branch (900 MHz). However, in this case, since the series inductance to bring the antenna booster to resonance at 900 MHz is high (46.6 nH), it is already blocking the high frequency. Hence, the notch filter may be avoided for the sake of simplicity. Similarly, a broadband matching network is included after the series inductor $L_{s\ low}$.

To validate the concept, a prototype is built (Figure 6.32). Two 5 mm × 5 mm × 5 mm antenna boosters are located at the corner of a 100 mm × 40 mm ground plane with a gap of 2 mm. Optimising the scheme of Figure 6.31, *S*-parameters are tested, showing a first frequency range at 810–964 MHz and a second region at 1770–2170 MHz with isolation below 13 dB at the high-frequency region (Figure 6.33). If needed, the isolation can be improved with a notch filter. Note that isolation at the low-frequency region is better than 17 dB thanks to the notch filter at 889 MHz, although the antenna boosters are 2 mm spaced (0.006λ at 900 MHz).

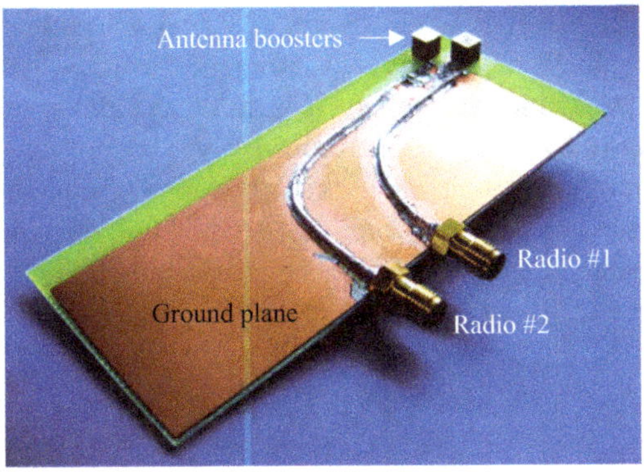

Figure 6.32 *Two closely spaced (2 mm) 5 mm × 5mm × 5 mm antenna boosters on a 100 mm × 40 mm ground plane. Radios #1 and #2 are centred at 900 MHz and 2000 MHz using the scheme shown in Figure 6.31.*

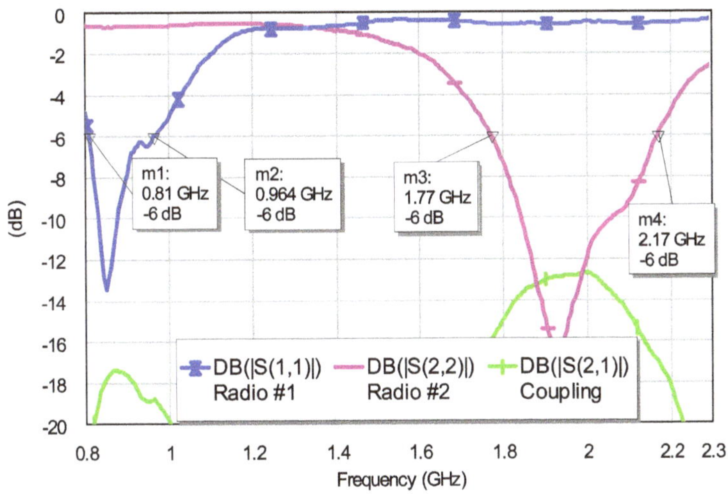

Figure 6.33 *Measured S-data. Although the antenna boosters are closely spaced at 900 MHz, the isolation is below 17 dB thanks to the notch filter.*

Multi-radio with *N* closely spaced antenna boosters

Each antenna booster provides operation at the frequency region corresponding to a radio. Since the antenna boosters are closely spaced to obtain a compact antenna booster system, mutual coupling is mitigated with reject filters.

The measured total efficiencies are tested in an anechoic chamber with 3D pattern integration. When testing the total efficiency of radio #1, the port at radio #2 is loaded with 50 Ω, and vice versa. As a result, the average total efficiency at radios #1 and #2 is 46.0% and 60.1%, respectively.

Hybrid solutions for multi-radio can combine the N-booster technique explained in this section with the teachings studied in the previous section. For example, in the present example, one booster can be exploited into two ports with a diplexing structure to provide triple radio operation [1].

6.8 Conclusions

Table 6.3 answers the objectives given at the beginning of the chapter. It also points to relevant figures to consolidate the concepts learnt. It is also advisable to refresh all essentials included in the grey boxes distributed along the chapter.

Table 6.3 Key aspects to take away

Topic	Review and take away
Multi-radio	When a wireless device operates on several radio protocols, a multi-radio antenna booster avoids using as many antenna booster components as radio protocols. Not only the number of antenna booster components is reduced, but more space is available to integrate other electronic components (Figure 6.1). However, as far as radio performance is concerned and space is not a constraint, the route with different antenna booster components is usually preferred since mutual coupling can be minimised by spacing the antenna booster components as much as possible. Hybrid approaches are also candidates in the design process where one antenna booster component groups several radios, and a single antenna booster component is reserved to cover the most challenging case.
Self-diplexed architectures	Self-diplexed architectures based on reject filters enable multi-radio (Figures 6.2 and 6.6). For a dual-radio operation centred at f_1 and f_2, two reject filters at f_1 and f_2 are connected to the antenna booster. Then, matching networks at f_1 and f_2 provide impedance matching. High isolation between ports is ensured thanks to the reject filters. This scheme can be scaled to include more radios (Figures 6.24 and 6.25).
Multiplexers	Multiplexers can also be employed to design multi-radio systems with antenna boosters. The antenna booster system is first matched with a multiband matching network (Figures 6.29 and 6.30). The design of such a multiband matching network can be addressed with the teachings of Chapter 5. This route requires existing multiplexers, whereas in the previous approach, the designers had more freedom to adjust the filters to the frequencies of operation. Both techniques should be compared when designing a multi-radio system to decide which is best in terms of performance, simplicity, and cost.

6.9 Exercises

A 3D electromagnetic software and a circuit simulator, including matching network synthesis, are needed to solve the following exercises.

6.9.1 Exercise 1

Consider the architecture of Figure 6.19. Repeat the results, varying the LC product of the filters. Do it for the reject filter at the low band (L = 22.9 nH, C = 1.48 pF) using as an example the following pairs: L(nH), C(pF) = {(10, 3.38), (15. 2.25), (20, 1.69), (25, 1.35), (30, 1.13)}. Note that all pairs keep the same original LC product = 22.9 nH × 1.48 pF.

6.9.2 Exercise 2

Consider Figure 6.6 but with a 7 mm × 3 mm × 1 mm antenna booster and a 40 mm × 5 mm ground clearance. Design a self-diplexing architecture to provide operation at two radios: one for 2.4–2.5 GHz and the other for 3–5 GHz. Compute the reflection coefficient at each port and the isolation between ports. Tune the reject filters to analyse how the notch can shift the isolation between ports. Compute the total efficiency considering real components.

6.9.3 Exercise 3

Repeat the previous exercise, 6.9.2, using a multiband-diplexer architecture. Compare results in terms of total efficiency, design time, cost of the components, and complexity.

6.9.4 Exercise 4

Consider the architecture of Figure 6.19. Remove the reject filter for the high-frequency region with a matching network that creates a high impedance at the low-frequency range (2.4–2.5 GHz) and simultaneously provides matching for the high-frequency region (3–5 GHz).

6.9.5 Exercise 5

Consider the architecture in Figures 6.13 and 6.14. Remove the broadband matching circuit and compute the reflection coefficient. Compare bandwidths with and without the broadband matching circuit.

6.9.6 Exercise 6

With the PCB of Figure 6.6 but a ground plane of 79 mm × 70 mm, design a dual radio system to provide operation at 863–928 MHz and 1560–1606 MHz. Design two-notch filters with lumped elements and their corresponding matching networks. Compute the reflection coefficient, isolation, and total efficiency at each port. Run a tolerance analysis considering 2% and 5% component tolerances and compare robustness.

6.9.7 Exercise 7

Repeat the previous Exercise 6.9.6 with lumped components having a worse Q, for example, Q about 30. Analyse the impact in terms of reflection coefficient, isolation, and total efficiency.

6.9.8 Exercise 8

Repeat Exercise 6.9.6 considering a dual-band 863–928 MHz and 1560–1606 MHz matching network and a diplexer. Use an off-the-shelf diplexer with measured S-parameters. To design the dual-band matching network, follow the teachings of Chapter 5.

6.9.9 Exercise 9

Repeat Exercise 6.9.6, considering a 59 mm × 40 mm ground plane. Which of the two frequency bands is more affected? Why?

6.9.10 Exercise 10

For Exercise 6.9.9, design a broadband matching network at the port operating at low frequencies (~900 MHz). Refer to the teachings of Chapter 4.

6.9.11 Exercise 11

With the PCB of Figure 6.6 but a ground plane of 79 mm × 70 mm, design a dual-radio system to provide operation at 863–928 MHz and 2400–2500 MHz. Integrate two notch filters with lumped elements and their corresponding matching networks. Compute the reflection coefficient, isolation, and total efficiency at each port.

6.9.12 Exercise 12

Repeat Exercise 6.9.11 with a dual-band 863–928 MHz and 2400–2500 MHz matching network and an off-the-shelf diplexer. To design the dual-band matching network, refer to the teachings of Chapter 5. Compare efficiency with the previous Exercise 6.9.11.

6.9.13 Exercise 13

Consider a 12 mm × 3 mm × 2.4 mm (height) antenna booster on an 80 mm × 40 mm ground plane spaced 3 mm from the edge of the ground plane. With the antenna booster at the corner, employ a feeding line of 3 mm × 1 mm. Design a multiband matching network for 1.5–1.6 GHz and 2.4–2.5 GHz. Add an integrated diplexer, including its real S-parameters. Compute S_{11}, S_{22}, S_{21}, and total efficiency. Repeat the exercise, considering a pair of reject filters instead of an integrated diplexer.

6.9.14 Exercise 14

Consider an 80 mm × 50 mm ground plane. Place a 7 mm × 3 mm × 1 mm antenna booster at the corner of the ground plane, spaced 3 mm from the edge (use a feeding line of 3 mm × 1 mm to feed the antenna booster). Add another antenna booster of the same size, 2 mm from the first. Add a second feeding line of the same size. Compute S_{11}, S_{22}, and S_{12} between 1.5 GHz and 2.5 GHz. With the scheme shown in Figure 6.31, design a dual-radio antenna booster system to provide operation at a first radio #1 between 1.5 GHz and 1.6 GHz and for a second radio #2 between 2.4 GHz and 2.5 GHz. Note that broadband matching networks are optional. Compute the total efficiency at each port.

References

[1] J. Anguera, A. Andújar, C. Puente, and R. Mateos, "Modular Multi-stage Antenna System and Component for Wireless Communications," *Patent US* 11,482,772 B2, Oct. 2022.

[2] J. Anguera and A. Andújar, and C. Puente, "Compact Antenna Technology for Wireless Communications," *Patent App.* WO 2020 0120589A1, June 2020.

[3] S-Y. Suh, V. K. Nair, D. Souza, and S. Gupta, "High Isolation Antenna for Multi-Radio Antenna System Using a Complementary Antenna Pair," *2007 IEEE Antennas and Propagation Society International Symposium*, pp. 1229–1232, 2007, doi: 10.1109/APS.2007.4395723.

[4] K. R. Boyle, M. Udink, A. de Graauw, and L. P. Ligthart, "A Dual-Fed, Self-Diplexing PIFA and RF Front-End," *IEEE Transactions on Antennas and Propagation*, vol. 55, no. 2, pp. 373–382, 2007, doi: 10.1109/TAP.2006.889852.

[5] S. Yang, C. Zhang, H. K. Pan, A. E. Fathy, and V. K. Nair, "Frequency-Reconfigurable Antennas for Multiradio Wireless Platforms," *IEEE Microwave Magazine* vol. 10, no. 1, pp. 66–83, 2009, doi:10.1109/MMM.2008.930677.

[6] A. Boukarkar, X. Q. Lin, Y. Jiang, and Y. Q. Yu, "A Tunable Dual-Fed Self-Diplexing Patch Antenna," *IEEE Transactions on Antennas and Propagation*, vol. 65, no. 6, pp. 2874–2879, 2017, doi:10.1109/TAP.2017.2689035.

[7] J. Anguera, C. Picher, A. Andújar, and C. Puente, "Concentrated Antenna-less Wireless Device Providing Operability in Multiple Frequency Regions," *Patent US* 9,379,443 B2, June 2016.

[8] C. Picher, J. Anguera, A. Bujalance, and A. Andújar, "Ground Plane Booster Antenna Technology Using a Self-Diplexed Matching Network for Multi-band Operation," *Microwave and Optical Technology Letters*, vol. 58, no. 2, pp.453–461, 2016.

[9] T. Bauer, C. Eggs, K. Wagner, and P. Hagn, "A Bright Outlook for Acoustic Filtering: A New Generation of Very Low-Profile SAW, TC SAW, and BAW Devices for Module Integration," *IEEE Microwave Magazine*, vol. 16, no. 7, pp. 73–81, 2015, DOI:10.1109/MMM.2015.2429512.

[10] A. Hagelauer, G. Fattinger, C. C. W. Ruppel, M. Ueda, K. Hashimoto, and A. Tag, "Microwave Acoustic Wave Devices: Recent Advances on

Architectures, Modeling, Materials, and Packaging," *IEEE Transactions on Microwave Theory and Techniques*, vol. 66, no. 10, pp. 4548–4562, 2018, DOI:10.1109/TMTT.2018.2854160.

[11] J. Anguera, A. Andújar, G. Mestre, J. Rahola, and J. Juntunen, "Design of Multiband Antenna Systems for Wireless Devices Using Antenna Boosters," *IEEE Microwave Magazine*, pp.102–114, 2019.

[12] A. Diallo, C. Luxey, P. Le Thuc, R. Staraj, and G. Kossiavas, "Study and Reduction of the Mutual Coupling between Two Mobile Phone PIFAs Operating in the DCS1800 and UMTS Bands," *IEEE Transaction on Antennas and Propagation*, vol. 54, no. 11, part I, pp. 3063–3074, 2006

[13] A. Lhilali, A. Diallo, C. Luxey, and P. Brachat, "Multiband Multi-antenna System for MIMO WLAN Box," *2010 14th International Symposium on Antenna Technology and Applied Electromagnetics & the American Electromagnetics Conference*, 2010, pp. 1–4, doi:10.1109/ANTEM.2010.5552504.

[14] L. Sun, Y. Li, and Z. Zhang, "Wideband Decoupling of Integrated Slot Antenna Pairs for 5G Smartphones," *IEEE Transaction on Antennas and Propagation*, vol. 69, no. 4, pp. 2386–2391, 2021.

[15] W. Zeng, Q. Chu, and B. Lu, "Two-Port Antenna Pair Decoupling by Modal Control for 5G Mobile Devices," *2021 13th International Symposium on Antennas, Propagation and EM Theory (ISAPE)*, pp. 1–3, 2021, doi: 10.1109/ISAPE54070.2021.9753138.

[16] A. Habashi, J. Nourinia, and C. Ghobadi, "Mutual Coupling Reduction Between Very Closely Spaced Patch Antennas Using Low-Profile Folded Split-Ring Resonators (FSRRs)," *IEEE Antennas and Wireless Propagation Letters*, vol. 10, pp. 862–865, 2011.

[17] Z. Du, X. Zhang, P. Qin, Y. Pu, and X. Xi, "Intercoupling Suppression of Very Closely Spaced MIMO Antenna Based on Current Cancellation Method," *IEEE Access*, doi: 10.1109/ACCESS.2022.3224461.

Chapter 7

Reconfigurable multiband architectures for wireless devices embedding antenna boosters

Wireless devices, such as smart meters, trackers, and sensors, need connections at multiple frequency bands with low power consumption and thus require multiband and efficient antenna systems. At the same time, antennas should be small to easily fit in the scarce space in wireless devices. Small, multiband, and efficient operation is addressed here with antenna boosters featuring volumes less than 90 mm^3 for operating at 698–960 MHz and some bands in a higher frequency range comprising 1710–2690 MHz. As studied in previous chapters, antenna boosters excite currents on the ground plane of the wireless device and do not rely on shaping complex geometric shapes to obtain multiband behaviour but on designing a multiband matching network. This approach results in a simpler, easier, and faster method than designing a new antenna for every device. Furthermore, since the multiband operation is achieved through a matching network, frequency bands can be configured and optimised with a reconfigurable matching network. Two kinds of reconfigurable multiband architectures with antenna boosters are presented. The first includes a digitally tunable capacitor (DTC) and a second one with radio-frequency switches, also called SPNT (Single Pole N-Throws). Results show that antenna boosters with reconfigurable architectures feature multiband behaviour and tiny sizes compared with other prior-art techniques.

The chapter has the following objectives:

1. Understand when reconfigurable matching networks are required instead of passive solutions
2. Know two types of microwave devices that allow tunability: the DTC and the SPNT
3. Design tunable architectures with DTC and SPNT with antenna boosters for achieving reconfigurable frequency bands
4. Understand how reconfigurable antenna booster architectures can provide multi-radio performance

7.1 Introduction

With the advent of the everything-connected era, antennas play a significant role. Every wireless device needs at least one antenna to transmit and receive information from smart meters, intelligent trackers, sensors, and any IoT (Internet of Things) device. In response, the wireless industry is moving fast to launch new devices into the market. This fact pushes RF/microwave and wireless engineers to design radiofrequency chains smartly using off-the-shelf devices such as filters, amplifiers, diplexers, front-end modules, etc. At the end of this radiofrequency chain, an antenna is needed to transmit and receive electromagnetic waves efficiently. Accordingly, an off-the-shelf antenna component would be attractive for simplifying the entire design process. It would streamline the antenna design phase and the device manufacturing process, reducing costs and being compatible with pick-and-place machines for mass production.

Besides off-the-shelf antenna components, miniature size and multiband performance are two other demanding requirements for wireless antenna systems. Common techniques to create small and multiband antennas for wireless devices rely on complex geometries where the several resonant modes of the antenna determine the frequency bands of operation [1–11]. On the one hand, this procedure requires a high level of expertise to shape the antenna geometry correctly to achieve acceptable behaviour to operate at the frequency bands of interest. This complexity of the antenna geometry shaping process makes RF/microwave and wireless engineers reluctant to design their antennas for wireless systems. On the other hand, some currently available off-the-shelf antennas are mainly narrowband, while others, although multiband, are difficult to adjust due to their self-resonance nature [12]. The off-the-shelf antenna boosters proposed here overcome these shortcomings. An antenna booster can be integrated inside any wireless device for operating at any frequency band by adequately adjusting a matching network (in the present book, from 400 MHz to 10 GHz; however, as explained in Chapter 1, frequency scaling can be applied to operate at other frequencies). The antenna booster can be easily adjusted, not by modifying its geometry, but through a proper matching network design. This is a straightforward and more familiar method for RF/microwave and wireless engineers acquainted with matching networks at every stage of a telecommunication system, for example, filters, amplifiers, etc. Therefore, designing a multiband antenna system with antenna boosters is more related to a microwave circuit design than an antenna geometry design, thus making antenna engineering much more like a filter design than a conventional antenna design.

The antenna booster technology relies on a non-resonant element that is electrically short in terms of the operating wavelength ($\sim\lambda/30$). Its correct location on the ground plane of a wireless device enables the excitation of efficient radiating modes on the ground plane [12–18]. Therefore, the term antenna booster is adopted here to refer to this non-resonant element. Antenna boosters are miniature, surface-mount (SMT), off-the-shelf components that replace conventional customised PIFA/Printed antennas for mobile/wireless connectivity applications (Figure 7.1). However, since the antenna booster is electrically small, its impedance is mainly reactive, and, thus, it is poorly matched at most of the bands of operation in the frequency range from 0.7 GHz up to 3 GHz, where many communication systems are allocated, such as cellular, short-range wireless, etc. This mismatch issue is overcome by adding a multiband matching network, resulting in a multiband

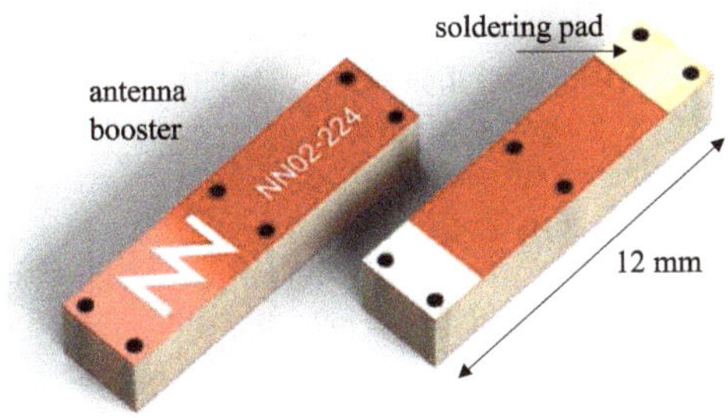

Figure 7.1 A commercial antenna booster (12 mm × 3 mm × 2.4 mm (height)). Two soldering pads on the bottom side allow the feeding of the antenna booster by either one of those pads [34] and fix it to the PCB (Printed Circuit Board). This is an example of an off-the-shelf antenna booster, which has been used extensively throughout the book. As a reference, when the antenna booster operates at 900 MHz, the electrical size is $\sim\lambda/28$.

antenna system with a small antenna element and competitive efficiency values [12–18].

Examples of antennas combined with matching networks appear in the literature [19–33]. However, those designs rely on adding intelligence to the antenna geometry. In addition, the antenna still represents a considerable volume in all the above cases. Although the volume of antennas has continuously shrunk, they still rely on complex geometries (Figure 7.1). On the contrary, antenna boosters can still operate at multiple frequency bands while maintaining a small antenna volume [12–18].

This chapter discusses the utilisation of antenna boosters in conjunction with reconfigurable matching networks, which incorporate both lumped passive components (such as capacitors and inductors) and active devices, enabling multiband operation. In the first case, a digitally tunable capacitor is presented, and in the second case, a radiofrequency switch-based solution. This results in reconfigurable matching networks that are helpful in tuning the frequency bands of operation. It is important to emphasise that a passive matching network can provide multiband operation. However, adding reconfigurability adds a new degree of freedom to expand the number of bands beyond those achieved with a passive matching network. Furthermore, another benefit of reconfigurability is its ability to effectively adjust the frequency bands of operation in wireless devices when encountering various challenging environments, including conductive surfaces, concrete, wood, plastics, and more, which can alter the antenna's frequency response. Finally, multiband behaviour is challenging for small form factor devices ($<\sim0.2\lambda \times \sim0.2\lambda$). Therefore, a reconfigurable antenna-booster architecture is convenient because an antenna booster is tiny enough to be embedded into the small device, and reconfigurability provides connection at multiple frequency bands. The reconfigurable architectures shown in this chapter are based on open-loop, which can adapt to prefixed states instead of closed-loop, requiring more electronics, such as integrated circuits, for sensing the impedance. For closed-loop schemes, the system can adapt in real time to any external change affecting the antenna system performance [35,36]. For the sake of validation, open-loop configurations are discussed.

The chapter is structured as follows (Figure 7.2): introduction to antenna booster technology is revisited in Section 7.2. Multiband reconfigurable matching networks using digitally tunable capacitors (DTC) and RF switches are described in Sections 7.3 and 7.4, respectively. A comparison with prior-art multiband reconfigurable antennas is discussed in Section 7.5. Conclusions are drawn in Section 7.6. A list of exercises are included in Section 7.7 and references given at the end.

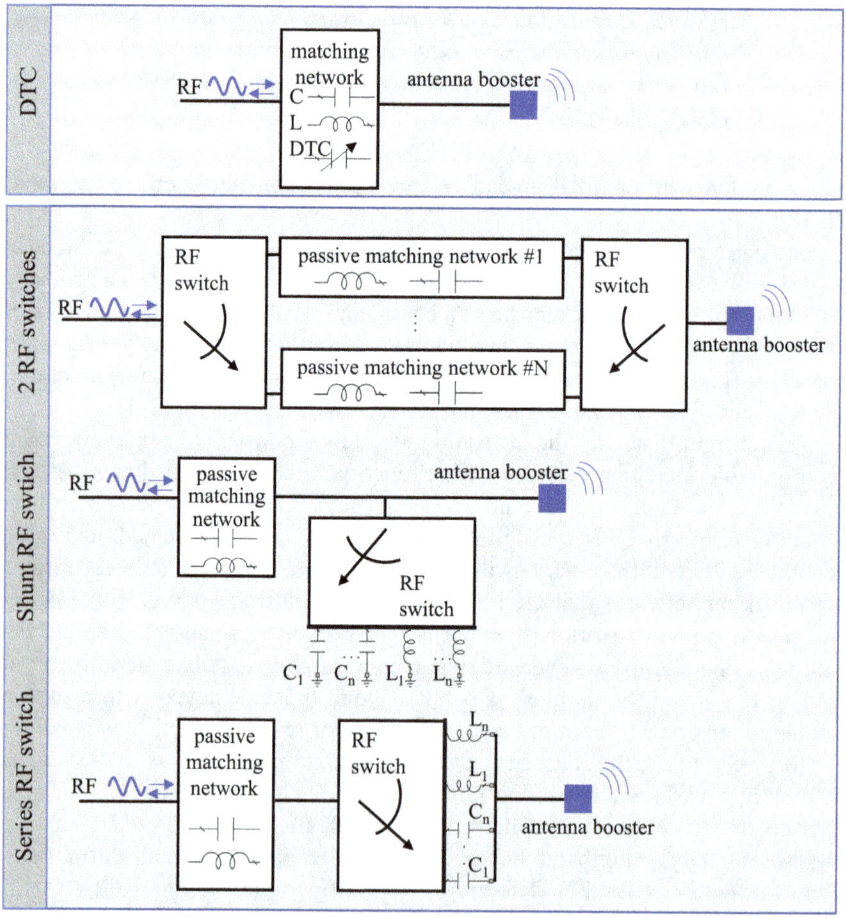

Figure 7.2 Reconfigurable matching network architectures with antenna boosters using DTC and RF switches

7.2 Antenna boosters

Antenna booster-based technology relies on miniature non-resonant elements called antenna boosters to provide operation in a wireless device (Figures 7.1 and 7.3). The same antenna booster as the one depicted in Figure 7.3 can operate at different bands depending on a matching network. This approach differs from conventional antenna designs, where the antenna geometry must be adjusted case by case, thus increasing the design time and cost. In addition, the same antenna booster can operate for different platform sizes since only the matching network changes from one platform to another [34]. Therefore, this is a faster and easier approach to designing multiband antenna systems than creating an antenna from scratch.

To illustrate the non-resonant nature of an antenna booster, an antenna booster (a 3D brass-made parallelepiped in this example) of 12 mm × 3 mm × 2.4 mm (height) is analysed, representing a maximum size of only $\lambda/35.8$ at 698 MHz. The non-resonant element is integrated into a 120 mm × 60 mm ground plane (Figure 7.4). The said antenna booster is in charge of exciting current modes in the ground plane, becoming a relevant part of the radiation process. This fact justifies the name antenna booster since it boosts currents on the ground plane [13].

The input impedance is reactive, particularly at the low bands (698–960 MHz), where the impedance is mainly capacitive (Figure 7.5). A passive matching network can match the antenna system shown in Figure 7.6, covering from 824–960 MHz to 1710–2690 MHz, as shown in [37]. However, the frequency range from 698 MHz to 824 MHz must be adequately matched since this frequency range

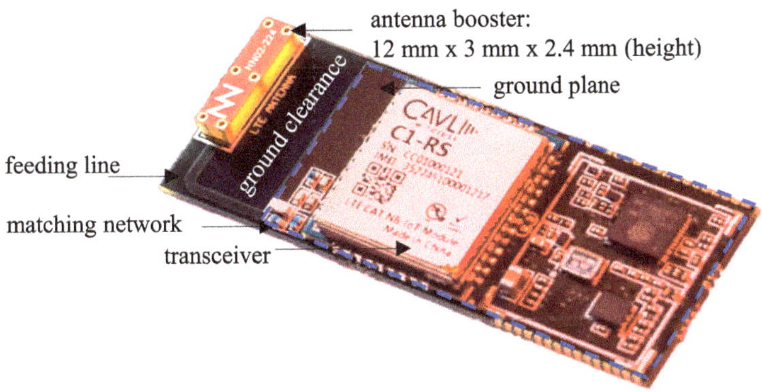

Figure 7.3 An antenna booster (in red) is located on the PCB of a wireless device. Its tiny size (e.g., 12 mm × 3 mm × 2.4 mm (height)) allows easy integration of the antenna booster within the device. The architecture comprises an antenna booster, typically at the corner of the device, a matching network, and a ground plane. The ground plane comprises a continuous conductive layer placed on top and bottom of the PCB substrate.

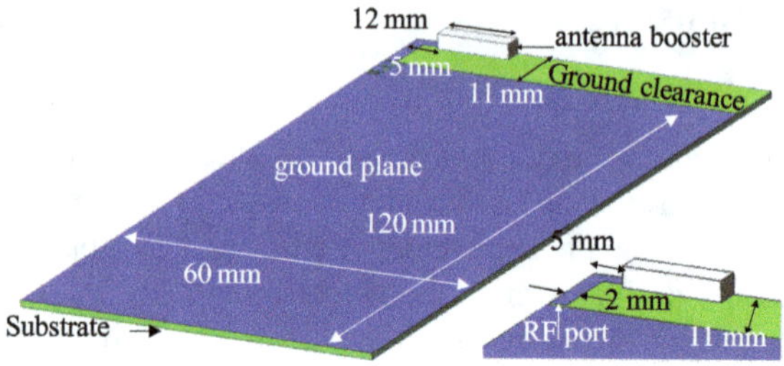

Figure 7.4 Antenna booster on a 120 mm × 60 mm ground plane over a dielectric substrate (ε_r = 4.15, tanδ = 0.017, and h = 1 mm). The antenna booster is a conductive σ = 5.9 × 10^7 S/m parallelepiped of 12 mm × 3 mm × 2.4 mm (height).

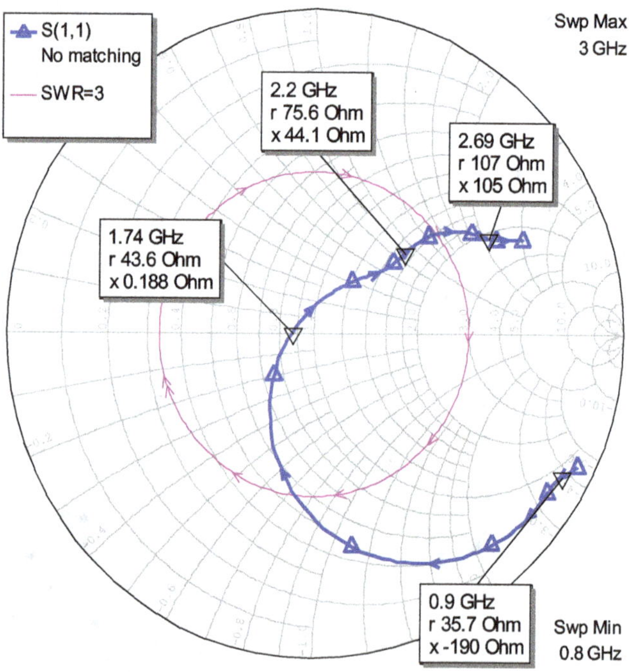

Figure 7.5 Simulated input impedance showing a reactive capacitive behaviour at the low-frequency region (698–960 MHz)

is required in many wireless devices. This frequency range represents a 16% bandwidth, which implies a significant challenge since it is located at the lower frequencies of operation. To enhance performance across the frequency spectrum ranging from 698 to 960 MHz, extending up to 1710 to 2200 MHz/2690 MHz, the subsequent sections introduce configurations integrating reconfigurable

(a)

(b)

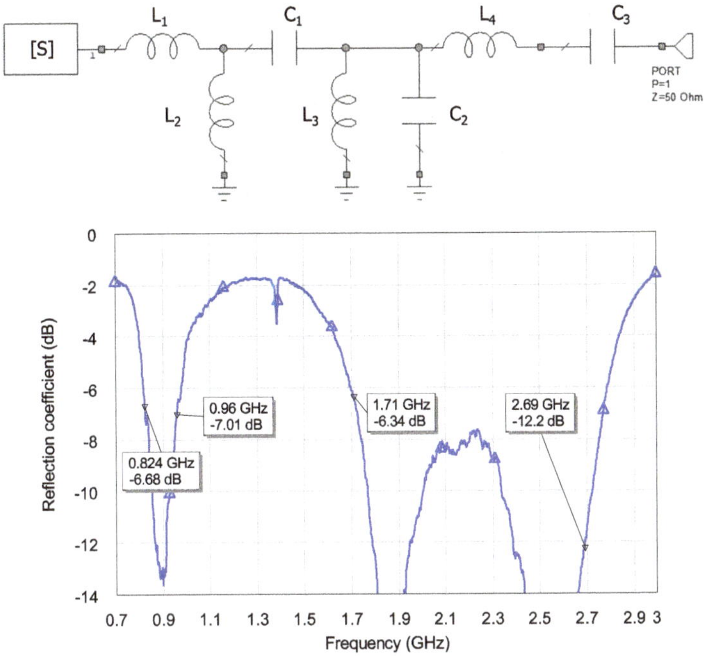

Figure 7.6 (a) A passive matching network for achieving multiband behaviour $L_1 = 4.3$ nH, $L_2 = 18$ nH, $C_1 = 0.9$ pF, $L_3 = 13$ nH, $C_2 = 1$ pF, $L_4 = 4.5$ nH, $C_3 = 2$ pF. (b) Reflection coefficient resulting from adding the matching network of (b) to the antenna system shown in Figure 7.4.

matching architectures alongside antenna boosters. Sections 7.3 and 7.4, present the utilisation of digitally tunable capacitors (DTC) and radiofrequency switches, respectively.

7.3 Digitally tunable capacitors for antenna boosters

The bandwidth of an antenna system cannot be infinitely enlarged by adding more components to a matching network, owing to the Fano limit [38]. To circumvent these limitations and expand bandwidth to accommodate additional frequency bands, a tunable capacitor is strategically integrated into the passive matching network, thereby creating a reconfigurable matching network. Hence, the reconfigurable antenna booster system comprises lumped capacitors, inductors, and a digitally tunable capacitor [39].

This section presents two architectures with DTC, one in series and another in shunt, to achieve a reconfigurable frequency band antenna system. The advantages of a passive matching network are also discussed.

7.3.1 Digitally tunable capacitors

A digitally tunable capacitor (DTC) is a microwave device where the capacitance can be digitally changed within a range from C_{min} to C_{max}. The tuning ratio (C_{max}/C_{min}) is intrinsically related to the DTC quality factor (Q); the larger the tuning ratio, the lower the Q [40]. For this example, the MEMS (micro-electro-mechanical system) technology DTC from [41] presents a Q = 700 and a C_{max} = 1 pF at 700 MHz with a tuning ratio of 2.5. Thanks to the low intrinsic losses of MEMS technology, the DTC features a high Q with a small footprint. In the present case, the DTC area is only 1.4 mm × 1.4 mm (Figure 7.7).

Besides 1.8 V, RF-ground, and RF-input connections, it supports the MIPI (Mobile Industry Processor Interface) and RFFE (Radio Frequency Front-End) interfaces with one clock and one series of data. With a 5-bit resolution in the digital control, 32 programmable capacitor states range from 0.4 pF to 1.0 pF with a constant step size of 18 fF. The MIPI RFFE is compatible with other radiofrequency components and baseband processors in the system, which can select the state of the DTC based on the desired use condition. In the described prototype, a laptop application controls a MIPI master interface to determine the state. Once the state is set, the MIPI digital cable bus is disconnected in order not to interfere with impedance and radiation measurements. An onboard battery and voltage regulator provide power to the DTC.

7.3.2 Multiband antenna system with a series DTC

This section illustrates how the frequency bands can be expanded compared to a passive multiband matching network. The antenna system comprises an antenna booster of 12 mm × 3 mm × 2.4 mm (height) on a 120 mm × 60 mm ground plane (Figure 7.4). The antenna system without a matching network is matched with an $S_{11} < -6$ dB from 1.547 GHz to 2.385 GHz but is very poorly matched at 824–960 MHz, presenting a highly capacitive impedance. By optimising a matching network with seven lumped components, the antenna booster is matched from

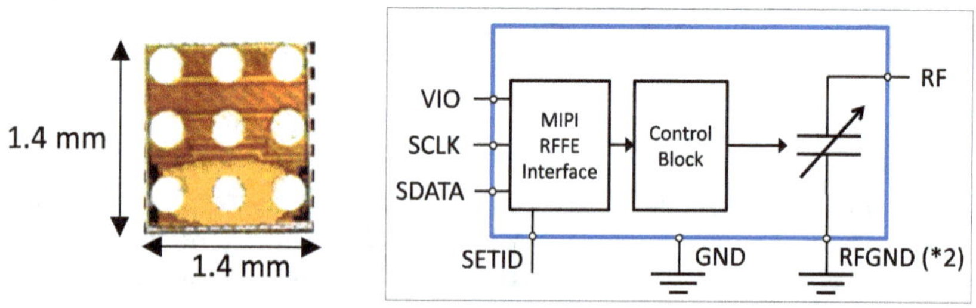

Figure 7.7 (Left) The footprint of the DTC. (Right) Block diagram of the DTC where the capacitor can be tuned from 0.4 pF to 1.0 pF. ESR (Equivalent Series Resistance) ~0.3 Ω at C_{max}. 5-bit resolution (32 capacitor states). SRF (Self-Resonant Frequency) > 15 GHz. RF power handling: +39 dBm. High Linearity. MIPI RFFE interface. Small size: ~2 mm^2. Low power: 180 μW typical.

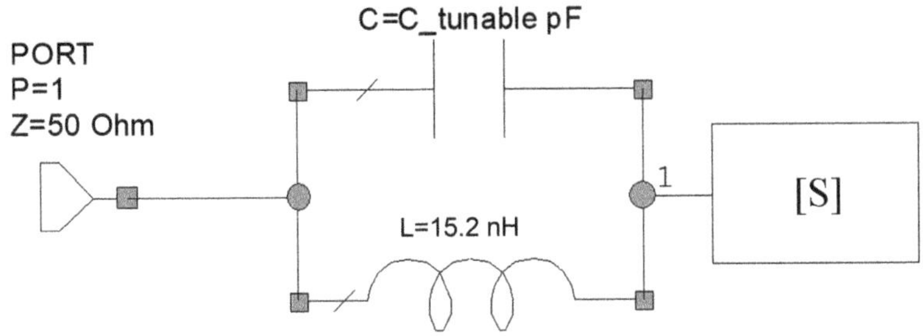

Figure 7.8 A reconfigurable matching network comprises a lumped inductor and a DTC represented by $C_{tunable}$. The S-parameter box represents the S_{11} data for the antenna booster in Figure 7.4.

824 to 960 MHz and 1710 to 2690 MHz (Figure 7.6). Chapter 5 explains that the matching network is obtained through matching network synthesis.

As far as the reconfigurable matching network is concerned, a scheme comprising a shunt LC in series is designed (Figure 7.8). As explained in Chapter 5, when the frequency is below the resonant frequency of the resonator f_0, the equivalent impedance is inductive. Furthermore, modifying the resonant frequency of the resonator can change the value of such inductance, such that when the frequency of operation is near but below f_0, the inductance becomes high. Therefore, the tunable capacitor controls the value of the equivalent inductance. For matching an impedance at a low frequency, for example, 750 MHz, the resonant frequency should be close to and above 750 MHz to provide a high inductance since the impedance at 750 MHz is highly capacitive. However, a lesser equivalent inductance is needed at 920 MHz since the impedance is not as capacitive. In this case, the resonant frequency f_0 should be above 920 MHz. For the high-frequency region ($f > 1.71$ GHz), the equivalent impedance is capacitive if $f_0 < 1.71$ GHz; therefore, it helps to match the frequency region above 1.71 GHz. Illustrating with numbers, the reactance at 750 MHz is $-252j$, requiring a series inductance of 53.4 nH (Figure 7.9). This value can be obtained with a $C_{\text{tunable}} = 2.1$ pF in a shunt with an $L = 15.2$ nH (the value for selecting the inductance value is discussed later). Similarly, to cancel the $-197j$ at 925 MHz, a series inductor of 6.3 nH is needed, which can be obtained with a $C_{\text{tunable}} = 1$ pF. In conclusion, to nullify the inductive impedance of 82j at a high frequency of 2.5 GHz, a series capacitor of 0.77 pF is necessary, achievable with a C_{tunable} of 1 pF, which is also used for matching at 925 MHz.

A shunt LC resonator behaves as an L and as a C

An LC resonator with a resonant frequency f_0 provides an inductive impedance if $f < f_0$ and a capacitive impedance if $f > f_0$. The impedance of the equivalent inductance is high when the frequency is below f_0 and approaches f_0. Analogously, the impedance of the equivalent capacitive is low if $f > f_0$ and close to f_0.

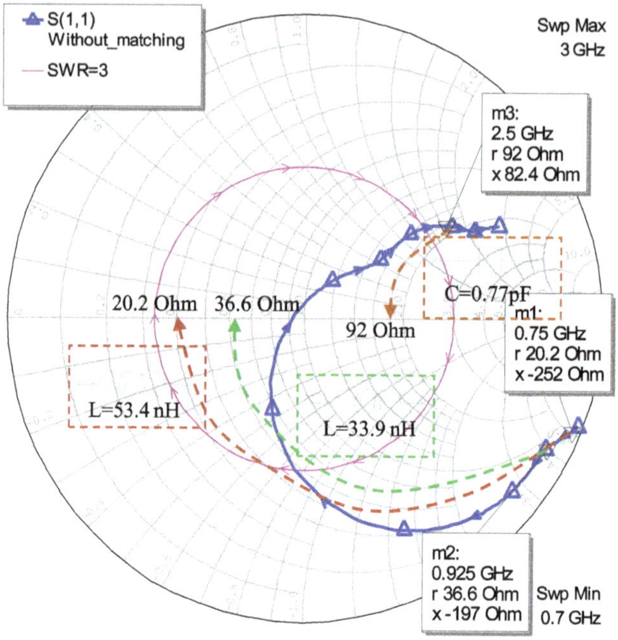

Figure 7.9 The impedance locus represents the impedance of the antenna booster in Figure 7.4 without any matching circuits, showing three representative frequencies (two for the low-frequency region – $f < 1$ GHz and one for the high-frequency region). The dashed lines represent the impedance movements after adding a series lumped inductor for the 0.75 GHz and 0.925 GHz frequencies and a capacitor for the 2.5 GHz. These different impedance values can be achieved by a reconfigurable matching circuit, such as the one shown in Figure 7.8.

What role does the inductance of the resonator play? In the previous example, it was fixed to 15.2 nH. What occurs if the inductance is 5 nH or 25 nH? By a simple circuit analysis, the L is fixed to 5 nH, 15.2 nH, and 25 nH, and for each case, the $C_{tunable}$ is calculated to cancel the reactive impedance at three different frequencies of 750 MHz, 925 MHz, and 2500 MHz (Figure 7.10). Results conclude that for an L = 5 nH, a ratio of $C_{max}/C_{min} = 5$ is needed, which is feasible to achieve with a DTC [34]. For L = 15.2 nH and L = 25 nH, C_{max}/C_{min} is acceptable. However, to minimise losses, a low inductance is preferred. In effect, the Q at 2500 MHz of an L = 15.2 nH and 25 nH is 126 and 75, respectively.

It is remarkable that with only one lumped component and the $C_{tunable}$, it is possible to match from 741 MHz to 1110 MHz and from 1640 MHz to 3220 MHz. As reflected in Figure 7.11, the S_{11} is better than the passive solution in a broader

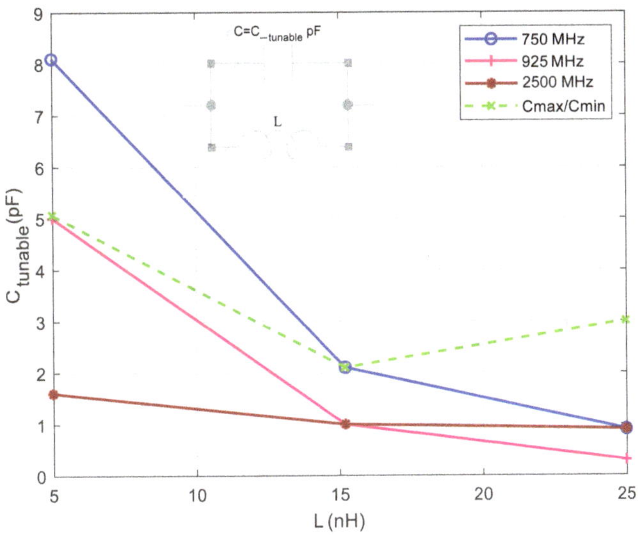

Figure 7.10 *Values of the $C_{tunable}$ and L to cancel the reactive impedance at 750 MHz, 925 MHz, and 2500 MHz. The value of L of the resonator comprising the DTC should be selected to be as low as possible to keep a high Q to minimise losses but at the same time to guarantee a practical C_{max}/C_{min} for the DTC.*

frequency range (Figure 7.6(b)): for the low-frequency region, the passive solution achieved a bandwidth BW = 16.6% compared to the 40% with the tunable solution. For the high-frequency region, the passive solution reached BW = 49.5%, whereas the tunable solution was BW = 64%.

The previous example needed a C_{max} = 3 pF and a C_{min} = 0.4 pF. For some DTCs, the range of C_{max}/C_{min} is lower than 7.5. A way to reduce the dynamic range involves adding a series inductance between the antenna booster and the LC filter. This extra inductance is adjusted empirically to obtain a C_{max}/C_{min} within the limits of the DTC. The following example shows C_{max}/C_{min} = 4.8 (Figures 7.12 and 7.13).

A reconfigurable frequency solution is attained with a reduced C_{max}/C_{min} = 4.8. However, the addition of the series inductance has mismatched the high-frequency region ($f >$ 1.71 GHz); therefore, less bandwidth is obtained. For this configuration, the bandwidths at the first and second frequency regions are 37% and 57%, respectively, whereas, for the solution with C_{max}/C_{min} = 4.8, the tunable bandwidths at the first and second frequency regions are 40% and 64%, respectively.

To summarise, a series DTC is combined in a shunt with an inductor for an antenna booster. Since the real part of the input impedance of the antenna booster without a matching network is within an SWR $\leq$ 3, the series tunable scheme is enough for cancelling out the reactive impedance both at low frequencies ($f <$ 1 GHz) and high frequencies ($f >$ 1.71 GHz) and thus obtaining matching at

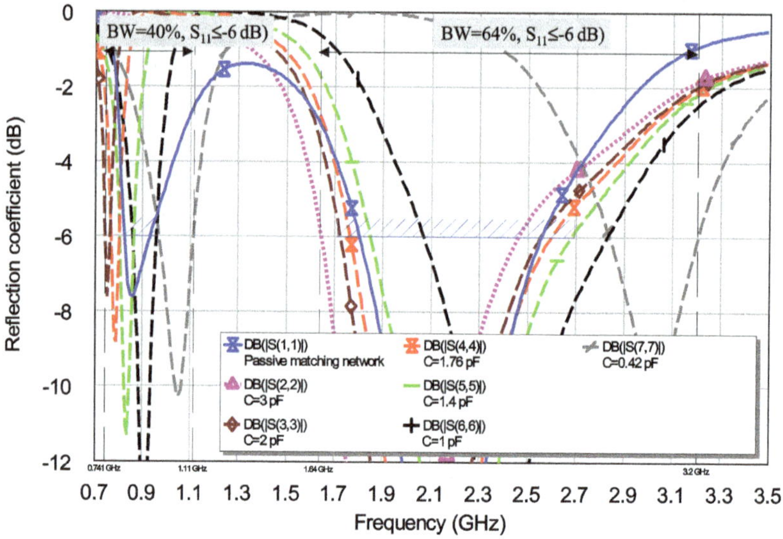

Figure 7.11 Several S_{11} states as a function of the capacitance for the scheme are shown in Figures 7.4 and 7.8. The capacitance range goes from 3 pF down to 0.4 pF.

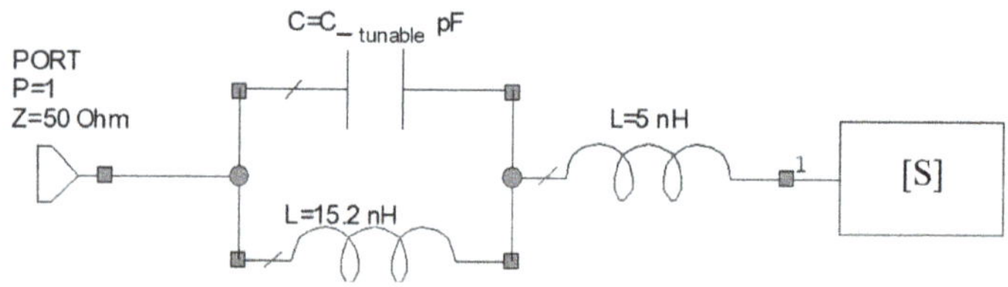

Figure 7.12 A reconfigurable matching network comprises a lumped inductor and a DTC represented by $C_{tunable}$. The S-parameter box represents the S_{11} data for the antenna booster in Figure 7.4. A series inductor of 5 nH is added to reduce the ratio C_{max}/C_{min}.

several frequency bands. For other ground plane sizes, the antenna booster impedance changes; hence, a different scheme should be adopted. In the end, the passive matching network will help modify the impedance so that the DTC tunes the frequency bands. In this sense, two different strategies are discussed next. The first one uses a shunt DTC, and the second one uses a network synthesis procedure that can be used either for a series or a shunt DTC.

7.3.3 Multiband antenna system with a shunt DTC

A shunt configuration is illustrated for the same antenna booster on the same ground plane size. The shunt DTC is combined with a passive matching network.

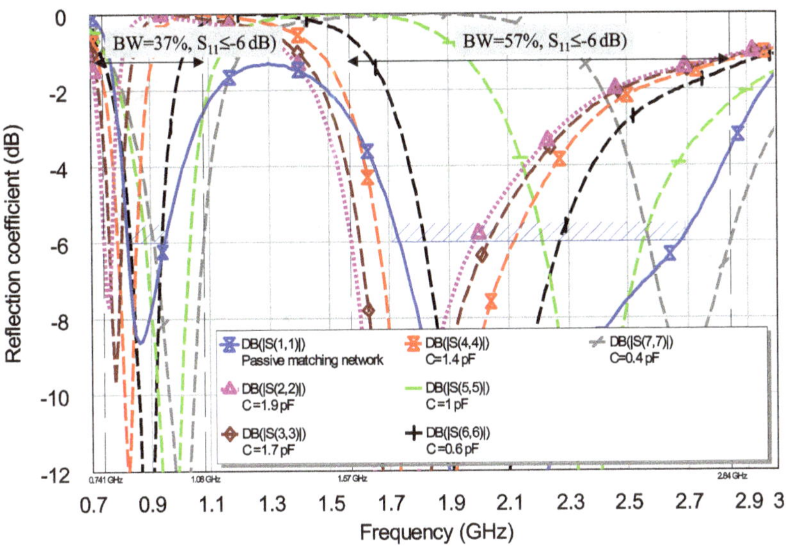

Figure 7.13 Several S_{11} states as a function of the capacitance for the scheme are shown in Figures 7.4 and 7.12. The capacitance range goes from 1.9 pF down to 0.4 pF.

As shown, the number of lumped components is large (six) compared to the only one included in the series case. However, the purpose is to physically validate an antenna booster with a DTC combined with a passive matching network with several lumped elements. This matching network, combined with the DTC, can also be helpful for other ground plane sizes.

A small antenna booster (86.4 mm^3) is proposed for multiband operation in combination with a reconfigurable matching network [42,43]. The antenna booster is placed at the corner of a 120 mm × 60 mm ground plane (Figure 7.14).

The said reconfigurable matching network uses the DTC implemented with MEMS described in the previous section. The RFFE digital signals for capacitance selection are provided through a removable ribbon cable. For the present case, only four states are enough to cover the frequency range of 698–960 MHz up to 1710–2690 MHz.

An analysis is done using an electromagnetic simulation of the PCB, including the antenna booster. The S-parameters calculated through electromagnetic simulation are subsequently brought into a circuit simulator (Cadence AWR Microwave Office) to formulate the matching network, which includes a mix of lumped capacitors, inductors, and a variable capacitor. Once the simulation shows a feasible result, a reference board is built (Figure 7.14).

The proposed reconfigurable matching network comprising a shunt DTC is obtained by engineering experience (Figure 7.15) after trying several matching networks and finding by trial-and-error which capacitor can tune the frequency

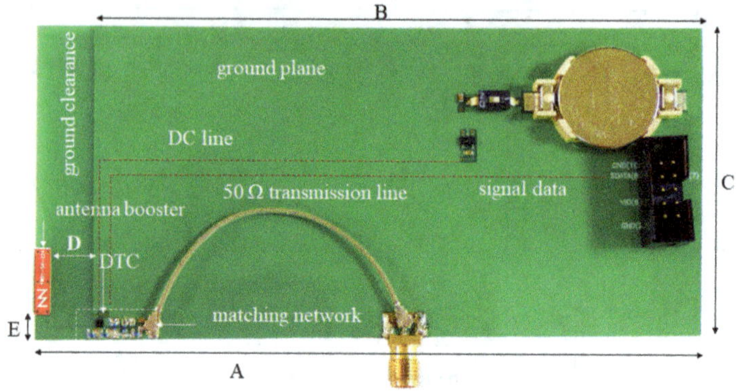

Figure 7.14 *The 12 mm × 3 mm × 2.4 mm (h) antenna booster with the reconfigurable matching network and its control circuit. The control lines from the battery and RFFE interface to the tunable capacitor go underneath the ground plane (B × C), shown in the top layer of the board. A = 131 mm, B = 120 mm, C = 60 mm, E = 5 mm, and D = 8 mm.*

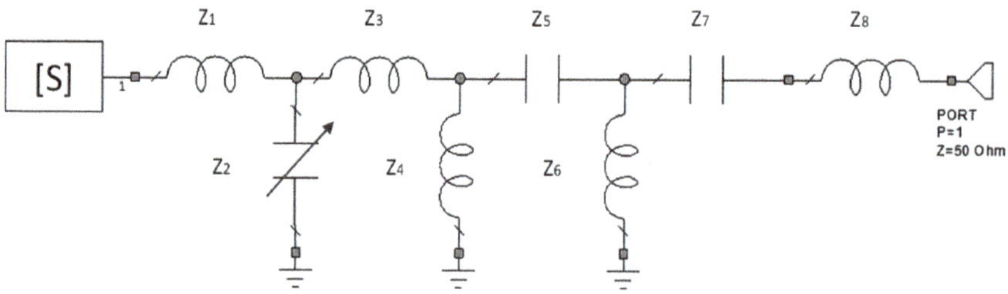

Figure 7.15 *A reconfigurable matching network where Z_2 is the Cavendish Kinetics DTC. Z_1 = 8 nH (Q > 60 with the frequency range 0.7–2.69 GHz), Z_3 = 7.3 nH (Q > 60), Z_4 = 11 nH (Q > 64), Z_5 = 0.8 pF (Q > 244), Z_6 = 13 nH (Q > 55), Z_7 = 1.9 pF (Q > 125), and Z_8 = 2.5 nH (Q >60). All passive components are SMD 0402, with high Q and tight tolerance (2%) [44].*

bands. An alternative method using a network synthesis tool is presented in the next section.

The programmable states for the DTC are S00 (0.40 pF), S02 (0.44 pF), S08 (0.55 pF) and S27 (0.92 pF).

The state of the DTC is software-controlled with a parallel interface connection at the end of the ground plane. Since the ground plane is a relevant element for the radiation process, the following set-up has been prepared to avoid an undesired effect of the interface connection. The DTC is DC-powered with a small battery on the ground plane (Figure 7.14). Then, a particular DTC state of the DTC is selected with the interface connection attached. Once the DTC is operating at a given state, the interface connection is removed. Since the DTC receives DC voltage from the battery, the state is maintained. Thus, both S_{11} and total efficiency measurements can be carried out without the impact of the interface connection. The typical power consumption of the DTC is 180 μW (∼100 μA static current on 1.8 V), representing about 2% of the power of a transceiver.

After optimising the matching network, some relevant states are chosen to cover from 698 MHz to 960 MHz up to 1710 MHz to 2690 MHz (Figure 7.16).

It is shown from the VSWR measurements that a particular state (S27) can cover some low bands of 700 MHz, those from 698 MHz to 746 MHz. From 746 MHz up to 824 MHz, the state S08 can be selected. Remarkably, four states are almost enough to obtain tolerable VSWR at the whole frequency range (Figure 7.16).

To analyse radiation performance, total efficiency (η_t) is measured inside an anechoic chamber using 3D pattern integration (Satimo Stargate-32). Total efficiency considers both radiation efficiency (η_r) and mismatch losses as follows: $\eta_t = \eta_r \times (1-|S_{11}|^2)$ (Figure 7.17).

To summarise the results, the average total efficiency values for several representative communication systems, namely 698–746 MHz, 746–798 MHz, 824–894 MHz, 880–960 MHz, 1710–1880 MHz, 1850–1990 MHz, 1920–2170 MHz, 2305–2360 MHz, and 2496–2690 MHz are gathered in Table 7.1. Information in Table 7.1 helps to choose the best state for a specific frequency band, as shown in the dark colour. For example, regarding the low-frequency region, if operation is needed at 699–746 MHz, the state S27 should be selected, providing an average efficiency value of 41.5%; if 746–798 MHz is desired, then state S08 should be established; for 824–960 MHz, S00 could be chosen. The same criteria apply to the high-frequency region. For example, to cover

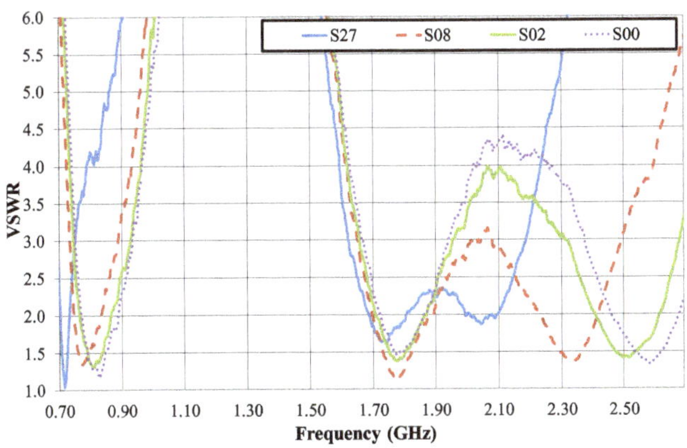

Figure 7.16 Measured VSWR for four states: S27, S08, S02, and S00

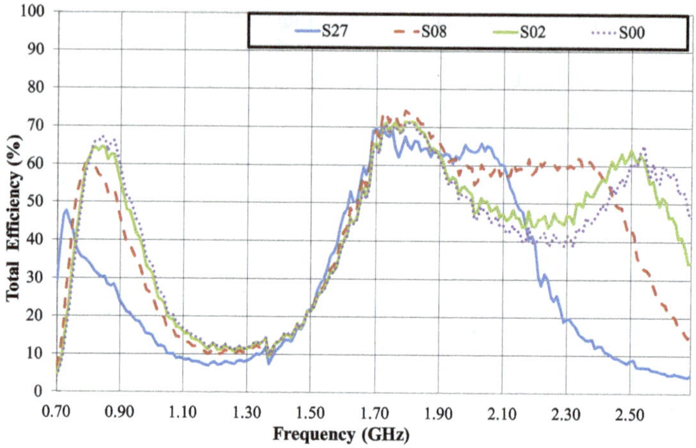

Figure 7.17 Measured total efficiency for the same four states as shown in Figure 7.16

Table 7.1 Measured total efficiency for four states and compared with the case with a passive matching network. Reference is the passive solution, as in Figure 7.6

State	698–746	746–798	824–894	880–960	1710–1880	1850–1990	1920–2170	2305–2360	2496–2690
S27	41.5	37.7	29.0	32.4	65.8	65.3	59.9	17.2	6.0
S08	19.0	53.6	54.0	41.5	71.5	63.8	59.1	60.4	26.2
S02	13.0	46.8	62.5	50.2	69.0	59.6	50.1	48.8	50.6
S00	11.5	44.1	65.4	53.7	68.4	58.5	47.7	42.5	57.2
Ref	11.8	23.1	69.9	72.0	77.1	79.0	74.9	72.4	78.4

Numbers at the first row indicate MHz.

1710–1880 MHz and 2305–2360 MHz, state S08 can be selected; state S27 can be chosen for covering 1850–1990 MHz, 1920–2170 MHz, and finally, state S00 can be set for covering 2496–2690 MHz. Due to the limitations of the anechoic chamber below 700 MHz, only measurements up to state S27 have been carried out. However, tuning can go below 700 MHz since more states are available.

It is worth pointing out that the reconfigurable antenna system exceeds the efficiency of the passive matching network case (Figure 7.6) at the low-frequency region from 699 MHz to 798 MHz (Table 7.1), which is the challenge frequency region among all frequency regions of operation. For example, at 699–746 MHz, the reconfigurable architecture reaches 5.4 dB more than the passive (reference case in Table 7.1) and 3.6 dB more at 746–798 MHz.

To sum up, a small antenna booster of 86.4 mm^3 and a reconfigurable matching network allow coverage in the frequency bands from 698 MHz to

960 MHz and 1710 MHz to 2690 MHz. The reconfigurable matching network is designed to achieve competitive performance using a MEMS DTC combined with lumped inductors and capacitors. The measured average total efficiency is above 40% and 57% at the 698–960 MHz and 1710–2690 MHz frequency ranges, respectively, making this architecture attractive for multiband wireless devices.

7.3.4 Matching network synthesis procedure for antenna boosters with DTCs

In contrast to the two previous examples where the matching networks were obtained by engineering knowledge, this section proposes a different design approach. In this case, a network synthesis tool is utilised to obtain the passive matching network topology in combination with a DTC. From an engineering standpoint, both methods are effective. However, automatic synthesis is simpler and faster.

The matching network synthesis is based on the same procedure explained in Chapter 5. However, a new component, the DTC, needs to be optimised in this case. The DTC comprises a set of several capacitors to be chosen by the optimiser, and the passive matching network can include as many components as desired. For practical purposes, the passive matching network is limited to up to four lumped elements. The decision of the topology depends upon the matching network synthesiser based on the total efficiency target (Figure 7.18(a)). In this example, the

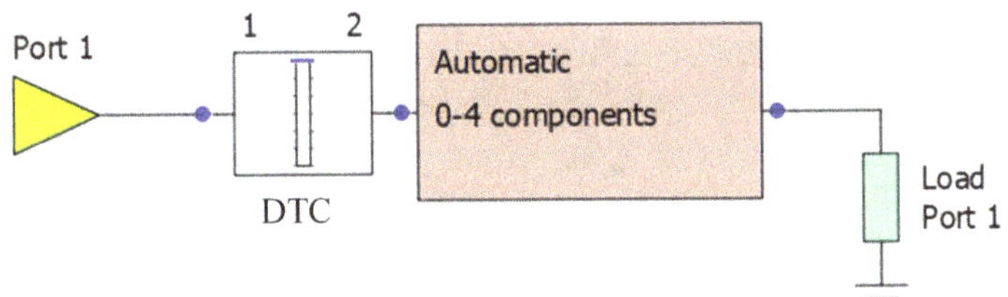

(a)

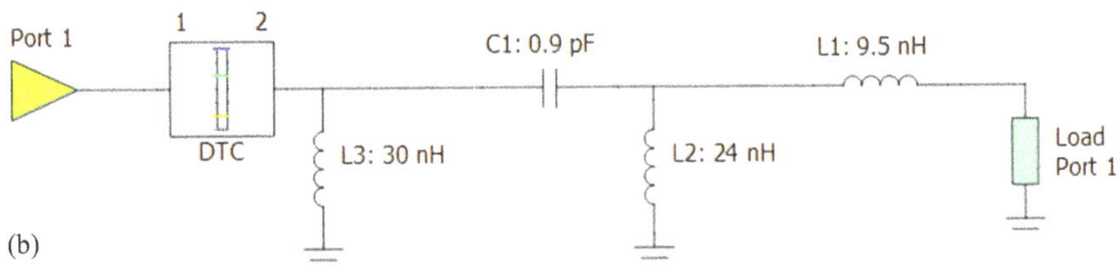

(b)

Figure 7.18 (a) A DTC in series with a passive matching network. The load represents the antenna booster in Figure 7.4. (b) Synthesized circuit where the DTC comprised three states (10 pF, 2 pF, and 1 pF) and a four-component matching network (real S-parameters of the lumped components are considered in the optimisation).

target has been fixed to optimise total efficiency to be as high as possible (100%) in the frequency range 750–960 MHz and 1710–2690 MHz.

The designer must evaluate whether the DTC is in series or shunt configuration and evaluate which option is optimal. Additionally, they should consider the placement of the passive matching network in relation to the feeding port and the DTC or between the DTC and the antenna booster (depicted as a load in Figure 7.18), among other potential configurations. For instance, one can split the passive matching network into two impedance blocks. A first impedance block between the feeding port and the DTC, and a second impedance block between the DTC and the antenna booster. For the current example, the selected candidate comprises a series DTC and a four-component matching network (Figure 7.18(b)).

After the synthesis procedure, the passive matching network needs four lumped components and achieves an $S_{11} < -6$ dB with an average total efficiency of 69.1% and 83.1% for 750–800 MHz and 1710–1880 MHz for state 1 (10 pF). For the state 2 (2 pF), the average total efficiency results in 77.6% and 81.2% for 800–890 MHz and 1850–1990 MHz, respectively. Finally, for state 2 (1 pF), the average total efficiency results in 81.2.8% and 83.1% for 880–960 MHz and 2200–2690 MHz (Figure 7.19, Table 7.2). Note that the matching network synthesis should be carried out considering the pad layout addressed in Chapter 9. In the end, pads alter the value of the inductors and capacitors, having a negligible impact on losses. Therefore, for the sake of simplicity, no pads are considered in the designs.

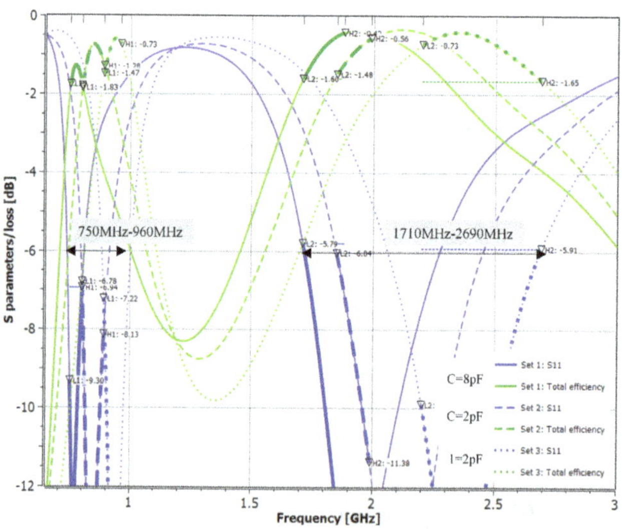

Figure 7.19 Simulated S-parameters and total efficiency for three different states of the DTC. The radiation efficiency of the antenna booster on its ground plane (Figure 7.4) has been included in the total efficiency computation

Table 7.2 Simulated total efficiency comparison between schemes at Figure 7.18

	750–800 MHz	800–890 MHz	890–960 MHz	1710–2200 MHz	2200–2690 MHz
DTC+4-component	69.1	77.6	81.2	85.1	83.2
DTC+Lshunt	54.9	66.0	81.2	91.2	83.1

The matching network synthesis procedure can also be applied to the configurations explained in Sections 7.3.2 and 7.3.3.

In conclusion, a DTC combined with a passive matching network is a valuable mechanism for tuning the frequency bands of operation of a wireless device comprising an antenna booster. A DTC provides several states of different capacitance values (typically 32 states) from C_{min} to C_{max}. Besides the tuning capability given by C_{max}/C_{min}, insertion loss needs to be considered to keep a competitive total efficiency. Finally, the power consumption of a DTC is hundreds of microwatts (180 μW in the example used in this chapter) and should be taken into the budget power consideration for implementing low-power devices.

7.4 RF switches for antenna boosters

DTC-based matching networks for antenna boosters shown in the previous section are convenient when only one component (a capacitor) of the matching network is enough to tune the frequency bands of operation. However, in other situations, changing only the value of a capacitor may not be enough to achieve satisfactory performance in the frequency bands of interest. As an alternative, the following example illustrates a small platform (50 mm × 50 mm) using RF switches to operate at several cellular frequency bands from 698 MHz to 960 MHz and 1710 MHz to 2200 MHz, as well as at 1575 MHz for GPS satellite geolocalisation (Figure 7.20).

In the present case, to gain more freedom, the multiband operation is addressed with simple matching networks comprising one or two lumped components designed for matching a particular frequency band. In this regard, a set of matching networks are connected using Single-Pole N-Throw switches (SPNT), as shown in the coming sections.

7.4.1 SPNT switch

SPNT switches permit the connection of a common radiofrequency input (pole) to several output ports (throws) (Figure 7.21). Besides the number of throws, insertion loss, and isolation are essential figures of merit since they impact both transmitted and received signal quality [45,46]. For the present case, the insertion loss of the SP8T switch is 0.43 dB and 0.56 dB at 698 MHz and 2170 MHz, respectively, with a power handling of 32 dBm.

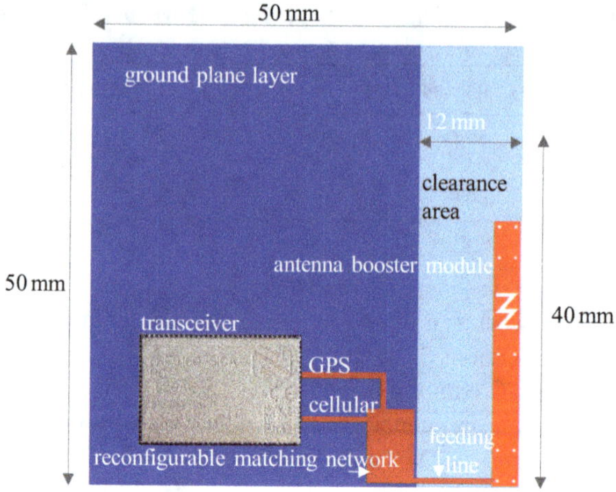

Figure 7.20 *An antenna booster module or loaded booster (in red) on a PCB (50 mm × 50 mm) connected to a reconfigurable matching network with two SPNT switches connected to a transceiver for cellular and GPS*

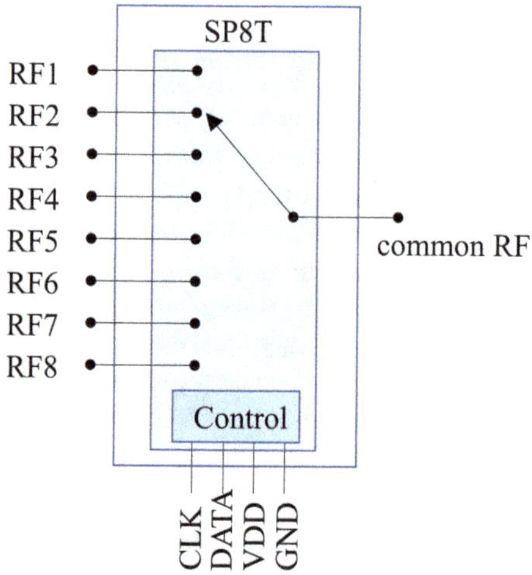

Figure 7.21 *An SP8T switch includes a common RF port, eight throws, and a control interface to set a connection between an RFi and the common RF*

For an SPNT switch, N typically adopts values from 2 up to 12, where each state can be controlled either with a GPIO (General Purpose Input Output) or an RFFE interface. The GPIO interface comprises control lines to set the state of the switch based on a simple DC voltage-driven scheme. For example, for an SP8T, three control lines are needed to control eight states: RF1 using 001, RF2 010, and so on. RFFE simplifies the number of control lines (1 clock and 1 data) and employs single-ended CMOS input/output ports, minimising power consumption [47].

SPNT switches are included in the following sections to provide operation at both cellular and satellite geolocalisation bands using an antenna booster module of only 90 mm^3 following the scheme presented in Figure 7.2 (2 RF SPNT).

7.4.2 Multiband operation with two SP8T switches

A multiband reconfigurable architecture comprising an antenna booster module (modular booster, Chapter 5) with a bunch of matching networks is proposed for operation at cellular bands as well as satellite geolocalisation (Figures 7.2 and 7.22) [43,48,49]. In this case, a larger antenna booster component than the one employed in the previous section is selected. The reason is that the ground plane size is small (38 mm × 50 mm), and moreover, the frequency of operation covers 698 MHz and above. Therefore, to obtain better efficiencies at said low frequency in the small ground plane, an antenna booster module of 30 mm instead of 12 mm is selected.

A transceiver [50] automatically controls the cellular frequency bands of an operation via a GPIO interface, although the RFFE interface can also serve this purpose. Such a GPIO interface controls the SP8T switches with three control lines (CLT1-3). In this case, six matching networks designed with a circuit simulator software [51] are used to match different cellular bands as well as satellite geolocalisation, including the measured S-parameters of the switches as well as the *S*-parameters of the SMD components of the matching network. For each frequency band, a simple L-type matching network using SMD components is enough. For example, if the transceiver decides to operate at 698–748 MHz, control lines set SP8T #1 and #2 to connect the common port to the matching network designed

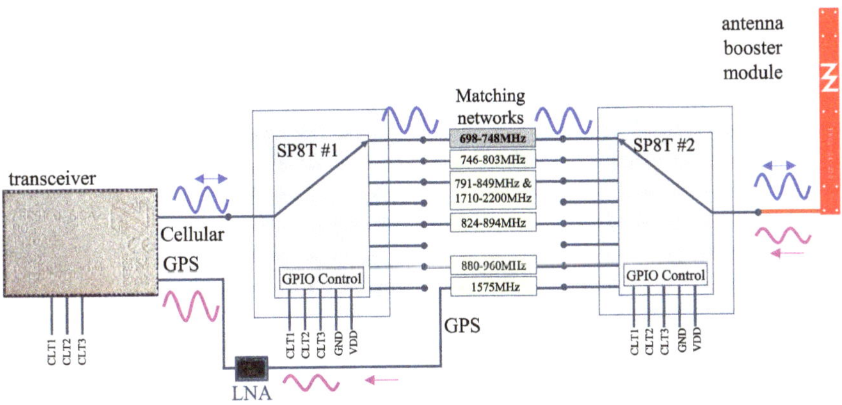

Figure 7.22 A transceiver (handling both cellular and GPS signals) governs the SP8T switches via lines CLT1-3, which regulate the reconfigurable matching network system consisting of lumped SMD components such as capacitors and inductors. An LNA (Low Noise Amplifier) amplifies the GPS signal before entering the transceiver

to match said frequency range. The same process is followed to operate at the other frequency regions. For 791–849 MHz, good matching is also obtained at the 1710–2170 MHz frequency region. For GPS reception, the transceiver controls the SP8T #2 to allow the GPS signal to reach the matching network for GPS. The LNA amplifies the signal before it enters the receiver's input GPS port.

The SP8T consumes low power: 240 μW (supply voltage and current of 3 V and 80 μA, respectively) [52].

It is interesting to note that there are still two states free at SP8T #2 (Figure 7.22). That state could be used for other radio protocols as long as no simultaneous communication is needed with the other two radios existing in the current solution.

An electromagnetic solver is used to obtain the *S*-parameter considering the ground plane printed on a substrate (ε_r = 4.15, tanδ = 0.017), the main device components such as the battery, and the top cover (ε_r = 2.9, tanδ = 0.01) (Figure 7.23). From a simulation point of view, the antenna booster module can be considered a 3D conductive body of 30 mm × 3 mm × 1 mm (height) with an LC (L = 15 nH, C = 0.5 pF) filter at 5 mm from the edge, according to the teachings of the loaded antenna booster studied in Chapter 5. In fact, this is the first step in the design process before the design of the matching networks. Once the filter is optimised to maximise potential bandwidth across the frequency range of interest, the next step consists of designing each matching network.

Besides the S-parameter of the RF switches, the lumped inductors and capacitors of the matching networks are considered in the electromagnetic simulation (Figure 7.24). As explained in Chapter 9, the pads for allocating the lumped

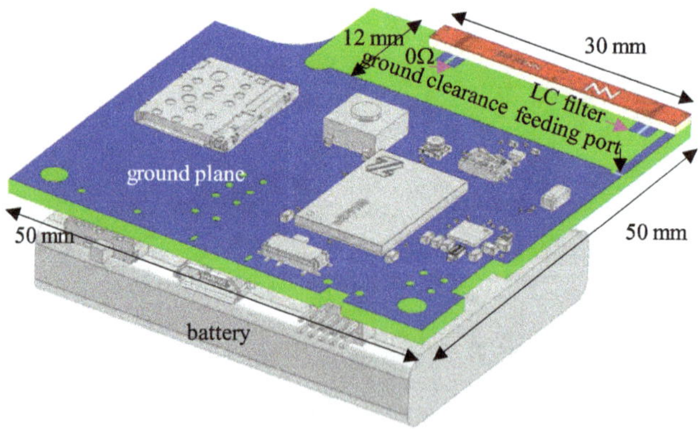

Figure 7.23 The 30 × 3 mm × 1 mm antenna booster-based component on a 50 mm × 38 mm ground plane printed on an FR4 substrate (1 mm thick, ε_r = 4.15, tanδ = 0.017). The main components, such as the battery, are taken into account for the simulation with Microwave Studio Suite [53]. The LC filter comprises a L = 15 nH and a C = 0.5 pF.

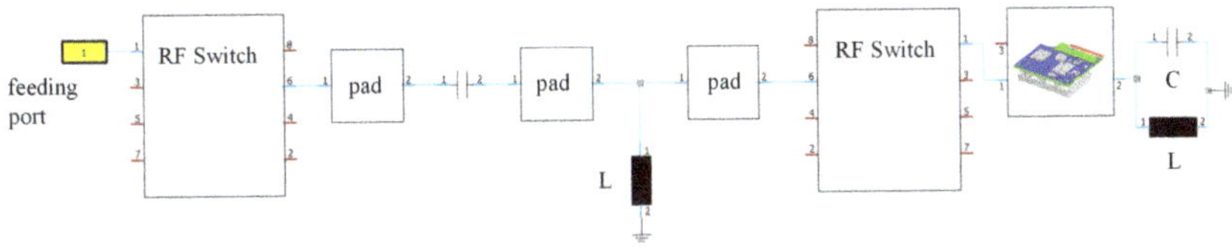

Figure 7.24 Circuit schematic for modelling the reconfigurable antenna system. The LC shows the filter used in the antenna booster element in Figure 7.23

STATE	Frequency band	Matching Network		
		Component	**Value**	$\mathbf{Q_{min}}$
RF5	698-748MHz & 1710-2200MHz	Z_1	5.1nH	40 at 698MHz
		Z2 (C10)	5.5pF	76 at 2200MHz
		Component	**Value**	$\mathbf{Q_{min}}$
RF7	746-803MHz& 1710-2200MHz	Z_1	4.7nH	40 at 746MHz
		Z_2	0Ω	-
		Component	**Value**	$\mathbf{Q_{min}}$
RF8	791-849MHz& 1710-2200MHz	Z_1	9.2pF	166 at 849MHz
		Z_2	6.2nH	40 at 791MHz
		Component	**Value**	$\mathbf{Q_{min}}$
RF3	824-894MHz	Z_1	1.5pF	461 at 894MHz
		Component	**Value**	$\mathbf{Q_{min}}$
RF1	880-960MHz	Z_1	2.5pF	402 at 960MHz
		Z_2	Open circuit	-
		Component	**Value**	$\mathbf{Q_{min}}$
RF4	GPS (1575MHz)	Z_1	2.2nH	103 at 1575MHz
		Z_2	2.5pF	242 at 1575MHz
RF2&6	available for other bands	empty		

Figure 7.25 Matching networks for cellular and GPS used in the electric scheme shown in Figure 7.22

components of the matching networks have also been considered. Those pads can be included as short transmission lines or by modelling the full layout with a full-wave simulator.

To keep the solution as compact as possible at the cellular bands, SMD 0201 (0.5 mm × 0.25 mm) lumped capacitors and inductors have been used for the matching networks, selecting the highest quality factor (Q) possible. For the 1575 MHz matching network (GPS), SMD 0402 lumped components have been used to minimise losses (Figure 7.25). From a behaviour perspective, 0402 SMD inductors provide a larger Q than the 0201 SMD inductors since the 0402 SMD size is larger. As a practical guideline, for the case presented here, the average total efficiency at low-frequency bands (698–960 MHz) will be 1 dB more if 0402 SMD inductors are used.

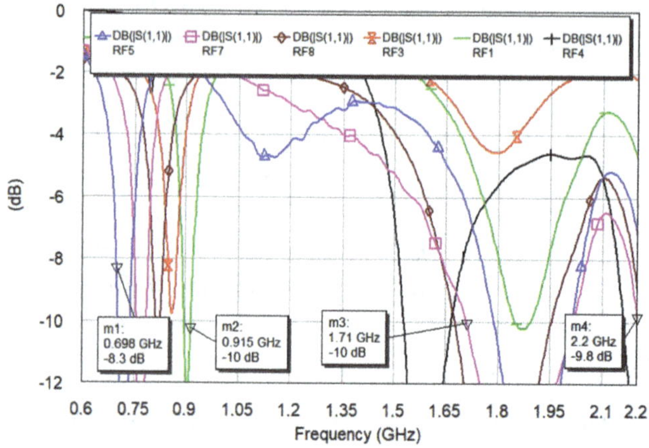

Figure 7.26 Simulated S_{11} for several states covering cellular and GPS bands. Each state corresponds to one matching network, as shown in Figure 7.25.

Once the S_{11} at the antenna port is obtained, the matching network architecture, including the passive matching networks and the RF switches, is designed. Each matching network provides operation at a single band (RF1, RF3, RF4) or dual-band (RF5, RF7, RF8) with an S_{11} that is optimised within the band (Figure 7.26).

Radiation patterns are computed from the entire 3D model, including all components (the top and bottom covers are not illustrated for clarity purposes in Figure 7.27). As observed, since the ground plane is small in terms of wavelength ($\sim 0.15\lambda$ at 900 MHz), radiation patterns are quasi-isotropic, which is convenient since signal reception does not depend upon the orientation of the device.

When the design is feasible in the simulation domain, the solution is implemented into an actual cellular IoT device (Figure 7.28). The device comprises a transceiver that drives the needed state of the switches to optimise performance with a suitable matching network. Besides the radiofrequency components, the device hosts many sensors, such as an accelerometer, light and colour, temperature, humidity, air quality, and air pressure.

For passive testing (Chapter 1), the radiofrequency module is removed and substituted by two microcoaxial 50 Ω transmission lines, one for testing the cellular bands and the other for testing the GPS band. Each transmission line should be grounded to the RF ground of the device at least at two points, one close to the end and the other close to the connector (Figure 7.29). For different situations where the transmission line is larger, a third ground point is suggested to avoid undesired resonances when testing either the reflection coefficient or the total efficiency.

A manual switch powered by the device battery controls the state of the switches. In this case, the interface to control the switches is GPIO, different from the RFFE utilised in the DTC case. In both cases, there are no external cables to set up the states of the switches, which could affect the passive measures.

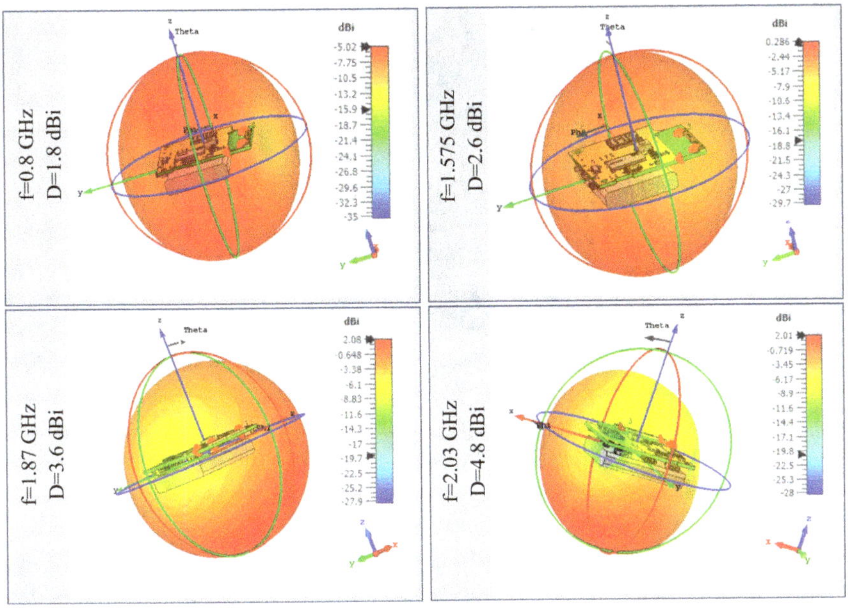

Figure 7.27 Simulated 3D radiation patterns. Radiation patterns are represented with a 30 dB dynamic range.

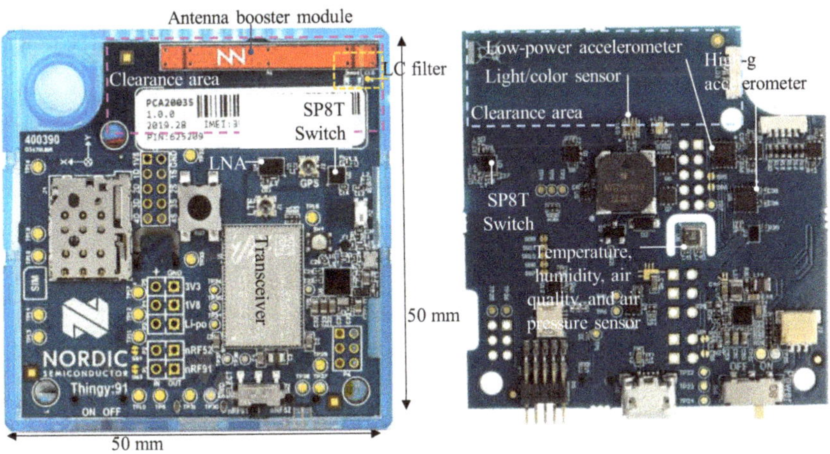

Figure 7.28 An IoT open platform device 50 mm × 50 mm (Thingy: 91 by Nordic Semiconductor [50], including the reconfigurable antenna system of Figure 7.22 by Ignion [54]. Besides cellular and GPS operation, thanks to the reconfigurable antenna system, sensors for IoT applications are included, such as temperature, accelerometer, humidity, air quality and pressure, and light data [48]. The LC filter is L = 15 nH, C = 0.5 pF.

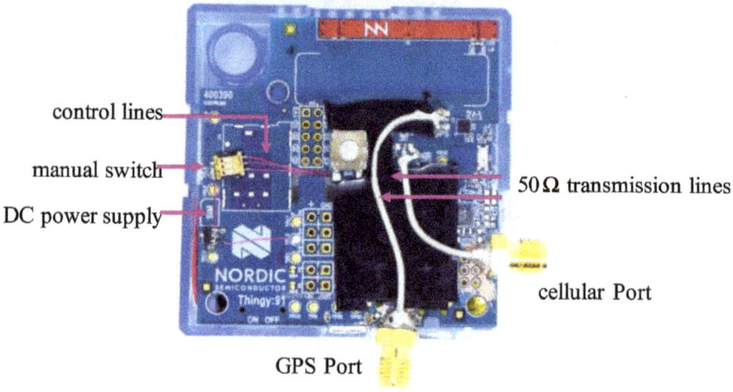

Figure 7.29 *Set-up for the passive testing of reflection coefficient and total efficiency*

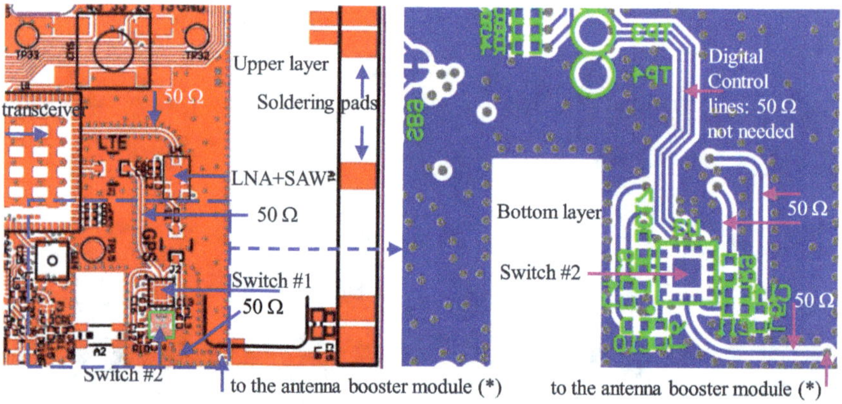

Figure 7.30 *(Left) Top layer; (right) bottom layer. The right part is magnified.*

Reconfigurable architecture versus a passive multiband matching network

The current reconfigurable matching system with 2-SPNT achieves 6.8 dB more total efficiency than a passive matching network in the 698–960 MHz frequency range, ensuring more battery life, connectivity, and range.

It is interesting to illustrate some guidelines for a correct layout of the reconfigurable matching system on a wireless device. In this case, one switch is placed on the top ground and the other on the bottom ground of the multilayer PCB (Figure 7.30). Starting from the antenna, there is a 50 Ω CPW line. Several vias

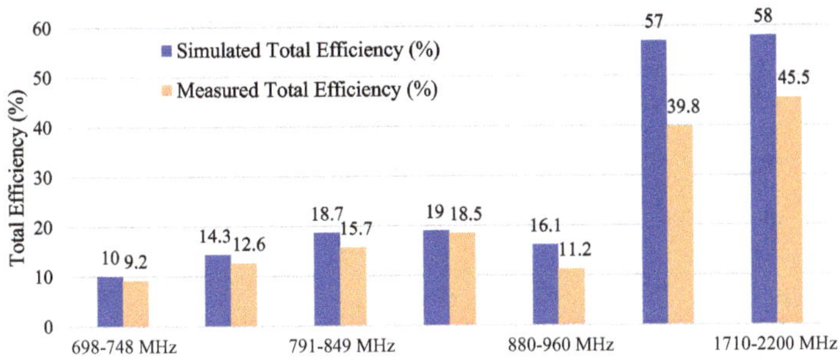

Figure 7.31 Simulated and measured total efficiency. Both the simulation and measurements include all components as well as the top and bottom covers of the device

surround the CPW line to mitigate interference with other RF components. Additionally, vias along the line surround the other 50 Ω CPW lines connecting one switch to another. Pads for hosting the SMD lumped inductors and capacitors are small (0.3 mm × 0.3 mm) to pack all the matching networks in the small space.

The reflection coefficient is measured to test each state, and the matching network is adjusted if needed. Finally, the total efficiency is measured in an anechoic chamber using 3D pattern integration and compared with the simulated data. As stated before, total efficiency considers mismatching and losses on the system, such as those of the antenna system, but also nearby components (e.g., the battery under the PCB) as well as top and bottom plastic covers (Figure 7.31). As reflected in the results, the difference between simulated and measured total efficiency is 0.7 dB and 1.3 dB on average in the low-frequency range (699–960 MHz) and high-frequency range (1575 MHz and 1710–2170 MHz), respectively.

At this point, it is remarkable to emphasise that the efficiency that a passive matching network can achieve to cover the same frequency range is much lower in the frequency range from 698 MHz to 960 MHz. The difference between the simulated total efficiency of the RF switches compared with the passive matching network is 6.8 dB. This difference considers losses from all RF components (RF switches, lumped inductors, and capacitors). This huge difference shows the advantage of using a reconfigurable matching system over a passive matching network. In the high-frequency range, total efficiency is practically the same.

The next section illustrates another reconfigurable architecture improving the performance of the solution, including the two SP8T. However, the present solution provides the flexibility of adding independent marching networks and also providing multi-radio features, which should also be considered when designing wireless devices.

7.4.3 Multiband operation with a single SP4T switch

With the objective of a compact solution, one SP4T switch is utilised instead of two.

A single SPNT switch can be combined with an antenna booster to provide a reconfigurable frequency system similar to the DTC. However, an SPNT switch offers extra degrees of freedom since not only a capacitor can be tuned but also an inductor (Figure 7.2). The advantage of a DTC is that it allows a fine-tuning of a broad set of capacitors. For example, the DTC explained in Section 7.3 provides 32 states of capacitance. However, if there is no need for a fine-tuned frequency, an RF switch offers more flexibility due to the combination of capacitors and inductors. At the same time, it is not limited by a maximum C_{max}/C_{min} of the DTC.

The SP4T switch follows the configuration of the SP8T in the previous section (Figure 7.21) but with four extra shunt switches ($SW_{5–8}$, Figure 7.32). Those switches can connect the output RF_i (i = 1 to 4) to the ground plane. With this configuration, around 2^8 combinations are possible. However, those combinations with SW1=SW2=SW3=SW4=OFF are not useful. As observed, this configuration provides many degrees of freedom, 65 to be precise. For example, if SW_1=ON and the remaining switches are OFF, Z_1 is in series. If SW_1=ON and SW_6=ON, the matching network is an L-type with Z_2 in shunt with the antenna booster and Z_1 in series. If SW_1=SW_2=ON, Z_1 and Z_2 are both connected in shunt and in series with the antenna booster. If SW_1=SW_2=SW_8=ON, the matching network Z_4 is in shunt with the antenna booster and Z_1 and Z_2 (both combined in shunt) in series. Additionally, a common matching network may be added at the common port (RFC) for fine-tuning purposes [55].

The architecture is applied to a 50 mm × 50 mm PCB where $Z_1 = 2.1$ pF, $Z_2 = 6$ pF, $Z_3 = 0\ \Omega$, and $Z_4 = 3.7$ nH. Finally, the common matching network is a 3.7 nH in shunt (Figure 7.33). As shown, only six states are enough to cover the 698–960 MHz and 1710–2170 MHz frequency ranges.

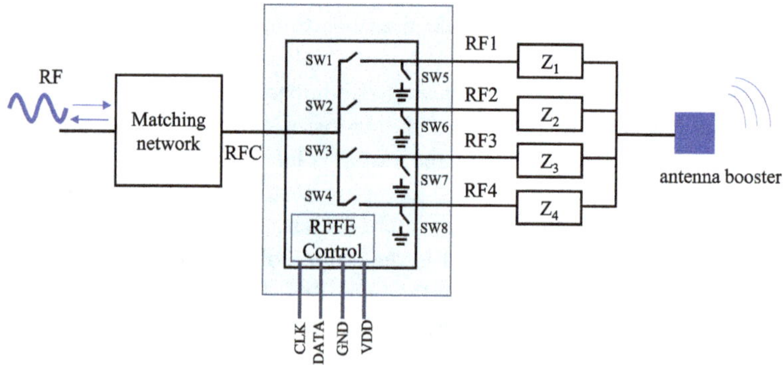

Figure 7.32 Architecture for multiband operation comprising an SP4T and an antenna booster. The antenna booster corresponds to the one shown in Figure 7.23 with an LC filter L = 15 nH, C = 0.3 pF. RFC stands for common RF port. The SP4T consumes low power: 72 μW (supply voltage and current of 1.8 V and 40 μA, respectively) [56].

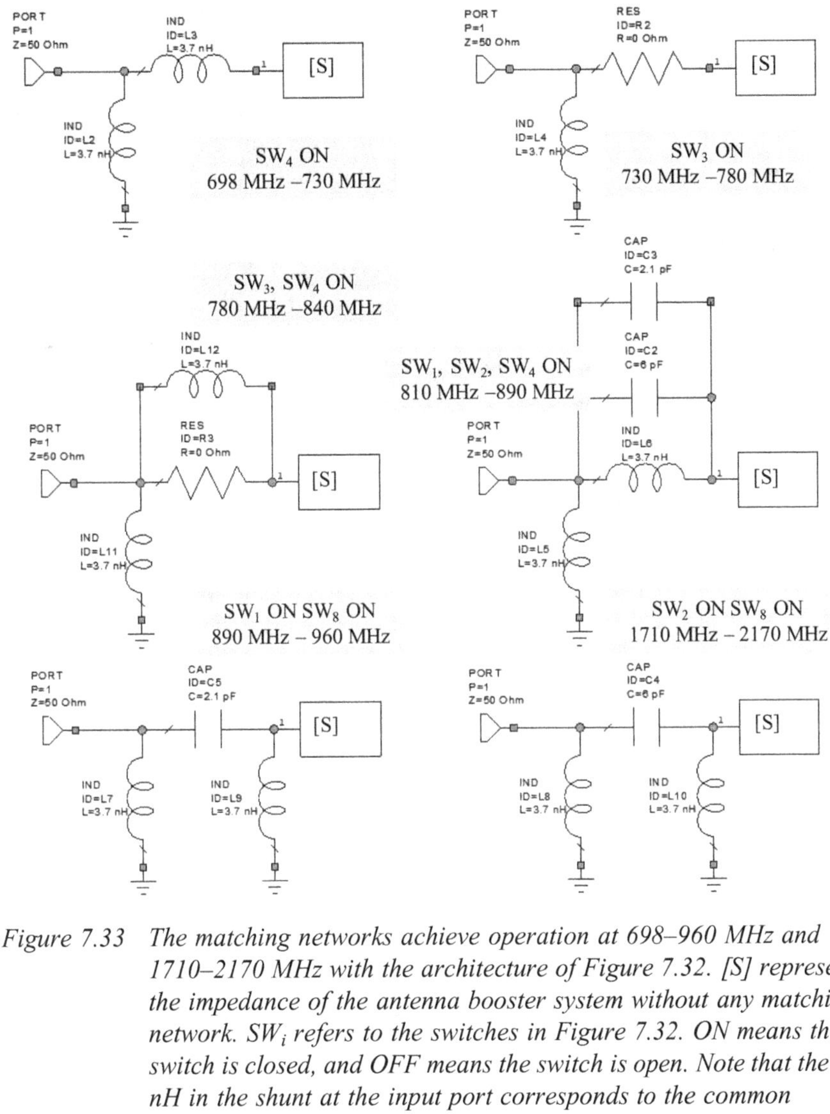

Figure 7.33 The matching networks achieve operation at 698–960 MHz and 1710–2170 MHz with the architecture of Figure 7.32. [S] represents the impedance of the antenna booster system without any matching network. SW_i refers to the switches in Figure 7.32. ON means the switch is closed, and OFF means the switch is open. Note that the 3.7 nH in the shunt at the input port corresponds to the common component shown in Figure 7.32. The antenna booster module includes an LC filter with $L = 15$ nH and $C = 0.3$ pF.

The matching networks illustrated in Figure 7.33 do not include the short transmission lines needed from the feeding line to the ports of the SP4T. These short transmission lines introduce inductances that can be easily taken into account with a full-wave simulation. For example, the SW_3 and SW_4 ON result in a shunt combination of 0 Ω and 3.7 nH. Ideally, this is a 0 Ω. However, the short transmission lines used at both sides of the 0 Ω introduce an inductance. As a result, the shunt combination of 0 Ω and 3.7 nH is not 0 Ω. In fact, this state is useful for operating at 780–840 MHz, whereas the 0 Ω matching (SW_3 ON) matches from 730 MHz to 780 MHz [57].

It is important to note that the SPNT used in this architecture should present a high impedance (C_{OFF} = 145 fF) when switches are in the OFF state. Some SPNTs are connected to a load (absorptive), which is inconvenient in this architecture. For example, let us assume SW_2=ON since only Z_2 in series is needed for operation. If other switches present high impedance in the OFF state, there is no issue. However, if switches are connected to a load, the matching network is no longer Z_2 in series but a combination of Z_2 with a load, which will cause losses in the system. This situation, however, is not critical for the two SPNT architectures presented in Section 7.4.2 because the different outputs of the SPNT are not interconnected.

The design of the matching networks can be addressed from an engineering perspective or by using matching network optimisation. In the first situation, the impedance of the antenna booster system without any matching network must be analysed. Therefore, one can start by matching the lowest frequency part by designing a series inductor. Note that an antenna booster system is non-resonant, at least at the lowest frequency of operation. Therefore, a series inductor is typically needed. Note that a first series is employed to match the 698–730 MHz range (Figure 7.33). In the present design, the real part is not 50 Ω. Therefore, the series inductance is designed in such a way that it is not on the real axis of the Smith chart but still in the capacitive zone. Therefore, a shunt inductor (the common inductor) is added with a value of L = 3.7 nH. The process is iterated for various frequency ranges while maintaining the same common inductor. Consequently, the aim is to achieve a matching where the impedance falls within the capacitive zone on the g = 1 circle of the Smith chart at the central frequency of each band. Note that the 730–780 MHz does not need any series inductor since the reactive part is less than the one at the 698–730 MHz. This tuning process should be carried out considering the full layout model since any short transmission line to connect the components of the matching network introduces inductance.

The second method relies on using a matching network synthesiser, as explained in Chapter 5. In this case, however, the optimisation problem is more processor-hungry since the process has to consider many combinations.

In this case, the control interface is RFFE, which relies on synchronous input with a clock and data line. As the previous section explains, the SP4T is monitored with an external kit (Figure 7.34). The kit allows the SP4T to be configured thanks to a software application and provides voltage to feed the SP4T. Once a particular state is set, the interface is removed. The reason is that such an interface alters the ground plane and, thus, impacts the S_{11} and efficiency. Hence, a battery is embedded within the PCB under examination to sustain the status of the SP4T. Consequently, the interface can be detached to accurately assess S_{11} (Figure 7.35) and efficiency (Figure 7.36).

The prototype is tested for S_{11} (as illustrated in Figure 7.35) using the matching network components obtained via electromagnetic simulation, taking into account

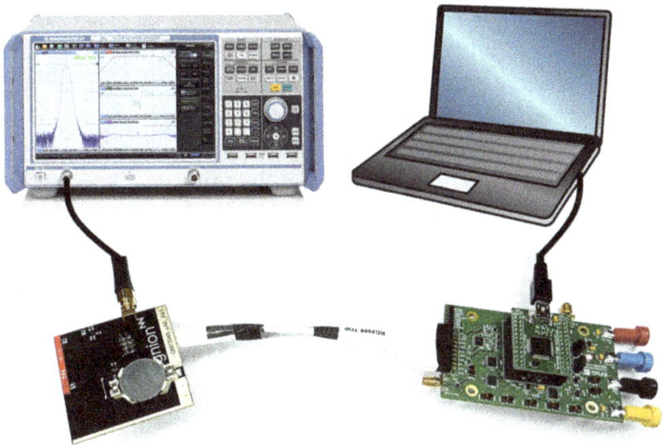

Figure 7.34 To turn ON/OFF the eight switches, a kit interface connected to a laptop is required. To avoid the effect of the interface between the kit and the PCB, a battery has been included in the PCB that allows the interface to be disconnected when testing S-parameters and efficiency [58].

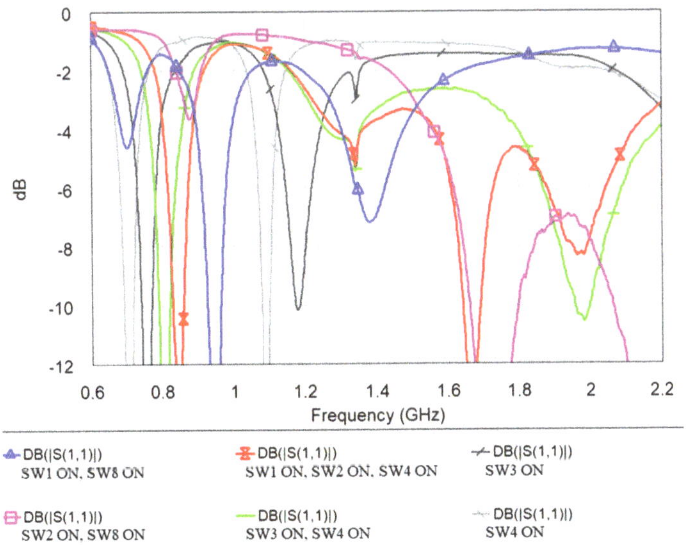

Figure 7.35 Measured reflection coefficient for achieving operation at 698–960 MHz and 1710–2170 MHz

the entire layout, and considering the states provided in Figure 7.33. Five states are needed to cover the low-frequency range, whereas only one state (SW_2 ON, SW_8 ON) is needed for the high-frequency region.

Once validated with the S_{11}, the total efficiency is tested in the anechoic chamber for each state (Figure 7.36). For comparison purposes, the total efficiency is compared with a passive solution with a multiband matching network following the teachings of Chapter 5. As demonstrated, the passive matching network cannot meet the bandwidth requirement at the low-frequency region (698–960 MHz). Hence, total efficiency is only comparable at 698–748 MHz, being much less at the other bands.

In terms of simplicity and efficiency, the 1-SP4T is a preferred architecture (Figure 7.32) over the 2-SP8T (Figure 7.22). The efficiency results of the 1-SP4T (Figure 7.36) outperform those of the 2-SP8T and even more compared to a passive matching network. The passive matching network only provides an efficiency comparable to the reconfigurable solution at some frequency bands (698–748 MHz and 1710–2170 MHz). The reason is that the matching network can only match some frequency bands as expected. The advantage of the reconfigurable solution is notorious. For example, at 880–960 MHz, the total efficiency for the 1-SP4T is 17.6 dB more.

Although the 2-SP8T presents less efficiency than the 1-SP4T, it allows the matching networks with several components to include not only series or shunt series as the 1-SP4T but also Π-type or even with more than three components. Besides, the 2-SP8T allows multi-radio. As disclosed in Figure 7.22, five input/output ports of the SP8T #2 are dedicated to covering cellular bands, and another port is for GPS. Still, two ports are free, which can be used for another two communication protocols. As a result, there is more design freedom. The designer should have in mind both routes to finally pick the most convenient in terms of simplicity, performance, and cost.

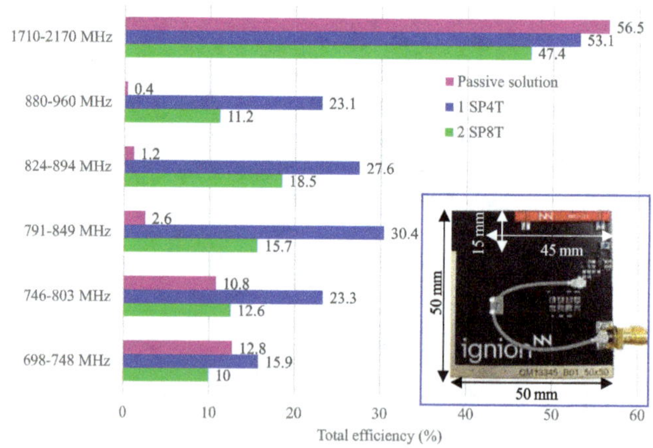

Figure 7.36 Measured total efficiency in the anechoic chamber for the 50 mm × 50 mm PCB, including the antenna booster module, compared to the solution with 2-SP8T (Figure 7.28) and a passive matching network

The 1-SP4T scheme uses only 7 states out of the 65 available. Those remaining states could be used to match other frequency bands or be used to compensate for the impact of the environment where the wireless device is located, such as, for example, concrete or a metal body [59].

7.5 Discussion

This section discusses several reconfigurable antenna solutions that appear in industry and scientific literature [2,31,60–73]. The selection criteria include reconfigurable antenna solutions for multiband behaviour at two frequency regions. The first is from 698 MHz to 960 MHz, and the second is from 1.71 GHz to at least 2.17 GHz. The first frequency region is the most challenging for designers since antennas are electrically smaller in this first frequency region than in the second. Consequently, squeezing bandwidth from the first frequency region is challenging.

Antennas are classified by length and width. In those antennas having irregular shapes, length and width have been selected as the minimum rectangle comprising the antenna. Once all antennas are put on a single map, a length and width average can be calculated. That single average antenna defines four quadrants: the best, the worst, and two intermediate quadrants, one having a length below the average but widths above, and the other way around for the other intermediate quadrant (Figure 7.37).

Regarding the active component for the matching network, not only DTC and SPNT are common, but some solutions include PIN diodes as single components [74,75] to open or short one or more conductive parts of an antenna and varactor diodes (variable capacitors) to obtain multiband performance [76,77]. For these solutions, extra electronics to separate the radiofrequency from DC must be included, whereas the encapsulated DTC/SPNT is embedded in the package,

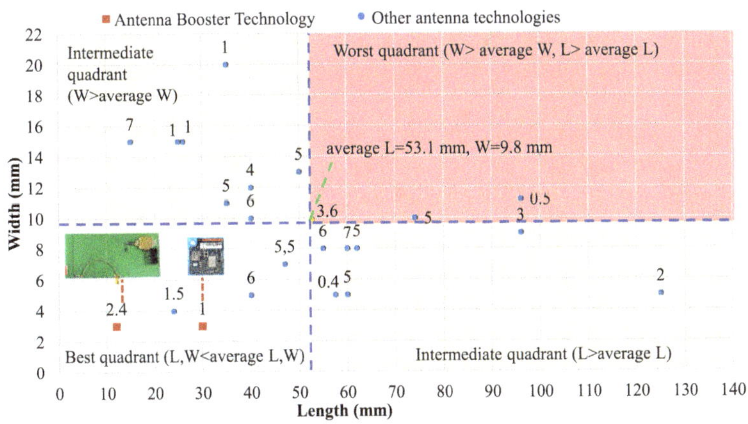

Figure 7.37 Classification of several reconfigurable multiband antenna techniques

making the integration easier and ready to use. Besides, the power consumption of the PIN diode in the ON-state is about ~1 mW compared to only 0.18 mW of the DTC used in this chapter. Generally, PIN diodes consume more power than MEMS-based components [78,79].

For comparison purposes, the two-antenna booster reconfigurable-based solutions presented in this chapter are included in the plot: the one using an antenna booster of 12 mm × 3 mm × 2.4 mm for operation from 698 MHz to 960 MHz and 1710 MHz to 2690 MHz, and the second one featuring an antenna booster module of 30 mm × 3 mm × 1 mm for operation at 698–960 MHz and 1710–2170 MHz as well as in GPS using the same reconfigurable matching circuit. Besides lengths and widths, antenna height is included in the plot, with a number indicating height in millimetres.

As clearly shown in Figure 7.37, the antenna booster reconfigurable-based solutions are in the best quadrant and the best location within that area; in particular, the one with 2.4 mm height is the best among all samples, featuring a small volume of only 86.4 mm^3. And in second place, the antenna booster module with 30 mm × 3 mm × 1 mm features a volume of 90 mm^3. This solution has a smaller volume than the 1.5 mm height, with a volume of 144 mm^3. As a final remark, antenna booster technology features the smallest footprint and volume while providing performance across many cellular and geolocalisation bands.

7.6 Conclusions

Two multiband reconfigurable antenna architectures for wireless devices using antenna boosters less than 90 mm^3 have been presented for operation at 698–960 MHz and 1710–2690 MHz. Combining DTC and RF switches with antenna boosters achieves multiband reconfigurability. On the one hand, DTC provides reconfigurability thanks to a tunable capacitor that, combined with a multiband matching network, allows for reconfiguring the frequency bands. In other situations where changing a capacitor is insufficient for providing enough freedom to tune the frequency bands, an architecture with RF switches like the ones presented here can be adopted.

A DTC and RF switch use interfaces that can be easily connected with transceivers. One interface uses as many *n* control lines to set 2^n states. The other one uses a digital data line and, thus, enables control of any number of states with such a line and a clock line.

Three solutions have been implemented, the one with DTC and an antenna booster of 86.4 mm^3 and the one with SPNT and an antenna booster module of 90 mm^3, showing competitive efficiency values for the wireless industry. These antenna booster architectures can be quickly adopted in wireless devices since the antenna booster does not need geometry customisation since the multiband operation is adjusted by the matching network, making it a fast and easy design process. Finally, an antenna booster can be easily embedded due to its small size and off-the-shelf nature, allowing one to pick and place it on a PCB.

In those situations, such as small ground planes or when the number of frequency bands is large, reconfigurable solutions provide much more efficiency compared to a passive matching network.

Finally, Table 7.3 answers the objectives given at the beginning of the chapter. It also points to relevant figures and equations to consolidate the concepts learnt. It is also advisable to refresh all essentials included in the grey boxes distributed throughout the chapter.

Table 7.3 Key aspects to take away

Topic	Review and take away
Reconfigurable matching networks for antenna boosters	When the ground plane is small in terms of the wavelength ($L < 0.2\lambda$), or the number of frequency bands is large, or the device is affected by external conditions (conductive materials, human body, etc.), a passive matching network may not meet the specifications. In this situation, reconfigurable matching networks provide the flexibility to cover the requested frequency bands.
DTC	Architectures combining DTCs and antenna boosters provide tunability to adjust the frequency bands, particularly at the frequencies where the ground plane is electrically small ($L < 0.2\lambda$). For those frequencies, the bandwidth is limited and, thus, not achievable with a passive matching network. Due to the nature of an antenna booster, a series DTC results in a simple reconfigurable architecture (Figures 7.8, 7.12, and 7.13). A DTC can also be connected in a shunt (Figure 7.15). Due to the small steps of the DTC, the frequency tunability is fine, whereas it may be coarse when using SPNT. The design of tunable matching networks with DTCs can also be addressed with matching network synthesis (Figure 7.18).
SPNT	SPNT also provides frequency band tunability. Architectures with two SP8T and a bunch of single matching networks to cover small frequency bands allow multiband operation (e.g., 698–960 MHz and 1710–2170 MHz) in a small PCB (Figure 7.22). The whole frequency band is divided into sub-sets where a simple matching network of 1 or 2 components is enough (Figures 7.25 and 7.26). Antenna boosters can also be bundled with one SPNT (Figure 7.32), resulting in a compact, tunable architecture. The architecture with 2-SPNT provides more flexibility in including matching networks with more than two components and also multi-radio operation. Both routes should be considered when designing reconfigurable antenna booster systems.

(Continues)

Table 7.3 (Continued)

Topic	Review and take away
Prototyping	From a proof-of-concept perspective, any interface connected to the PCB may alter the currents in the ground plane and, thus, modify the reflection coefficient, total efficiency, and radiation patterns, for example. Therefore, it is advisable to provide a DC voltage not from the interface but from an embedded battery on the PCB (Figures 7.14 and 7.36). Once a particular state has been configured, the interface can be removed, and the embedded battery will power the DTC or SPNT.
Consumption	For battery-powered wireless devices, consumption should be minimised. Low-power tunable circuits, such as those based on MEMS switches and DTC, are competitive candidates. As a reference, the DTC used in this chapter consumes 180 μW and the SP4T 72 μW.

7.7 Exercises

A 3D electromagnetic software and a circuit simulator are needed to solve the following exercises.

7.7.1 Exercise 1

Consider the antenna booster system described in Figure 7.4 and the reconfigurable matching network in Figure 7.12. Optimise the values of the inductors and determine the range of the tunable capacitor to obtain an $S_{11} < -6$ dB across 698–960 MHz and 1710–2690 MHz. Consider real components with a finite Q. Compute the total efficiency.

7.7.2 Exercise 2

Repeat the previous exercise, 7.7.1, changing the size of the ground plane to 60 mm × 60 mm. Analyse the frequency range that can be covered using the $S_{11} < -6$ dB criteria. Why does the 120 mm × 60 mm ground plane provide better performance? To answer the question, use the potential bandwidth that can be achieved for each scenario, as explained in Section 3.5, Chapter 3.

7.7.3 Exercise 3

Design a reconfigurable matching network using two switches for the 60 mm × 60 mm ground plane of the previous exercise. Consider the insertion losses of the switches using measured S-data found in datasheets. What SPNT do you need to cover from 698 MHz to 960 MHz and 1710 MHz to 2690 MHz?

7.7.4 Exercise 4

Repeat the previous Exercise 7.7.3 with the 60 mm × 60 mm ground plane using only one series switch with the architecture given in Figure 7.32. Compare results and analyse the pros and cons of each solution in terms of performance, simplicity, and cost.

7.7.5 Exercise 5

For the antenna booster in Figure 7.4, simulate the impact on S_{11} and efficiency when the antenna system is close to an infinite conductive plane ($\sigma = 5.9 \times 10^7$ S/m) for the following distances: $h = 20$ mm, $h = 10$ mm, and $h = 5$ mm. Analyse the impedance and think if a reconfigurable matching network can be used to have an $S_{11} < -6$ dB for each case.

7.7.6 Exercise 6

Benchmark different DTCs by comparing the tunability range (C_{min}, C_{max}), quality factor, self-resonant frequency, power consumption, control interface, R_{ON}, and C_{OFF}.

7.7.7 Exercise 7

Benchmark different SPNTs by comparing insertion loss, number of states, isolation, power consumption, control interface, R_{ON}, and C_{OFF}.

7.7.8 Exercise 8

Consider the antenna booster system in Figure 7.4 but with a 109 mm × 60 mm ground plane. Design a reconfigurable matching network with an SP2T to provide a multi-radio operation at 863–928 MHz and 1.71–2.7 GHz. Compute S_{11} and total efficiency considering a real SP2T and lumped components for the matching network.

7.7.9 Exercise 9

For exercise 7.7.8, compare the results regarding total efficiency and design simplicity with a diplexer-based solution described in Chapter 6, Figure 6.25.

7.7.10 Exercise 10

Consider a 12 mm × 3 mm × 2.4 mm antenna booster on a 60 mm × 60 mm ground plane (use Figure 7.4 as a reference). Compare the following architectures (Figure 7.38) in terms of S_{11} and total efficiency for multi-radio operation at radio #1 800–900 MHz and radio #2 1.7–3 GHz. Note that MN #1 should provide operation at radio #1, MN#2 at radio #2, and MN#3 should be matched at radios #1 and #2. Which of both architectures is easy to scale to more than two radios?

7.7.11 Exercise 11

For the SP4T of Figure 7.32, demonstrate that 65 different matching networks can be obtained.

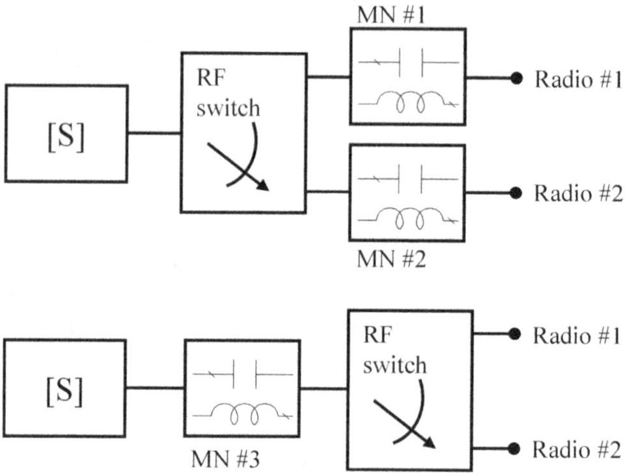

Figure 7.38 The [S] corresponds to the input impedance of the antenna booster shown in Figure 7.4. MN stands for Matching Network

References

[1] J. Anguera, A. Andújar, M. C. Huynh, C. Orlenius, C. Picher, and C. Puente, "Advances in Antenna Technology for Wireless Handheld Devices," *International Journal on Antennas and Propagation*, vol. 2013, 2013. Article ID 838364.

[2] M. Zheng, H. Wang and Y. Hao, "Internal Hexa-Band Folded Monopole/Dipole/Loop Antenna with Four Resonances for Mobile Device," *IEEE Transactions on Antennas and Propagation*, vol. 60, no. 6, pp. 2880–2885, 2012. DOI: 10.1109/TAP.2012.2194687

[3] Byoung-Nam Kim, Seong-Ook Park, Jae-Ho Lee, Jeong-Kun Oh, Kyung-Joon Lee and Gwan-Young Koo, "Hepta-band Planar Inverted-F Antenna with Novel Feed Structure for Wireless Terminals," *2007 IEEE Antennas and Propagation Society International Symposium*, Honolulu, HI, pp. 1257–1260, 2007.

[4] Y. L. Ban, Y. F. Qiang, Z. Chen, K. Kang, and J. L. W. Li, "Low-Profile Narrow-Frame Antenna for Seven-Band WWAN/LTE Smartphone Applications," *IEEE Antennas Wireless Propagation Letters*, vol. 13, pp. 463–466, 2014.

[5] Z. X. Chen, Y. L. Ban, Z. Chen, K. Kang, and L. W. Li, "Two-Strip Narrow-Frame Monopole Antenna with a Capacitor Loaded for Hepta-Band Smartphone Applications," *Progress in Electromagnetics Research*, vol. 145, no. 2, pp. 31–38, 2014.

[6] Y. L. Ban, C. L. Liu, J. L. W. Li, and R. Li, "Small-Size Wideband Monopole with Distributed Inductive Strip for Seven-Band WWAN/LTE Mobile

Phone," *IEEE Antennas and Wireless Propagation Letters*, vol. 12, pp. 7–10, 2013.

[7] C. Deng, Y. Li, Z. Zhang, and Z. Feng, "A Novel Low-Profile Hepta-Band Handset Antenna Using Modes Controlling Method," *IEEE Transactions on Antennas and Propagation*, vol. 63, no. 2, pp. 799–804, 2015.

[8] Y. Wang and Z. Du, "Wideband Monopole Antenna with Less Nonground Portion for Octa-Band WWAN/LTE Mobile Phones," *IEEE Transactions on Antennas and Propagation*, vol. 64, no. 1, pp. 383–388, 2016.

[9] D. Huang, Z. Du and Y. Wang, "An Octa-band Monopole Antenna with a Small Nonground Portion Height for LTE/WLAN Mobile Phones," *IEEE Transactions on Antennas and Propagation*, vol. 65, no. 2, pp. 878–882, 2017.

[10] D. Wu, S. W. Cheung, and T. I. Yuk, "A Compact and Low-Profile Loop Antenna with Multiband Operation for Ultra-Thin Smartphones," *IEEE Transactions on Antennas and Propagation*, vol. 63, no. 6, pp. 2745–2750, 2015.

[11] Zi-Qiang Xu, Qiang-Qiang Zhou, Yong-Ling Ban, and Simon S. Ang, "Hepta-Band Coupled-Fed Loop Antenna For LTE/WWAN Unbroken Metal-Rimmed Smartphone Applications," *IEEE Transactions on Antennas and Propagation*, vol. 17, no. 2, pp. 311–814 2018.

[12] J. Anguera, A. Andújar, G. Mestre, J. Rahola, and J. Juntunen, "Design of Multiband Antenna Systems for Wireless Devices Using Antenna Boosters," *IEEE Microwave Magazine*, Dec. pp. 102–114, 2019.

[13] J. Anguera, A. Andújar, C. Puente, and J. Mumbrú, "Antennaless Wireless Device capable of Operation in Multiple Frequency regions," *Patent US*. 8,237,615 B2. Aug. 2012.

[14] J. Anguera, A. Andújar, C. Puente, and J. Mumbrú, "Antennaless Wireless Device" *Patent Application* WO 2010/015365, February 11, 2010.

[15] A. Andújar, J. Anguera, and C. Puente, "Ground Plane Boosters as a Compact Antenna Technology for Wireless Handheld Devices," *IEEE Transactions on Antennas and Propagation*, vol. 59, no. 5, pp. 1668–1677, 2011.

[16] J. Anguera, A. Andújar, C: Puente, "Wireless Handheld Devices, Radiation Systems and Manufacturing Methods," *Patent US* 9,331,389 B2, May 2016.

[17] J. Anguera, C. Picher, A. Bujalance, and A. Andújar, "Ground Plane Booster Antenna Technology for Smartphones and Tablets," *Microwave and Optical Technology Letters*, vol. 58, no. 6, pp. 1289–1294, 2016.

[18] J. Anguera, A. Andújar, and C. Puente, "Antenna-Less Wireless: A Marriage Between Antenna and Microwave Engineering," *Microwave Journal*, vol. 60, no.10, pp. 22–36, 2017.

[19] J. de Mingo, A. Valdovinos, A. Crespo, D. Navarro, and P. García, "An RF Electronically Controlled Impedance Tuning Network Design and Its Application to an Antenna Input Impedance Automatic Matching System," *IEEE Transactions on Microwave Theory and Techniques*, vol. 52, no. 2, pp. 489–497, 2004.

[20] S. Yang, C. Zhang, H. K. Pan, A. E. Fathy, and V. K. Nair, "Frequency-reconfigurable Antennas for Multiradio Wireless Platforms," *IEEE Microwave Magazine*, pp. 66–83, 2009.

[21] Y. W. Chi and K. L. Wong, "Quarter-wavelength Printed Loop Antenna With an Internal Printed Matching Circuit for GSM/DCS/PCS/UMTS Operation in the Mobile Phone," *IEEE Transactions on Antennas and Propagation*, vol. 57, no. 9, pp. 2541–2547, 2009.

[22] V. Iyer, S. N. Makarov, D. D. Harty, F. Nekoogar, and R. Ludwig, "A Lumped Circuit for Wideband Impedance Matching of a Non-Resonant, Short Dipole or Monopole Antenna," *IEEE Transactions on Antennas and Propagation*, vol. 58, no. 1, pp. 18–26, 2010.

[23] Y. Li, Z. Zhang, J. Zheng, Z. Feng, and M. F. Iskander, "A Compact Hepta-Band Loop-Inverted F Reconfigurable Antenna for Mobile Phone," *IEEE Transactions on Antennas and Propagation*, vol. 60, no. 1, pp. 389–392, 2012.

[24] Ying Liu, Yan-Min Zhou, Gui-Feng Liu, and Shu-xi Gong, "Heptaband Inverted-F Antenna for Metal-Rimmed Mobile Phone Applications," *IEEE Antennas and Wireless Propagation Letters*, vol. 15, pp. 996–999, 2016.

[25] Hang Xu, Hanyang Wang, Steven Gao, Hai Zhou, Yi Huang, Qian Xu, and Yujian Cheng, "A Compact and Low-Profile Loop Antenna with Six Resonant Modes for LTE Smartphone," *IEEE Transactions on Antennas and Propagation*, vol. 64, no. 9, pp. 3743–3751, 2016.

[26] F. A. Asadallah, J. Constantine, and Y. Tawk, "A Multiband Compact Reconfigurable PIFA Based on Nested Slots," *IEEE Antennas and Wireless Propagation Letters*, vol. 17, no. 2, pp. 331–334, 2018.

[27] Y. Chen, R. Martens, R. Valkonen, and D. Manteuffel, "Evaluation of Adaptive Impedance Tuning for Reducing the Form Factor of Handset Antennas," *IEEE Transactions on Antennas and Propagation*, vol. 63, no. 2, pp. 703–710, 2015.

[28] C. Hoarau, N. Corrao, J.D. Arnould, P. Ferrari, and P. Xavier, "Complete Design and Measurement Methodology for a Tunable RF Impedance-matching network," *IEEE Transactions on Microwave Theory and Techniques*, vol. 56, pp. 2620–2627, 2008.

[29] Y. Chen and D. Manteuffel, "A Tunable Decoupling and Matching Concept for Compact Mobile Terminal Antennas," *IEEE Transactions on Antennas and Propagation*, vol. 65, no. 4, pp. 1570–1578, 2017.

[30] I. Ida, J. I. Takada, T. Toda, and Y. Oishi, "An Adaptive Impedance Matching System and Its Application to Mobile Antennas," *Proc. IEEE Region 10 Conf.*, 2004, pp. 543–546.

[31] J. Ilvonen, R. Valkonen, J. Holopainen, and V. Viikari, "Multiband Frequency Reconfigurable 4G Handset Antenna with MIMO Capacity," *Progress in Electromagnetics Research*, vol. 148, pp. 233–243, 2014.

[32] P. Bahramzy, P. Olesen, P. Madsen, J. Bojer, S. Caporal Del Barrio, A. Tatomirescu, P. Bundgaard, Arthur S. Morris III, and G. F. Pedersen, "A Tunable RF Front-End with Narrowband Antennas for Mobile Devices,"

IEEE Transaction on Microwave Theory and Techniques, vol.63, no. 10, pp. 3300–3310, 2015.

[33] S. Padmanathan *et al.*, "Compact Multiband Reconfigurable MIMO Antenna for Sub-6GHz 5G Mobile Terminal," *IEEE Access*, vol. 10, pp. 60241–60252, 2022, doi 10.1109/ACCESS.2022.3180048.

[34] J. Anguera, N. Toporcer, and A. Andújar, "Slim Bar Boosters for Electronic Devices," *Patent US* 9,960,478 B2, May 2018.

[35] J. L. Hilbert, *Tunable RF Components and Circuits: Applications in Mobile Handsets,* CRC Press, 2016.

[36] Q. Gu and A. S. Morris, "A New Method for Matching Network Adaptive Control," *IEEE Transactions on Microwave Theory and Techniques*, vol. 61, no. 1, pp. 587–595, 2013, doi: 10.1109/TMTT.2012.2230022.

[37] A. Andújar, J. Anguera, and R. Mª Mateos, "Multiband Non-Resonant Antenna System with Reduced Ground Clearance," *European Conference on Antennas and Propagation*, EUCAP 2017, Paris, France, April 2017.

[38] R.M. Fano, "Theoretical Limitations on the Broad-Band Matching of Arbitrary Impedances," *Journal of the Franklin Institute*, vol. 249, pp. 57–83, and pp. 139–154, 1950, 1950.

[39] J. Anguera, A. Andújar, J. L. Leiva, C. Schepens, R. Gaddi, and S. Kahng, "Multiband Antenna Operation with a Non-Resonant Element Using a Reconfigurable Matching Network," *European Conference on Antennas and Propagation*, EUCAP 2018, London, UK, April 2018.

[40] M. P. J. Tiggelman, K. Reimann, F. Van Rijs, J. Schmitz, and R. J. E. Hueting, "On the Trade-off between Quality Factor and Tuning Ratio in Tunable High-Frequency Capacitors," *IEEE Transactions on Electron Devices*, vol. 56, no 9, pp. 2128–2136, 2009.

[41] Cavandisk-Kinetics. https://www.cavendish-kinetics.com/

[42] J. Anguera and A. Andújar, "Wireless Device," *Patent US* 10,122,403 B2, Nov. 2018.

[43] J. Anguera, A. Andújar, J. L. Leiva, O. Massó, J. Tonnesen, E. Rindalsholt, R. Brandsegg, and R. Gaddi, "Reconfigurable Multiband Operation for Wireless Devices Embedding Antenna Boosters," *Electronics*, vol. 10, p. 808, 2021.

[44] Murata. https://www.murata.com/en-us

[45] D. Wang, R. Wolf, A. Joseph, A. Botula, P. Rabbeni, M. Boenke, J. Dunn, "High Performance SOI RF Switches for Wireless Applications". *2010 10th IEEE International Conference on Solid-State and Integrated Circuit Technology*. doi:10.1109/icsict.2010.5667307.

[46] H. Feng, J. Zhao, C. Zhou, and M. Song, "Design and Analysis of the Capacitive RF MEMS Switches with Support Pillars," *Sensors*, vol. 22, p. 8864, 2022. https://doi.org/10.3390/s22228864

[47] "Enabling the IoT Opportunity," *MIPI Alliance*, White Paper, September 2020

[48] J. Anguera, A. Andújar, J. L. Leiva, Y. Cobo, J. Tønnessen, E. Rindalsholt, and R. Brandsegg, "Virtual Antenna™ Provides Mobile and GPS Connection in the Thingy:91 Cellular IoT Module," *Everything RF*, May 2020.

[49] J. Anguera, "Embedded Antennas for IoT," *Webinar* organised by Nordic Semiconductor in collaboration with Ignion (Fractus Antennas in the video due to the old branding), October 2019. Available at https://www.youtube.com/watch?v=rjPd0sFuAyU
[50] Nordic Semiconductor. https://www.nordicsemi.com/
[51] Microwave office, Cadence. https://www.cadence.com/en_US/home.html
[52] Qorvo. www.qorvo.com. Datasheet QM12038.
[53] CST Studio Suite. https://www.3ds.com/products-services/simulia/products/cst-studio-suite/
[54] Ignion. www.ignion.io
[55] J. Anguera, A. Andújar, and C. Puente, "Multiband Antenna Booster Architecture with a Single Switch," *US Patent Application* US17749989; and European Patent Application EP22174591.2.
[56] Qorvo. www.qorvo.com. Datasheet QM13345.
[57] E. García, A. Andújar, J. L. Pijoan and J. Anguera, "Reconfigurable Antenna Booster Element for Multiband Operation in IoT Devices with an SP4T," *2023 17th European Conference on Antennas and Propagation (EuCAP)*, Florence, Italy, 2023, pp. 1–4, doi:10.23919/EuCAP57121.2023.10133063.
[58] Qorvo. www.qorvo.com
[59] A. Fernández, A. Andújar, J. L. Pijoan and J. Anguera, "Reconfigurable Multiband Antenna Booster Architecture for Different Environments," *IEEE Access*, vol. 11, pp. 85537–85548, 2023, doi:10.1109/ACCESS.2023.3303259.
[60] L. H. Trinh, F. Ferrero, L. Lizzi, R. Staraj and J. Ribero, "Reconfigurable Antenna for Future Spectrum Reallocations in 5G Communications," *IEEE Antennas and Wireless Propagation Letters*, vol. 15, pp. 1297–1300, 2016, DOI:10.1109/LAWP.2015.2505669.
[61] S. Caporal del Barrio, E. Foroozanfard, A. Morris and G. F. Pedersen, "Tunable Handset Antenna: Enhancing Efficiency on TV White Spaces," *IEEE Transactions on Antennas and Propagation*, vol. 65, no. 4, pp. 2106–2111, 2017, DOI:10.1109/TAP.2017.2662221.
[62] B. Mun, C. Jung, M. Park and B. Lee, "A Compact Frequency-Reconfigurable Multiband LTE MIMO Antenna for Laptop Applications," *IEEE Antennas and Wireless Propagation Letters*, vol. 13, pp. 1389–1392, 2014, DOI:10.1109/LAWP.2014.2339802.
[63] Y. Ban, S. Sun, P. Li, J. L. Li and K. Kang, "Compact Eight-Band Frequency Reconfigurable Antenna for LTE/WWAN Tablet Computer Applications," *IEEE Transactions on Antennas and Propagation*, vol. 62, no. 1, pp. 471–475, 2014, DOI: 10.1109/TAP.2013.2287522.
[64] C. Peng, I. Chen, and C. Liu, "Multiband Printed Asymmetric Dipole Antenna for LTE/WLAN Applications," *International Journal of Antennas and Propagation*, vol. 2013, pp. 1–6, 2013.
[65] C. Zhang, S. Yang, S. El-Ghazaly, A. E. Fathy and V. K. Nair, "A Low-Profile Branched Monopole Laptop Reconfigurable Multiband Antenna for Wireless Applications," *IEEE Antennas and Wireless Propagation Letters*, vol. 8, pp. 216–219, 2009, DOI: 10.1109/LAWP.2008.2007224.

[66] C. Hu, D. Huang, H. Kuo, C. Yang, C. Liao and S. Lin, "Compact Multi-branch Inverted-F Antenna to Be Embedded in a Laptop Computer for LTE/WWAN/IMT-E Applications," *IEEE Antennas and Wireless Propagation Letters*, vol. 9, pp. 838–841, 2010, DOI:10.1109/LAWP.2010.2069079.

[67] H. Li, J. Xiong, Y. Yu and S. He, "A Simple Compact Reconfigurable Slot Antenna with a Very Wide Tuning Range," *IEEE Transactions on Antennas and Propagation*, vol. 58, no. 11, pp. 3725–3728, 2010, DOI: 10.1109/TAP.2010.2071347.

[68] Y. Li, Z. Zhang, J. Zheng and Z. Feng, "Compact Heptaband Reconfigurable Loop Antenna for Mobile Handset," *IEEE Antennas and Wireless Propagation Letters*, vol. 10, pp. 1162–1165, 2011, DOI: 10.1109/LAWP.2011.2171311

[69] Y. Ren, "Ceramic Based Small LTE MIMO Handset Antenna," *IEEE Transactions on Antennas and Propagation*, vol. 61, no. 2, pp. 934–938, 2013, DOI: 10.1109/TAP.2012.2231660.

[70] I. Dioum, A. Diallo, S. M. Farssi and C. Luxey, "A Novel Compact Dual-Band LTE Antenna-System for MIMO Operation," *IEEE Transactions on Antennas and Propagation*, vol. 62, no. 4, pp. 2291–2296, 2014, DOI: 10.1109/TAP.2014.2301151.

[71] A. Singh *et al.*, "Compact active antenna for mobile devices supporting 4G LTE," *Loughborough Antennas and Propagation Conference* (LAPC), Loughborough, 2014, pp. 525–529, DOI:10.1109/LAPC.2014.6996441.

[72] M. Choi, H. Wi, B, Mun, Y. Yoon, H. Lee, and B. Lee, "A Compact Frequency Reconfigurable Antenna for LTE Mobile Handset Applications," *International Journal of Antennas and Propagation*, vol. 2015, pp. 1–10, 2015.

[73] T. Houret, L. Lizzi, F. Ferrero, C. Danchesi and S. Boudaud, "TC-Enabled Frequency-Tunable Inverted-F Antenna for IoT Applications," *IEEE Antennas and Wireless Propagation Letters*, vol. 19, no. 2, pp. 307–311, 2020, DOI:10.1109/LAWP.2019.2961114.

[74] I. Masroor, J. Ansari, A. Pandey, and P. Kumar Mishra, "Design and Analysis of Multiband Fractal Reconfigurable Antenna Using PIN Diodes for Smart Wireless Communications," *Progress In Electromagnetics Research C*, vol. 120, 209–222, 2022. doi:10.2528/PIERC22041402

[75] C. Luxey, I. Dussopt, J. L. Le Sonn, and J. M. Laheurte, "Dual-Frequency Operation of CWP-Fed Antenna Controlled by PIN Diodes," *IEE Electronics Letters*, vol. 36, no. 1, pp. 2–3, 2000.

[76] S. Oh, H. Yoon and S. Park, "A PIFA-Type Varactor-Tunable Slim Antenna with a PIL Patch Feed for Multiband Applications," *IEEE Antennas and Wireless Propagation Letters*, vol. 6, pp. 103–105, 2007, doi:10.1109/LAWP.2007.893096.

[77] N. Behdad and K. Sarabandi, "Dual-band Reconfigurable Antenna with a Very Wide Tunability Range," *IEEE Transactions on Antennas and Propagation*, vol. 54, no. 2, pp. 409–416, 2006, doi:10.1109/TAP.2005.863412.

[78] J. Bartolic, "MEMs, FET and PIN RF Switching Devices and Circuits for Reconfigurable Antennas," *The Future of the Antenna Technology in the Communications and Observation Systems*, Navarra, Spain, 2008.

[79] L. Santamaria, F. Ferrero, R. Staraj and L. Lizzi, "Electronically Pattern Reconfigurable Antenna for IoT Applications," *IEEE Open Journal of Antennas and Propagation*, vol. 2, pp. 546–554, 2021, doi:10.1109/OJAP.2021.3073104.

Chapter 8

Wireless devices embedding antenna boosters near conductive bodies

The integration of small, multiband antennas into wireless devices has become increasingly appealing. However, when these devices are located in proximity to conductive bodies, such as a smart meter with a metallic enclosure, their performance in terms of bandwidth and efficiency can be significantly compromised. This chapter presents a novel technique aimed at addressing this issue and achieving optimal bandwidth and efficiency values for multiband wireless devices situated near conductive bodies. By implementing this technique, wireless devices can overcome the negative effects caused by their close proximity to the conductive body, thereby ensuring competitive performance. In particular, the analysis focuses on the 824–960 MHz and 1710–2690 MHz frequency ranges for two wireless platforms with 120 mm × 60 mm and 60 mm × 60 mm ground planes, both being electrically close (less than $\lambda/30$ at 824 MHz) above a conductive flat body.

The chapter has the following objectives:

1. Know the impact of a conductive body electrically close to a wireless device on the reflection coefficient and total efficiency
2. Mitigation techniques to minimise the impact of detuning and efficiency reduction due to the conductive body

8.1 Introduction

Horizontal antenna elements with respect to an electrically conductive body are attractive due to their thin profile. However, they suffer from mismatching and low efficiency when electrically close to such conductive bodies [1]. The problem is the out-of-phase image current, which tends to cancel out the radiation. Replacing the electric conductive body with a magnetic conductive body solves the problem, as the image antenna shifts from being out-of-phase to in-phase [2]. A magnetic conductor does not exist in nature, but its performance can be synthesised by the so-called artificial magnetic conductors (AMC) or high-impedance surfaces (HIS) [3–19]. In practice, an artificial magnetic conductor comprises subwavelength structures following periodic patterns, where its behaviour is evaluated using the reflection phase method [3]. From a practical point of view, when the artificial surface presents a reflection phase within ±90°, the surface behaves as an AMC/HIS, which helps design low-profile antennas over said surface. In this sense, in [3,4], a horizontal dipole is placed close to a perfect electric conductor (PEC), including mushroom-like structures forming an HIS occupying a specific surface far beyond the antenna footprint. In [5–9], the cells forming the AMCs are optimised using genetic algorithms and fractal shapes. A dipole over an AMC is designed at 0.04λ concerning the AMC [9]. In [12], a monopole antenna over a small high-impedance surface is analysed, where the monopole has been placed at 0.08λ with respect to the high-impedance surface. Circularly polarised antennas using AMC are proposed in [14]. Active elements are combined with HIS to reconfigure the frequency band of operation [15].

Another technique presented in the literature to obtain reasonable efficiencies when an antenna is close to a PEC consists of adding multiple arms to a dipole antenna [20,21]. However, this technique presents limited bandwidth.

With the advent of the IoT (Internet of Things) devices, wireless devices may be close to a conductive body. For example, a wireless device operating at cellular bands on a fridge, a microwave, a washing machine, on the roof of a vehicle, on a container, etc. Such a wireless device usually includes a PCB (Printed Circuit Board) and an antenna. Therefore, when the PCB is electrically near a conductive body, the induced current on the conductive body minimises the radiation because it is out-of-phase with respect to the currents on the PCB. One possible solution is using an artificial magnetic conductor between the PCB and the conductive body. However, this may be a bulky solution. In this chapter, we propose a simple technique that disturbs the induced currents on the conductive body so that, in the end, the electromagnetic performance of the wireless device above the conducive body presents adequate bandwidth and efficiency. Moreover, the method does not need to add elements beyond the antenna system footprint, as AMC/HIS does.

The chapter aims to illustrate a technique that mitigates the adverse effects on bandwidth and efficiency when a wireless device is electrically close to a conductive body. For example, a metallic container may need a wireless device for tracking purposes. Other devices made out of conductive parts, such as washing machines, fridges, and heating machines, may be wirelessly connected for smart-home applications (Figure 8.1). Placing a wireless device close to such bodies in all cases may severely impact the bandwidth and efficiency.

Figure 8.1 A container, a washing machine, a fridge, and a heater comprise a conductive body. Placing wireless devices electrically close to a conducive body is challenging for obtaining bandwidth and efficiency.

To reach the objective, the chapter is divided as follows: a multiband antenna system comprising a PCB and an antenna booster for operating at 824–960 MHz and 1710–2690 MHz is presented in Section 8.2. The performance is analysed when the antenna system is placed at several distances over a finite PEC. Section 8.3 discusses an approach to address the limitation of PEC's proximity to the antenna system, which involves connecting the antenna system's ground plane to the PEC using vias. Section 8.4 presents a prototype for validating the electromagnetic analysis, showcasing measured SWR and total efficiency under various conditions: without PEC, with PEC, and with PEC utilising the proposed method of connecting the antenna system's ground plane to the PEC. Further analysis for a smaller ground plane is presented in Section 8.5. Conclusions are drawn in Section 8.6. Finally, a list of exercises are included in Section 8.7 and references at the end.

8.2 The multiband antenna system

First, the multiband antenna system is presented. In this case, the wireless device comprises a PCB and an antenna booster (Figure 8.2), and a multiband matching network that provides operation at 824–960 MHz and 1710–2690 MHz in free-space [22,23]. Such a 12 mm × 3 mm × 2.4 mm antenna booster is a 3D conductive parallelepiped ($\sigma = 5.9 \times 10^7$ S/m).

The multiband behaviour is not achieved by shaping a complex antenna geometry but by designing a multiband matching network using lumped elements such as capacitors and inductors (Figure 8.3). For a given input impedance without any matching network (the [*S*] box in Figure 8.3), a matching network can be synthesised using a microwave circuit simulator, given the total efficiency as a goal [24].

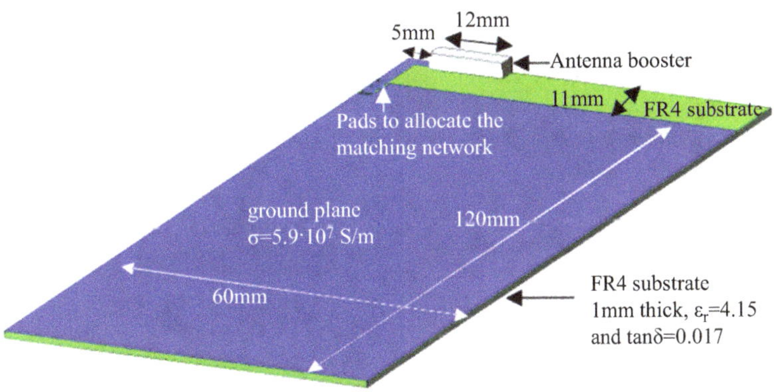

Figure 8.2 A 12 mm × 3 mm × 2.4 mm (height) antenna booster on a 120 mm × 60 mm ground plane printed over an FR4 substrate ($\varepsilon_r = 4.15$, tan$\delta = 0.017$). The antenna booster is modelled as a 3D conductive box with a conductivity of $\sigma = 5.9 \times 10^7$ S/m.

Total efficiency considers the power delivered to the load over the available power. This way, this total efficiency takes into account return loss as well as losses of the matching network as follows (Equation (8.1)):

$$\eta_t = \eta_r \cdot \left(1 - |S_{11}|^2\right) \tag{8.1}$$

where η_r is the radiation efficiency, which includes the losses of the antenna system, including the antenna booster, the PCB, and the losses due to the matching network components (inductors and capacitors). For the present case, the values of inductors L_i and capacitors C_i of the matching network have been optimised to obtain an S_{11} less than −6 dB across 824–960 MHz and 1710–2690 MHz using real SMD (Surface Mounted Device) 0402 components.

Without the PEC, the S_{11} of the antenna system of Figure 8.2, in combination with the matching network shown in Figure 8.3, presents a satisfactory S_{11} of less than −6 dB for 824–960 MHz and 1710–2690 MHz.

It is interesting to analyse the impact on electromagnetic performance when the antenna system is close to a conductive body like a PEC (Figure 8.4). This situation may represent an antenna system near a conductive body, such as a conductive device (a fridge, a container, a car roof, etc.).

In this regard, several cases have been analysed (Figure 8.4) where the distance h from the antenna system to the PEC varies from h = 30 mm, 20 mm, to 10 mm. Since 30 mm and 20 mm feature practically the same results, only 20 mm is shown

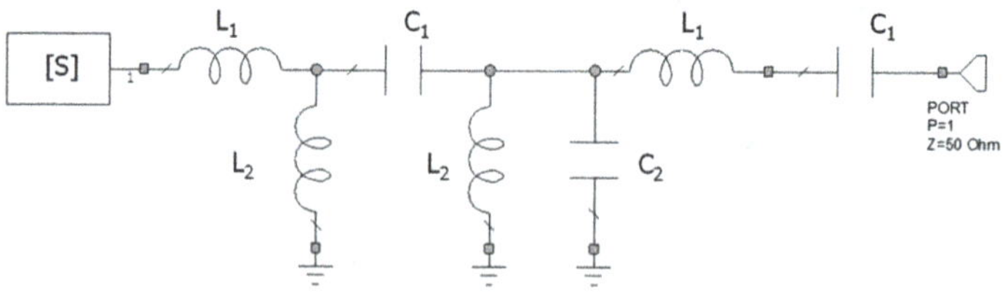

Figure 8.3 Matching network included in the antenna system of Figure 8.2. The [S] represents the impedance of the antenna booster on the PCB without any matching network.

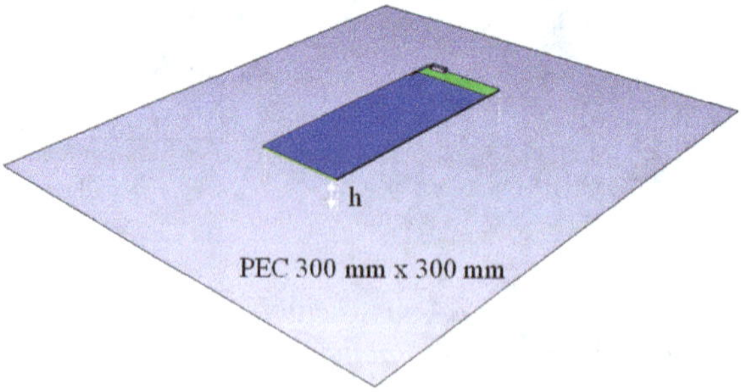

Figure 8.4 Antenna system of Figure 8.2 on a 300 mm × 300 mm PEC at height h

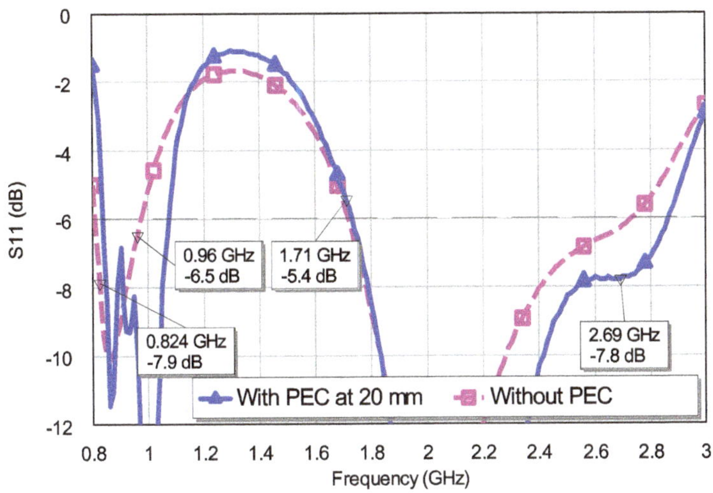

Figure 8.5 *Simulated S_{11} without and with PEC at h = 20 mm. Without PEC: L_1 = 4.5 nH, L_2 = 14 nH, C_3 = 1 pF, C_4 = 1.3 pF, L_5 = 6.1 nH, C_6 = 2.9 pF, and L_7 = 0 nH (see topology in Figure 8.3). With PEC at 20 mm: L_1 = 3.2 nH, L_2 = 12 nH, C_3 = 1 pF, C_4 = 1.5 pF, L_5 = 5.3 nH, C_6 = 2.6 pF, L_7 = 0 nH.*

(Figure 8.5). It is observed that for h = 20 mm, the S_{11} without and with PEC are practically the same, even better at the low bands with the PEC. It is important to outline that the matching network is optimised for each scenario for the bands of interest (824–960 MHz and 1710–2690 MHz). The optimisation is carried out by constraining an $S_{11} < -6$ dB across the band of interest using a genetic algorithm that finds the best L and C values to reach, if possible, the goal of $S_{11} < -6$ dB. This optimisation procedure is done using a microwave simulator (Microwave Office).

However, for h = 10 mm, the S_{11} at low bands is severely altered (Figure 8.6). This is aligned with the well-known phenomenon of horizontal wire antennas close to conductive bodies [1, 3]. For this case, 10 mm represents 0.027λ at 824 MHz. This S_{11} alteration is a significant drawback if the antenna system must operate near a conductive body. It is, therefore, of practical interest to mitigate the adverse influence of the PEC over the antenna system. The following section proposes a simple method to achieve it.

It is worth pointing out that for the h = 10 mm case where bandwidth at the low bands is poor, one may ask if the matching network shown in Figure 8.3 is convenient or if we need to change the topology to get better results in terms of S_{11}. A new strategy is followed to solve this question: the optimisation is carried out by not imposing a predetermined matching network but letting the microwave simulator synthesise the best topology. Even in this situation, the results fail to achieve a better response than the one reported in Figure 8.6. This result, however, can be theoretically predicted without the need to test matching networks. This is addressed in the next section with a method that estimates the bandwidth without

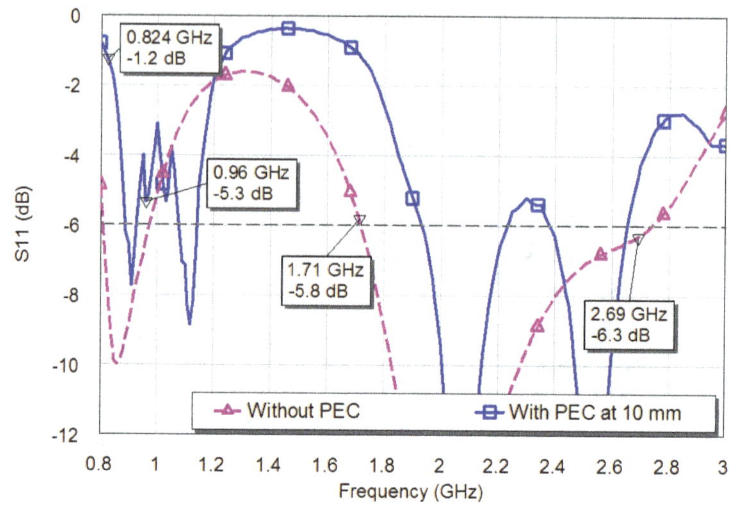

Figure 8.6 Simulated S_{11} without and with PEC at $h = 10$ mm. Without PEC: $L_1 = 4.5$ nH, $L_2 = 14$ nH, $C_3 = 1$ pF, $C_4 = 1.3$ pF, $L_5 = 6.1$ nH, $C_6 = 2.9$ pF, and $L_7 = 0$ nH (see topology in Figure 8.3). With PEC at 10 mm: $L_1 = 0.4$ nH, $L_2 = 13.7$ nH, $C_3 = 0.6$ pF, $C_4 = 0$ pF, $L_5 = 10$ nH, $C_6 = 1.6$ pF, $L_7 = 1.3$ nH.

the need to design a matching network. Such an approach is used to compare the bandwidth of the antenna system above the PEC with the proposed technique to improve the bandwidth at the low-frequency bands (<1 GHz).

8.3 The proposed solution: electromagnetic analysis

The proposed solution comprises a conductive means such as a conductive wire or a via between a PCB of a wireless device and a conductive body to modify the induced current to alter out-of-phase currents (Figure 8.7). The original situation in Figure 8.7 represents a horizontal current above a PEC, and the equivalent problem is based on image theory, where an out-phase image current appears. If the distance is short in terms of the operating wavelength, the impedance tends to be shorted out, and the radiation efficiency decreases [1]. The proposed solution consists of empirically adding conductive wire-made vias (vias hereafter) to alter the image currents and minimise out-of-phase currents [25]. Figure 8.7 illustrates qualitatively that the vias modify the out-of-phase currents by adding some in-phase currents. The analysis is addressed by electromagnetic simulation using an MoM code (IE3D). Several cases have been analysed: with one via, two vias, and three vias connecting the ground plane of the antenna system with the PEC. For each case, several positions have been tested, with the three vias case being the one that gives positive results (Figure 8.8).

It is shown that in terms of S_{11}, the solution with three vias provides almost the same S_{11} as the one without the PEC (Figure 8.9). Again, for this case, the values of

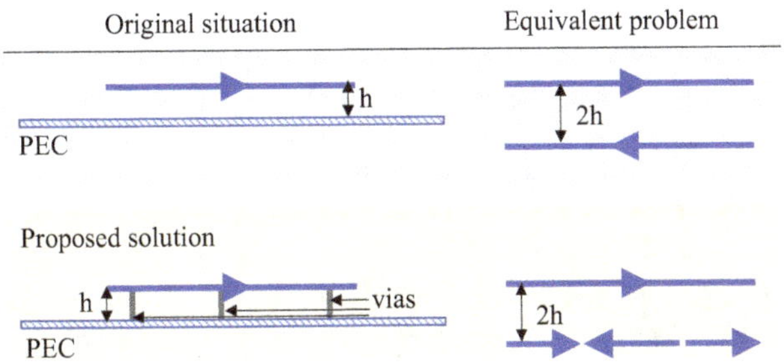

Figure 8.7 *Illustration showing an ideal situation of a horizontal current radiating over a PEC at a distance h and the proposed solution including three vias. The equivalent problem obtained using image theory shows the out-of-phase image current for the case without the three vias and a hypothetical case showing how out-of-phase currents are minimized.*

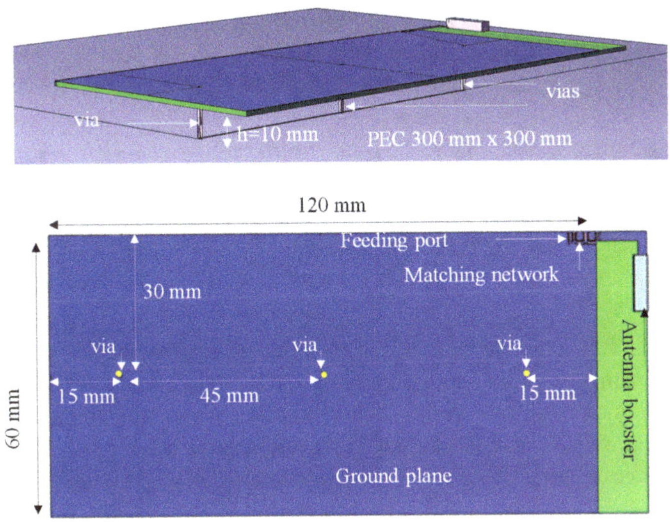

Figure 8.8 *The antenna system of Figure 8.2 over a PEC of 300 mm × 300 mm at a height h = 10 mm, including three vias (radius = 0.5 mm) electrically connecting the ground plane of the antenna system and the PEC underneath*

the lumped elements in the matching network have been optimised to achieve the best S_{11} across the 824–960 MHz and 1710–2690 MHz frequency regions.

It is interesting to deeply analyse the impact of the PEC on the antenna system in terms of bandwidth. The bandwidth of the antenna system with the PEC ($h = 10$ mm) and with the PEC ($h = 10$ mm), including the three vias, is computed without any matching network. This consideration removes the matching network

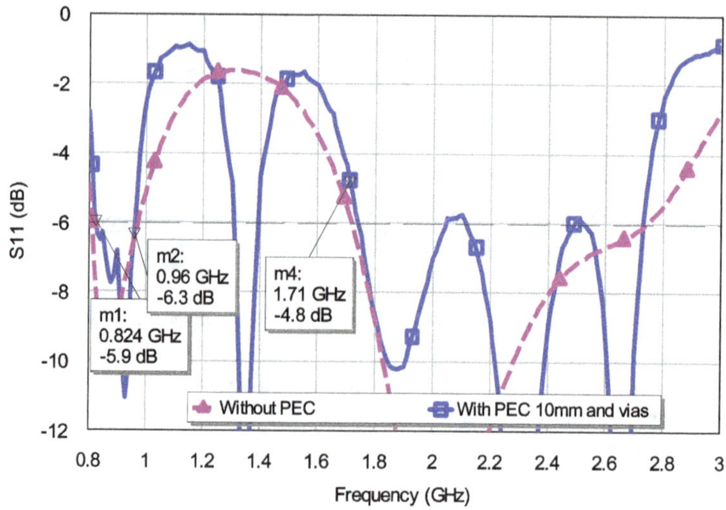

Figure 8.9 *Simulated S_{11} for the antenna system of Figure 8.4 (no vias connecting the ground plane and the PEC) and Figure 8.8 (with the three vias connecting the ground plane and the PEC). Each case uses the matching network topology shown in Figure 8.3 with the optimised L and C values to obtain performance ($S_{11} < -6$ dB) at 824–960 MHz and 1710–2690 MHz. Without PEC: $L_1 = 4.5$ nH, $L_2 = 14$ nH, $C_3 = 1$ pF, $C_4 = 1.3$ pF, $L_5 = 6.1$ nH, $C_6 = 2.9$ pF, and $L_7 = 0$ nH (see topology in Figure 8.3). With PEC 10 mm and vias: Without PEC: $L_1 = 4.5$ nH, $L_2 = 14$ nH, $C_3 = 1$ pF, $C_4 = 1.3$ pF, $L_5 = 6.1$ nH, $C_6 = 2.9$ pF, $L_7 = 0$ nH.*

effect on the bandwidth calculation. Using the input impedance data, the bandwidth is estimated using a quality factor approach [26], using Equations (8.2) and (8.3), where R and X are the real and imaginary parts of the input impedance of the antenna system without a matching network; SWR is the reference SWR where the bandwidth is computed, being SWR=3 ($S_{11} = -6$ dB) for this case.

$$Q_a(\omega) = \frac{\omega}{2R(\omega)}\sqrt{\left[\frac{dR(\omega)}{d\omega}\right]^2 + \left[\frac{dX(\omega)}{d\omega} + \left|\frac{X(\omega)}{\omega}\right|\right]^2} \tag{8.2}$$

$$BW = \frac{f_2 - f_1}{f_0} = \frac{SWR - 1}{Q_a \cdot \sqrt{SWR}} \tag{8.3}$$

Although this procedure has some limitations when the impedance presents small impedance loops, it is simple, fast, and provides a good bandwidth estimation without the need to design a matching network. This formulation is now employed to compare the two situations, the one without the three vias and the other including them (Figure 8.10). A different approach to calculating this bandwidth is known as potential bandwidth [27,28]. The name potential bandwidth is also adopted here.

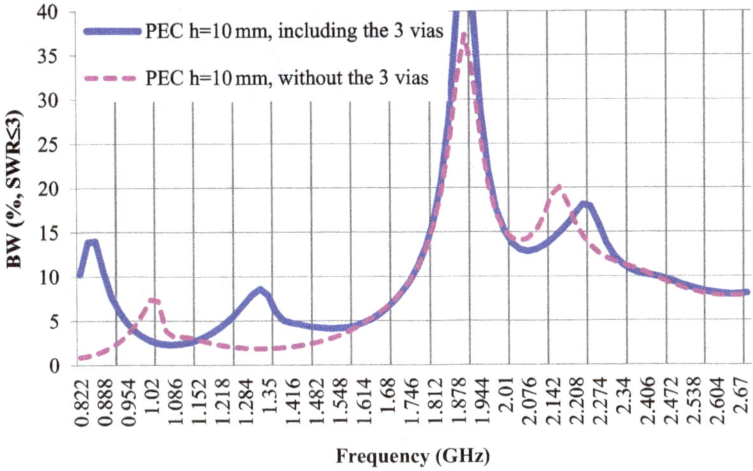

Figure 8.10 Potential bandwidth for the antenna system of Figure 8.2 is 10 mm over a PEC of 300 mm × 300 mm with and without including the three vias

The wording *potential* refers to the bandwidth one could expect when matching the impedance with a simple matching network of one or two lumped components.

The analysis clearly demonstrates that the potential bandwidth of the proposed architecture significantly exceeds that of the design without the inclusion of three vias, especially evident in the lower frequency range ($f < 0.96$ GHz) where the proximity of the PEC to the antenna system's ground plane is notable (Figure 8.10). As frequency rises, the electrical separation between the PEC and the antenna system's ground plane increases, resulting in a nearly identical potential bandwidth to that without the PEC. Thus, the impact of the PEC on impedance becomes negligible. This result is aligned with the results given in Figures 8.6 and 8.9, where it shows how the bandwidth at the 824–960 MHz region is less for the case without the three vias. In conclusion, while the potential bandwidth remains nearly identical in the high-frequency range (1.71–2.69 GHz), the actual bandwidth, after designing a matching network, is greater for the configuration that incorporates vias (with a bandwidth of 45% in the high-frequency range, as depicted in Figure 8.9) compared to the configuration without vias (with a bandwidth of 29%, as shown in Figure 8.6). The reason is that the matching network synthesis for the case without vias forces bandwidth at the low-frequency region and, at the same time, pushes to achieve bandwidth at the high-frequency region. Therefore, less bandwidth is obtained than predicted by the potential bandwidth formulation.

At this point, it is worth analysing the initial idea of minimising the out-of-phase currents over the PEC by adding the three vias. Since the significant impact occurs at the low-frequency bands, the frequency of 850 MHz is used as a reference to compute the current on both the ground plane of the antenna system and the PEC underneath (Figure 8.11).

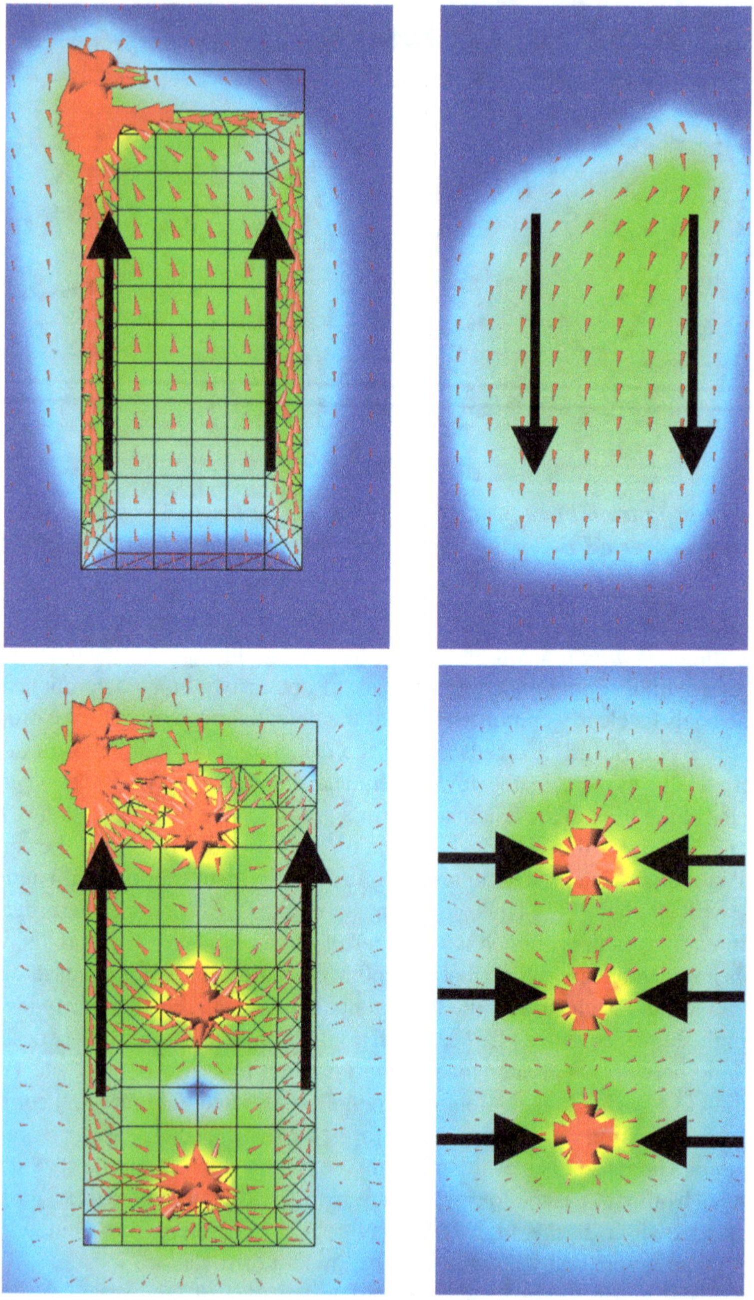

Figure 8.11 Simulated current distribution at 850 MHz for the case without the three vias (top) and including the three vias (bottom). Black arrows have been intentionally added to emphasise the most relevant current amplitude

For the case without vias (Figure 8.11 top), the currents on the ground plane and the PEC are mainly out-of-phase (note that the opposite direction of the arrows indicates that currents are out-of-phase). The black arrows have been intentionally added to indicate the main current distribution. Since the electric distance between the ground plane of the wireless device and the PEC is only 0.028λ at 850 MHz, the out-of-phase current impacts the impedance, as observed in the S_{11} results (Figure 8.6).

Nearby objects
Any object near the wireless device, such as conductive bodies, human/animal tissues, concrete, wood, and earth, impacts the impedance and radiation properties. If the device is in a fixed position, the matching network must be designed according to the real scenario. For moving devices, a tunable solution may be considered. Otherwise, a compromise must be made.

However, for the scenario including the three connecting vias (Figure 8.11 bottom), the current on the PEC presents mainly currents that are orthogonal to the currents on the ground plane. As a result, the impedance is like the case without the PEC (Figure 8.9). This makes the present technique simple to use to design antenna systems near conductive bodies.

8.4 Physical validation

Once it has been proven in the simulation domain that the proposed mechanism is adequate to achieve sufficient bandwidth and efficiency when the antenna system is near a conductive body, a prototype is built to corroborate the numerical analysis (Figure 8.12).

The prototype follows the same scheme shown in the simulation: the antenna system comprises the PCB and the antenna booster with the matching network. The matching network is designed using lumped 0402 and 0603 SMD components. The reason for using 0603 in some cases (L_2, see Figures 8.3 and 8.12) has been chosen to obtain the maximum component Q for the component. For instance, $L_2 = 18$ nH has a minimum $Q = 80$ across the 824–960 MHz and 1710–2690 MHz frequency regions, whereas the same component using the 0402 SMD type has, across the same frequency range, a minimum $Q = 57$. With this selection, the losses due to the matching network are minimised.

First, the antenna system without a PEC is measured in terms of SWR and total efficiency as a reference (Figure 8.13). Total efficiency has been calculated using measured 3D-pattern integration with the Satimo Stargate-32 anechoic chamber. The shaded area represents the 824–960 MHz and 1710–2690 MHz frequency regions. Measured total efficiencies are, on average, 69.4% and 75.6% for the lowest and highest frequency regions, respectively.

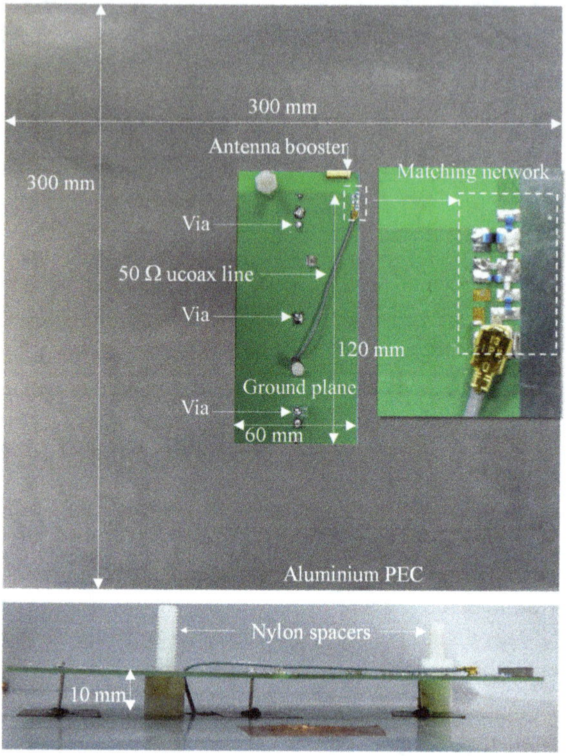

Figure 8.12 *Built prototype. Nylon screws are placed to separate 10 mm of the antenna system from the 300 mm × 300 mm PEC. A 50 Ω microcoaxial line feeds the antenna system, which is terminated through an SMA connector under the PEC.* $L_1 = 4.3\ nH$, $L_2 = 18\ nH$, $C_3 = 0.9\ pF$, $C_4 = 1\ pF$, $L_5 = 13\ nH$, $C_6 = 2\ pF$, *and* $L_7 = 4.5\ nH$ *(see topology in Figure 8.3).*

When the antenna system is placed 10 mm (0.027λ at 824 MHz) above a 300 mm × 300 mm PEC, the SWR and total efficiency change, especially at the low-frequency region, as predicted by the simulations (Figure 8.14). The average efficiencies are now 15.4% and 59.5% for the low- and high-frequency regions, respectively. This represents a drop of 6.5 dB in efficiency at the low-frequency region compared to the case without PEC.

However, when the proposed solution using the three vias is applied to the antenna system above the PEC, both SWR and total efficiency change positively, improving the values of both SWR and total efficiency (Figure 8.15). In this case, the measured total efficiencies are, on average, 62.8% and 60.8% for the lowest and highest frequency regions, respectively. The improvement at the low-frequency region is huge: 6.1 dB (62.8%/15.4%). This increase is equivalent to the decrease due to the case without PEC and with it. This means that the proposed method

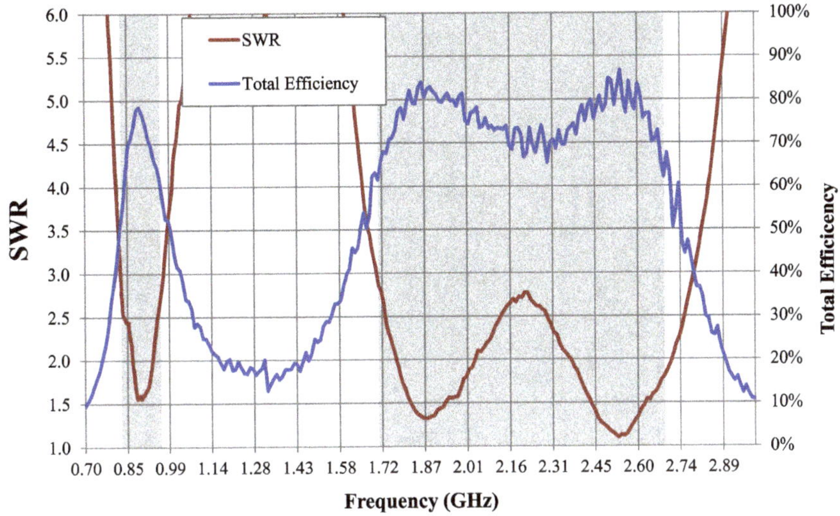

Figure 8.13 Measured SWR and total efficiency without the PEC

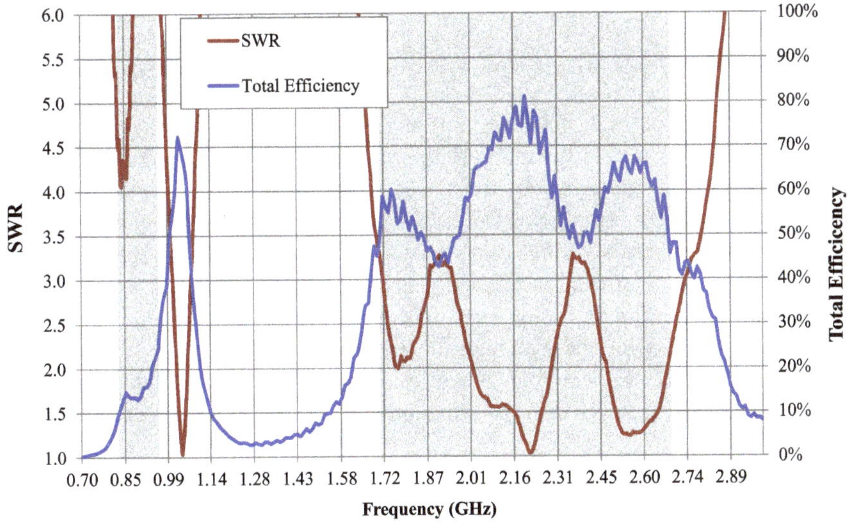

Figure 8.14 Measured SWR and total efficiency for the wireless device 10 mm above a PEC of 300 mm × 300 mm. PCB is at the centre of the PEC

effectively minimises out-of-phase currents induced on the PEC, so a similar efficiency is obtained in the case without the PEC and with the PEC having the three vias connecting the ground plane of the antenna system and the PEC. In this regard, it is essential to note that a simple approach when a wireless device is close to a conductive body consists of impedance matching to maximise the delivered power

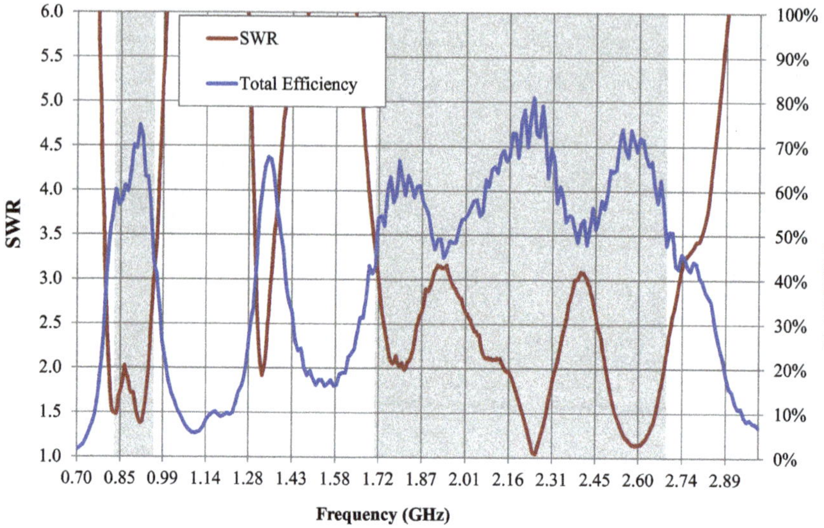

Figure 8.15 Measured SWR and total efficiency for the wireless device 10 mm above a 300 mm × 300 mm PEC and the three vias connecting the ground plane of the wireless device and the PEC. PCB is at the centre of the PEC.

to the antenna. In this case, however, not only is impedance matching relevant but, more importantly, it also helps mitigate the impact of the conductive body.

Additionally, an unexpected result was found: a new band of operation at 1.32 GHz with a peak efficiency of 67.0%.

Finally, further study of the improvement in the low-frequency region is relevant. In this regard, the radiation efficiency is computed based on the measured data of SWR and total efficiency using Equation (8.1) (Figure 8.16). The results reflect that even after improving the matching for the case of the antenna system over the PEC, the total efficiency would be much less than the proposed solution since the radiation efficiency is lower than the radiation efficiency of the proposed solution.

Improving the efficiency of a wireless device embedding antenna boosters when it is electrically close to a conductive body

Bandwidth and radiation efficiency decrease when the distance between the ground plane and the PEC underneath is less than $\sim 0.03\lambda$. This negative impact is mitigated by connecting the ground plane of the device to the PEC below using vias. For the PCB evaluated in this section, three vias are enough to obtain bandwidth and efficiencies, like in the case without the PEC. The inclusion of three vias results in a notable enhancement of total efficiency,

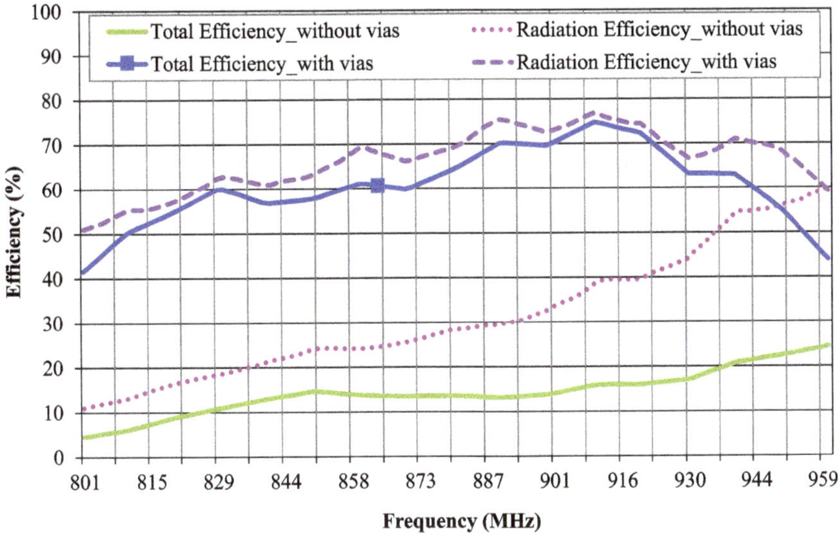

Figure 8.16 Comparison of the measured total and radiation efficiency for the antenna system (10 mm over the 300 mm × 300 mm PEC without and with the three vias)

achieving an improvement of 6.1 dB within the low-frequency operational range (824–960 MHz). This significant enhancement makes the technique particularly advantageous for situations where a wireless device must be positioned near a conductive body.

It is worth establishing the separation distance at which the radiation efficiency of a wireless device over a PEC can be neglected when no vias are considered. For example, from Figure 8.16, the radiation efficiency without vias and vias is the same at 960 MHz, representing a separation of 0.032λ. Therefore, as a practical rule, if the separation of a device is less than $\sim 0.03\lambda$, the present technique is attractive to minimise the bandwidth and efficiency reduction due to the conductive body below.

To generalise the design method, a smaller PCB size is studied in the next section to determine if the connecting mechanism also helps to mitigate the bandwidth and efficiency drop of the conductive body underneath the wireless device.

8.5 Application to other platforms

To examine the potential of the solution on other platforms, a different ground plane size is evaluated. In this case, a 60 mm × 60 mm ground plane above 10 mm of a PEC without vias, two and three vias is studied in terms of bandwidth and efficiency (Figures 8.17 and 8.18). In all cases, the same 12 mm × 3 mm × 2.4 mm (height) antenna booster as the previous section is utilised. Likewise, the size of the ground plane is 300 mm × 300 mm, and the PCB is placed at the centre of the PEC in all cases. In this analysis, no matching networks are designed since the comparisons with and without vias are made using potential bandwidth and radiation efficiency (Equations (8.1)–(8.3)). Finally, the vias connecting the ground plane and the PEC have a radius of 0.5 mm.

First, potential bandwidth using Equations (8.2) and (8.3) is computed from 0.6 GHz up to 3 GHz (Figure 8.19). It is clearly shown that the case without vias presents a poor bandwidth below 1 GHz. For the case with a via at the centre of the ground plane, bandwidth remains practically the same except for a little increase at

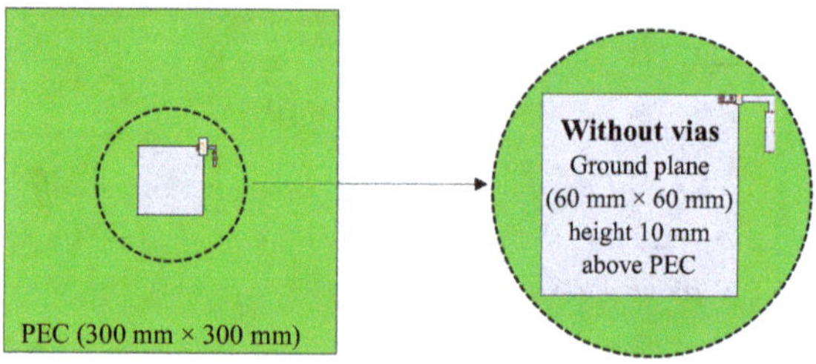

Figure 8.17 A wireless device including a 60 mm × 60 mm ground plane and a 12 mm × 3 mm × 2.4 mm antenna booster placed at 10 mm over a 300 mm × 300 mm ground plane

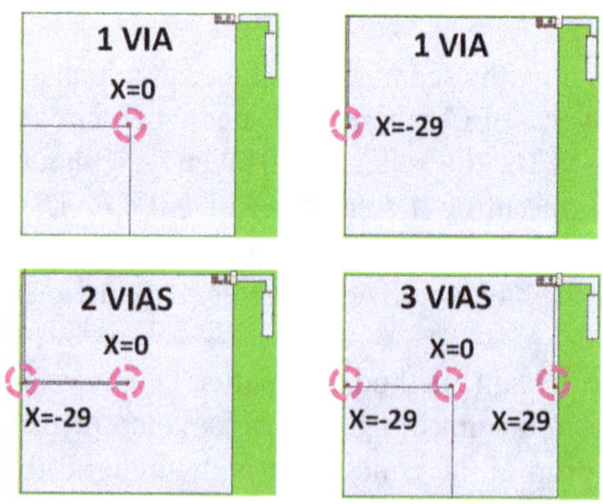

Figure 8.18 The circles indicate the location of the via connecting the ground plane and the PEC below. All ground planes are spaced 10 mm from the PEC below, having a 300 mm × 300 mm size. Antenna booster is 12 mm × 3 mm × 2.4 mm (height).

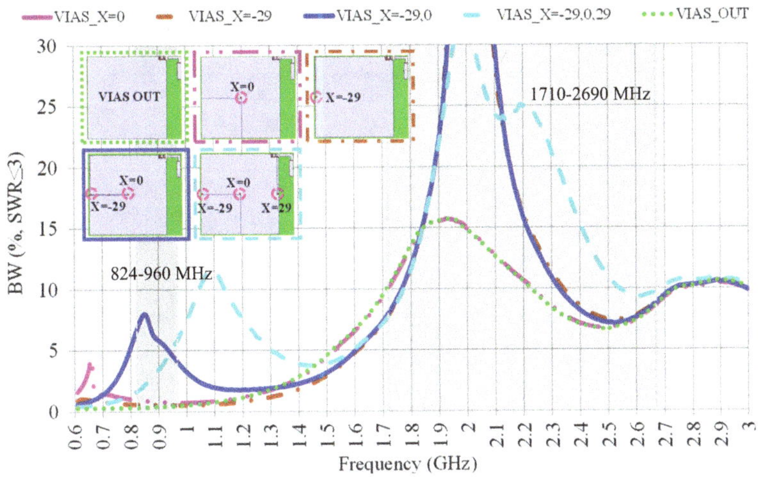

Figure 8.19 Potential bandwidth without vias, with one via at the centre and edge, two vias, and three vias

0.65 GHz. The one via case at the edge ($X = -29$ mm) provides a similar bandwidth below 1 GHz again but increases in the frequency range between 1.85 GHz and 2.3 GHz with respect to no vias. However, a significant change occurs when adding two vias, one at the centre and the other at the edge. For this situation, bandwidth below 1 GHz is much larger than the no vias case, as well as in the high-frequency region above 1.7 GHz. For the three vias case, however, bandwidth is larger than the no vias case but centred around 1.1 GHz, whereas the case with two vias is centred in the frequency range of 824–960 MHz.

Besides bandwidth, the radiation efficiency is also examined (Figure 8.20). All cases with one, two, and three vias improve the radiation efficiency compared to the no vias case in the frequency region from 0.6 GHz to 1 GHz. The case with two vias provides an increment of 6 dB approximately at 0.8 GHz, which is a considerable increment, and at the same time, bandwidth is enhanced compared with no vias cases. Once again, connecting vias between the ground plane and the PEC below provides a positive effect in those scenarios where the wireless device, including a ground plane, is electrically close to a conductive body.

To gain further control in tuning where the bandwidth and efficiency improvements occur, the via can be combined with an electronic component such as an inductor, capacitor, switch, filter, or the like [25] (Figure 8.21). Accordingly, the radioelectric performance of the radiating system may be improved with different adjustments corresponding to other situations or frequencies. Such elements thus provide additional ways for configuring and optimising the performance of the radiating system and, therefore, the device when the radiating system is electrically close to the electrically conductive body.

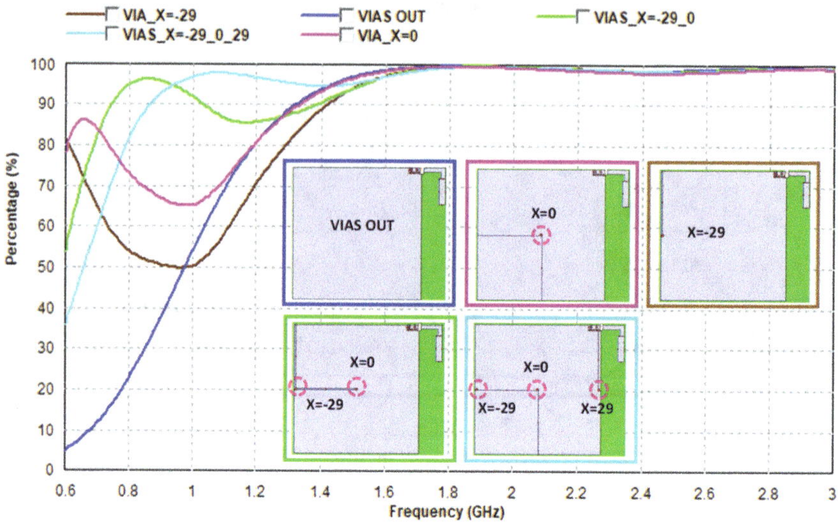

Figure 8.20 Simulated radiation efficiency showing the radiation efficiency improvement when including the vias. For the 2-via case at x = 0 and x = −29, not only radiation efficiency improves compared to no vias case, but also the potential bandwidth.

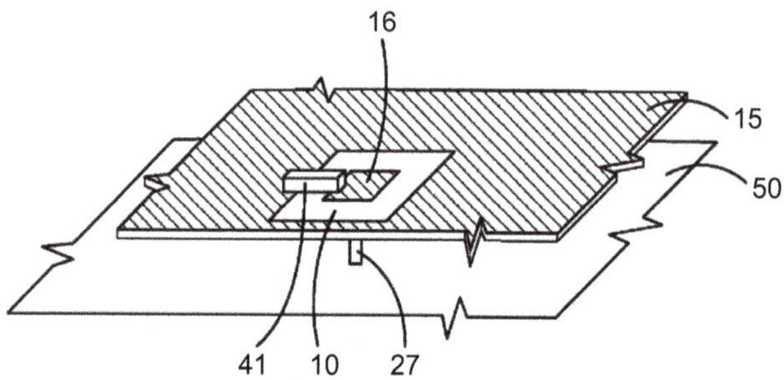

Figure 8.21 A wireless device comprising a ground plane (15), a soldering pad (16) isolated with respect to the ground plane through (10), and a via (27) connected to the PEC (50). (41) represents an SMD component such as an inductor which is connected to the via (27) [25].

8.6 Conclusions

A simple technique has been presented to mitigate the negative impact on bandwidth and efficiency of a horizontally placed antenna booster system over a conductive body. The method uses vias between the ground plane of the PCB and the conductive body. The number of vias and their position have been empirically obtained. Two ground plane cases have been discussed, a large one with

120 mm × 60 mm and a small one with 60 mm × 60 mm. In both scenarios, the introduction of vias between the PCB's ground plane and the underlying PEC improves radiation efficiency at frequencies where the wavelength is relatively short, while also enhancing bandwidth compared to configurations lacking vias. It is concluded that heights less than 0.03λ suffer from efficiency degradation, and therefore, the proposed technique enhances such radiation efficiency to efficiency levels comparable to those obtained without the PEC.

The advantage of the proposed technique relies on its simplicity since there is no need to add complex structures such as artificial magnetic or high impedance surfaces based on periodic structures, which are much more difficult to integrate than the proposed design method.

This technique has its field of application on those wireless devices close to conductive bodies, such as the roof of a vehicle, smart meters with metallic covers, etc., which is interesting for IoT devices.

Table 8.1 answers the objectives given at the beginning of the chapter. It also points to relevant figures and equations to consolidate the concepts learnt. It is also advisable to refresh all essentials included in the grey boxes distributed throughout the chapter.

Table 8.1 Key aspects to take away

Topic	Review and take away
Impact of a PEC	When a PEC is close to a wireless device, its performance in terms of reflection coefficient and total efficiency is altered. As a practical rule, when the distance is less than 0.03λ, the PEC affects the initial response without it (Figures 8.5 and 8.6). For much larger PECs (size $\sim 3\lambda$), the PEC starts to affect separations around 0.1λ. The PEC impact may be frequency-detuned. In such a case, the matching network may recover the original response. However, when the PEC is electrically close to the wireless device, the radiation efficiency decreases (the closer the PEC, the worse). In such a case, total efficiency can only be improved by recovering the reflection coefficient. The negative impact of a PEC is due to the short-circuit effect of the out-of-phase currents induced on the PEC (Figures 8.7 and 8.11).
Mitigation technique	Shorting vias between the ground plane of the wireless device and the PEC modifies the induced currents on the PEC to mitigate its negative impact (Figure 8.8). The position has been empirically determined to be where two or three vias are enough to recover a similar performance without the PEC. An improvement of 6 dB in total efficiency has been achieved when comparing the efficiency of a multiband wireless device embedding an antenna booster with the PEC and the PEC, including vias (Figure 8.16). The total efficiency of the case, including vias, is similar to the total efficiency of the wireless device without the presence of the PEC.

(Continues)

Table 8.1 (*Continued*)

Topic	Review and take away
	The positive effect on the radiation efficiency of including vias is clearly shown in Figure 8.20, where the radiation efficiency is much improved with vias when the wireless device is close to the PEC.
Radiation patterns	Depending on the size of the PEC, the radiation patterns of the wireless device will change. As a limit, if the PEC is electrically large in terms of wavelength, no radiation will be below the PEC. In this regard, the resulting radiation pattern of a wireless device with a PEC underneath will become more directive with less radiation under the PEC.
Not expected positive results	A non-expected positive result when adding vias is a new frequency band with high efficiency (Figure 8.15).
Further degrees of freedom	To gain flexibility, besides the vias, lumped components such as capacitors, inductors, or even tuneable components such as digitally tuneable capacitors may be used (Figure 8.21).

8.7 Exercises

A 3D electromagnetic software and a circuit simulator, including matching network synthesis, are needed to solve the following exercises.

8.7.1 Exercise 1

Design a 120 mm × 60 mm ground plane printed over a dielectric 1 mm thick with $\varepsilon_r = 4.15$ and $\tan\delta = 0.017$ and a 12 mm × 3 mm × 2.4 mm (height) antenna booster like the one shown in Figure 8.2. Design a matching network (see an example in Figure 8.3 or use the techniques explained in Chapter 5) to provide operation at 824–960 MHz and 1710–2690 MHz. Compute S_{11} and total efficiency. Repeat the exercise when the wireless device is 5 mm over a finite 300 mm × 300 mm PEC. Compare S_{11} and total efficiency without the PEC. Add three vias according to the teachings of this chapter to obtain a similar S_{11} and efficiency as in the case without PEC. Repeat the experiment when the PEC is only at 2.5 mm.

8.7.2 Exercise 2

Repeat Exercise 8.7.1 with a wireless device having a 90 mm × 60 mm ground plane to operate at 824–890 MHz and 1700–2200 MHz.

8.7.3 Exercise 3

Repeat Exercise 8.7.1, analysing the impact of the PEC size. For example, use a 200 mm × 200 mm PEC. Repeat for a big ground plane of 1 m × 1 m. Place the device centred with respect to the PEC below.

8.7.4 Exercise 4

For each previous case, compute the radiation patterns at 850 MHz and 2000 MHz. Compare those with the case without a PEC.

8.7.5 Exercise 5

Based on Exercise 8.7.1 and the scheme shown in Figure 8.21, study the benefits of adding a lumped inductor or capacitor at each of the three vias.

8.7.6 Exercise 6

Using the same wireless device size as in Exercise 8.7.1, design a new matching network with the teachings of Chapter 4 to provide operation from 698 MHz to 824 MHz. Investigate the benefits of adding vias connecting the ground plane to a 300 mm × 300 mm PEC 10 mm below the wireless device. Compute S_{11}, total efficiency, and radiation patterns.

8.7.7 Exercise 7

Reproduce the results of Figures 8.19 and 8.20 for the case using a via at the centre and another via with an $x = -29$ mm. Design a matching network with lumped elements to provide operation at 860–930 MHz. Compute S_{11} and total efficiency with and without vias.

8.7.8 Exercise 8

Investigate Exercise 8.7.7 with a PEC at 5 mm. If needed, use a reactive element in series with the two vias, as in Figure 8.21.

8.7.9 Exercise 9

Based on the antenna booster structure given in Figure 8.2, design a matching network for operating at 860–930 MHz, including a 300 mm × 300 mm PEC 10 mm below, including connecting vias. Compute S_{11} and total efficiency with and without vias. Simulate the current distribution at 900 MHz with and without vias to determine the effect of vias.

8.7.10 Exercise 10

Repeat Exercise 8.7.9 when the size of the PEC is 200 mm × 200 mm and 130 mm × 130 mm. In both cases, the PCB should be centred with respect to the PEC below.

8.7.11 Exercise 11

Repeat the experiment in Figures 8.18 and 8.19, considering a circular ground plane with the same area as the 60 mm × 60 mm square ground plane.

8.7.12 Exercise 12

For the 60 mm × 60 mm ground plane shown in Figure 8.17, design a matching network at 900 MHz for the case without vias and with the vias at $x = 0$ and $x = -29$ (see Figure 8.19). Compute the bandwidth at $S_{11} = -6$ dB for both cases and compare it with the bandwidth potential. Simulate total efficiency when considering lumped components in a matching network with a finite Q. For this purpose, use inductors and capacitors, considering measured S-data.

References

[1] C. A. Balanis, "*Antenna Theory: Analysis and Design*," Second Edition: Wiley, 1996.

[2] R. C. Hansen, "Effects of a High-Impedance Screen on a Dipole Antenna," *IEEE Antennas and Wireless Propagation Letters*, vol. 1, pp. 46–49, 2002.

[3] D. Sievenpiper, L. Zhang, R. F. J. Broas, N. G. Alexopoulos, and E. Yablonovitch, "High-impedance Electromagnetic Surfaces with a Forbidden Frequency Band," *IEEE Trans. Microwave Theory Technology*, vol. 47, pp. 2059–2074, 1999.

[4] R. F. J. Broas, D. F. Sievenpiper and E. Yablonovitch, "A High-Impedance Ground Plane Applied to a Cellphone Handset Geometry," *IEEE Transactions on Microwave Theory and Techniques*, vol. 49, no. 7, pp. 1262–1265, 2001.

[5] J. McVay, A. Hoorfar, and N. Engheta, "Applications of Hilbert Artificial Magnetic Conductor Ground Planes to Performance Enhancement of Electrically Small Antennas," in *Proc. 2003 IEEE Sarnoff Symp. Advanced Wired and Wireless Communications*, pp. 297–30, 2003.

[6] D. H. Werner and P. L. Werner, "The Design Optimization of Miniature Low-Profile Antennas Placed in Close Proximity to High-Impedance Surfaces," *IEEE Antennas and Propagation Society International Symposium*, Columbus, US., vol. 1, pp. 157–160, 2003.

[7] J. McVay, A. Hoorfar, and N. Engheta, "High Impedance Metamaterial Surfaces Using Hilbert-Curve Inclusion." *IEEE Microwave and Wireless Components Letters*, vol. 14, no. 3, pp. 130–132, 2004.

[8] J. McVay, A. Hoorfar, and N. Engheta, "Small Dipole Antenna Near Peano High-Impedance Surfaces," *IEEE Antennas and Propagation Society International Symposium*, vol. 1, 2004.

[9] Jong Ryul Sohn, Heung-Sik Tae, Jae-Gon Lee, and Jeong-Hae Lee, "Comparative Analysis of Four Types of High-Impedance Surfaces for Low Profile Antenna Application," *IEEE Antennas and Propagation Society International Symposium*, vol. 1A, Washington, DC, pp. 758–761, 2005.

[10] E. Carrubba, S. Genovesi, A. Monorchio, and G. Manara, "AMC-based Low-profile Antennas for 4G Communication Services," *2007 IEEE Antennas and Propagation Society International Symposium*, Honolulu, HI, pp. 3364–3367, 2007

[11] G. S. A. Shaker, D. Busuioc and Safieddin Safavi-Naeini, “Small Antenna Over an Engineered Ground Plane,” *2007 IEEE Antennas and Propagation Society International Symposium*, pp. 4072–4075, 2007, doi:10.1109/APS. 2007.4396435.

[12] N. M. Mohamed-Hicho, E. Antonino-Daviu, M. Cabedo-Fabrés, M. Ferrando-Bataller, “A Novel Low-Profile High-Gain UHF Antenna Using High-Impedance Surfaces,” *European Conference on Antennas and Propagation*, EuCAP 2014, pp. 919–922.

[13] I. Liberal, Igor S. Nefedov, I. Ederra, R. Gonzalo, and S. A. Tretyakov, “Reconfigurable Artificial Surfaces Based on Impedance Loaded Wires Close to a Ground Plane,” *IEEE Transactions Antennas and Propagation* vol. 60, no. 4, pp. 1921–1930, 2012.

[14] F. Yang and Y. Rahmat-Samii, “Reflection Phase Characterizations of the EBG Ground Plane for Low Profile Wire Antenna Applications,” *IEEE Transactions on Antennas and Propagation*, vol. 51, no. 10, pp. 2691–2703, 2003.

[15] F. Costa, A. Monorchio and G. P. Vastante, “Tunable High-Impedance Surface with a Reduced Number of Varactors,” *IEEE Antennas and Wireless Propagation Letters*, vol. 10, pp. 11–13, 2011.

[16] Y. Chen and T. Ku, “A Low-Profile Wearable Antenna Using a Miniature High Impedance Surface for Smartwatch Applications,” *IEEE Antennas and Wireless Propagation Letters*, vol. 15, pp. 1144–1147, 2016.

[17] M. Li, S. Xiao, J. Xiong and B. Wang, “Horizontal Dipole Located Close to Ground Plane with Bidirectional Endfire Radiation,” *IEEE Antennas and Wireless Propagation Letters*, vol. 13, pp. 1144–1147, 2014.

[18] F. Yang and Y. Rahmat-Samii, “Curl Antennas over Electromagnetic Band-gap Surface: A Low Profiled Design for CP Applications,” *IEEE Antennas and Propagation Society International Symposium. 2001* Digest. Boston, MA, pp. 372–375 vol. 3, 2001.

[19] M. Elayachi, J. Ribero and P. Brachat, “Planar Low Profile and Gain Enhancement of Printed Antennas Using EBG Structures,” *2008 IEEE Antennas and Propagation Society International Symposium*, San Diego, CA, pp. 1–4, 2008.

[20] S. Best, “Improving the Performance Properties of a Dipole Element Closely Spaced to a PEC Ground Plane,” *IEEE Antennas and Wireless Propagation Letters*, vol. 3, pp. 359–363, 2004.

[21] M. Polívka, M. Švanda, and P. Černý, “Multiple-Arm Folded Monopole Antenna Operating Extremely Close to a Conductive Plane,” *2008 14th Conference on Microwave Techniques*, Prague, 2008, pp. 1–5.

[22] J. Anguera, N. Toporcer, and A. Andújar, “Slim Radiating Systems for Electronic Devices,” *Patent US* 9,960,478 B2, May 2018.

[23] A. Andújar, J. Anguera, and R. Mª Mateos, “Multiband Non-Resonant Antenna System with Reduced Ground Clearance,” *European Conference on Antennas and Propagation*, EUCAP 2017, Paris, France, 2017.

[24] J. Anguera, A. Andújar, J. Rahola, J. Juntunen, "Synthesis of Multiband Matching Networks for Non-Resonant Antennas," *Computing and Electromagnetics* CEM'17, Barcelona, Spain, 2017.

[25] J. Anguera and A. Andújar, "Devices with Radiating Systems Proximate to Conductive Bodies," *Patent US* 11,532,877 B2, Dec. 2022.

[26] S. R. Best, "The Inverse Relationship Between Quality Factor and Bandwidth in Multiple Resonant Antennas," *IEEE Antennas and Propagation Society International Symposium*, pp. 623–626, 2006.

[27] J. Rahola, "Bandwidth Potential and Electromagnetic Isolation: Tools for Analyzing the Impedance Behavior of Antenna Systems," in *Proceedings of the EuCAP 2009 Conference*, Berlin, 2009.

[28] J. Anguera, A. Andújar, G. Mestre, J. Rahola, and J. Juntunen, "Design of Multiband Antenna Systems for Wireless Devices Using Antenna Boosters," *IEEE Microwave Magazine*, pp. 102–114, 2019.

Chapter 9
Matching network synthesis considering pad layout

This chapter addresses the design of multiband matching networks for antenna boosters from a practical perspective. This chapter considers the pad layout when synthesising a matching network. Pads to allocate the SMD (Surface Mounted Device) components of the matching network alter the impedance since pads are short transmission lines. Therefore, the matching network should be designed considering the pad layout to minimise changing the values of the components of the matching network when it is physically implanted in a wireless device.

The chapter has the following objective:

1. To understand and put into practice the matching network synthesis from the electromagnetic simulation domain as well as from the physical domain, including the pad layout to allocate the components of the matching network.
2. To know which components impact a particular frequency region to simplify the fine-tuning of the matching network if needed.

9.1 Introduction

Two methods to design matching networks for antenna boosters considering the pad layout are presented (Figure 9.1). The first considers the design from the simulation domain, and the second method from the physical domain. In the first method, the antenna system, comprising an antenna booster, the pad layout, the ground plane, and other components, is simulated. The electromagnetic simulation results in a [*S*] matrix with as many ports as components the matching can have, plus an extra port for the radiofrequency source. Then, matching network synthesis is carried out, where the algorithm decides how many components are needed and their value. Besides, the synthesis can be run considering real components in the same manner as in Chapters 4 and 5. The second method is suitable when there is a full-wave electromagnetic simulator available but only a physical device and a vector network analyser. If the electromagnetic simulation of the device is accurate, the first method can provide a matching network that can be directly mounted on the physical device. However, the reality is variable, and not all devices can undergo an accurate simulation phase because some components are not well characterised in the microwave domain. In these situations, the second method is

	SIMULATION DOMAIN	PHYSICAL DOMAIN
Method	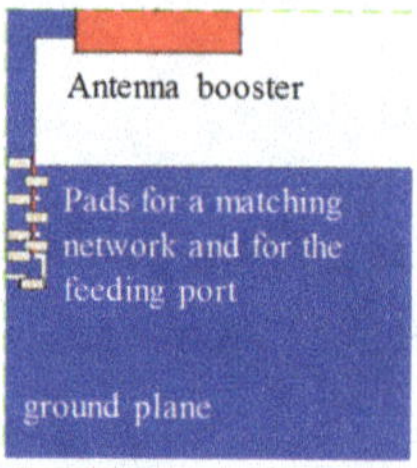	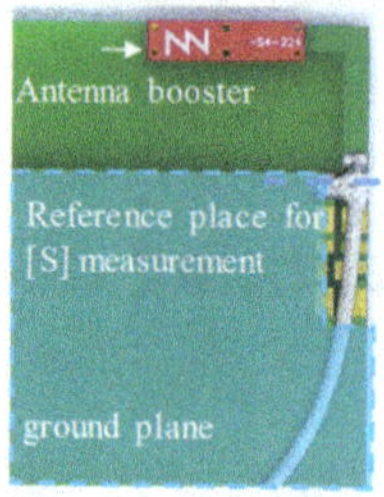
Description	• Parasitic effects of pads and considered in the full-wave electromagnetic simulation. • Needs: 1) a full 3D electromagnetic simulator to obtain the [S] data. 2) A Network Synthesis Tool 3) Need to know the exact dimensions of the PCB, pads layout, and antenna booster to obtain the [S] data.	• Parasitic effects of pads can be modelled by circuit simulation with short transmission lines • A full 3D simulator is not needed, only: 1) The physical device 2) Microwave simulator with a Network Synthesis Tool 3) A Vector Network Analyzer
Procedure	1) Full-wave simulation to obtain the [S] matrix 2) Import the [S] matrix to a network synthesis tool to obtain the matching network. 3) Specify targets 4) Network synthesis	1) Measurement of S_{11} of the antenna booster in the real device without any matching network 2) Import the [S] matrix to a network synthesis tool to obtain the matching network. Pads are modelled by short transmission lines 3) Specify targets 4) Network synthesis

Figure 9.1 Methods to synthesise a matching network considering the pad layout

very appropriate since it works directly with the input impedance of the antenna booster on the device, including all electronics and casings. The matching network obtained with this method can be directly mounted into the real device to be ready to operate without further human intervention. Both ways are detailed in the following sections.

It is interesting to analyse what would happen if a matching network is synthesised without considering the pad layout or if it is implemented into a PCB (Printed Circuit Board), including the layout (Figure 9.2).

Pads are conductive traces and can be modelled as short transmission lines. As the phase delay of a transmission line is proportional to frequency, the impact of the pad layout on the impedance is typically more severe at higher frequencies. This results in a mismatch of the original response. For example, a multiband matching network for an antenna booster achieves an S_{11} below −6 dB across 824–960 MHz and 1710–2690 MHz. However, when the same matching network considers the pad layout, the S_{11} changes. In this example, the S_{11} is degraded to −3.3 dB at 2.69 GHz (Figure 9.2). The degradation is mainly a frequency shift that can be easily compensated if the matching network synthesis considers the pad layout in the design process.

The chapter is divided as follows: after introducing the two methods for synthesising matching networks above, first from the simulation domain and second

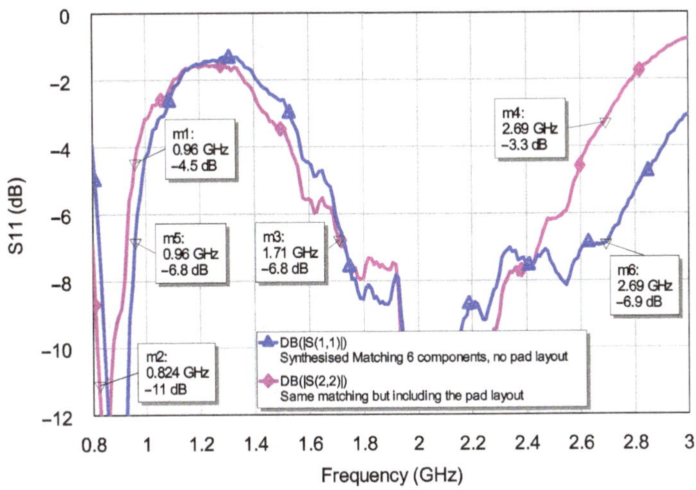

Figure 9.2 *If the pad layout is not considered in the matching design process, the response will be altered due to the impact on impedance due to the layout*

from the physical domain, both approaches are intensely detailed in Sections 9.2 and 9.3. In Section 9.4, the effect of a series/shunt inductor/capacitor is discussed to provide a fine-tuned impedance adjustment when the procedures are not accurate. Section 9.5 concludes. Exercises to consolidate the concepts are included in Section 9.6. Finally, References section contains extra reading.

9.2 Matching network synthesis, including the entire layout from the simulation domain

Three PCBs representative of typical wireless devices illustrate the design process (Figure 9.1): platform #1 with a 145 mm × 130 PCB, platform #2 with a 131 mm × 60 mm PCB, and platform #3 with a 76 mm × 60 mm PCB. In addition, all platforms comprise a 12 mm × 3 mm × 2.4 mm (height) antenna booster.

The design process comprises the following general steps:

1. Generation of the [S] matrix
   Electromagnetic simulation of the antenna booster on the PCB, including pads to allocate a matching network component: lumped inductors and capacitors. The output of this step is an $[S]_{N\times N}$ frequency-dependant matrix having $N \times N$ dimensions, where $N-1$ is equal to the number of components for the matching network, and the one additional port is the feeding port. Note that there are as many $[S]_{N\times N}$ matrices as the number of frequencies used in the simulation.
2. Definition of the target specifications
   Once the [S] matrix is obtained, the next step in the design procedure is to select real components, such as SMD inductors and capacitors, that meet the

target specifications across the frequency bands of interest. This may be done manually or automatically by software [1]. Target specifications can be, for example, a specific total efficiency across frequency bands.

3. Computation
   The design procedure gives several matching network candidates.
4. Implementation and measurement
   Each case is implemented, and VSWR and efficiency measurements are carried out to validate the design process.

It is important to note that no optimisation of radiation patterns is included. This step is unnecessary since the radiation pattern will be determined mainly by the platform size rather than the antenna booster. Also, the dimensions of the devices are not large in terms of wavelength, around 0.2λ to 1λ. For example, a device with a 120 mm length is 0.32λ at 824 MHz and 0.88λ at 2.2 GHz, so the resulting radiation patterns are low directive patterns suitable for wireless communication. For mobile/IoT applications, low directivities and stable radiation patterns across the bands are preferred since the end wireless device may be located at any position or orientation. Therefore, optimising the radiated power to space matters here, which is achieved by optimising the total efficiency (Figure 9.3, Equation (9.1) or Chapter 4 for more details). Details for each step are given next.

$$\eta_t = \frac{P_r}{P_{avs}} = \eta_r \cdot \left(1 - |S_{11}|^2\right) \quad \eta_r = \frac{P_r}{P_{load}} \tag{9.1}$$

It is crucial to outline that the synthesis procedure considers a given pad layout with enough flexibility to allocate several matching network components to ensure the requested multiband performance. However, a different route can be taken regarding the matching network synthesis without any pad layout. Once the matching network is synthesised, pads are allocated. Since pads will detune the S_{11} response, a second iteration is needed, which optimises the value of each component of the matching network to compensate for the pad effect. The following section uses the first method, which can easily be applied to the second route.

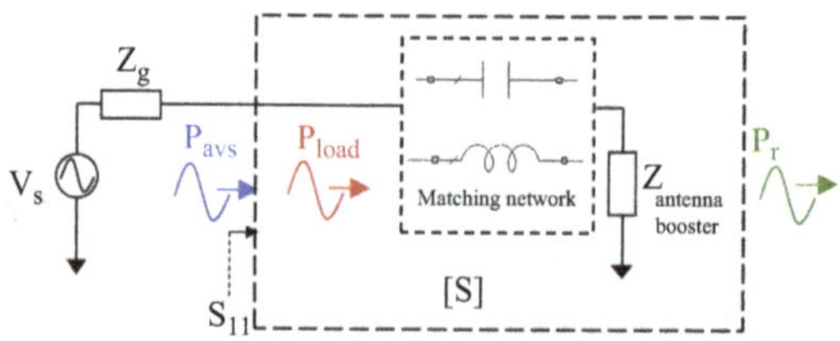

Figure 9.3 Total efficiency is defined as the power delivered to space (P_r) over the available power of the generator (P_{avs}). Z_g represents the internal impedance of the generator, and the $Z_{antenna\ booster}$ is the impedance of the antenna booster without any matching network

9.2.1 Generation of the [S] matrix

An electromagnetic simulator is needed to generate the $[S]_{N\times N}$ matrix. In the present case, an MoM-based commercial code was used (Figure 9.4). Each PCB has a ground plane printed on a 1 mm thick substrate layer made from FR4 ($\varepsilon_r = 4.15$, $\tan\delta = 0.017$). Pads to allocate a matching network with lumped elements are also considered. In this case, pads help allocate both 0603 and 0402 SMD components (inductors and capacitors) (Figure 9.5). Pads can also be designed to allocate small components such as 0201. In these examples, N is equal to eight, meaning that the matching network can have up to seven SMD components: four series components (at ports #1, #3, #5, and #7) and three shunt components (at ports #2, #4, and #6). Port #8 is left for the feeding source. Therefore, it is interesting that this method considers the parasitic effect of the pads to allocate the SMD lumped components. Smaller pads, such as those used in real devices, can be used too. In the present case, pads are 2 mm × 2 mm to simplify the soldering process in the lab.

For the present analysis, the number of pads has been selected to allocate seven lumped components. In some cases, such as those requiring a single band or not many bands, the number of pads can be reduced. Another possibility is to use open circuits or short circuits so the number of components is less than seven, as in the examples shown next.

It is interesting to observe the behaviour of the antenna booster when there is no matching network, that is, when ports #1, #3, #5, and #7 are short-circuited, and ports #2, #4, and #6 are open circuit (Figure 9.6). In the absence of the matching network, most of the frequency range of interest is poorly matched. For example,

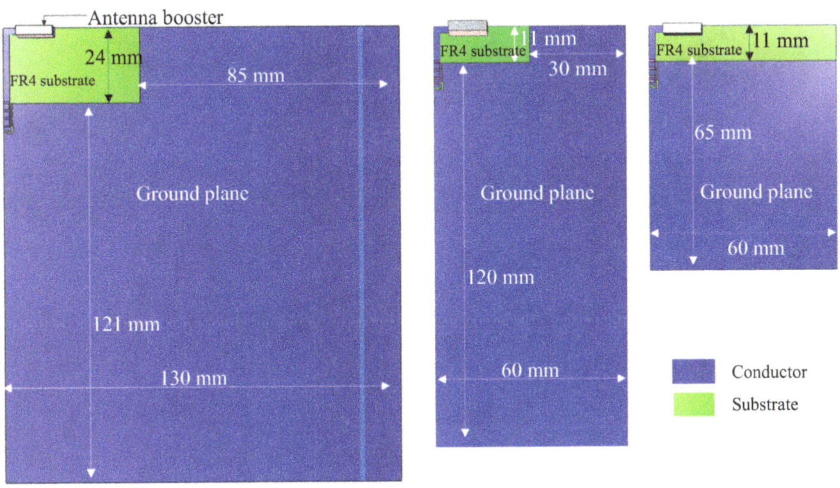

Figure 9.4 Representative platforms of wireless devices: (left) platform #1; (middle) platform #2; (right) platform #3. FR4 substrate is 1 mm thick, $\varepsilon_r = 4.15$, $\tan\delta = 0.017$. The antenna booster is 12 mm × 3 mm × 2.4 mm (height)

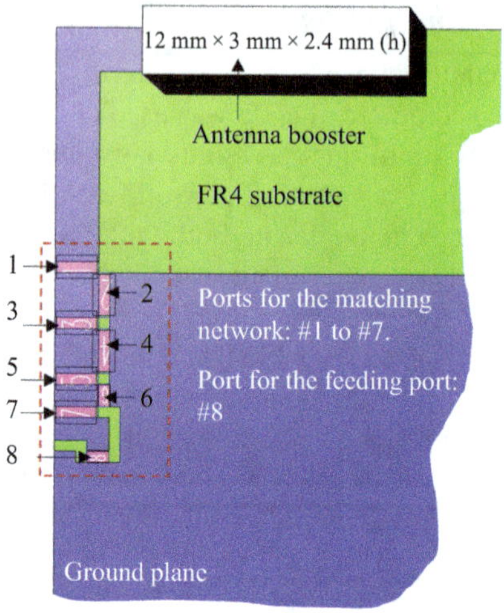

Figure 9.5 Pads for allocating the matching network. Each pad is 2 mm × 2 mm with 0.5 mm gaps between them and the ground plane. Ports #1, 3, 5, and 7 are for series components; ports # 2, 4, and 6 are for shunt components; port #8 is the feeding port.

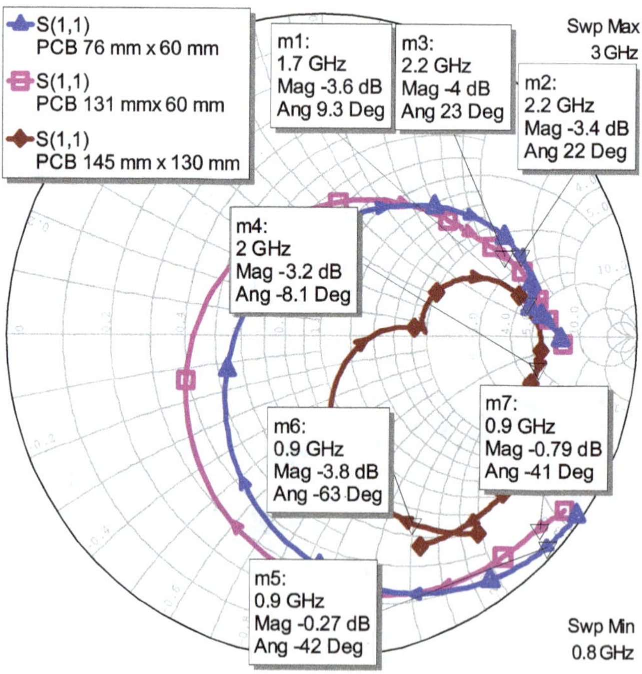

Figure 9.6 Simulated input impedance without any matching network

the frequency range from 698 MHz to 960 MHz has a poor S_{11} (<-1 dB for platforms #2 and #3), which would result in an antenna system of poor total efficiency due to severe mismatch loss. Also, at the higher bands, from 1.71 GHz up to 2.69 GHz, S_{11} is still poor (around -3 dB), presenting a severe mismatch. Note that a $S_{11} = -1$ dB results in only 20% of the available power delivered to the antenna system (the term $1-|S_{11}|^2$ in Equation (9.1)). For $S_{11} = -3$ dB, the delivered power is still poor, being 50% of the available power.

In all three cases, the antenna booster is placed at the corner of the ground plane, enabling the excitation of currents in the ground plane. This fact can be easily tested with an electromagnetic simulator by computing radiation efficiency (η_r). Since radiation efficiency is independent of antenna matching, it is a good figure of merit to judge if the chosen placement for the antenna booster is correct. If the efficiency is low, a different location should be selected to attain competitive efficiency. Bear in mind that placing the antenna booster is crucial. Typically, the corners of the ground plane are preferred to maximise bandwidth and efficiency as long as the antenna booster is of electric nature (Chapter 3).

9.2.2 Definition of the target specifications

Target specifications are given regarding power delivered to the impedance box ($Z_{\text{antenna booster}}$ in Figure 9.3) rather than S_{11}. This should be taken into account since optimising S_{11} may provide a solution with a good S_{11}, not because the power is delivered to the antenna booster but instead delivered to the resistive part of the lumped elements of the matching network due to its finite quality factor (Q). It should be considered that the optimisation is done using real components that possess certain losses (finite Q). Therefore, optimising the total power delivered to the antenna booster is paramount, not S_{11}. This ensures that the optimised matching network provides the best possible total efficiency, defined as the power delivered to the antenna booster over the available power. This definition equals the transducer power gain used in two-port network systems such as amplifiers (Figure 9.3). Total efficiency targets (Table 9.1) are included in the optimisation procedure of the matching networks [1–6].

Once the target specifications are defined, the matching network set-up is needed. In this case, all ports from #1 to #7 will have an SMD component, either an inductor or a capacitor. In some cases, a port may be a combination of both, such as

Table 9.1 Total efficiency target specifications

Platform #1: PCB 145 mm × 130 mm	**Platform #2: PCB 131 mm × 60 mm**	**Platform #3: PCB 76 mm × 60 mm**
824–960 MHz: −1 dB (79%) 1710–2170 MHz: −0.5 dB (89%)	824–960 MHz: −2.5 dB (56%) 1710–2690 MHz: −1.5 dB (70%)	824–960 MHz: −6 dB (25%) 1710–2690 MHz: −1.5 dB (70%)

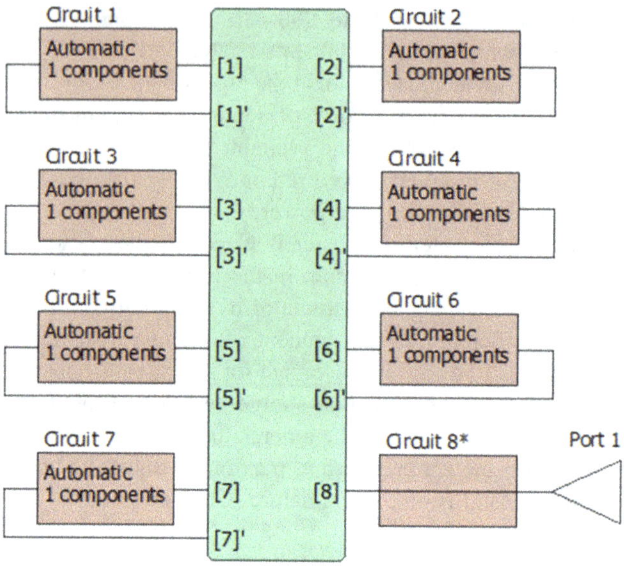

Figure 9.7 The $[S]_{8\times 8}$ matrix obtained from the simulation. From port #1 to #7, a lumped element is included. Port #8 is reserved for the source. Ports #1, 3, 5, and 7 are for series components (inductor, capacitor, or 0 Ω); ports # 2, 4, and 6 are for shunt components (inductor, capacitor, or open circuit).

shunt LC or series LC. In this example, inductors and capacitors have high Q, tight tolerance, and SMD types. Finally, the feeding port is connected to port #8 (Figure 9.7). The best matching topology and component values are obtained after finishing the synthesis process, as explained in the next section.

It is worth noting that the synthesis process is fast (typically minutes), taking into account the huge space of possibilities. For example, if we consider the search space between L_{min} = 0 nH, L_{max} = 20 nH at 0.2 nH steps and C_{min} = 0 pF, C_{max} = 20 pF at 0.2 pF steps, the total number of possible matching networks to explore results in ~13.000 billion matching networks. This clearly demonstrates that a brute-force search method is not feasible and highlights the advantages of optimisation algorithms.

9.2.3 Computation

The matching circuit component types (inductor or capacitor) and values must be determined to obtain the target total efficiency. For each component position, the optimisation process aids in selecting the component type (inductor or capacitor). The number of possible combinations grows exponentially with the number of ports. In addition, some of the ports for components may be short-circuited or left open. The optimisation goal is set for fixed component types to maximise the total efficiency, considering impedance mismatch, component losses, and antenna

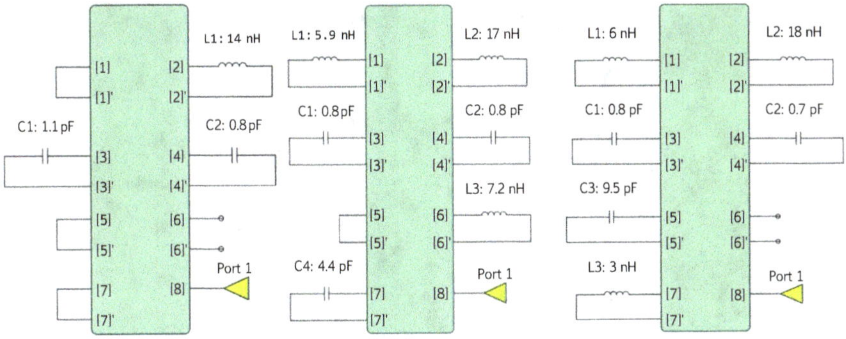

Figure 9.8 Synthetised matching networks: (a) platform #1 (PCB 145 mm × 130 mm), (b) platform #2 (PCB 131 mm × 60 mm), (c) platform #3 (76 mm × 60 mm). Note that some ports are left open (shunt) or shorted (series) since no L or C is needed.

radiation efficiency. Various optimisation algorithms (search-based, discrete, gradient, genetic, etc.) may be used to find the optimum component values, considering any losses and parasitic reactance of inductors and capacitors. Therefore, high Q components should be selected as part of the search space to minimise losses in the matching network.

As a first step, all ports from #1 to #7 have a component where optimisation will determine the appropriate option: inductor, capacitor, or open/short circuit. If the resulting matching networks have components close to a short circuit or an open circuit, then the matching network should be simplified to a smaller number of components. After optimising the cases under study, the matching networks use three, six, and six lumped components for platform #1, platform #2, and platform #3, respectively (Figure 9.8). Each component is modelled with its S-parameter given by a selected vendor. In this example, the simulations take less than 5 min on an 11-core computer.

This strategy considers a fixed pad layout. However, a different approach may consist in first synthesising a matching network without any pad layout. Once the matching network is obtained, the pad layout and, thus, the value of the components can be optimised to compensate for the pad effect. This process involves two steps (synthesis and optimisation) but ends with a compact layout. In the current process, we must consider an initial layout that may ultimately require fewer components than the layout supports, resulting in unnecessary pads.

9.2.4 Measurements

After completing the simulation process, implement the same matching networks obtained through simulation on identical PCBs used in the electromagnetic simulation. In this case, a 50 Ω microcoaxial transmission line terminated with an SMA connector is employed (Figures 9.9 and 9.10). A practical tip should be considered. The microcoaxial transmission line should be grounded at least at three points: one

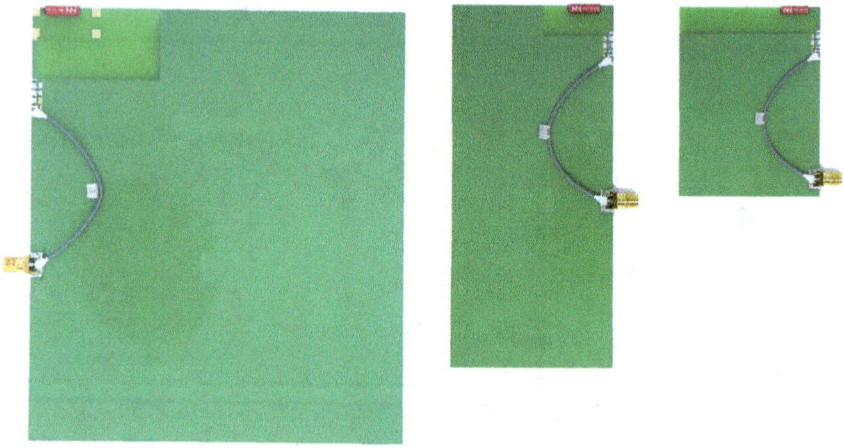

Figure 9.9 Implemented PCB with the dimensions as simulated in Figure 9.4

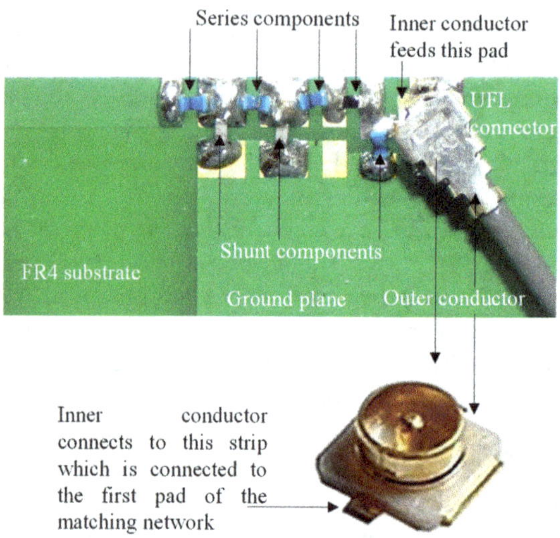

Figure 9.10 Detail of the transition from the coaxial cable to the input port

close to the first pad of the matching network, another in the middle, and another at the microcoaxial-to-SMA transition. The reason is to avoid a stub effect when the microcoaxial transmission line is not grounded correctly. In an actual device, the transmission line between the transceiver and the matching is typically a printed transmission line as short as possible ($<0.015\lambda$).

The voltage standing wave ratio for each case is measured with a vector network analyser, and total efficiency is measured in a SATIMO Stargate-32 anechoic chamber. Total efficiency is measured using 3D pattern integration. It takes into account both the radiation efficiency of the antenna system and mismatching, as in Equation (9.1). In this manner, losses in the matching network are also considered (Figures 9.11–9.13).

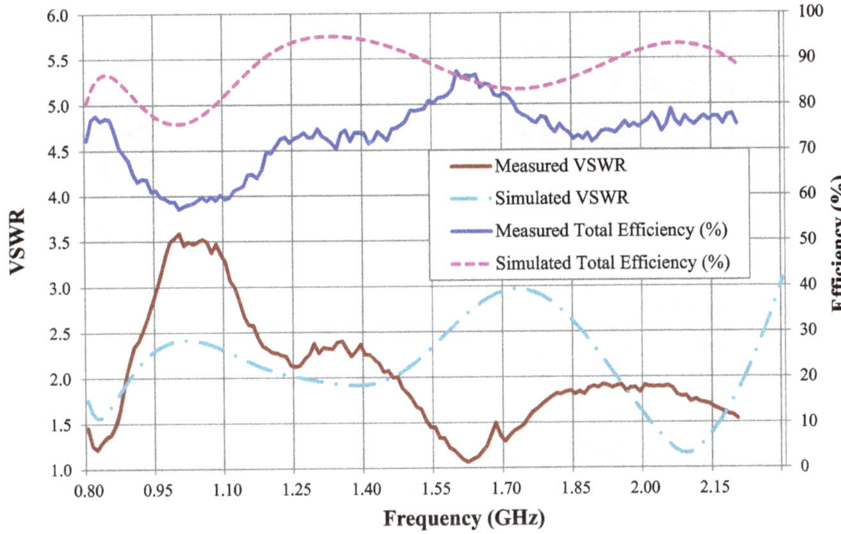

Figure 9.11 Measured and simulated VSWR and total efficiency for platform #1 (PCB 145 mm × 130 mm). The shaded area indicates 824–960 MHz and 1710–2170 MHz.

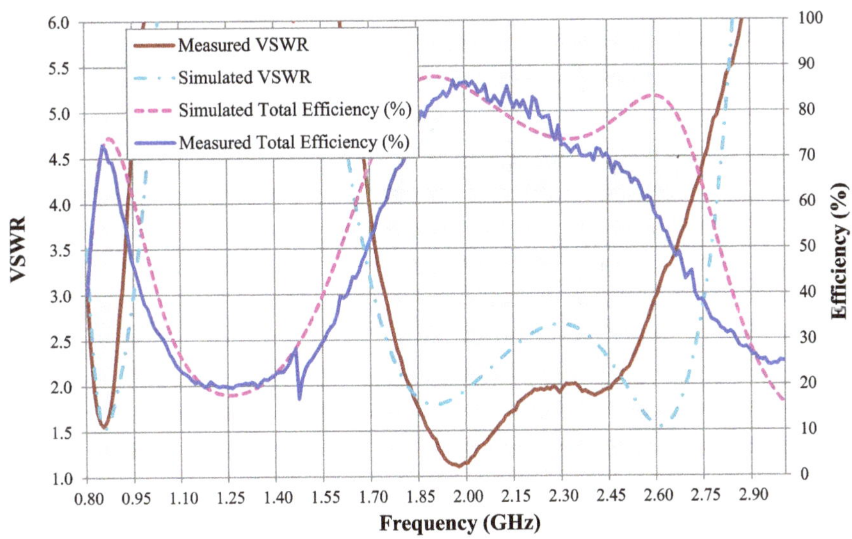

Figure 9.12 Measured and simulated VSWR and total efficiency for platform #2 (PCB 131 mm × 60 mm). The shaded area indicates 824–960 MHz and 1710–2690 MHz.

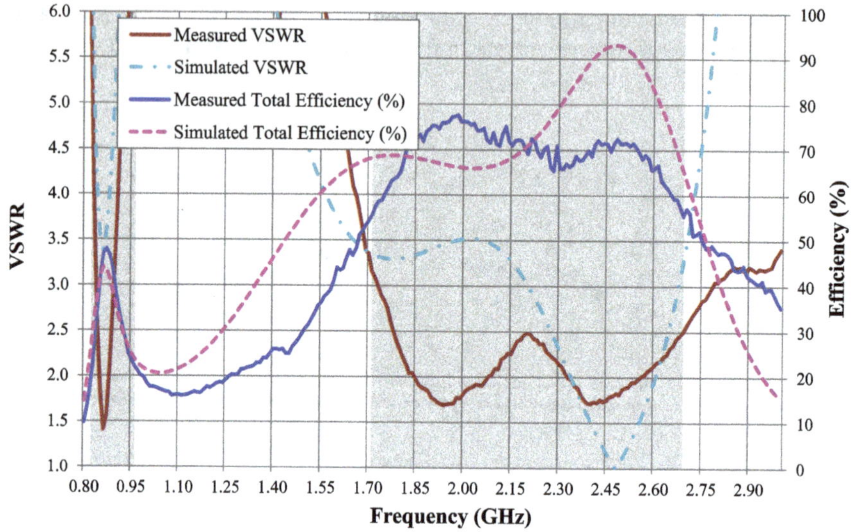

Figure 9.13 Measured and simulated VSWR and total efficiency for platform #3 (PCB 76 mm × 60 mm). The shaded area indicates 824–960 MHz and 1710–2690 MHz.

Matching network synthesis

The design of wireless devices with antenna boosters is simple since it is based on designing a matching network. The matching network provides the flexibility to make the device operate from single-band to multiband.

The design of matching networks for antenna boosters can be addressed by matching network synthesis tools. These tools obtain the matching network topology as well as the values of the components given the impedance of the antenna booster in the device. Furthermore, the synthesis can also take into account real components, and thus, the resulting matching network can be directly implemented on the wireless device.

Besides measured VSWR and total efficiency, it is worth noting that measured total efficiency (which takes into account radiation efficiency and mismatch loss) agrees well with the simulated total efficiency, in particular for platforms #2 and #3.

After these results show multiband performance for all three platforms, one can conclude that a multiband antenna system can be efficiently designed not by shaping the antenna element but by developing a multiband matching network, thanks to the antenna booster. Since a multiband matching network can be easily and quickly designed using the antenna booster procedure presented here, it opens a new paradigm for designing multiband antenna systems for wireless devices, which is fast and straightforward.

Table 9.2 Comparison of measured total efficiencies through computer simulation (computer) and obtained based on engineering experience (human)

Platform #1 (PCB 145 mm × 130 mm)						
Measured total efficiency (%)	824 MHz	960 MHz	Average	1710 MHz	2170 MHz	Average
Computer	76.0	59.7	67.8	81.7	75.6	75.7
Human-based	63.0	76.0	67.7	77.0	67.0	72.8
Platform #2 (131 mm × 60 mm)						
Measured total efficiency (%)	824 MHz	960 MHz	Average	1710 MHz	2690 MHz	Average
Computer	59.5	43.2	61.0	54.0	43.4	72.6
Human-based	51.8	60.9	66.6	70.2	59.9	76.2
Platform #3 (76 mm × 60 mm)						
Measured total efficiency (%)	824 MHz	960 MHz	Average	1710 MHz	2690 MHz	Average
Computer	20.3	22.2	34.9	55.4	56.6	69.5
Human-based	18.8	34.7	40.5	71.5	61.0	71.9

9.2.5 Matching network human-based design versus machine-based design

An interesting question is how well the implemented matching network performs compared to a human-based design. This section compares the three cases with matching networks designed by an experienced antenna designer (Table 9.2).

The results are similar, making the computer-based method an efficient automatic approach to designing multiband antenna systems for wireless devices embedded antenna boosters. Besides, the human-based process took days to find a matching network after evaluating several candidates, whereas the computer-based method is easy and fast, taking a couple of hours from the simulation of the antenna booster on the PCB considering the layout, running the matching network synthesis, implementing the matching network, and testing.

9.3 Matching network synthesis given the measured input impedance

The previous matching network synthesis method is convenient if one is provided with a full-wave simulator. However, no electromagnetic model is available in other situations, and the designer relies only on the physical wireless device (Figure 9.1). In this new scenario, the following method proposes a matching network synthesis method for antenna boosters to design from single-band to multiband antenna systems. In this case, a vector network analyser is required since the matching network synthesis needs the input impedance of the antenna booster on the real device without any matching network.

The first method synthesises a matching network given the measured impedance of an antenna booster on a PCB without any matching network (Section 9.3.1). Once the matching network is obtained, pads are added according to the number of components needed in the matching network. Once pads are added, each component is optimised to compensate for the pad effect [7].

In the second method (Section 9.3.2), a fixed layout is considered first; thus, the matching network synthesis considers the pad layout from the beginning. This process is faster since fewer iterations are needed. However, it may end with some extra pads that are not required.

It is important to outline that both methods can be followed to design the matching network not only for a bare PCB as shown in the illustrative examples below but, more importantly, for an entire device populated with electronics, batteries, casings, etc. In this case, the S_{11} should be measured, considering the entire device. As detailed next, the initial S_{11} should be taken regarding the correct phase delay to remove the delay introduced by the microcoaxial used for testing purposes.

9.3.1 Considering the pad layout after the matching network synthesis

As demonstrated in Chapter 4, a 12 mm × 3 mm × 2.4 mm (height) antenna booster achieves multiband performance from 824 MHz to 960 MHz up to 1710 MHz to 2690 MHz. The antenna booster is integrated into a wireless device, including a 120 mm × 60 mm ground plane (Figure 9.14). Said antenna booster is responsible for exciting currents in the ground plane, making the ground plane an essential part of the radiation process [8,9].

Since the antenna booster is small in terms of the operating wavelength ($\lambda/30$ at 824 MHz), its input impedance is mainly reactive across the low-frequency region from 824 MHz to 960 MHz. Therefore, a multiband matching network is designed. As explained next, such a design is addressed through a network synthesis tool based on genetic algorithms [10].

The input impedance derived from the configuration depicted in Figure 9.14 serves as the input data for the algorithm, which aims to ascertain an appropriate matching network based on specified objectives and constraints. Such S_{11}

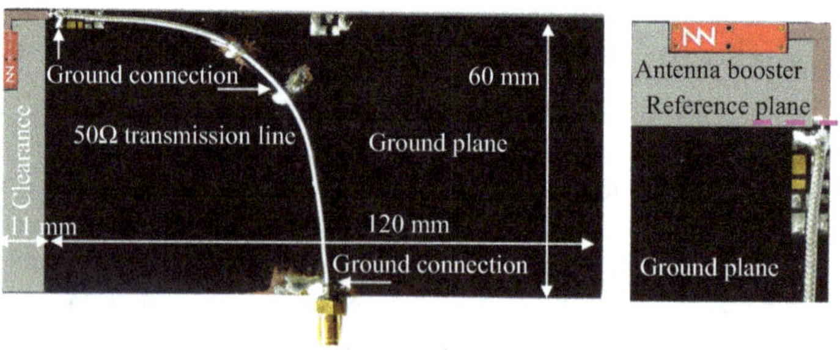

Figure 9.14 A 12 mm × 3 mm × 2.4 mm antenna booster. The ground plane is 120 mm × 60 mm printed over an FR4 substrate 1 mm thick (ε_r = 4.15, tanδ = 0.017). The reference plane indicates where the S_{11} is measured. Note that such measurements do not include the pad layout.

measurements must consider the reference plane at the first gap to allocate the first component of the matching network. That gap is used for the reference plane when testing the S_{11} (Figure 9.14).

The goal is to obtain an $S_{11} < -6$ dB across 824–960 MHz and 1710–2690 MHz. The constraint is to keep the number of components for the matching network up to six inductors or capacitors. Also, given the frequency region of operation, the inductors L are limited to varying from 0 nH to 20 nH and capacitors from 0 pF to 20 pF. This reduces the search space, and consequently, the genetic algorithm finds a set of possible matching networks in less than 2 min. Among the several matching network candidates, the selected one has a cost closest to zero, where zero means that the matching network will provide an $S_{11} < -6$ dB across the specified frequency range. The algorithm may find a solution with a smaller number of components. Otherwise, the maximum number of components should be increased. From a practical point of view, the maximum number of components is recommended to be less than eight. Note that the synthesis process can also use commercially available components instead of ideal components.

The synthesised matching network comprises six lumped components and achieves a simulated $S_{11} < -6$ dB at the frequency regions of interest – 824–960 MHz and 1710–2690 MHz (Figure 9.15).

Pads are now considered to design the matching network (Figure 9.16). These pads are needed to allocate the lumped elements. In this case, SMD 0603 and 0402 type components are selected to obtain the best Q for each component.

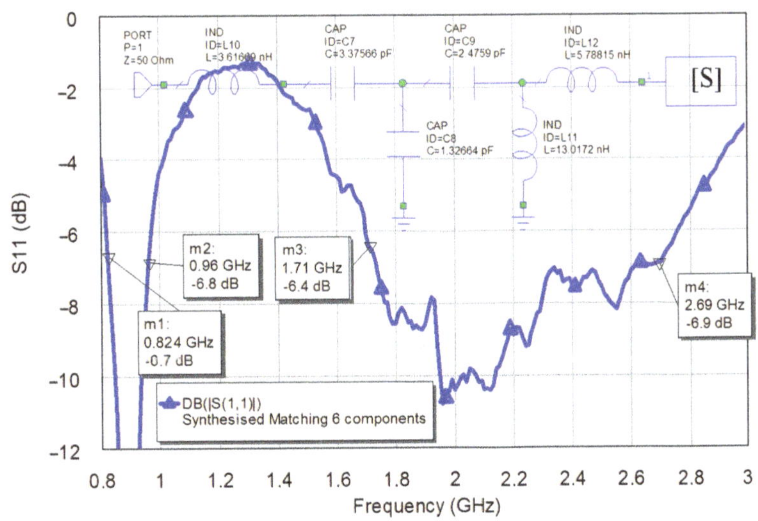

Figure 9.15 A multiband matching network synthesised with the genetic algorithm tool [10]. [S] represents the measured S_{11} for the 12 mm × 3 mm × 2.4 mm (h) antenna booster on a 120 mm × 60 mm ground plane without any matching network.

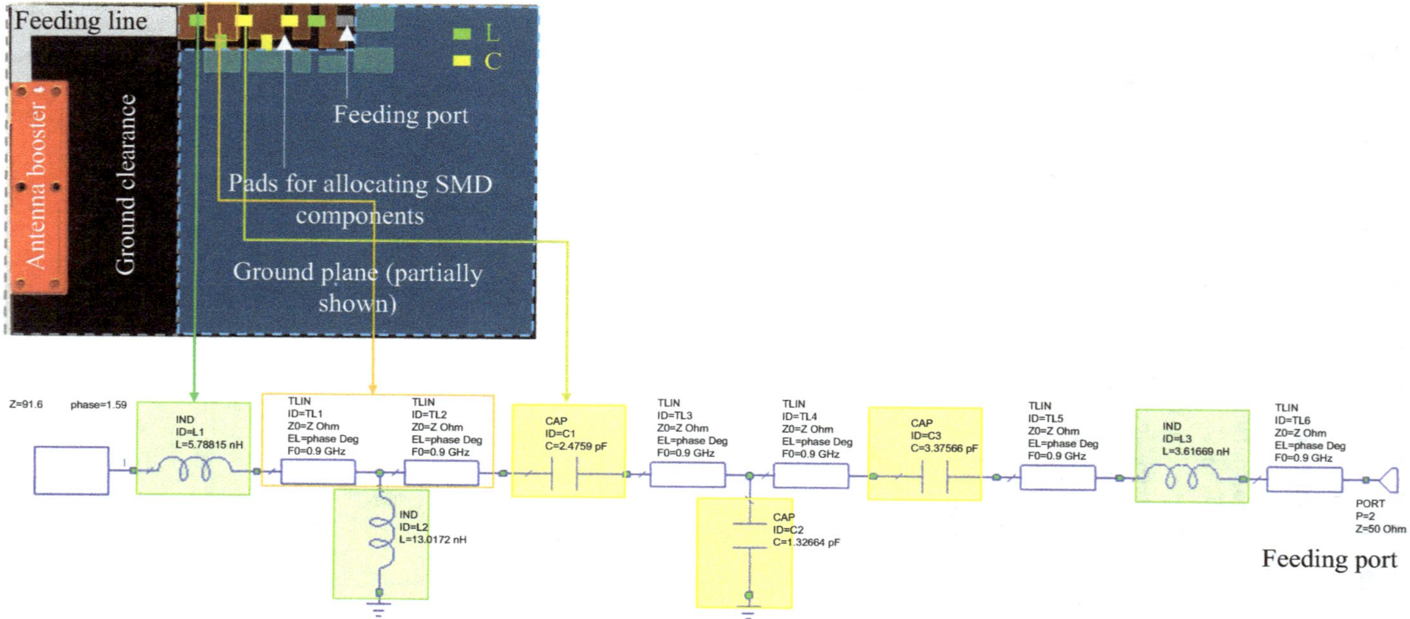

Figure 9.16 *Matching network, including pad effects. The box at the left represents the measured input impedance of the antenna system shown in Figure 9.14. Z corresponds to the equivalent characteristic impedance of pads, and 1.59° is the phase delay of half the length of a pad (note that a shunt component is placed in the middle of a pad).*

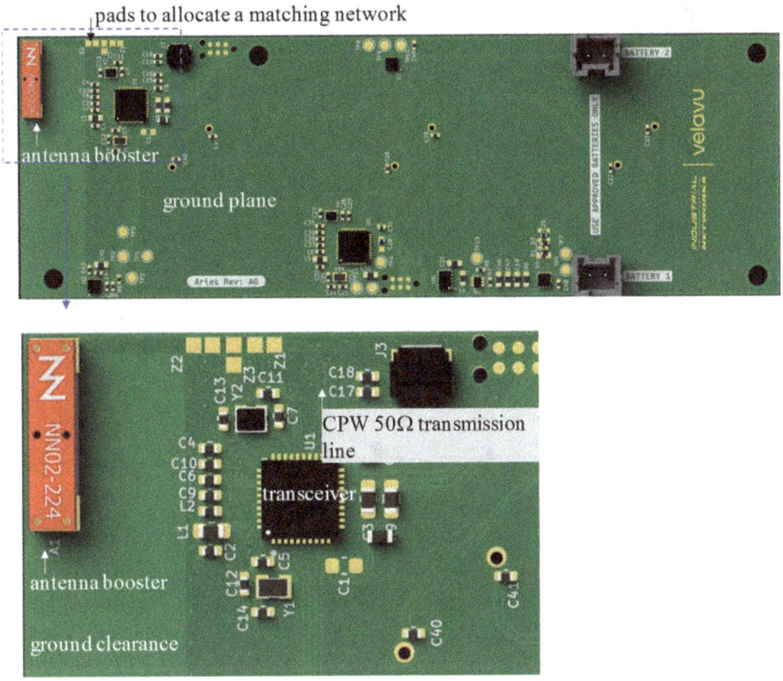

Figure 9.17 A 12 mm × 3 mm × 2.4 mm (height) antenna booster on a wireless device, including a pad layout for a series (Z_2), shunt (Z_3), and series (Z_1) matching network [11]. Thanks to Velavu.

Short transmission lines can model the effect of pads. In this case, pads are modelled by short asymmetric CPW lines (only a coplanar ground plane is on one side). The characteristic impedance and phase delay can be calculated with a circuit simulator.

For illustrative purposes, a real device, including a pad layout for a series-shunt-series matching network, is shown in Figure 9.17.

Once they are included in the model, the S_{11} is analysed, and as expected, there is a slight shift toward lower frequencies (Figure 9.2). Fortunately, this frequency shift can be compensated by changing the values of the L and C of the matching network. A genetic algorithm is run to optimise the L and C values to get an $S_{11} < -6$ dB across the frequency range. Once the genetic algorithm finds the L and C, the new S_{11} is analysed, where it can be observed that S_{11} can be below −6 dB again (Figure 9.18).

Finally, each L and C are substituted with commercial values to implement the matching network (Figure 9.19).

The advantage of this step is that the S-parameters of L and C are included in the simulation, giving a more precise S_{11} result (Figure 9.20). The final SMD components used for building the feasibility prototype are shown in Table 9.3. The

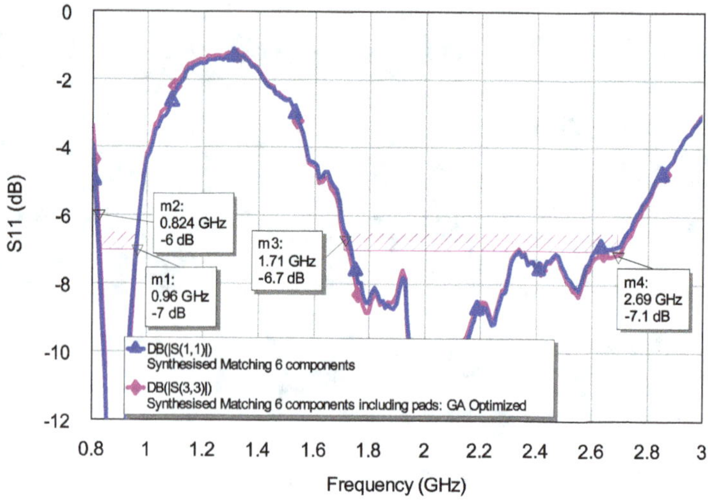

Figure 9.18 *Comparison of the original S_{11} without pads and the S_{11} optimised when pads are considered*

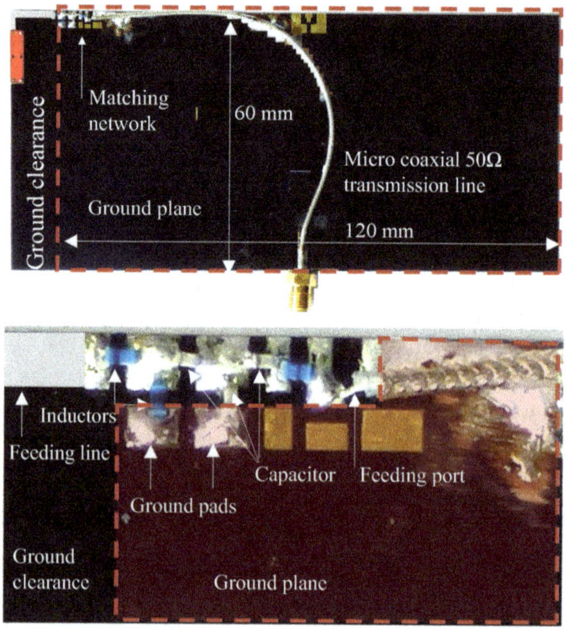

Figure 9.19 *Prototype with the 12 mm × 3 mm × 2.4 mm (h) antenna booster on a 120 mm × 60 mm ground plane, including a matching network with six lumped components (three inductors and three capacitors)*

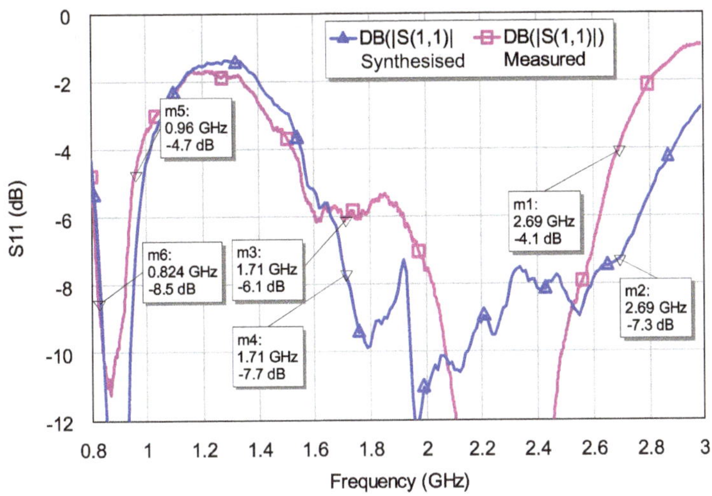

Figure 9.20 *Comparison of the simulated S_{11} obtained after the matching network synthesis procedure and the measured S_{11}*

Table 9.3 *Values of the feasibility prototype*

Component	Value	Minimum *Q* within 824–2690 MHz
L_1	4.7 nH	58
L_2	14 nH	62
C_1	1.9 pF	127
C_2	0.9 pF	228
C_3	3.3 pF	82
L_3	2.9 nH	63

terminology for each component corresponds to the matching network shown in Figure 9.17.

Once the matching network is implemented with the values given in the simulation, it can be observed that it is in good agreement (Figure 9.20). Although some differences exist, the most important conclusion is that the matching network has been obtained using a fully automated method without any adjustment. This opens the window to facilitate the design of multiband antenna systems for wireless designers who are not antenna experts.

Finally, the total efficiency (Equation (9.1)) of the prototype is measured inside an anechoic chamber (Figure 9.21). It is measured through 3D pattern integration with the Satimo Stargate-32 anechoic chamber. The average efficiency at 824–960 MHz is 60.6%, and at 1710–2690 MHz is 70.5%.

To sum up, a systematic multiband antenna system design has been proposed. The method relies on using an antenna booster ($\lambda/30$ at 824 MHz) with the addition of a multiband matching network. The matching network design is carried out by a

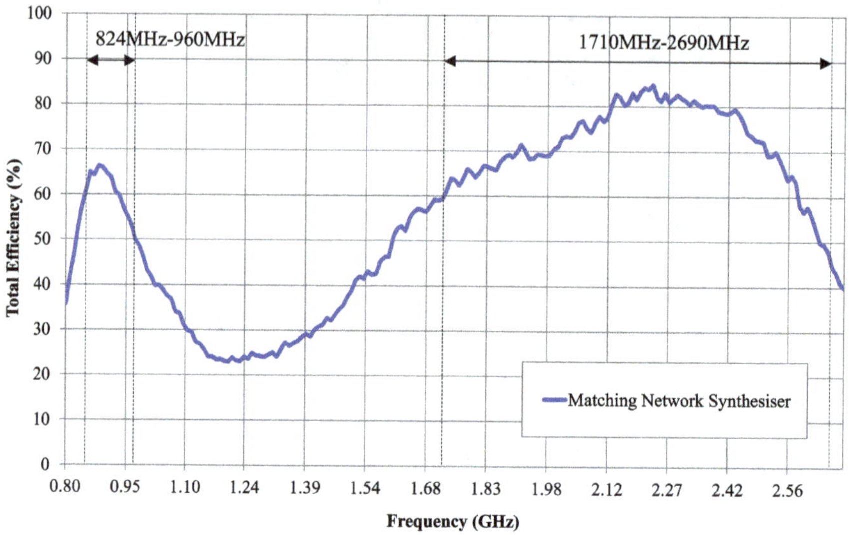

Figure 9.21 Measured total efficiency for the antenna system shown in Figure 9.19 with the synthesised matching network

network synthesis tool based on genetic algorithms. To implement the matching network in a device, the pad layout to host the SMD components of the matching network has been taken into account. Those pads shift the original S_{11} response, but this shift can be compensated by adjusting the inductors and capacitors of the matching network using genetic-based optimisation. Once the values are optimised considering the pads, the ideal values are substituted by real commercial values used for prototyping.

After the proposed design process, the measured total efficiency shows an average efficiency of 60.6% at 824–960 MHz and 70.5% at 1710–2690 MHz (Figure 9.21).

This multiband antenna design method with an embedded antenna is fast, simple, easy, and attractive for wireless designers in charge of designing the entire RF chain, including antenna boosters.

9.3.2 Considering the pad layout during the matching network synthesis

The same 12 mm × 3 mm × 2.4 mm (height) antenna booster is used to achieve multiband performance from 824–960 MHz to 1710–2690 MHz, thanks to a multiband matching network. However, in this case, the antenna booster is integrated on a 145–130 mm ground plane, including a 45–24 mm ground clearance (Figure 9.22).

Since the antenna booster is small in terms of the operating wavelength ($\lambda/30$ at 824 MHz), its input impedance is mainly reactive across the low-frequency region from 824 to 960 MHz (Figure 9.23). At the high-frequency region (1710–2690 MHz),

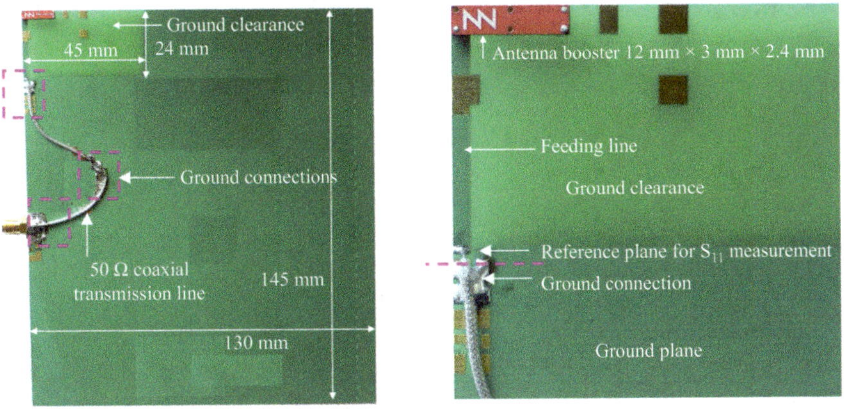

Figure 9.22 *(Left) A 12 mm × 3 mm × 2.4 mm antenna booster. The ground plane is 145 mm × 130 mm printed over an FR4 substrate 1 mm thick (ε_r = 4.15, tanδ = 0.017) with a 45 mm × 24 mm ground clearance; (right) reference S_{11} plane positioned on the first component from the antenna booster (end of the coaxial line).*

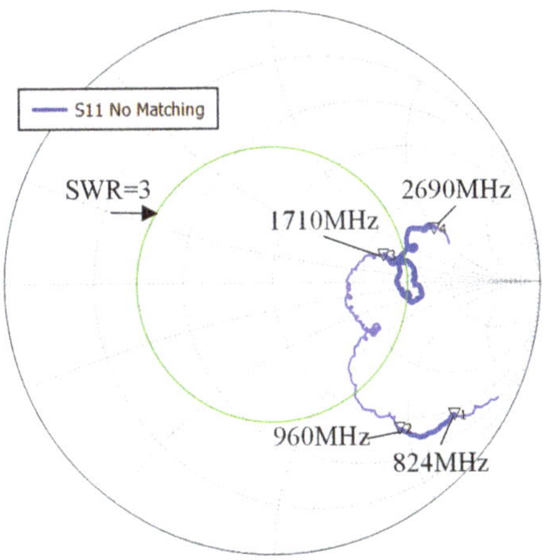

Figure 9.23 *Input impedance of the antenna booster is shown in Figure 9.22 without any matching network. The reference plane for the [S] measurement has been corrected at the end of the coaxial 50 Ω transmission line. The green circle represents SWR = 3.*

the impedance is over the SWR < 3 ($S_{11} < -6$ dB) green circle on the Smith chart. Therefore, a multiband matching network is designed to have an SWR < 3 across the frequency regions. Such design is addressed through a matching circuit synthesis tool by maximising the system efficiency [1–6]. As elaborated further, the synthesis of the matching network involves the consideration of pads for allocating Surface-Mounted

Device (SMD) components within the matching network, based on the measured impedance of the antenna booster on the PCB in its original state without any matching network.

In this case, the synthesis is carried out from a measured S_{11} for the antenna booster on a PCB, as in the previous case. Therefore, this method is suitable for wireless engineers directly designing the physical device rather than in the simulation domain, as in Section 9.2.

In this example, the synthesis is directly addressed with commercially available components.

The input impedance depicted in Figure 9.23 serves as the measured input data for synthesising the matching network, aiming to achieve a total efficiency of greater than -1 dB across both frequency ranges. This impedance must be tested without considering the pads and having the reference plane at the end of the coaxial line where the first component of the matching network will be mounted (Figure 9.22 right). This critical aspect determines the method's accuracy [12–14]. If the phase is not adequately corrected, there will be a disagreement between the simulated S_{11} and the measured S_{11}. As Figure 9.22 (right) shows, the input impedance of the antenna booster on the PCB must be tested at the point where the first component will be placed (either an inductor, capacitor, or 0 Ω if no component is needed in the first place). A short circuit is added at this point to shift the reference plane so the vector network analyser can calculate the reference plane. The short circuit can be implemented by adding a 0 Ω to short the first pad to the ground. Once the reference plane is set, the short circuit is removed. Another simple preferred option is to characterise the phase delay of the transmission line with an open circuit [14,15].

The goal is to optimise total efficiency η_t across 824–960 MHz and 1710–2690 MHz. Such total efficiency considers the losses of the SMD components of the matching network, which considers the real measured data of the lumped capacitors and inductors. Besides, the constraint is to keep the number of components for the matching network as low as possible. In this case, SMD components 0402 and 0603 are selected to get the best Q. Pads are modelled by short, asymmetrical CPW lines (only a coplanar ground plane is on one side). In this case, pads are modelled by short asymmetric CPW lines (only a coplanar ground plane is on one side). A circuit simulator can calculate the characteristic impedance and phase delay (Figure 9.24). For example, the first 2 mm × 2 mm pad from the feeding line is modelled as two transmission lines since we can connect a shunt component, either an inductor or a capacitor, in the middle. The election will be conducted by the matching network synthesiser tool. For modelling the pads, several considerations need to be taken. First, the characteristic impedance is given by the characteristic impedance Z_0 of a CPW line. In this case, since the CPW has only a ground plane on one side, Z_0 needs to be multiplied by the square root of two. The reason is that symmetric CPW lines have two capacitance effects, one between the line and the ground plane on one side and the other between the line and the ground plane on the opposite side. However, in this asymmetrical case, there is only one capacitance effect that increases Z_0 by

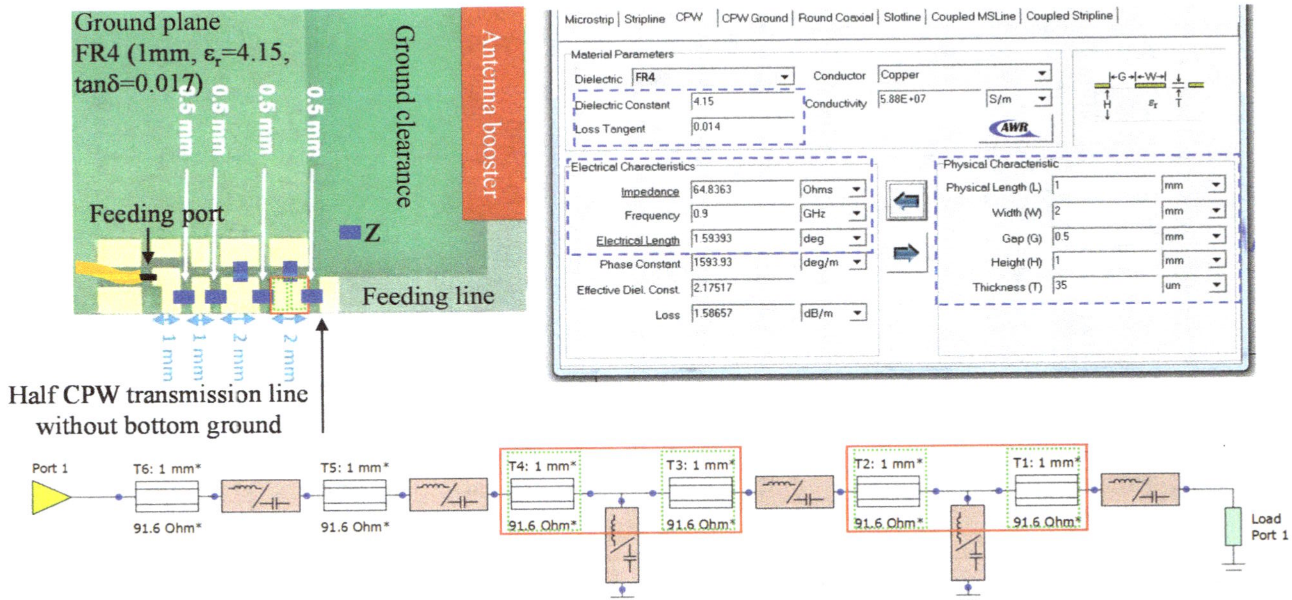

Figure 9.24 *Pads are modelled by transmission lines, considering the characteristic impedance Z_0 and length. Besides, effective relative permittivity must be introduced as well.*

the square root of two since Z_0 is defined as the square root of the inductance over the capacitance of the transmission line.

Second, the length of the pad needs to be modelled. Depending on the simulators, a phase delay needs to be introduced. In others, the effective relative permittivity. Using any transmission line calculator, both are easily computed; for example, in a 1 mm pad over a 1 mm thin substrate having $\varepsilon_r = 4.15$, the phase delay at 0.9 GHz results in 1.59°, where the effective relative permittivity is 2.17.

Considering the pad layout, the network synthesis tool allows obtaining the impedance values to minimise a human fine-tuning of the matching network values. Note that this method obtains a matching network with a fixed pad layout. An alternative approach is to run the synthesis without any pad effect to let the algorithm find a matching network. Once the matching network is obtained, the pad layout is included to allocate the matching network components. Afterwards, the given matching network must be optimised to consider the pad effects.

It is crucial to outline that although the pad layout is fixed from the beginning, the method is flexible since if the matching network synthesis finds that the first series component is not needed, it will simply neglect to add any reactance at that position. This is the situation illustrated next, where the first component of the matching network is not a series reactance but a shunt reactance (Figure 9.25). In other words, each block in Figure 9.24 showing an *L* or *C* inside can finally be an *L*, a *C*, or a short circuit for the series case, or an open circuit for the shunt case.

The synthesised matching network comprises three lumped components (Figure 9.25) with commercial values and achieves a simulated $S_{11} < -6$ dB at the frequency regions of interest – 824–960 MHz and 1710–2690 MHz (Figure 9.26).

The advantage of using commercial values is that the *S*-parameters of the *L* and *C* are included in the simulation, giving a more precise S_{11} result. The commercial values correspond to high-*Q* tight tolerance components.

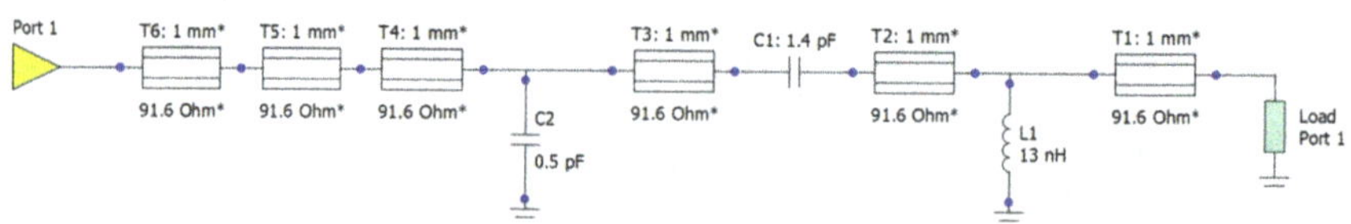

Figure 9.25 Matching networks, including pad effects and lumped components, with commercial values. The box at the right (Load port 1) represents the measured input impedance of the antenna system shown in Figure 9.22 right. 91.6 Ω corresponds to the equivalent characteristic impedance of pads, and 1 mm corresponds to half the pad's length (note that a shunt component is placed in the middle of a pad). L_1 is an SMD0603 with $Q > 85$, whereas C_1 ($Q > 165$) and C_2 ($Q > 330$) are 0402. All are high-Q tight tolerance (2%) components

Design with only a vector network analyser
With a single S_{11} and a correct modelling of the pad layout, a single/multi-band matching network is easily obtained with matching network synthesis.

Once the matching network is implemented with the commercial values given in the simulation (Figure 9.26), it can be observed that the simulation and implementation are in good agreement. Moreover, the matching network has been obtained without human adjustments, only using a systematic matching network method. This provides an opportunity for wireless designers to easily and rapidly design multiband antenna systems.

Finally, the total efficiency (η_t) of the prototype is measured inside an anechoic chamber (Figure 9.26). It is measured by integrating the 3D pattern with a Stargate-32 anechoic chamber. Average efficiency is 68.3% and 81.6% at 824–960 MHz and 1710–2690 MHz, respectively.

The radiation efficiency without the matching network has been numerically computed and included in the software simulator. Hence, losses include those of the matching network and the losses of the antenna system (PCB and antenna booster). This provides more realistic data, which leads to a good agreement between simulated and measured total efficiency.

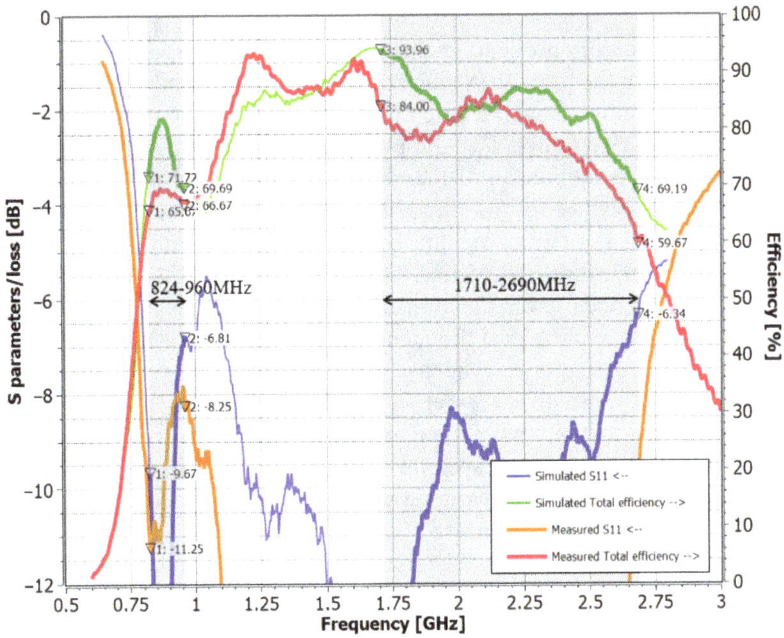

Figure 9.26 Comparison of the simulated S_{11} and total efficiency and the measured S_{11} and total efficiency

It is worth mentioning that this platform has a 145 mm × 130 mm PCB, the same as studied in Section 9.2. The synthesised matching networks have the same topology (three components) (Figures 9.8 and 9.25). In this situation, the high-frequency region is larger, covering from 1710 MHz to 2690 MHz. However, the antenna booster is the same. Only the matching network has been redesigned for this new situation. This illustrates that an antenna booster can be used for any frequency band (if there is bandwidth) and any platform (as shown in Section 9.2).

Further improvements may be adopted in the previous models, including the capacitance between pads and pads to the ground. Despite the capacitance between pads being on the scale of tens of femtofarads, incorporating their impact into the model enhances accuracy, leading to improved alignment between simulation and measurement [14].

Both methods rely on modelling the pad layout with transmission lines. An alternative method involves modelling the pad layout with a full-wave electromagnetic simulator [15]. In this case, the parasitic effects due to capacitances between pads and pads to the ground plane are taken into consideration.

9.4 Fine-tune the reflection coefficient response by changing the lumped components of the matching network

When implementing the previous methods, the final S_{11} response may be shifted. As discussed in the previous sections, the accuracy depends on how well the pad layout is modelled as well as the characterisation of the materials. For the method based on a measured S_{11}, the accuracy also depends on the reference plane as well as the calibration of the vector network analyser. Although both methods feature accurate results, the reality is, in some cases, not as ideal as desired. Therefore, a practical method is proposed here to adjust the response, considering little microwave expertise. The consideration relies on the impact of each lumped component as a function of frequency. For example, consider a multiband situation with a first frequency region centred at f_1 and a second frequency region centred at f_2, where $f_2 > f_1$. A change in a series inductor will typically impact it more at f_2 than at f_1. On the contrary, a change in a series capacitor will impact it more at f_1 than at f_2. Analogously, a shunt inductor will affect more in f_1, and a shunt capacitor will impact more in f_2. However, this does not mean that by simply changing these values, the frequency response is directly shifted downward or upward.

To illustrate an example, a multiband matching network for an antenna booster is designed. The value of the shunt L_2 is changed from its original value of 13.5 nH to 11.1 nH, that is, a 17% reduction. For this particular example, the response is shifted upwards by 7.7% at the first frequency region and 4.0% at the second frequency region (Figure 9.27).

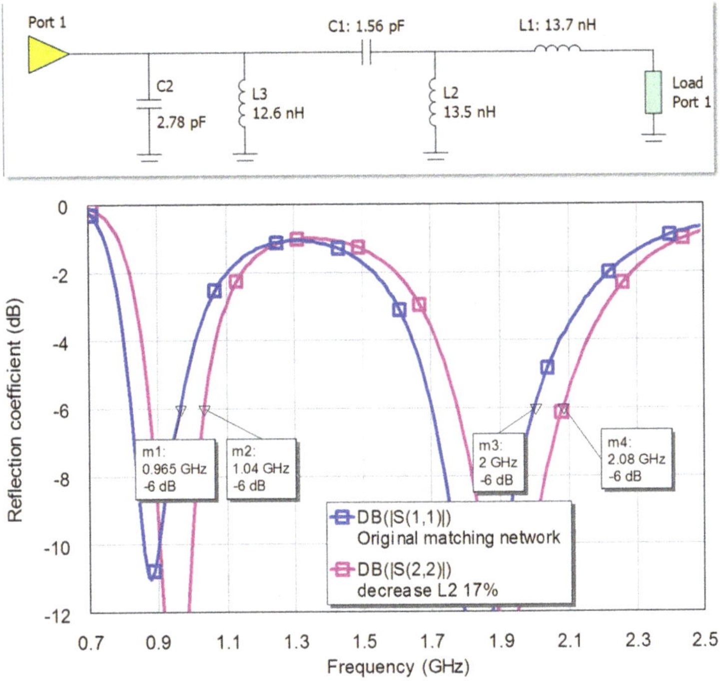

Figure 9.27 Impact on the reflection coefficient when changing the value of one component from L_2 = 13.5 nH to L_2 = 11.1 nH

9.5 Conclusions

The pad layout must be considered when designing wireless devices with antenna boosters (and for any antenna as well), including matching networks. This results in a more accurate design process. As a result, the matching network synthesised considering the pad layout can be directly built on a PCB, resulting in a good agreement between the simulated and measured S_{11} and total efficiency.

Two methods have been proposed to synthesise matching networks for antenna boosters, considering the pad layout. The first method uses the electromagnetic model of the antenna booster and the PCB, including the layout, to allocate the components. At each gap, a port is included. As a result, a $[S]$ matrix with N ports is obtained: $N-1$ ports allocate the $N-1$ components, and one port allocates the source. The matching network synthesis allows treating this $[S]$ matrix to get the desired matching network according to the target specification, which can be set in terms of total efficiency. The procedure can consider real components with their finite Q and tolerance. Suppose the electromagnetic model considers all details of the antenna system (antenna booster, feeding line, pad layout, electronics, casing, battery). In that case, the matching network obtained with this method can be built into the real device.

The second method relies on using the input impedance of the antenna booster on a real device. Again, the process needs the S_{11} with a proper reference plane. Once the input impedance is obtained, the pad layout is modelled using short transmission lines. As in the first method, real components can also be considered in the synthesis procedure. This method is suitable if no electromagnetic simulation is available, and the designers need to design the matching network given the physical device.

Both methods are easy and fast to implement and are software-driven, with minimum or no human intervention in adjusting the matching network once it is obtained after these procedures. Furthermore, since the matching network synthesis tool can provide several matching networks, the designer can select some of them to test in the lab to choose the one most appropriate to their needs.

Table 9.4 answers the objective given at the beginning of the chapter. It also points to relevant figures and equations to consolidate the concepts learnt. It is also advisable to refresh all essentials in the grey boxes distributed along with the chapter.

Table 9.4 Key aspects to take away

Topic	Review and take away
Full layout matching network synthesis	A wireless device, including the antenna booster and the pad layout, is simulated to obtain an $[S]_{N\times N}$ matrix where $N-1$ is the number of lumped elements for the matching network and one port is reserved for working as the RF source (Figures 9.5 and 9.7). The matching network is obtained with a matching network synthesiser to maximise the power delivered to the antenna booster system (Figure 9.8). From a practical perspective, up to eight components are enough (to keep low losses), and therefore, the pad layout is designed to host those $N-1$ SMD components in a ladder manner (series, shunt, series, etc.). The matching network synthesis tool will determine the number of components needed and add short circuits and open circuits when required. This route requires a full 3D electromagnetic simulator. The synthesis process can be carried out considering real components.
Including the layout with transmission lines	In those situations where a full 3D electromagnetic simulator is unavailable, the matching network synthesis can be addressed directly from the actual device, where the pad layout is considered as transmission lines. The input impedance without the pad layout should be carefully measured. This implies a precise phase plane correction for measuring the S_{11} of the antenna booster embedded in the real wireless device (Figures 9.14 and 9.22). Once the S_{11} is obtained, the synthesis is carried out by any of the two methods: (a) synthesise the matching network without any pad layout. Then, include the pad layout and optimise the values of the matching network components (Figures 9.15 and 9.16). (b) Include a pad layout with transmission lines. In this case, the number of maximum components is fixed from the beginning; typically, seven or eight is enough to keep low losses (Figures 9.24 and 9.25). The synthesis tool will finally determine how many components are needed. Both routes can include real components. In this regard, method *b* needs fewer interventions.

9.6 Exercises

A 3D electromagnetic software and a circuit simulator, including matching network synthesis, are needed to solve the following exercises.

9.6.1 Exercise 1

Simulate the input impedance of a 12 mm × 3 mm × 2.4 mm (height) antenna booster on a 120 mm × 60 mm ground plane (Figure 9.28) from 600 MHz to 3 GHz. For the antenna booster, consider a 3D conductive parallelepiped having a conductivity of $\sigma = 5.9 \times 10^7$ S/m. Put the ground plane on a 131 mm × 60 mm PCB with a substrate 1 mm thick, with $\varepsilon_r = 4.15$, $\tan\delta = 0.017$. Represent the S_{11} from 0.65 GHz to 3 GHz.

9.6.2 Exercise 2

For Figure 9.28, compute the potential bandwidth for SWR = 3 using Equations (3.2) and (3.3) (Chapter 3) from 0.6 GHz to 3 GHz. Use the potential bandwidth as a reference for the following exercises.

9.6.3 Exercise 3

For Figure 9.28, synthesise a matching network to achieve performance at 824–960 MHz and 1710–2690 MHz. Use as a target: total efficiency greater than 80% at both frequency regions. Use real components with high Q and tight tolerance. Compare the results when synthesising a matching network with 4, 5, 6, and 7 components.

9.6.4 Exercise 4

For Figure 9.28, select the matching network of Exercise 9.6.3 and include a pad layout modelled as transmission lines, as in Figure 9.24. Compute the S_{11} and compare it with and without the pad layout. Analyse the results. Optimise the value of the matching network components, including the pad layout.

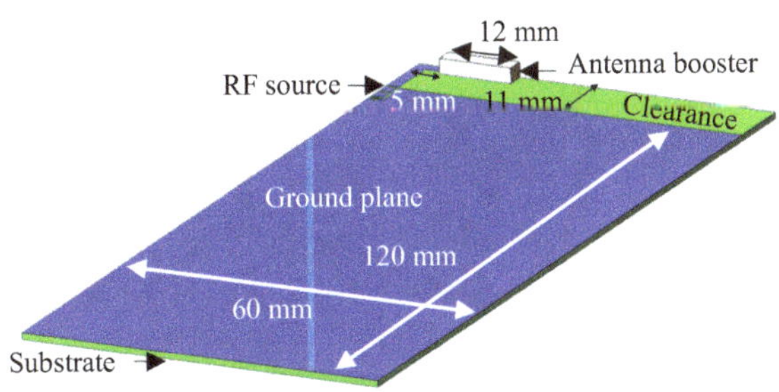

Figure 9.28 Ground plane and antenna booster (conductive parallelepiped with conductivity $\sigma = 5.9 \times 10^7$ S/m), the substrate is 1 mm thick, $\varepsilon_r = 4.15$, and $\tan\delta = 0.017$

9.6.5 Exercise 5

For the exercise in Section 9.6.4, model the pad layout with a full wave simulator. Compare the results with previous models.

9.6.6 Exercise 6

For the exercise in Section 9.6.4, consider a pad layout as shown in Figure 9.5. Run a full 3D electromagnetic simulation to obtain the $[S]_{8\times8}$ from 0.7 GHz to 3 GHz. Synthesise a matching network, as shown in Figure 9.7. Compare the results in terms of components of the matching network, S_{11} and total efficiency with the approach of the previous exercise.

9.6.7 Exercise 7

For the previous exercise in Section 9.6.6, compare the S_{11} with the substrate using the original $\varepsilon_r = 4.15$ with $\varepsilon_r = 3.8$ and $\varepsilon_r = 4.4$. Determine how critical the variation of the permittivity of the substrate is.

9.6.8 Exercise 8

For the previous exercise in Section 9.6.7, keep the substrate with $\varepsilon_r = 4.15$. Evaluate the matching network losses if $\tan\delta = 0.001$ or $\tan\delta = 0.03$.

9.6.9 Exercise 9

Considering the antenna booster system in Figure 9.28 and the pad layout in Figure 9.5, run a full 3D electromagnetic simulation to obtain the $[S]_{8\times8}$ from 0.65 GHz to 1.2 GHz. Design a broadband matching network centred at 0.9 GHz to cover two times the bandwidth given by the potential bandwidth at 0.9 GHz. Review the broadband principles explained in Chapter 4.

9.6.10 Exercise 10

Repeat the previous exercise in Section 9.6.9 reducing the pad size from 2 mm × 2 mm to 1 mm × 1 mm and the gap between pads and pads to the ground plane from 0.5 mm to 0.2 mm. Repeat the matching synthesis for the specs given in exercise in Section 9.6.3.

9.6.11 Exercise 11

Consider the antenna booster system where the pad layout has grounding at both sides (symmetric coplanar waveguide) (Figure 9.29). Synthesise a matching network for providing operation at 824–960 MHz.

9.6.12 Exercise 12

For exercise in Section 9.6.6, determine how each component may be helpful to fine-tune the frequency response. Take each component at a time and change it ±20% to determine how the reflection response changes.

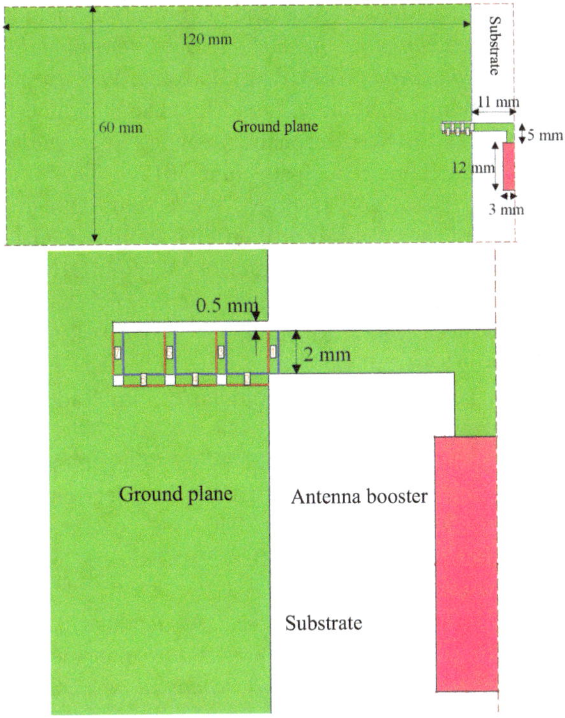

Figure 9.29 *(Top) Ground plane and 12 mm × 3 mm × 2.4 mm (height) antenna booster (conductive parallelepiped with conductivity $\sigma = 5.9 \times 10^7$ S/m), the substrate is 1 mm thick, $\varepsilon_r = 4.15$, and $\tan\delta = 0.017$. (Bottom) Detail of the coplanar line with 2 mm × 2 mm pads. Ports #1, 3, and 5 can be used to allocate series components, whereas ports #2, 4, and 6 are for shunt components. Port #7 is the feeding port.*

References

[1] J. Anguera, A. Andújar, G. Mestre, J. Rahola, and J. Juntunen, "Design of Multiband Antenna Systems for Wireless Devices Using Antenna Boosters," *IEEE Microwave Magazine*, Dec. pp. 102–114, 2019.

[2] J. Rahola, "Optimization of Matching Circuits for Antennas," *European Conference on Antennas and Propagation*, EuCAP 2011, Rome, April 2011.

[3] J. Rahola, "Power Waves and Conjugate Matching," *IEEE Transactions on Circuits and Systems—II: Express Briefs*, vol. 55, pp. 92–96, 2008.

[4] J. Rahola, "Estimating the Performance of Matching Circuits for Antennas," *European Conference on Antennas and Propagation*, EuCAP 2010 conference, Barcelona, April 12–16, 2010.

[5] J. Rahola, "Antenna Matching Circuit Optimization Including Stop Band Definitions," *Proceedings of the 2011 IEEE International Symposium on Antennas and Propagation* (APSURSI), Spokane, WA, July 3–8, 2011.
[6] J. Anguera, A. Andújar, J. Rahola, J. and Juntunen, "Synthesis of Multiband Matching Networks for Non-Resonant Antennas," *Computing and Electromagnetics CEM'17*, Barcelona, Spain, June 2017.
[7] J. Anguera, A. Andújar, D. Vye, and M. Wien, "Easy Design of IoT Devices with Antenna Boosters," Webinar at EDICON, Beijing, China, October 2020. Available https://www.youtube.com/watch?v=W3WpMD1MUw0&t=815s
[8] J. Anguera, A. Andújar, C. Puente, and J. Mumbrú, "Antennaless Wireless Device Capable of Operation in Multiple Frequency Regions," *Patent US.* 8,237,615 B2, Aug. 2012.
[9] J. Anguera, N. Toporcer, and A. Andújar, "Slim Radiating Systems for Electronic Devices," *Patent US* 9,960,478 B2, May 2018.
[10] J. Anguera, A. Andújar, D. Vye, and M. Wien, "Easy Design of IoT Devices with Antenna Boosters", Webinar at EDICON (Electronic Design Innovation) at Beijing, China, 23 October 2020. See presentation: https://www.youtube.com/watch?v=W3WpMD1MUw0&t=1038s
[11] Velavu. https://www.velavu.com/
[12] A. Fernández, J. Rahola, J. Juntunen, A. Andújar, J. L. Pijoan, and J. Anguera, "Embedding Antenna Booster in Smart-metering Platforms," *European Conference on Antennas and Propagation*, EUCAP 2021, Madrid, Spain, March 2022.
[13] J. Rahola, J. Lappalainen and J. Juntunen, "Requirements for Accurate Design of Matching Circuits at 5 GHz," *2020 14th European Conference on Antennas and Propagation (EuCAP)*, 2020, pp. 1–5, DOI:10.23919/EuCAP48036.2020.9135840.
[14] N. Ramos, "Automatic Implementation of Multiband Matching Networks with Antenna Boosters," *Master Thesis* at Telecommunication Engineering of Universitat Ramon Llull, Barcelona, Spain. 2022. Directed by J. Anguera and A. Andújar.
[15] M. Viñas, "Automatic Design of Matching Networks for Antenna Boosters," *Master Thesis* at Telecommunication Engineering of Universitat Politècnica de Catalunya, Barcelona, Spain. 2022. Directed by J. Anguera and A. Andújar.

Chapter 10

Antenna booster technology versus other antenna technologies

This chapter compares the performance of antenna boosters with self-resonant antennas, FPC (Flexible Printed Circuit) antennas, and microstrip patch antennas. The comparison is not only made in terms of passive parameters such as total efficiency but also in terms of TRP (Total Radiated Power), TIS (Total Isotropic Sensitivity), and finally, TTFF (Time To First Fix) in the case of satellite communications.

The chapter has the following objectives:

1. To evaluate the performance in terms of total efficiency, TRP and TIS of a wireless device embedding an antenna booster and a self-resonant antenna.
2. To assess the advantages of antenna booster technology compared to a monopole antenna for single-band operation in terms of bandwidth, efficiency, and reuse across different device sizes.
3. To compare the size of an antenna booster for multiband performance with other antenna techniques.
4. To show the advantages of antenna booster technology compared to FPC antennas.
5. To demonstrate the relevance of quasi-isotropic radiation patterns of wireless devices embedding antenna boosters compared to microstrip patch antennas. In particular, what is the impact of TTFF on the reception of satellite signals?

10.1 Introduction

As deeply explained throughout the previous chapters, the performance of an antenna booster system within a wireless device can be optimised by the proper location of the antenna booster on the ground plane. Such location enables the excitation of currents in the ground plane, and, as a result, bandwidth and efficiency are competitive. Since the nature of the antenna booster is reactive, a matching network is needed to provide operation at the frequency band or bands of operation. As shown in the book, this design process is simple, fast, and software-driven since matching network synthesis tools can address the design of the matching networks.

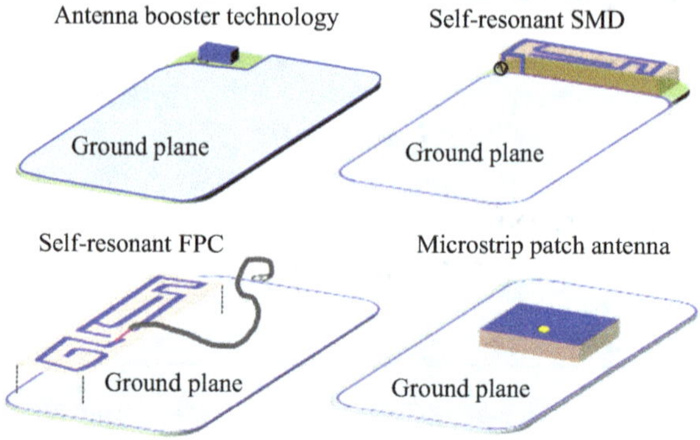

Figure 10.1 Antenna booster technology compared to self-resonant SMD, FPC, and microstrip patch antenna

This chapter compares antenna booster technology with self-resonant SMD (Surface Mounted Device) antennas, FPC, and microstrip patch antennas (Figure 10.1).

The chapter is divided as follows: In Section 10.2, antenna booster technology is compared to a self-resonant antenna in an actual device. Then, an antenna booster is compared to a monopole antenna for single-band operation in terms of bandwidth and reuse of the solution across different ground plane sizes in Section 10.3. In Section 10.4, a general comparison is made among several techniques based on self-resonant antennas. Section 10.5 deals with the advantages of antenna boosters compared to FPC antennas. A discussion on the benefits of antenna boosters versus microstrip patch antennas is detailed in Section 10.6. Section 10.7 gives practical takeaways. In Section 10.8, a list of exercises is proposed. Finally, the References section includes references for further reading.

10.2 Antenna boosters versus a self-resonant antenna in a real device

A 12 mm × 3 mm × 2.4 mm (height) antenna booster is proposed for multiband operation in a smartphone for operating 824–960 MHz, 1710–2170 MHz, and 2500–2690 MHz. The antenna booster is compared in a free-space scenario in terms of total efficiency, total radiated power (TRP), and total isotropic sensitivity (TIS), with the original antenna of the smartphone having a volume of 707 mm^3. In addition, field test measurements are carried out to compare the performance of both antenna solutions in real-life situations, particularly when a user holds the smartphone in the talking position.

10.2.1 Free-space comparison

As demonstrated in the book, an antenna booster is a small, non-resonant element able to excite currents on the ground plane, which cannot be considered an antenna but an exciter of currents in the ground plane, becoming the radiating element. Its position on the PCB matters, being the corner of the ground plane in the preferred position to obtain better bandwidth and efficiency (Chapter 3). Other small antenna technologies, such as PIFA (Printed Inverted-F Antennas) and monopoles, also employ the ground plane. However, this is pushed to its limits for the antenna booster technology since there is no need to use a ten times larger antenna such as PIFA. Thus, the contribution to the radiation process is primarily due to the ground plane using a tiny antenna booster instead of a classic antenna, which is much more prominent in size. One of the questions that this feature raises is: do those currents excited on the ground plane negatively impact the sensitivity of the antenna system? To answer the question, the following experiment compares a self-resonant multiband 707 mm^3 monopole antenna with an 86.4 mm^3 (12 mm × 3 mm × 2.4 mm height) antenna booster in terms of TRP and TIS (Chapter 1) [1,2].

First, both antenna systems are measured in terms of total efficiency regarding the original antenna and the antenna booster in an anechoic chamber (Satimo Stargate-32). Total efficiency (η_t) takes into account both radiation efficiency (η_r) and mismatch losses as follows: $\eta_t = \eta_r \cdot (1 - |S_{11}|^2)$. The results show that total efficiency is almost the same at the low bands, being slightly better in the case of the antenna booster element. For the high-frequency bands, the total efficiency with the antenna booster element is well above the original, particularly for the frequency range from 1710 to 2400 MHz. At 2500–2690 MHz, the total efficiency peak is better for the original antenna. However, the total efficiency peak for the antenna booster element is more balanced within the 2500–2690 MHz frequency range.

It is observed that efficiencies are better at the high bands ($f > 1.71$ GHz) than at the lower bands because the ground plane is electrically smaller at the lower bands (824–960 MHz) (Figure 10.2).

The antenna booster is integrated into the active smartphone based on these passive results, including the designed matching network. In this case, a micro-coaxial 50 Ω transmission line connects the input/output port of the front-end module with the input port of the matching network. TRP and TIS are measured inside an anechoic chamber, and the values are compared with the original antenna (Figure 10.3).

From the active results, several conclusions are drawn:

- At the low-frequency bands, TRP is almost the same, which is aligned with the total efficiency results. For TIS, the same trend is observed.
- For the high bands, TRP is around 1 dB better for the antenna booster. TIS is better in nearly all channels by 1 dB or 2 dB.

Based on this data, the smartphone operates similarly in the low bands but better in the high-frequency bands with the booster-based solution. It is important to emphasise that the volume of the original antenna is 707 mm^3, whereas it is only 86.4 mm^3 for the antenna booster system. At this point, it is interesting to analyse

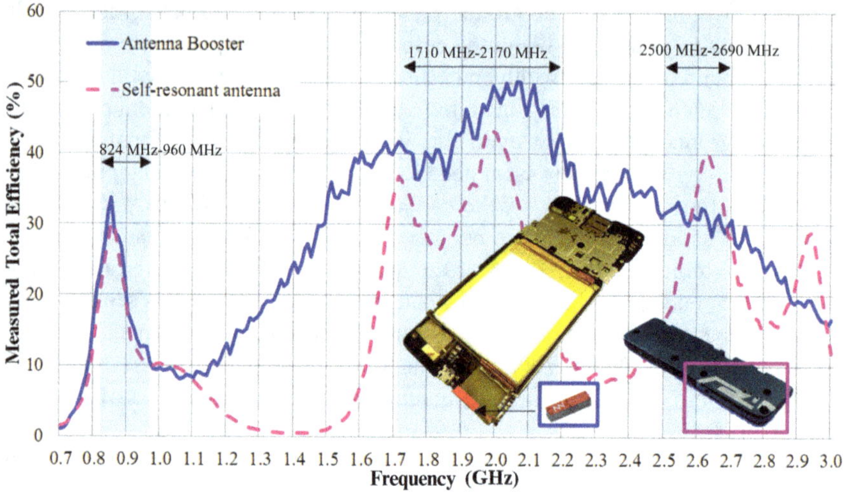

Figure 10.2 Measured total efficiency for the original self-resonant antenna and the antenna booster integrated into the real smartphone

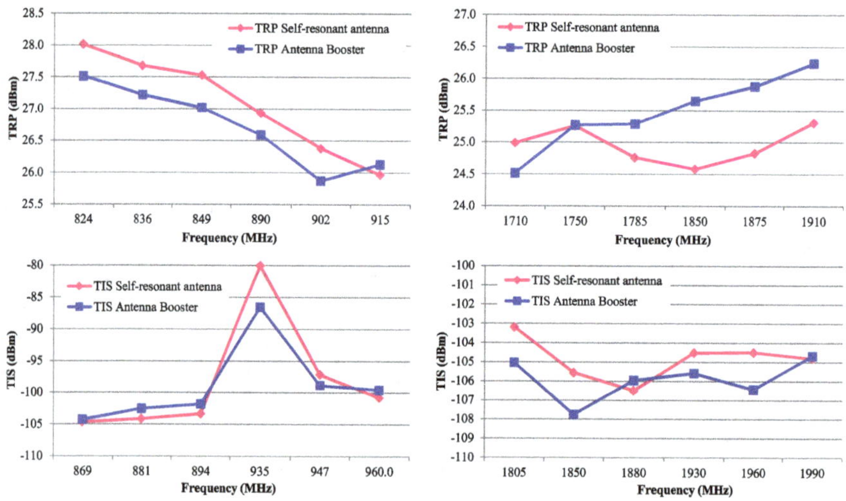

Figure 10.3 Comparison of the antenna booster with the original monopole antenna: top left is TRP for the 824–960 MHz and the top right for the 1710–1990 MHz. The bottom left is TIS for the low bands, and the right is for the high bands.

how the antenna booster system behaves under a real scenario regarding the talking position, where the head and hand effects are considered. The following section analyses both antenna solutions in a real-life scenario.

Excitation of currents in the ground plane and TIS

The excitation of currents in the ground plane due to an antenna booster has no negative implication on deteriorating TIS. On the contrary, it has been proven that TIS is comparable (slightly better) for an antenna booster than that of a self-resonant monopole.

10.2.2 Talking-position comparison

Besides evaluating both antenna solutions in terms of electromagnetic parameters such as total efficiency and TRP/TIS, another interesting and complementary method of characterisation consists of comparing both antenna systems in a real-life scenario. In this regard, a field test is carried out, which measures the received power when a user holds the phone in a talking position. To acquire the received power, an application called GSM field is used. This application generates a position vector and the received power for each position (Figures 10.4 and 10.5). A semi-urban scenario has been considered. Four case studies are analysed: one for 900 MHz, one for 2 GHz, and each one for the original antenna and the antenna booster system. For each case, the user is at walking speed, holding the smartphone

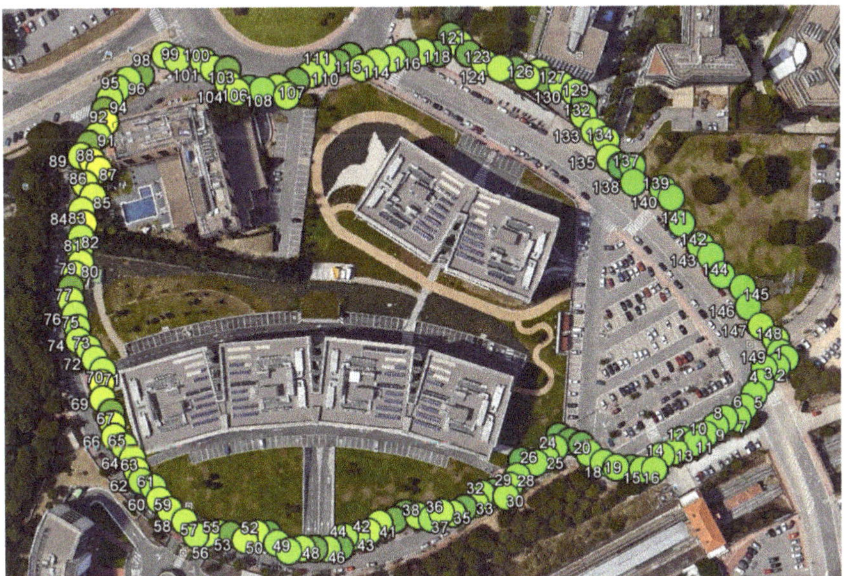

Figure 10.4 A software application allows obtaining the received power in field-test measurements in the real smartphone as a function of the user position. The field test comprises buildings, the main road, train tracks, and a forest. The whole path is 1,500 m.

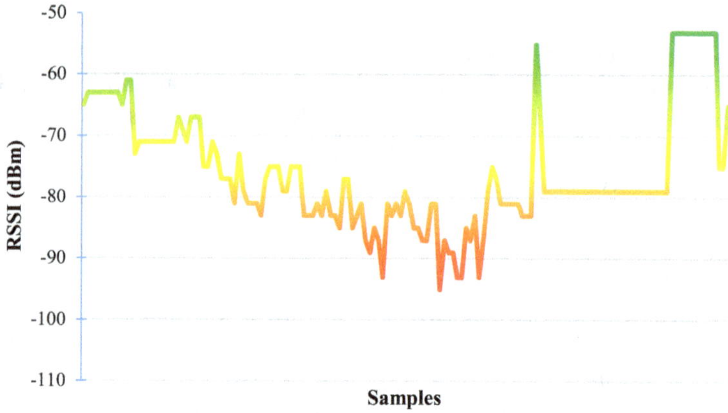

Figure 10.5 Received power as a function of space (close path shown in Figure 10.4) at GSM900 for antenna booster system

Table 10.1 Averaged received power

Average Rx. power (dBm)	Self-resonant antenna	Antenna booster
900 MHz	−64.0	−63.4
2 GHz	−81.6	−75.4

in a talking position. In this way, the human head-and-hand effect is considered in the measurement.

When the user walks in a closed path (Figure 10.4), the received power changes from levels close to −60 dBm down to −100 dBm (Figure 10.5). The average received power is calculated once the received vector is obtained (Table 10.1). The results conclude that the performance of the low-frequency band (900 MHz) is almost the same for both the original and antenna booster systems. However, at the high-frequency band (2 GHz), the received power is around 6 dB better for the antenna booster system than for the original self-resonant antenna. This is again aligned with the passive and active results, where the antenna booster element outperforms the original antenna (Figure 10.2).

A small antenna booster of only 12 mm × 3 mm × 2.4 mm (86.4 mm^3) has been proposed for multiband operation in a wireless device for 824–960 MHz, 1710–2170 MHz, and 2500–2690 MHz. The performance is compared with passive and active data, with the original self-resonant antenna having a much larger volume (707 mm^3). The passive and active results conclude that performance is similar at the lower frequency bands and better, both in efficiency, TRP, and TIS, at the high-frequency bands.

A field test was conducted where both the original self-resonant antenna and the antenna booster system were compared when a user performed a call and held the smartphone in a talking position. In this scenario, both the human head and hand are considered. The received power in the field test reveals that performance is similar in the low-frequency bands and 6 dB better in the high bands.

The results show that although antenna boosters excite currents on the ground plane, TIS is as competitive (and even better) as conventional monopole solutions.

10.3 Antenna booster versus a monopole antenna for 900 MHz

The objective is to compare the behaviour of an antenna booster with that of a monopole antenna under the same conditions. To do this, a reference 120 mm × 60 mm ground plane plus 40 mm × 11 mm with a small 20 mm × 11 mm ground clearance is recreated (Figure 10.6). On the one hand, the spiral monopole antenna is fine-tuned with a shunt inductor. On the other hand, the antenna booster is 12 mm × 3 mm × 2.4 mm (height). At 900 MHz, the antenna booster is non-resonant with a capacitive behaviour. Hence, a simple Lseries–Lshunt matching network is designed [3,4].

The results of the antenna designs are 863–928 MHz. However, the antenna booster bandwidth is larger (17.4%) than the monopole antenna (9.5%), making it attractive since it can allocate more radio protocols beyond the 863–928 MHz frequency range and becomes more robust due to detuning provoked by nearby components of wireless devices or due to changes in the relative permittivity of the substrate. Note that these results are aligned with those in Chapter 3, Section 3.8.1, Figure 3.27, where it was demonstrated that an antenna booster obtains more

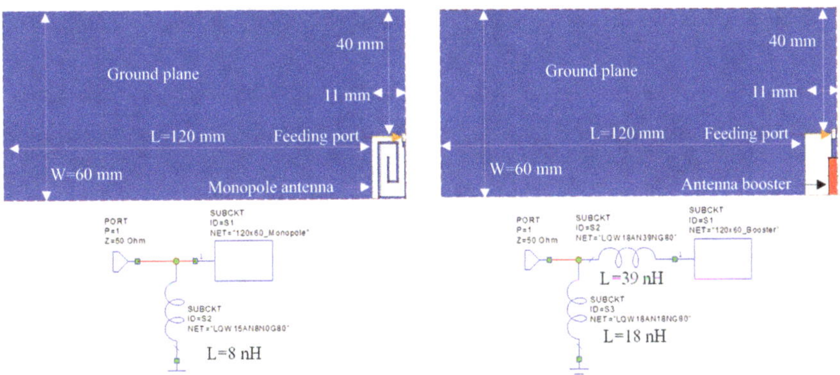

Figure 10.6 (Left) Monopole antenna on a small clearance area of 20 mm × 11 mm with its matching network connected to a 120 mm × 60 mm + 40 mm × 11 mm ground plane. (Right) 12 mm × 3 mm × 2.4 mm (h) antenna booster on the same small clearance as the monopole. The ground plane is printed on a FR4 substrate 1 mm thick, $\varepsilon_r = 4.15$, and $\tan\delta = 0.017$.

bandwidth and, thus, makes better and more efficient use of the ground plane to achieve bandwidth and efficiency.

The size of the ground plane is changed from a small one of 20 mm × 30 mm to 200 mm × 200 mm while keeping the same clearance area for both cases, the same as the one for the reference 120 mm × 60 mm ground plane. The total number of simulated ground planes is 342.

The worst reflection coefficient within the 863–928 MHz range is stored for each ground plane and then represented in a 2D plot. These results have been displayed in a thermal graph, where the warmer colours mean poor matching, and the cooler ones are better ($S_{11} \leq -6$ dB). Moreover, the edge with $S_{11} = -6$ dB is defined by a black line (Figure 10.8). The ground planes providing an $S_{11} \leq -6$ dB within 863–928 MHz are within the area inside the contours defined in Figure 10.8 called the *cluster*.

Remarkably, the antenna booster is well-matched ($S_{11} \leq -6$ dB) with many more different ground plane sizes than the monopole antenna (Figure 10.8): the antenna booster can be reused across 53.8% of cases, whereas the monopole antenna can only be reused 4.6%. For example, if the original 120 mm × 60 mm ground plane is reduced to 90 mm × 60 mm, the antenna booster still keeps an $S_{11} \leq -6$ dB across 863–928 MHz, whereas the monopole antenna degrades to −4 dB. It is clearly shown from the cluster shapes (Figure 10.8) that the antenna booster is robust to ground plane changes, whereas the monopole is sensitive to changes.

The shape of the antenna booster cluster is also intriguing, as it can be extrapolated to form a cross delimited by the −6 dB curve since the antenna is adapted within this zone. Moreover, as mentioned earlier, this antenna was designed on a 120 mm × 60 mm ground plane, which is why the coldest zones (better matching) are where L = 120 mm and W = 60 mm or vice versa, indicating that it also has symmetry. The worst ones are in the corners, except the top right corner.

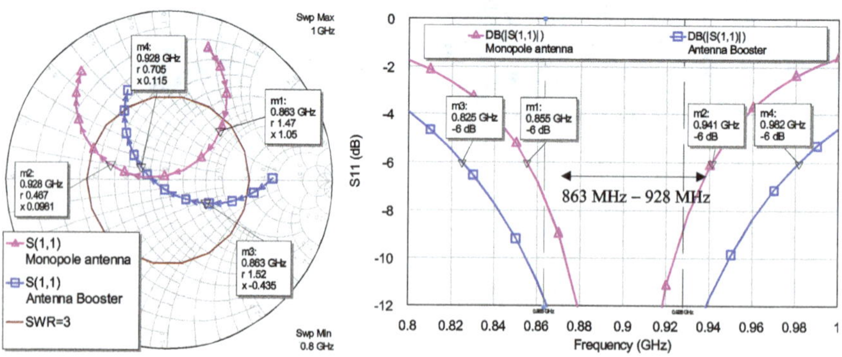

Figure 10.7 (Left) Simulated input impedance for the monopole antenna and antenna booster, each with its matching network, as shown in Figure 10.6. (Right) S_{11} shows that the antenna booster features 1.8 times more bandwidth.

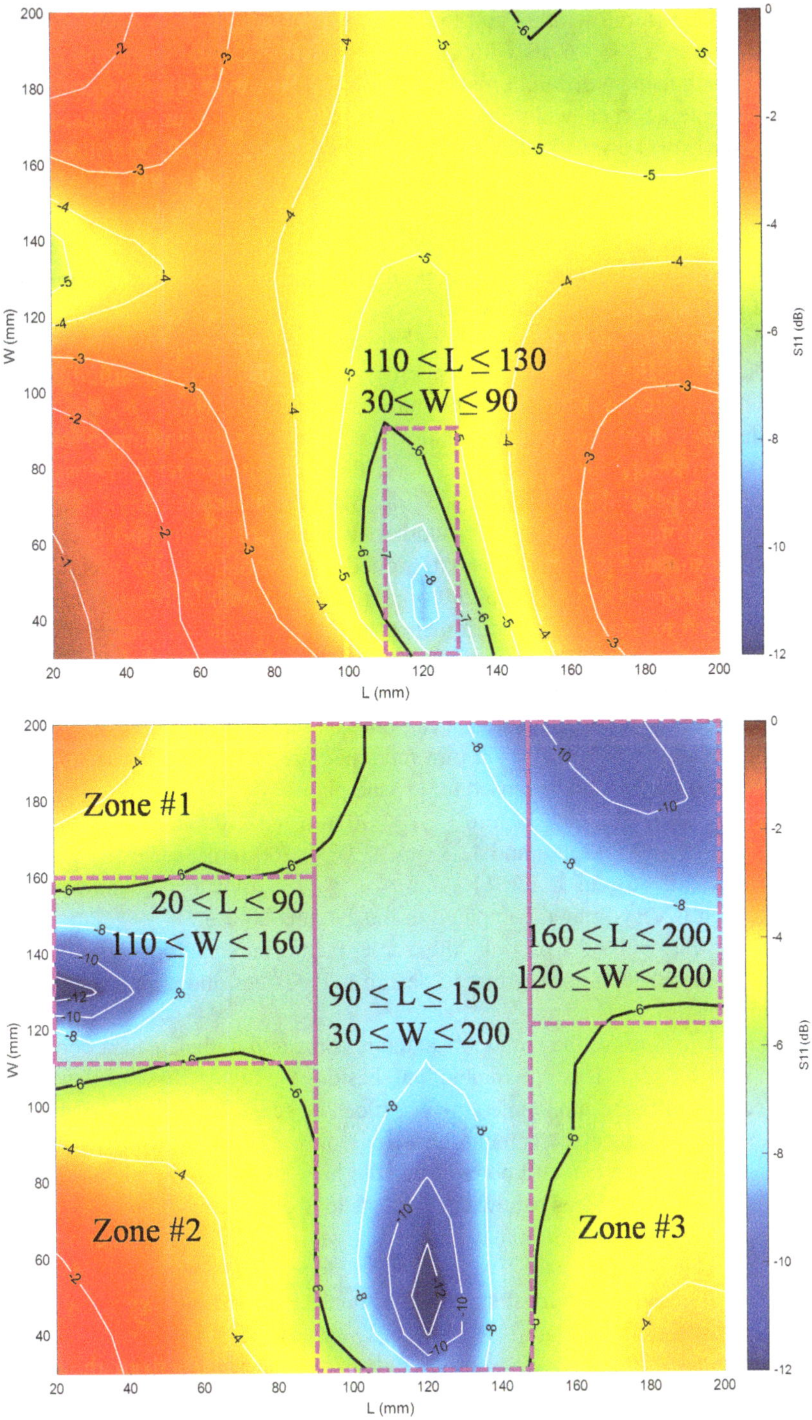

Figure 10.8 Worst S_{11} in the 863–928 MHz frequency range for the monopole antenna (top) and antenna booster (bottom) as a function of the length L and width W of the ground plane

The top left (large widths) and the bottom right part (large lengths), although not providing an $S_{11} < -6$ dB with the current Lseries–Lshunt matching network, can be matched using a different matching network. However, the most critical area is the bottom left corner, which includes electrically small ground planes. It is left as an exercise on how to match off-cluster areas (zones #1, 2, 3) following the teachings of Chapter 4). In particular, zones #1 and #3 need a different matching topology (antenna booster connected to a series L and shunt C). Zone #2 (small PCB) requires special treatment since bandwidth decreases when the ground plane becomes smaller. Therefore, broad banding mechanisms or reconfigurable solutions are needed to achieve bandwidth.

A monopole antenna and an antenna booster have been compared for operating at 863–928 MHz. Both solutions present an $S_{11} \leq -6$ dB on this specific frequency band on the reference 120 mm × 60 mm ground plane. However, the antenna booster features more bandwidth at 900 MHz: 17.4% for the antenna booster and 9.5% for the monopole antenna. Therefore, the antenna booster can absorb new radio protocols beyond those in the 863–928 MHz frequency band.

Besides having more bandwidth, another advantage of the antenna booster over the monopole antenna is its robustness to ground plane variations. 342 ground planes ranging from 20 mm × 30 mm to 200 mm × 200 mm with increments of 10 mm have been used to analyse the performance of the monopole antenna and antenna booster. In all cases, the antenna set-up is the same: same antenna element and same matching network. The analysis shows that the antenna booster can be reused at 53.8%, whereas the monopole antenna can only be reused at 4.6% (only for ground planes similar to the 120 mm × 60 mm reference ground plane). For example, the antenna booster system can be reused for ground planes with 90 mm $\leq L \leq$ 150 mm and 30 mm $\leq W \leq$ 200 mm, 20 mm $\leq L \leq$ 90 mm and 110 mm $\leq W \leq$ 160 mm, and 160 mm $\leq L \leq$ 200 mm and 120 mm $\leq W \leq$ 200 mm. However, the monopole antenna can be reused for 110 mm $\leq L \leq$ 130 mm and 30 mm $\leq W \leq$ 90 mm, which is very limited compared to the antenna booster.

Keeping the bill of materials the same (same antenna booster and same matching network) reduces the time to design IoT devices and simplifies procurement and logistics.

The main features of the monopole antenna compared to the antenna booster are gathered in Table 10.2. It is remarkable that, besides the higher reuse of the antenna booster compared to the monopole antenna, it can be reused if the ground clearance is further reduced. Since the antenna booster is only 12 mm × 3 mm × 2.4 mm, ground plane clearance can still be reduced. We have to keep in mind that reducing the ground clearance reduces bandwidth. Still, as long as the bandwidth is enough for the application, such a reduction may enable a larger integration of other components on the PCB. This is not the case for the monopole antenna. If the ground clearance is reduced, the design process must start from scratch. For example, when the ground clearance is reduced from 20 mm × 11 mm (220 mm^2) to 15 mm × 9 mm (135 mm^2) for the reference 120 mm × 60 mm ground plane, the bandwidth of the antenna booster decreases from 17.4% to 11.5%. Although smaller, it is still better than the monopole antenna on the original clearance.

Table 10.2 Comparison between the monopole antenna and the antenna booster. The red and black arrows on the current distribution for the monopole antenna have been intentionally added to emphasise the out-of-phase currents. Ground plane clearance for both cases is 20 mm × 11 mm.

	Monopole antenna	**Antenna booster**
Current distribution at 900 MHz (The ground plane is partially shown; the ground plane is 120 mm × 60 mm for both cases)	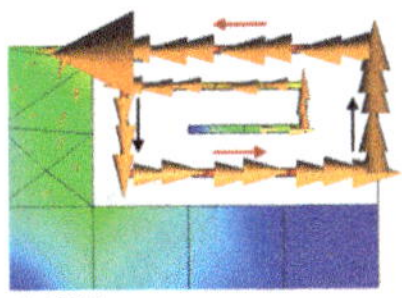	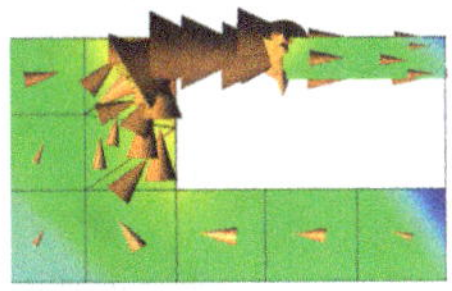
Reuse ($S_{11} < -6$ dB)	4.6%	53.8%
Total efficiency	Lower due to out-of-phase current portions	Higher. e.g., 1.2 dB more for a 60 mm × 60 mm ground plane
BW at 900 MHz for a 120 mm × 60 mm ground plane	9.5%	17.4%
Reducing the clearance area	Redesign the antenna geometry	Only change the matching network
Impact of the PCB substrate on frequency shifting when changing the ε_r = 4.15–4.7	2.3%	1.3%
Design process	Requires an iterative process	Fast: the antenna booster is an off-the-shelf component and also due to matching network synthesis

The total efficiency is superior for the antenna booster case. For example, for a 20 mm × 80 mm ground plane size, the antenna booster is 2.3 dB above the efficiency of the monopole.

Besides the spiral monopole, the antenna booster has also been compared to a meander and a fractal-shaped antenna inspired by the Hilbert curve. The results show the same trend: reuse for meander and fractal-based is 0.6% and 10.5%, respectively compared to 53.8% for the antenna booster.

10.4 Antenna boosters versus multiband self-resonant antennas

Now that the entire design process of a multiband antenna system that combines an antenna booster with the systematic design of matching networks has been explained (Chapter 5), it is worth comparing the advantages of the proposed antenna booster design method with a traditional approach based on self-resonant antennas relying on complex geometries (Table 10.3 and Figure 10.9). The comparison considers the same frequency operation for both cases: 824–960 MHz and 1710–2690 MHz and passive solutions; that is, no reconfigurability is involved.

Table 10.3 Comparison between a self-resonant antenna and an antenna booster design. (*)An antenna engineer required 20 simulations [5].

824–960 MHz and 1710–2690 MHz		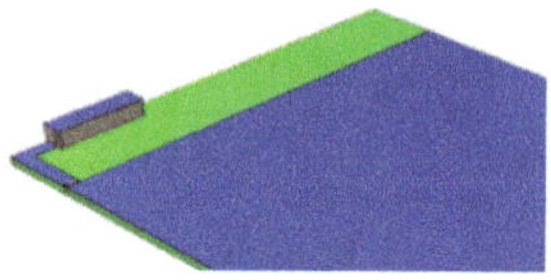
Size	35 mm × 11 mm × 2.4 mm	12 mm × 3 mm × 2.4 mm
Volume	924 mm^3	86.4 mm^3
Simulation design time	1. 20$^{(*)}$ full 3D electromagnetic simulations to find a geometry to match the frequency bands: 100 minutes 2. Matching network for fine-tuning: 5 minutes Total simulation time: 105 min	1. One full 3D electromagnetic simulation to obtain [S] parameters: 5 min 2. Matching optimisation: 5 minutes (for a multiband matching network like the ones shown here) Total simulation time: 10 min
Lab design time	Slow (days). Need to tune the antenna by adding and removing conductive portions. Once the geometry is obtained, a new antenna prototype must be fabricated.	Fast (hours). Only need to tune the matching network. If the simulation is accurate enough, there is no need to adjust the matching network.
Knowledge to design	Antenna engineer	Wide variety of engineers: not exclusively antenna engineers but also wireless engineers, RF engineers, microwave engineers
Can it be reused for any frequency band?	No, a new geometry must be designed. Only can be reused for frequencies close to the original design.	Yes, only the matching network must be designed.
Easy to integrate	Complex: If the device's width is less than 30 mm, it becomes difficult due to the antenna size. Also, due to its size, it is more complicated.	Easy, its small footprint makes it flexible to place in small devices.
Usable for MIMO	Limited to two antennas due to their large size.	Easy for higher-order MIMO due to its small size [6].

First, in terms of size, the antenna booster for multiband operation is less than a tenth the size of a customised antenna. It is also interesting to analyse other bespoke antennas operating at the same bandwidth. For instance, from several representative antennas that appeared in references [7–16], the average volume for operation at 824–960 MHz and 1710–2690 MHz is 971 mm^3, demonstrating that the customised antenna design presented here with a volume of 924 mm^3 is very representative (Figure 10.9). The size L, W is calculated as the smallest L, W of an imaginary parallelepiped enclosing the antenna.

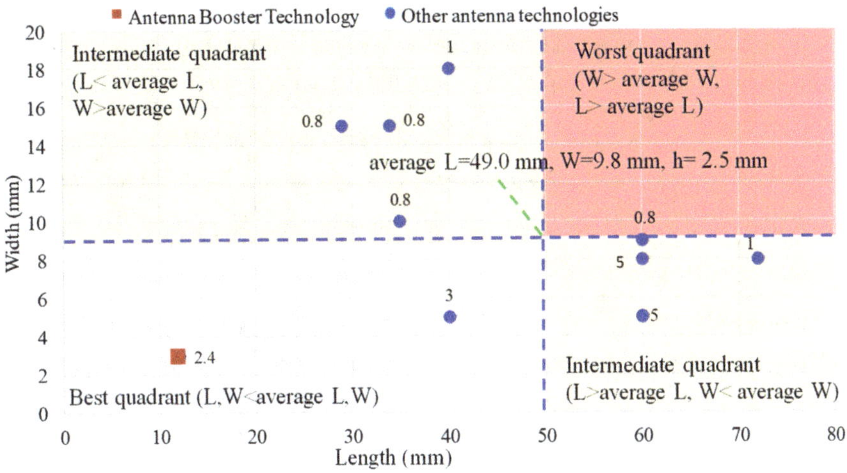

Figure 10.9 Size comparison of a 12 mm × 3 mm × 2.4 mm antenna booster with other antennas in the prior art for operating approximately at 824–960 MHz and 1710–2690 MHz. The number associated with each dot represents the height of the antenna element.

As observed in Figure 10.9, as far as size is concerned, the 12 mm × 3 mm × 2.4 mm antenna booster presents a length and width much below the average length L = 49.0 mm and average width W = 9.8 mm. Even in the best quadrant features, the 40 mm × 5 mm × 3 mm is more than three times larger than the antenna booster.

Regarding design in the simulation domain, a customised antenna must be designed from scratch to create several resonances at the frequency bands of interest. This requires several steps to adjust the antenna geometry carefully. Twenty steps were needed to obtain the geometry for the resonant antenna presented here. However, only a single electromagnetic simulation is required for the antenna booster to generate the S_{11}; then, all the work to design the matching network is carried out automatically. This is also translated to the laboratory domain since the geometry of the self-resonant antenna needs to be adjusted, either by cutting some conductive traces or adding them. This also takes time, and a final prototype must be fabricated again. This procedure, however, is not required for the antenna booster since, once the chip is made, the impedance adjustment is made only with the matching network, making the design process faster and more cost-effective.

Small, easy design and easy to use

Antenna boosters are tiny elements that simplify the embedding into a wireless device compared to self-resonant antennas.

Embedding an antenna booster is also easy: typically, the antenna booster should be at the corner of the ground place to obtain more bandwidth/efficiency (for magnetic boosters, other locations such as the centre of the ground plane edge can be used as well, Chapter 3, Figure 3.23). Then, matching can be easily designed to freely select the bands of operation regardless of the form factor of the wireless device and the frequency of operation.

With respect to expertise, a self-resonant antenna requires antenna engineering knowledge. This is a bottleneck for wireless/microwave engineers designing RF modules because they need more expertise to develop their antennas. They can easily use off-the-shelf filters, amplifiers, diplexers, front-end modules, and other components, but not antennas. However, the antenna booster is like another microwave component. Its operation is adjusted the same way a matching network is designed for an amplifier or a filter.

Once a self-resonant antenna relying on resonances is designed, it is difficult to reuse it for other platforms or frequency bands. However, this book proves that the same antenna booster can be easily used for different frequency bands and platforms.

Finally, as far as multiple-input/multiple-output (MIMO) is concerned, since the size of an antenna booster is small in terms of the operating wavelength ($\sim\lambda/30$), several antenna boosters can be allocated at different positions on the device to achieve high-order MIMO, as shown in Chapter 5.

It is worth emphasising the unique feature of the non-resonant nature of antenna boosters compared to the resonance behaviour of self-resonant antennas (Figure 10.10). As widely explained throughout the book, an antenna booster

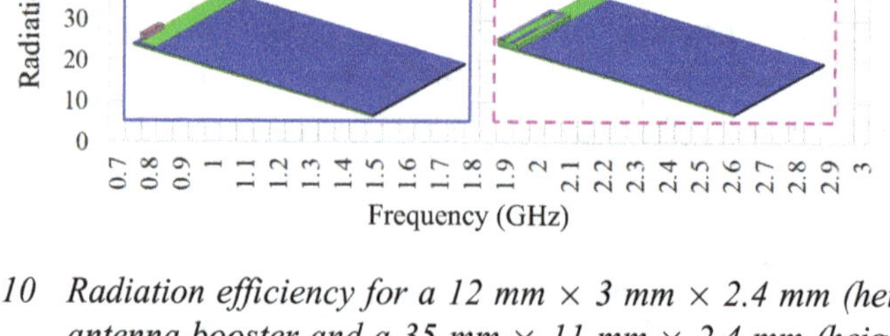

Figure 10.10 Radiation efficiency for a 12 mm × 3 mm × 2.4 mm (height) antenna booster and a 35 mm × 11 mm × 2.4 mm (height) self-resonant monopole with two branches both on a 120 mm × 60 mm ground plane

features a non-resonance impedance, at least at the lowest frequency of operation. This fact guarantees that radiation efficiency is quite flat over a wide range of frequencies. This allows the antenna to be efficient when matched at the desired frequencies of operation. However, self-resonant antennas feature resonance at the frequency of operation. The drawback of such an approach is clearly understood when plotting the radiation efficiency across frequencies. At resonance, the self-resonant antenna features maximum radiation efficiency. However, between those resonances, a valley appears. Therefore, the antenna can be efficient at the frequency regions where it was initially designed, but not out of resonance. As a consequence, the antenna cannot be reused, as is the case of the antenna booster. This is a clear differentiator factor that makes the antenna booster easily operative across many frequency bands, either with passive or reconfigurable matching networks. In the Figure 10.10 example, the antenna booster can operate across the whole 0.7 GHz to 3 GHz range. However, the self-resonant antenna is efficient at the resonances but about 3 dB less efficient in the antiresonance at 1.35 GHz.

Note that efficiency drops for classic self-resonant antennas have also been discussed in Chapter 3, Section 3.8, Figure 3.28, where self-resonant antennas with different geometries have been compared to an antenna booster, reaching the same conclusions about the alternate of maximum and drops in efficiency for the self-resonant antennas.

This example shows that antenna booster technology is future-proof since the frequency of operation of any coming protocol can be adopted by changing the matching network. On the contrary, for a self-resonant approach, a new antenna must be designed.

10.5 Antenna boosters versus FPC

Two different technologies are compared: antenna boosters and flexible printed circuit antennas. The comparison is addressed from measured results concerning efficiency, concluding that despite being the antenna booster more than fifty times smaller in the area, it provides better efficiency across the frequency range of 698–960 MHz and 1710–2690 MHz across three different printed circuit boards (PCB): a big PCB of 131 mm × 60 mm, a medium PCB of 95 mm × 42 mm, and a small PCB of 65 mm × 42 mm.

10.5.1 Antenna boosters versus off-PCB FPC antennas

The comparison is made at the frequency range of 698–960 MHz and 1710–2690 MHz, where most commercial wireless protocols are allocated.

The embedded antenna comprises a 12 mm × 3 mm × 2.4 mm (height) antenna booster mounted on a 131 mm × 60 mm PCB, including a 120 mm × 60 mm ground plane. An antenna booster is an electrically small ($\sim\lambda/30$) non-resonant element that is easily embedded on the PCB of a wireless device thanks to its SMD nature and small size (Figure 10.11 left). Being so small, an antenna booster is off-resonance, meaning it needs a matching network to provide a good impedance match to the RF

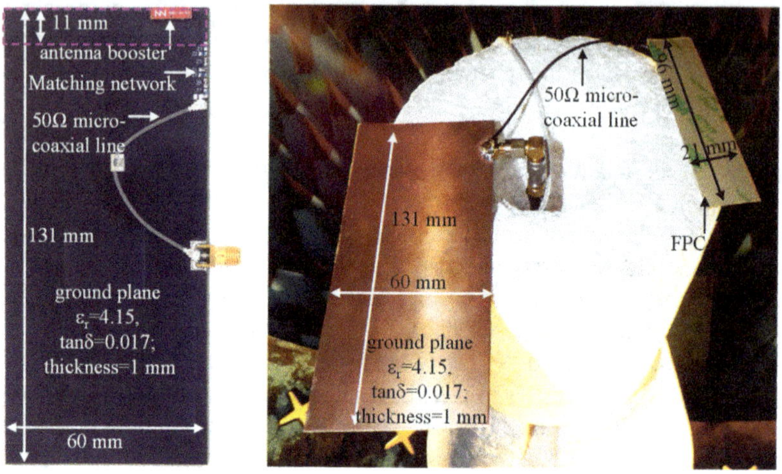

Figure 10.11 *Left: a 12 mm × 3 mm × 2.4 mm antenna booster on a 131 mm × 60 mm PCB comprising a 120 mm × 60 mm ground plane and a matching network with lumped SMD components for multiband performance at 698–960 MHz and 1710–2690 MHz. Ground clearance is 60 mm × 11 mm. Right: a 96 mm × 21 mm × 0.2 mm FPC antenna connected to the 131 mm × 60 mm PCB through a 50 Ω microcoaxial transmission line.*

front end [1]. By designing the proper matching network following network synthesis microwave tools [5], a multiband radiation performance is easily obtained from the booster-ground plane set.

For the present case, the antenna booster (a 3D conductive parallelepiped) is 12 mm × 3 mm × 2.4 mm (height), representing about $\lambda/35$ at 698 MHz.

The multi-band matching network for the antenna booster is easily designed using a network synthesis tool once the input impedance is measured. The network synthesis optimises total efficiency (η_t), which is the ratio between the power radiated to space and the available power from the generator. Optimising total efficiency and not return loss ensures that the selected lumped components of the matching network introduce low losses. The optimisation process considers available commercial high-Q SMD components. SMD 0402 and SMD 0603 high-quality factor components have been included (Figure 10.12).

The setup for the FPC antenna is simple: the FPC antenna is directly connected to the PCB through a 50 Ω microcoaxial, which is included with the FPC antenna (Figure 10.11 right). The feeding location on the PCB is maintained in the same position as the antenna booster. The FPC antenna is 96 mm × 21 mm, representing $\sim\lambda/5$ at the lowest frequency region (698 MHz). This maximum electrical size is about eight times larger than the size of the antenna booster. In terms of area, the antenna booster (12 mm × 3 mm) is 56 times smaller than the FPC antenna (96 mm × 21 mm).

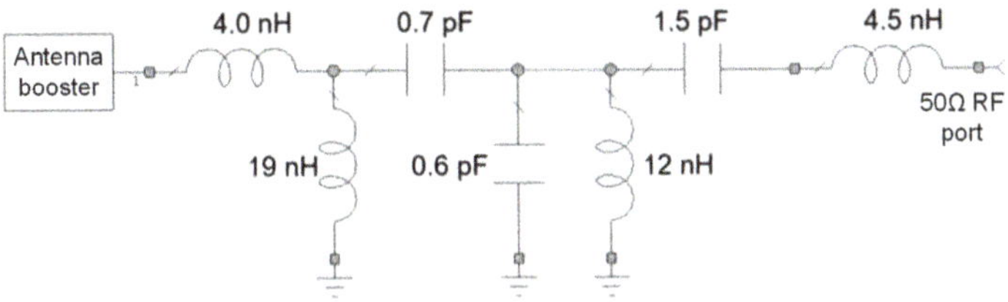

Figure 10.12 Multiband matching network using SMD inductors and capacitors. The minimum quality factor Q across the band for each component is: 4.0 nH Q_{min} = 63; 19 nH Q_{min} = 50; 0.7 pF Q_{min} = 260; 0.6 pF Q_{min} = 260; 12 nH Q_{min} = 82; 1.5 pF Q_{min} = 157; 4.5 nH Q_{min} = 56.

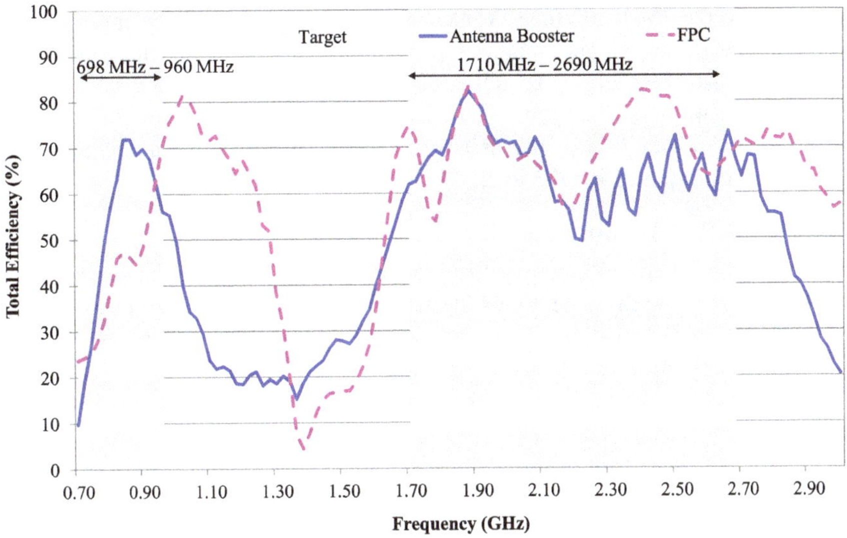

Figure 10.13 Measured total efficiency for the antenna booster and the FPC

Most FPC antennas rely on self-resonant structures. Therefore, there is no need to use a matching network [17]. After averaging seven multiband FPC antennas operating from 698 to 960 MHz and 1710–2690 MHz [18], the average length and width are 89.6 mm and 25.3 mm, respectively. Therefore, the FPC antenna analysed here is representative of an average FPC antenna.

To compare the performance of both antenna technologies, total efficiency is measured in an anechoic chamber (Satimo Stargate-32) using 3D pattern integration (Figure 10.13). Total efficiency considers both radiation efficiency η_r and impedance match as $\eta_t = \eta_r \cdot (1-|S_{11}|^2)$. Therefore, total efficiency is used as a figure of merit for comparison purposes.

Total efficiency shows similar average values (Table 10.4). For the FPC, several off-PCB locations have been tested to get the best performance, being the ones selected in the best-case scenario.

Table 10.4 Measured total efficiency for the FPC off-PCB and an antenna booster

Average total efficiency (%)	698 MHz–960 MHz	1710 MHz–2690 MHz
FPC	42.3	70.1
Antenna booster	51.3	65.7

The following section addresses the impact when the FPC antenna is mounted on a PCB.

10.5.2 Antenna booster versus on-PCB FPC antennas

Once the off-PCB results for the FPC have been presented, the FPC is embedded on the PCB, considering several heights and positions. In this regard, three situations have been considered: a big 131 mm × 60 mm PCB, a medium 95 mm × 42 mm PCB, and a small 65 mm × 42 mm PCB. Since wireless devices should be compact, the size of the device is typically as large as the PCB, thus not leaving space for having the FPC antenna off the PCB.

10.5.2.1 PCB 131 mm × 60 mm

Several heights of the FPC above the ground plane of the PCB have been considered: 2.4 mm (the same height as the antenna booster), 5 mm, 10 mm, 15 mm, and 20 mm (Figure 10.14 left). Since wireless devices should be small and present slim profiles, 20 mm is taken as the maximum height.

Besides height, different FPC antenna locations have also been analysed since the wireless designer has no degree of freedom to allocate the FPC antenna on the device. Three cases have been considered: top, middle, and bottom edge (Figure 10.14 right). Therefore, for the FPC antenna case, there are five height cases and three positions, giving, as a result, 15 different measurements. To simplify the results, the best location among the top, middle, and bottom edges for each FPC antenna height in terms of total efficiency is represented. As a result, there are five total efficiency measurements, one per FPC height and one total efficiency measurement for the antenna booster (Figure 10.15). This procedure is repeated for the medium and small PCBs.

When mounted on the PCB, the total efficiency results for the FPC antenna drastically change from the off-PCB case (Figure 10.15). The total efficiency for the FPC antenna at 2.4 mm height (the same height as the antenna booster) is only 0.8% in the low-frequency regions. In contrast, it is 52.0% for the antenna booster, representing a ratio of 18 dB better total efficiency for the antenna booster. Other values are summarised in Table 10.5, where it can be concluded that:

- The total efficiency of the antenna booster is always better at both frequency regions, even when the FPC antenna is 20 mm above the PCB.
- The efficiency of the FPC antenna improves as the height increases, from 0.8% at $h = 2.4$ mm to 14.8% at $h = 20$ mm in the low-frequency region and in a similar manner in the high-frequency region. These results are aligned with

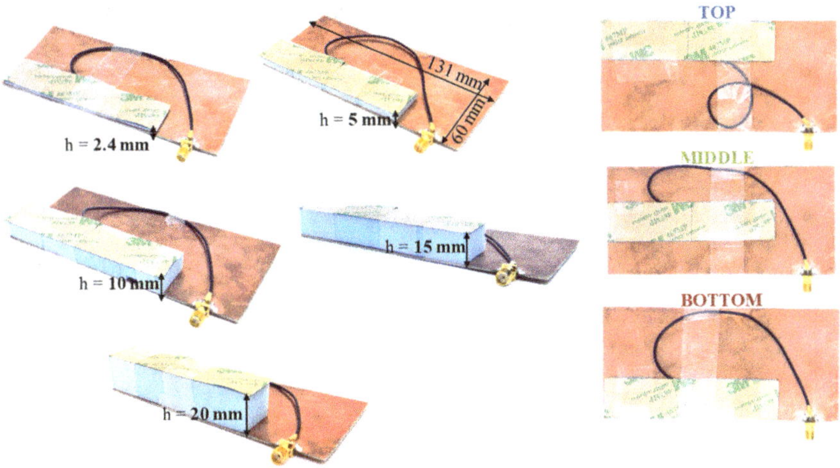

Figure 10.14 *(Left) The FPC antenna is mounted at different heights above the ground plane. A foam spacer is included. The PCB, including the ground plane, is 131 mm × 60 mm, which is the same size as the PCB size of the antenna booster. (Right) Three positions have been tested for each height shown on the left: top edge, middle edge, and bottom edge.*

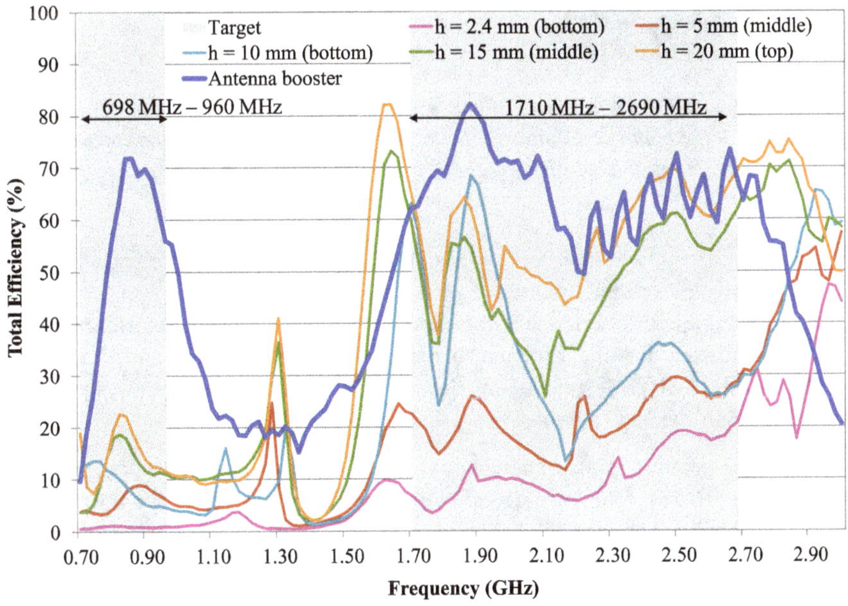

Figure 10.15 *Measured total efficiency for both the embedded antenna booster and the FPC antenna at different heights. Only the best efficiency among the three positions (top, middle, and bottom) is shown for each height of the FPC.*

Table 10.5 Measured total efficiency for the FPC antenna and the antenna booster

Average total efficiency (%)	698–960 MHz	1710–2690 MHz
FPC antenna h = 2.4 mm	0.8	10.9
FPC antenna h = 5 mm	5.7	21.3
FPC antenna h = 10 mm	9.0	34.5
FPC antenna h = 15 mm	11.7	47.5
FPC antenna h = 20 mm	14.8	56.4
Antenna Booster	52.0	65.7

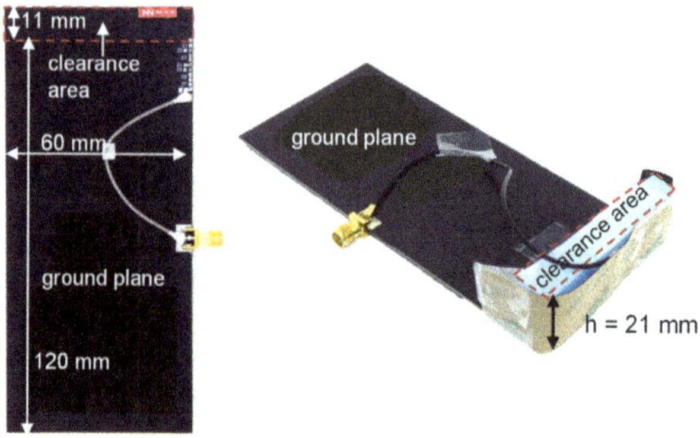

Figure 10.16 (Left) 12 mm × 3 mm × 2.4 mm antenna booster on a 131 mm × 60 mm PCB, including a matching network with SMD components. (Right) the 96 mm × 21 mm FPC antenna.

antenna theory, where a horizontal current element is not efficient when it is electrically close to a conductive plane [19].

- Comparing the FPC antenna at the same height as the antenna booster results in 18 dB better for the antenna booster at 698–960 MHz and 7.8 dB at 1710–2690 MHz, despite being the antenna booster eight times smaller in length than the FPC antenna.

The FPC antenna is also analysed when it is vertically oriented on the same PCB (Figure 10.16). The overall efficiency results once more demonstrate the poorest performance for the FPC antenna, particularly in the low-frequency range, where the antenna booster solution's measured average total efficiency differs by 4.3 dB (Figure 10.17 and Table 10.6). Furthermore, not only is the FPC antenna the worst, but it forces the device to be at least 21 mm in height, which goes against the direction of designing thin wireless devices. On the contrary, the antenna booster solution is only 2.4 mm in height, enabling very slim wireless devices.

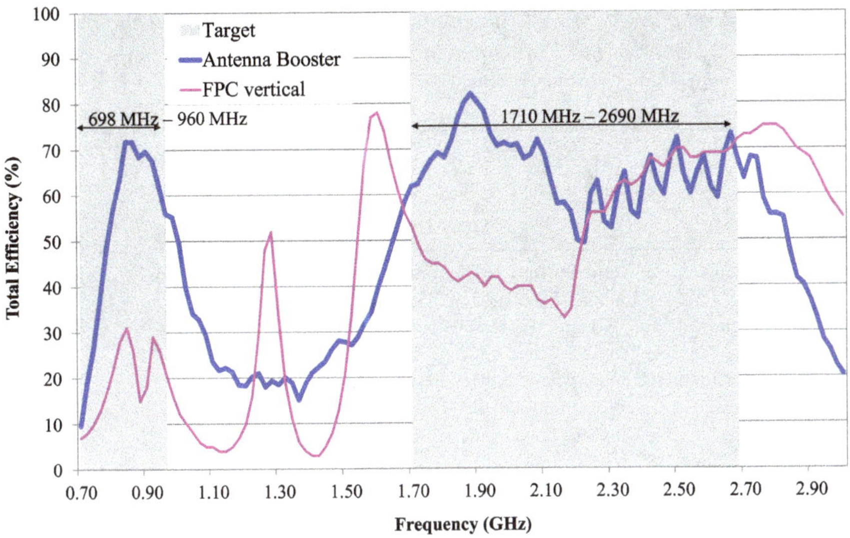

Figure 10.17 Measured total efficiency for the antenna booster and FPC solutions of Figure 10.16

Table 10.6 Measured total efficiency for the FPC and the antenna booster of Figure 10.16

Average total efficiency (%)	698–960 MHz	1710–2690 MHz
FPC antenna vertical position	19.3	52.6
Antenna Booster	52.0	65.7

The same analysis as the one in Figure 10.14 has been conducted for two other platforms: a medium size of 95 mm × 42 mm and a small one of 65 mm × 42 mm, considering the same setup as shown in the next section. For the 95 mm × 42 mm PCB case, the same matching network as the 131 mm × 60 mm PCB case has been reused. However, a new matching network has been synthesised for the 65 mm × 42 mm PCB, following the teachings of Chapters 5 and 9 [20].

10.5.2.2 Antenna booster versus FPC for different PCB sizes

As shown before, the FPC antenna changes its behaviour drastically from off-PCB conditions to on-PCB. Since wireless devices should be small and compact, the antenna is typically embedded within the device. On the one hand, the size of FPC antennas for multiband operation from 698 MHz to 2690 MHz is large. In the present case, the antenna area is 56 times larger (Table 10.7) than the antenna booster. Antenna boosters are off-the-shelf SMD components that are easily embedded in a PCB. Once its position on the PCB is optimised (usually the corner

Table 10.7 Comparison of FPC antennas with antenna booster technology. Schemes of the FPC antenna and antenna booster have the same scale.

	FPC antenna 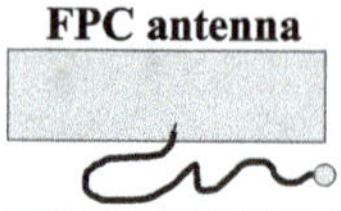	**Antenna booster**
Area	Big 96 mm × 21 mm Average size*: 89.6 mm × 25.3 mm	Small 12 mm × 3 mm
Automatic Pick and Place	No	Yes
Predictable performance	No, performance depends on microcoaxial and FPC antenna position	Yes
It fits in small wireless devices (<90 mm in length)	No	Yes
Enables thin devices	No, it needs to be about 20 mm over the PCB to obtain reasonable efficiency	Yes, the antenna booster height is only 2.4 mm
BoM	Adhesive, FPC, coaxial line and receptacles at the FPC antenna and PCB	Antenna booster and a matching network are all placed on the PCB

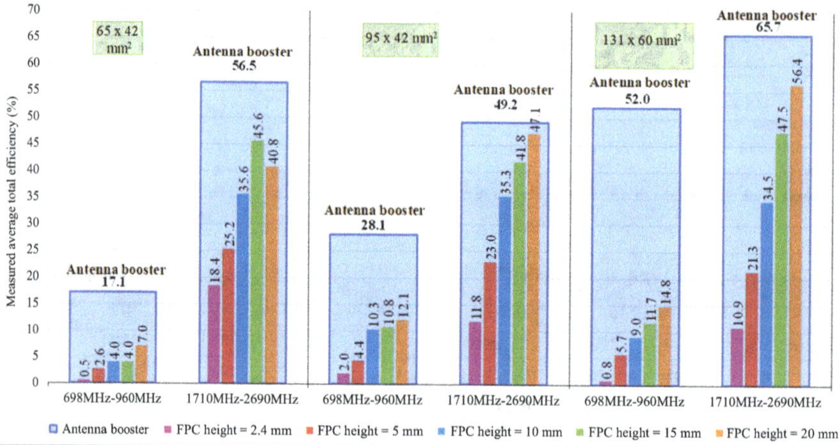

*The average size is obtained from seven antenna samples [18].

of the device provides the best bandwidth and efficiency [5]), the performance remains when mass-producing hundreds of thousands of devices. On the contrary, the FPC is more unpredictable since the position of the manually mounted FPC antenna impacts its performance.

From a bill of material (BoM) perspective, the antenna booster technology needs an antenna booster and lumped SMD components for the multiband matching network. On the other hand, the FPC needs an adhesive so the antenna can be attached to the casing of the device, a microcoaxial line with a connector, and a receptacle on the PCB to connect the antenna.

Finally, the performance is always better for the antenna booster despite being eight times smaller concerning its maximum length (or 56 times less area) for all PCB sizes and even when the FPC antenna is 20 mm over the ground plane PCB. The vast differences are significant in the low-frequency region (698–960 MHz). For example, taking as a reference the FPC antenna at 10 mm over the ground plane, the antenna booster is 6.3 dB, 4.3 dB, and 7.6 dB more efficient on average at the low-frequency region for the small, medium, and large PCB, respectively. In the high-frequency region (1710–2690 MHz), the difference is mild, but still, the antenna booster exceeds the performance of the FPC antenna.

As observed for both technologies, the ground plane impacts, especially in the low-frequency region. As the ground plane size increases, the total efficiency rises, too.

Based on these results, FPC antennas, like the one presented in this section, are large, and when placed above the ground plane of the device, the efficiency drastically drops. A further negative impact on the overall FPC performance can be expected when batteries are placed in the device since they are covered with a conductive material, thus decreasing the distance between the FPC and the conductive part of the battery. However, this falls outside the scope of this study.

Additionally, the performance is unpredictable since it depends on the ability of the human being to mount the antenna on the PCB of the device. On the contrary, embedded antenna boosters are convenient for designing wireless devices due to their small size, multiband behaviour, off-the-shelf nature, and compatibility with the pick-and-place assembly process, ensuring stability when mass-producing devices with embedded antennas.

In summary, when leaving the antenna selection for wireless devices to the end of the design cycle, by just putting a UFL connector on the PCB in the initial design stages, the risks of not being able to get acceptable performance increase materially, and the design effort of selecting and tuning an antenna booster is easily recovered given the price difference and manufacturing cost savings.

10.6 Linearly polarised antenna boosters versus circularly polarised microstrip patch antennas

The polarisation of an onboard satellite antenna typically features a circularly polarised antenna. Therefore, to minimise the PLF – Polarisation Loss Factor (see Chapter 1), the polarisation at the receiving antenna should also be circular. For wireless devices, a common way of achieving circular polarisation is using a microstrip patch. For a patch, two degenerate orthogonal 90º out-of-phase modes can be excited, and thus, a circular polarised is obtained. The nature of a microstrip patch makes a radiation pattern that is directed towards the perpendicular direction to the patch surface, and little radiation goes under the ground plane sustaining the patch antenna. This makes a patch antenna an ideal candidate when the antenna is pointing to the sky. However, what happens when the device with an embedded antenna is not pointing to the sky? What happens if the device is on the seat of a vehicle? What is the performance of both the circularly polarised patch antenna and antenna booster if the device is upside-down? Does the radiation pattern of a microstrip patch antenna, typically directive, become worse than that of a quasi-isotropic linearly polarised antenna booster when the orientation of the device is not known, or even worse, cannot be controlled? To address these questions, a comparison of circularly polarised microstrip patch antennas and linearly polarised antenna boosters is conducted to determine their performance, depending on the position and orientation of the wireless device inside a vehicle.

The antenna booster is 7.0 × 3.0 × 2.0 mm^3 and, centred at a frequency of 1575 MHz (Figure 10.18), radiates with a quasi-isotropic radiation pattern with linear polarisation [21].

The LNA (Low Noise Amplifier) included for these measurements is an external LNA attached to the device with SMA transitions, which introduces 0.3 dB losses. This LNA has a gain of 16.5 dB, a noise figure of 1.03 dB, and is centred at 1575 MHz.

Since the 7 mm × 3 mm × 2 mm GPS antenna booster is non-resonant at the frequency bands of operation ($\lambda/27$ at 1575 MHz), a matching network with lumped components (SMD 0402) is designed according to the principles studied in Chapter 3. Due to the narrow bandwidth needed for GPS (1.5%), the matching network was simple, comprising only three lumped components: series and shunt inductors (Table 10.8).

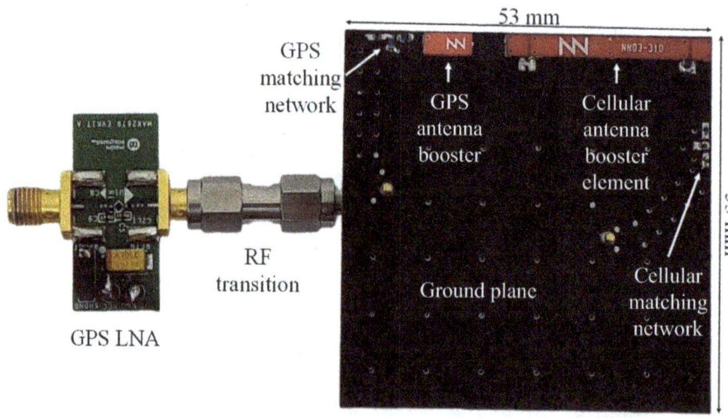

Figure 10.18 A 53 mm × 53 mm PCB with a linearly polarised GPS (Global Positioning System) antenna booster. A ground plane is printed over the FR4 substrate (1 mm thick, ε_r = 4.15, tanδ = 0.017). The chip antenna booster component on the right operates at 698–960 MHz and 1710–2170 MHz. The GPS antenna booster is 7 mm × 3 mm × 2 mm.

Table 10.8 Matching network for the GPS antenna booster

	GPS PORT	
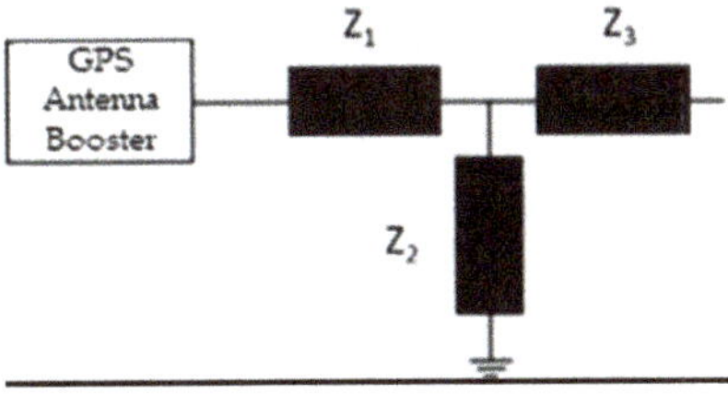	Z1	L = 15 nH, Q = 108
	Z2	L = 9.1 nH, Q = 117
	Z3	L = 2.2 nH, Q = 114

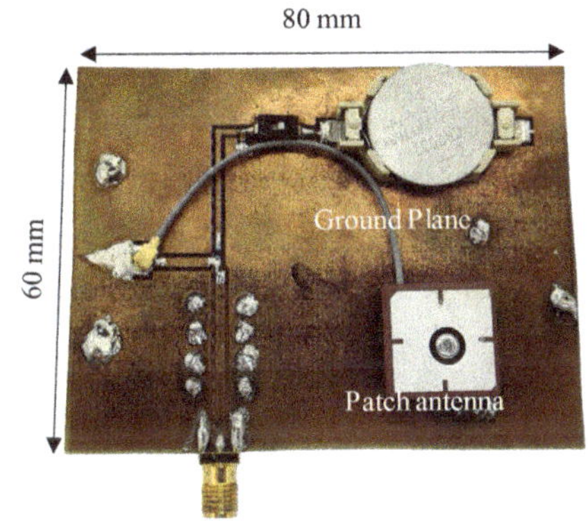

Figure 10.19 An 80 mm × 60 mm PCB with a circularly polarised microstrip patch antenna is employed for the study. The battery feeds the embedded LNA under the patch antenna.

Both evaluation boards are connected to two GPS receivers to carry out simultaneous measurements in different locations of a vehicle: on the front and rear dashboards and on top and under the seat. Besides the different locations, several orientations (up, upside down, and vertical) have been tested, resulting in 11 different scenarios for both antenna technologies.

A microstrip patch antenna with an embedded 16 dB gain LNA is utilised for the circularly polarised antenna case (Figure 10.19).

A 50 km route including highways, cities, and forests is selected to perform the averages and compare the two antenna technologies using the TTFF (Time To First Fix). Time to First Fix (TTFF) is a metric used in global navigation satellite systems (GNSS), such as GPS, to define the time it takes for a receiver to determine its initial position fix after being powered on or started. TTFF is typically measured from the moment the receiver is turned on or activated to the point when it calculates a valid position fix, which includes latitude, longitude, and often altitude.

The definition of TTFF can vary depending on the specific GNSS receiver and its capabilities, but in general, it represents the time taken for the receiver to acquire the necessary satellite signals, perform signal processing, and compute a reliable position fix. Shorter TTFF values are desirable because they indicate that the receiver can quickly determine its location, which is essential for applications like navigation and location-based services and minimises battery consumption [21].

The main conclusions are summarised next (Figures 10.20 and 10.21). Due to the quasi-isotopic nature of the radiation pattern of the antenna booster, the TTFF is less than that of the microstrip patch antennas at all positions. In Figure 10.20, the average TTFF is the average along the 50 km route, whereas the city average TTFF is the average along the city (wide avenues, narrow streets, including ten-story flats). As shown, the circularly polarised antenna is the worst (~10 s) compared to the linearly polarised antenna booster in urban scenarios.

According to the bill of materials, the matching network for the antenna booster technology requires lumped SMD components, an LNA, and an antenna booster. On the other hand, a ceramic patch, LNA, coaxial cable, and receptacles are necessary for the microstrip patch antenna (Table 10.9).

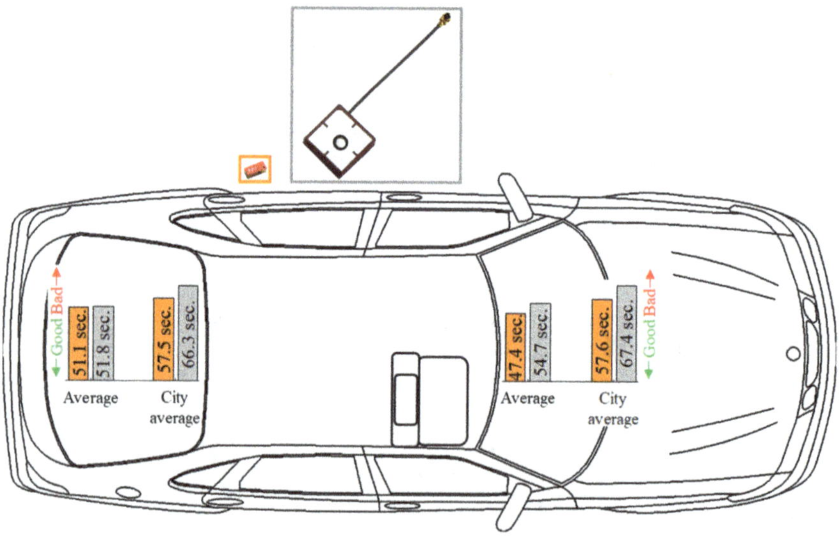

Figure 10.20 Average TTFF in seconds for the different positions for the circularly polarised microstrip antenna and the linearly polarised antenna booster. These results are represented on a schematic design of an automobile, indicating the position of the device. For each position, the results were averaged considering the different orientations at the rear dashboard (left) and front dashboard (right).

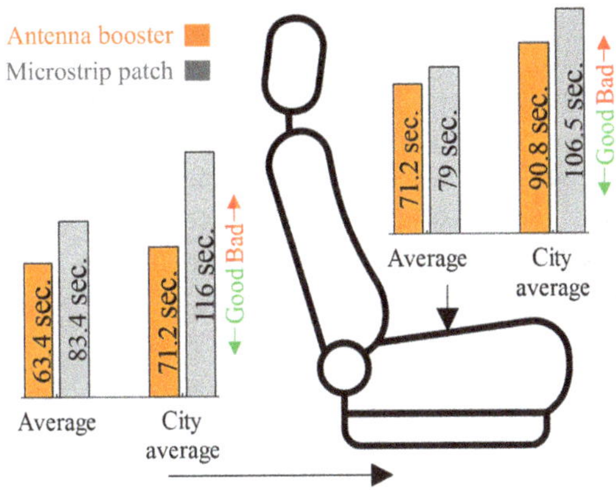

Figure 10.21 TTFF over the seat and under the seat

Table 10.9 Comparison of the microstrip patch antenna with antenna booster technology

	Circular polarised: microstrip patch antenna	**Linear polarised: antenna booster**
	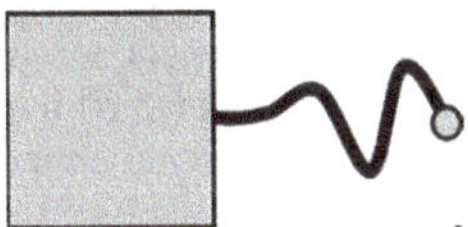	
Volume of the antenna part	$17.0 \times 17.0 \times 6.1$ mm^3	$7.0 \times 3.0 \times 2.0$ mm^3
Automatic pick and place	No	Yes
TTFF total average	64.9 s	57.8 s
Omnidirectional radiation pattern	No	Yes
BoM	Ceramic patch, LNA, coaxial line, and receptacles at the microstrip antenna and PCB	Antenna booster, LNA, and a matching network are all placed on the PCB

For unpredictable positions and orientations of a wireless device having a linearly polarised antenna booster, the theoretical loss of 3 dB due to a different polarisation than the circularly polarised incoming electromagnetic signal from the satellite is not relevant. Other antenna parameters, such as the radiation patterns, become more critical. This is the case for the antenna booster with quasi-isotropic radiation patterns that allow the device to be robust to reception at any orientation of the device, and thus, the TTFF is less, ensuring a longer battery duration.

Robust to orientations

Linearly polarised quasi-isotropic radiation patterns achieved through the use of antenna boosters offer enhanced resilience to satellite signal reception when compared to circularly polarised microstrip patch antennas. While the patch antenna is circularly polarised, the quasi-isotropic radiation pattern enables reception regardless of the wireless device's orientation. Consequently, the Time to First Fix (TTFF) is reduced with the antenna booster solution, leading to extended battery life.

10.7 Conclusions

Table 10.10 answers the objective given at the beginning of the chapter. It also points to relevant figures and equations to consolidate the concepts learnt. It is also advisable to refresh all essentials included in the grey boxes distributed along the chapter.

Table 10.10 Key aspects to take away

Topic	Review and take away
About TIS	TIS has been measured in a real wireless device for an antenna booster (86.4 mm^3) and a self-resonant antenna (707 mm^3), concluding that similar TIS performance, is slightly better for the antenna booster at the high-frequency bands (Figure 10.3). This fact confirms that exciting currents on the ground plane due to an antenna booster are an effective means for achieving bandwidth, efficiency, TRP, and TIS.
Future proof	Antenna boosters are ready to be used for any future protocol since their performance can be tuned by the proper location of the antenna booster in the device and the design of a matching network. However, self-resonant antennas operate at a particular frequency band or bands. Therefore, a new design, starting from scratch, is needed when a new radio protocol appears. This takes time and money until a new self-resonant antenna is available.
Antenna booster versus printed monopole antenna	According to the set-up in Figure 10.6, the antenna booster features more bandwidth: 1.8 times. Therefore, the antenna booster solution can allocate more radio protocols, or the antenna booster system size can be further reduced. In terms of simplicity (Figure 10.8), the antenna booster solution can be reused across a large number of form factors compared to the monopole antenna: the reuse ratio is 53.8%, whereas it is only 4.6% for the monopole antenna. The antenna booster is more robust to variations in the relative permittivity of the substrate compared to the monopole antenna (Table 10.2).
Size antenna booster versus self-resonant antenna	Antenna booster technology is characterised by the small size of the antenna boosters (Figure 10.9). On average, the size is ten times smaller. Besides, the clearance area can also be small, as shown in Chapter 3. This feature facilitates the integration of antenna boosters into thin wireless devices. Moreover, such a

(Continues)

Table 10.10 (Continued)

Topic	Review and take away
	small size enables the integration of several antenna boosters for MIMO or multi-radio applications.
Non-resonant nature	Due to the non-resonant nature of antenna booster technology, radiation efficiency is flat over a wide range of frequencies compared to self-resonant antennas. As proven, self-resonant antennas present maximum efficiency at resonance. However, there are efficiency drops between resonances, making them unusable at those frequencies. This implies that several self-resonant antennas are needed to cover different frequency ranges. However, an antenna booster is useful for any frequency band.
Easy to use	Antenna boosters are easy to use since the frequencies of operation can be easily adjusted by the design of the matching network. Such design is addressed by software tools that synthesise the topology of the matching network as well as the values of the components.
Off-PCB FPC versus antenna booster	The off-PCB FPC antenna performs similarly to the antenna booster (Figures 10.11 and 10.13). However, the performance drastically deteriorates when placing the FPC antenna in the device.
On-PCB FPC versus antenna boosters	The FPC antenna has been evaluated at several heights above the ground plane of a PCB, from 2.4 mm up to 20 mm, presenting in all cases worse efficiency values than the embedded booster antenna at both frequency regions 698–960 MHz and 1710–2690 MHz. When comparing the FPC antenna over the ground plane of the PCB at the same height (2.4 mm) as the antenna booster, total efficiency is 15.5 dB, 11.4 dB, and 18.1 dB more for the antenna booster at the low-frequency region for the small, medium, and large PCB, respectively. For the high-frequency region, the total efficiency is 4.8 dB, 6.2 dB, and 7.8 dB more for the antenna booster for the small, medium, and large PCB, respectively (Table 10.7).
FPC dependence on the mounting	The FPC antenna also presents dependent behaviour on its position on the PCB, where total efficiency can change by 10 dB. However, due to the SMD nature of the antenna booster, this situation is irrelevant for the embedded antenna booster since the position is fixed on the PCB and automatically mounted following a pick-and-place assembly process.
Linearly polarised antenna booster versus circularly polarised microstrip patch antenna	The quasi-isotropic radiation pattern of the wireless device, including the antenna booster, makes it more robust to the reception of satellite signals than a microstrip patch antenna (Figures 10.20 and 10.21). Therefore, the competitive advantage is not determined by polarisation but by the ability of the device to capture receiving signals from any direction. As a result, TTFF is better for the antenna booster solution than the microstrip patch, which is convenient for satellite reception applications to get the position faster. Less TTFF ensures more battery duration, which is relevant for battery-powered wireless devices.

10.8 Exercises

A 3D electromagnetic software and a circuit simulator, including matching network synthesis, are needed to solve the following exercises.

10.8.1 Exercise 1

Consider the 12 mm × 3 mm × 2.4 mm (height) antenna booster on a 120 mm × 60 mm ground plane (Figure 10.22). Using the teaching in Chapter 5, design a matching network to obtain an $S_{11} < -6$ dB for 863–928 MHz, 1560–1606 MHz, and 2400–2500 MHz. Include SMD components, taking into account finite Q. Note the design time.

Compute the bandwidth potential (Equations (3.2) and (3.3), Chapter 3) to check whether it is possible to achieve the proposed bandwidth. Use this figure of merit for the following Exercises: 10.8.2 to 10.8.8.

10.8.2 Exercise 2

Design a self-resonant antenna with the same ground plane size as the previous exercise to achieve the same S_{11} performance. The maximum antenna height cannot exceed the height of the antenna booster, which is 2.4 mm. Note the design time and compare it with the previous design. Compare the volume of the self-resonant antenna with the antenna booster. The width of the antenna can be as large as 60 mm if needed. Total volume should not exceed 60 mm × 11 mm × 2.4 mm (height).

10.8.3 Exercise 3

Repeat Exercise 10.8.1 considering a ground plane of 120 mm but now with only 20 mm width. Can you reuse the same matching network as Exercise 10.8.1, or do you need to design a new one? How long does it take to obtain the matching network?

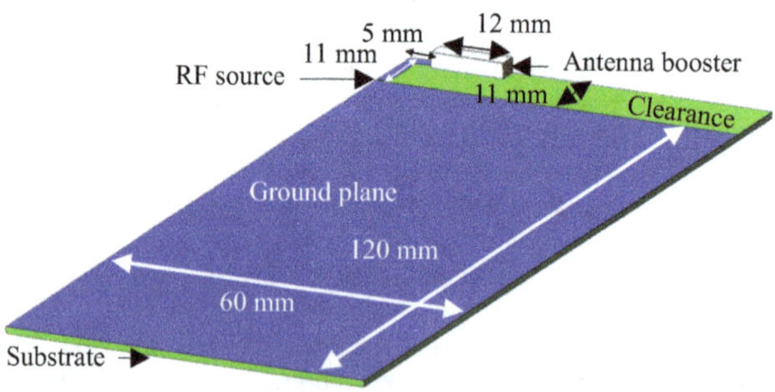

Figure 10.22 The antenna booster can be modelled as a conductive parallelepiped with a $\sigma = 5.9 \times 10^7$ S/m. The substrate is 1 mm thick, $\varepsilon_r = 4.15$ and $\tan\delta = 0.017$.

10.8.4 Exercise 4

Repeat Exercise 10.8.2, but with the same ground plane as Exercise 3. If the self-resonant antenna designed in Exercise 10.8.2 is wider than 20 mm, you should develop a new self-resonant for this device. Create a self-resonant antenna in the available 20 mm × 11 mm space. The maximum height cannot exceed the height of 2.4 mm, which is the height of the antenna booster.

10.8.5 Exercise 5

Repeat Exercise 10.8.3, considering a ground plane of 120 mm but now with only 12 mm width. Can you reuse the same matching network as exercise one, or do you need to design a new one? How long does it take to obtain the matching network?

10.8.6 Exercise 6

Repeat Exercise 10.8.4, but with the same ground plane as Exercise 10.8.5. You may design a new self-resonant for this device. Design a self-resonant antenna in the available 12 mm × 11 mm space. The maximum height cannot exceed the height of 2.4 mm, which is the height of the antenna booster. Compare bandwidth and efficiency with the antenna booster.

10.8.7 Exercise 7

Considering the device in Figure 10.22. Design a matching network to provide an $S_{11} < -6$ dB across 863–928 MHz, 1560–1606 MHz, and 2.5–3 GHz with the same antenna booster. Use the teachings of Chapter 5.

10.8.8 Exercise 8

Repeat Exercise 10.8.7 with a self-resonant antenna. Compare the design time with respect to the previous exercise and the total efficiency.

10.8.9 Exercise 9

Repeat Exercises 10.8.7 and 10.8.8, but with a 120 mm × 20 mm ground plane. Use the same antenna booster. Compare the design time. How many self-resonant antennas have you designed to satisfy the requirements given in all exercises? It is important to note that in all cases, you can employ the same antenna booster for every ground plane size and frequency band of operation.

10.8.10 Exercise 10

Using Figure 10.9 as a reference, consult the latest publications on antenna designs to achieve the same frequency bands. Compare the size of the new antenna solutions to that of antenna booster technology.

10.8.11 Exercise 11

Considering a ground plane of 70 mm × 50 mm. Design a 15 mm × 10 mm ground clearance at the corner to include a 7 mm × 3 mm × 2 mm antenna booster to operate at 1575 MHz. Design a matching network with finite Q components. Compute S_{11}, total efficiency, and radiation pattern. Design a 15 mm × 15 mm × 2 mm microstrip patch antenna with chamfered corners to achieve circular polarisation at 1575 MHz on a 70 mm × 50 mm ground plane. Place the patch antenna at the centre of such a ground plane. Select the appropriate substrate (ε_r and tanδ). Compute the total efficiency and radiation pattern. Compare the results with the antenna booster in terms of bandwidth, total efficiency, and radiation patterns.

10.8.12 Exercise 12

Considering the antenna booster in Figure 10.6, reduce the ground clearance until you achieve a bandwidth of 9% at $S_{11} = -6$ dB. How much smaller is the clearance area compared to the monopole antenna in Figure 10.6? Design a monopole antenna for such clearance. Compute bandwidth and total efficiency and compare them with the antenna booster. Analyse the reuse of both designs when the ground plane changes from 30 mm × 30 mm to 200 mm × 200 mm with steps of 10 mm.

10.8.13 Exercise 13

Considering the schemes in Figure 10.6, design a meander antenna and a fractal-shaped antenna and compare the results with the spiral monopole and the 12 mm × 3 mm × 2.4 mm antenna booster. Which ones provide more bandwidth, efficiency, and reuse across different ground plane sizes (use the same set as explained in Section 10.3)?

References

[1] J. Anguera, N. Toporcer, and A. Andújar, "Slim Radiating Systems for Electronic Devices," *Patent US* 9, 960, 478 B2, 2018.

[2] A. Andújar and J. Anguera, "Integration of a Non-Resonant Antenna in a Smartphone for Multiband Operation," *12th European Conference on Antennas and Propagation (EuCAP 2018)*, 2018, pp. 1–4, doi:10.1049/cp.2018.1110.

[3] J. Gui, A. Andújar, and J. Anguera, "On the Reuse of a Matching Network for IoT Devices Operating at 900MHz Embedding Antenna Boosters," *Electronics* vol. 11, p. 1267, 2022. https://doi.org/10.3390/electronics11081267.

[4] B. Oller, A. Andújar, and J. Anguera, "Antenna Booster Versus a Spiral Monopole Antenna for Single-Band Operation at 900 MHz," *Electronics* vol. 12, p. 2067, 2023. https://doi.org/10.3390/electronics12092067

[5] J. Anguera, A. Andújar, G. Mestre, J. Rahola, and J. Juntunen, "Design of Multiband Antenna Systems for Wireless Devices Using Antenna Boosters," *IEEE Microwave Magazine*, pp. 102–114, 2019.

[6] J. Anguera, A. Andújar, R. M. Mateos, and S. Kahng, "A 4 × 4 MIMO Multiband Antenna System with Non-Resonant Elements for Smartphone Platforms," *2017 11th European Conference on Antennas and Propagation* (EUCAP), Paris, France, 2017, pp. 2705–2708, doi:10.23919/EuCAP.2017.7928580.

[7] Y. L. Ban, Y. F. Qiang, Z. Chen, K. Kang, and J. L. W. Li, "Low-profile Narrow-frame Antenna for Seven-band WWAN/LTE Smartphone Applications," *IEEE Antennas and Wireless Propagation Letters*, vol. 13, pp. 463–466, 2014.

[8] Z. X. Chen, Y. L. Ban, Z. Chen, K. Kang, and L. W. Li, "Two-strip Narrow-frame Monopole Antenna with a Capacitor Loaded for Hepta-band Smartphone Applications," *Progress in Electromagnetic Research*, vol. 145, no. 2, pp. 31–38, 2014.

[9] C. Deng, Y. Li, Z. Zhang, and Z. Feng, "A Novel Low-profile Hepta-band Handset Antenna Using Modes Controlling Method," *IEEE Transactions on Antennas and Propagation*, vol. 63, no. 2, pp. 799–804, 2015.

[10] D. Wu, S. W. Cheung, and T. I. Yuk, "A Compact and Low-profile Loop Antenna with Multiband Operation for Ultra-thin Smartphones," *IEEE Transactions on Antennas and Propagation*, vol. 63, no. 6, pp. 2745–2750, 2015.

[11] Zi-Qiang Xu, Qiang-Qiang Zhou, Yong-Ling Ban, and S. S. Ang, "Hepta-Band Coupled-Fed Loop Antenna for LTE/WWAN Unbroken Metal-Rimmed Smartphone Applications," *IEEE Transactions on Antennas and Propagation*, vol. 17, no. 2, pp. 311–314, 2018.

[12] S. Yang, C. Zhang, H. K. Pan, A. E. Fathy, and V. K. Nair, "Frequency-reconfigurable Antennas for Multiradio Wireless Platforms," *IEEE Microwave Magazine*, pp. 66–83, 2009.

[13] Y. Li, Z. Zhang, J. Zheng, Z. Feng, and M. F. Iskander, "A Compact Hepta-Band Loop-Inverted F Reconfigurable Antenna for Mobile Phone," *IEEE Transactions on Antennas and Propagation* vol. 60, no. 1, pp. 389–392, 2012.

[14] A. Chen, M. Sun, Z. Zhang, and X. Fu, "Planar Monopole Antenna with a Parasitic Shorted Strip for Multistandard Handheld Terminals," *IEEE Access*, vol. 8, pp. 51647–51652, 2020, DOI:10.1109/ACCESS.2020.2979483.

[15] Y.-L. Ban, C.-L. Liu, J. L.-W. Li, and R. Li, "Small-Size Wideband Monopole with Distributed Inductive Strip for Seven-Band WWAN/LTE Mobile Phone," *IEEE Antennas and Wireless Propagation Letters*, vol. 12, pp. 7–10, 2013. doi:10.1109/LAWP.2012.2233706.

[16] P. Huang and L. Huang, "Design of an Antenna with Ground Plane Slots for LTE/WWAN Mobile Phones," *Progress In Electromagnetics Research Letters*, vol. 83, pp. 15–22, 2019. doi:10.2528/PIERL19012701.

[17] Y. Huang, *Antennas: From Theory to Practice*, 2nd Edition, John Wiley & Sons, 2021.

[18] "What Are Flexible Printed Circuit (FPC) Antennas?," *Everything RF*, Feb 16, 2021

[19] C. Balanis, *Antenna Theory and Design*, 3rd Edition, John Wiley & Sons, 2005.

[20] J. Anguera, A. Fernández, C. Puente, A. Andújar, and J. Groot, "Antenna Boosters Versus Flexible Printed Circuit Antennas for IoT Devices," *Signals*, vol. 3, pp. 326–340, 2022.

[21] J. Gui, J. L. Leiva, A. Andújar, J. Groot, J. L. Pijoan, and J. Anguera, "Linearly Polarized Antenna Boosters Versus Circularly Polarized Microstrip Patch Antennas for GPS Reception in IoT Devices," *Energies*, vol. 15, p. 9623, 2022. https://doi.org/10.3390/en15249623.

Chapter 11
From zero to hero in antenna booster technology

This chapter summarises the essentials of antenna booster technology.

The chapter has the following objectives:

1. Review the definition of an antenna booster and the architecture of an antenna booster system.
2. Refresh the basic steps to design a wireless device for single to multiband operation compared to a self-resonant antenna approach.
3. Revisit the role of the ground plane.
4. Consolidate the main techniques to design single, multiband, multi-radio, and reconfigurable solutions with antenna boosters.
5. Remind yourself of the effect of the pad layout for simplifying the integration of an antenna booster into a wireless device.
6. Summarise the benefits of antenna booster position, ground clearance, and ground plane length.
7. Present industrial examples of antenna booster products and wireless devices embedding antenna boosters.
8. Provide all you must know on one page.

11.1 Introduction

This chapter summarises the main takeaways of antenna booster technology. For a more detailed summary, the reader is encouraged to review all conclusions at the end of all chapters.

This chapter is divided as follows. First, the definition of an antenna booster is provided in Section 11.2. The main parts of an antenna booster architecture are revisited in Section 11.3. Then, the basic steps to design from single to multiband operation are presented and compared to a classic approach with self-resonant antennas (Section 11.4). The role of the ground plane is discussed in Section 11.5 since an antenna booster is in charge of exciting a ground plane that determines the electromagnetic performance of the wireless device. The most practical techniques to design single, multiband, multi-radio, and reconfigurable antenna booster systems are reviewed in Sections 11.6, 11.7, and 11.8. The pad layout is revisited from a practical perspective to simplify the embedding of antenna boosters (Section 11.9). Basic steps to boost the performance are provided by changing the antenna booster location, modifying the ground clearance, and changing the ground plane length (Section 11.10). To show the impact of antenna booster technology in the wireless industry, several antenna products are illustrated, as well as real wireless devices embedding antenna boosters (Section 11.11). A single-page summary is presented in Section 11.12, and the main takeaways are in Section 11.13. Finally, essential reading is gathered in the References section.

11.2 Antenna boosters

As extensively demonstrated throughout the book, antenna booster technology radically differs from the classic antenna approach. For the traditional antenna approach, the performance is mainly determined by shaping the geometry of an antenna from fractal-shaped structures, multi-branch antennas, and genetically optimised antennas [1–3]. However, on the contrary, antenna booster technology relies on using a non-resonant and electrically small element called an antenna booster [4,5]. The antenna booster efficiently excites currents in the ground plane of a wireless device. In this manner, the ground plane becomes the main radiating element, acting as an antenna. Furthermore, since the ground plane is larger than the size of any antenna within the device, it is an effective means for achieving bandwidth and efficiency. Therefore, an antenna booster, also called a ground plane booster, should not be considered an antenna but a boosting element to excite the ground plane. This combination allows an antenna booster to be tiny in terms of the operating wavelength λ, typically $\lambda/30$, and even smaller than other antenna techniques (Figures 11.1 and 11.2). Besides, an antenna booster has a very simple structure, as the one shown in the book, consisting of a simple 3D conductive parallelepiped. For industrial applications, other simpler designs exist [6].

Antenna booster
An electrically small (typically with a size $\leq \lambda/20$ at the lowest frequency of operation) non-resonant element at least at some frequencies band of operation (normally, the lower bands) can excite currents on a ground plane of a wireless device to transmit and/or receive electromagnetic signals.

Figure 11.1 A 7 mm × 3 mm × 2 mm antenna booster for dual-radio operation at 1.57 GHz and 2.5 GHz [7]. The electrical size at the lowest frequency of operation is ~$\lambda/27$.

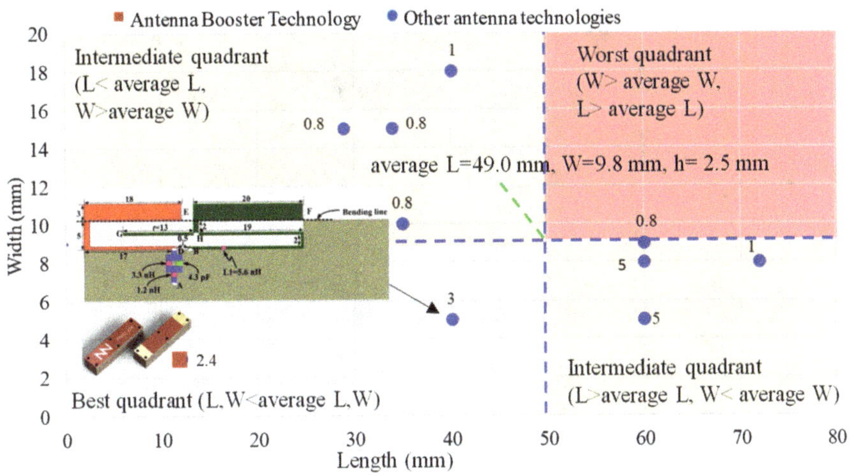

Figure 11.2 Antenna boosters are electrically smaller (typically ten times) than other antenna technologies. The 12 mm × 3 mm × 2.4 mm antenna booster is not only smaller than the 40 mm × 5 mm × 3 mm but also simple to manufacture.

In terms of size, antenna booster technology has been compared with self-resonant antennas and flexible printed circuit antennas (Chapter 10), concluding that antenna boosters are small and easy to use. For example, Figure 11.2 is revisited, illustrating the small dimension of an antenna booster for multi-band operation at 824–960 MHz and 1710–2690 MHz. In particular, a 12 mm × 3 mm × 2.4 mm (height) antenna booster is compared to other antenna techniques available in the literature. L and W are the sizes of the smallest parallelepiped enclosing the antenna structure. The L, W space is divided into three regions: the smaller one, where L and W are both smaller than the average L_{avg}, W_{avg}; an intermediate area where either $W > W_{avg}$ or $L > L_{avg}$; and the worst zone, where both L and W are larger than the average size. As shown, the antenna booster is in the best location, far away from the rest.

Since the antenna booster is non-resonant, typically at the lowest frequencies of operation, a matching network is needed to provide operation either at a single band or multiband operation (Figure 11.3). This is an enormous advantage and differentiator factor since the matching network offers a simple way to tune the antenna booster from single to multi-band operation for any frequency range between ∼ 400 MHz and 11 GHz (Section 11.6). Note that this is the preferred frequency range for the antenna boosters proposed in this book. However, the antenna booster principle can be scaled to other regions of the electromagnetic spectrum. For example, a 12 mm × 3 mm × 2.4 (height) antenna booster at 900 MHz can be scaled to provide operation at 30 GHz by reducing its dimensions by a factor of 0.9/30, resulting in 360 μm × 90 μm × 72 μm.

A matching network comprises lumped passive components: inductors and capacitors. In those situations, such as those requiring a large number of frequency

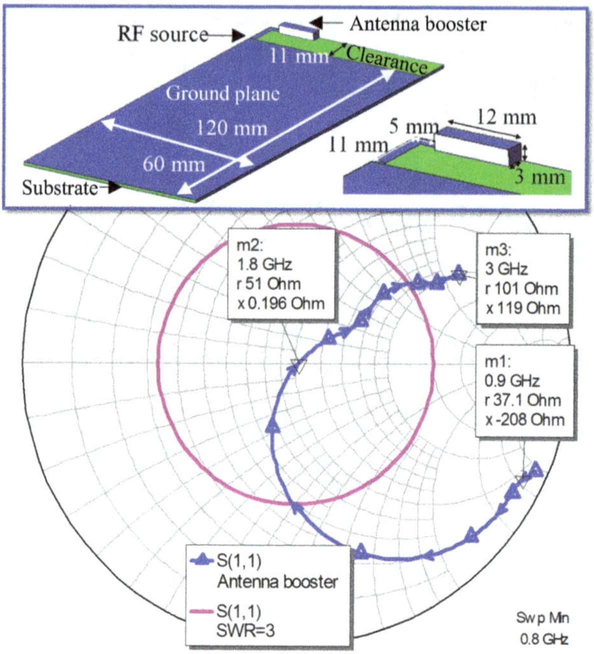

Figure 11.3 An antenna booster showing a non-resonance behaviour (capacitive impedance) at a low-frequency region of ~900 MHz

bands or featuring electrically small ground planes $L < \lambda/\pi$ (Equation (1.26), Chapter 1), a tuneable or reconfigurable matching network is preferred if a passive matching network cannot provide enough bandwidth (Section 11.7).

11.3 The architecture

The architecture of an antenna booster system comprises three main elements: first, a tiny element, generically called an antenna booster; second, a ground plane conductor usually embedded in the printed circuit board of the device; and third, a matching circuit that ultimately shapes the frequency response of the antenna system to the specific needs of a wireless device (Figure 11.4).

Properly engineering the antenna booster and its position in a wireless device enables the excitation of currents on the ground plane to become the crucial element in the radiation process. The matching circuit provides operation at the frequency bands where the wireless device will radiate. Doing so removes all the complexity in the antenna part, putting the intelligence in the matching circuit design. This makes the antenna booster ten times smaller than any other antenna parts while flexibly adapting to any wireless protocol and device. Therefore, the antenna booster can be utilised for any band and any device, simplifying and speeding up the whole design process.

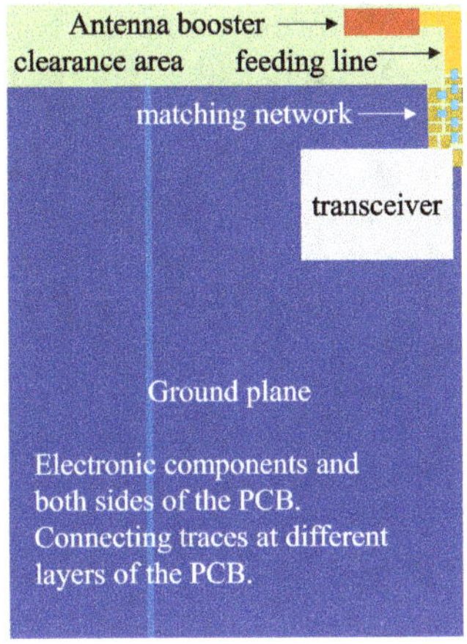

Figure 11.4 Architecture of an antenna booster system: the antenna booster, the matching network, and the ground plane of the PCB (Printed Circuit Board) of the wireless device

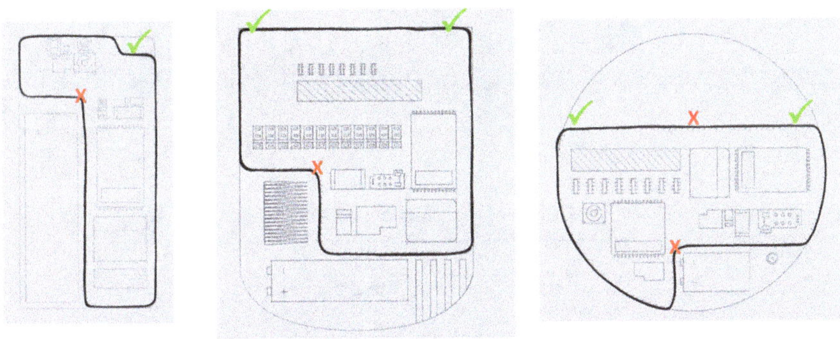

Figure 11.5 The position of an antenna booster matters. The corners of the ground plane of the wireless device are typically preferred to get the best bandwidth. However, other positions may work as well if bandwidth is not a significant constraint.

As deeply analysed in Chapter 3, the corners of the ground plane are the preferred locations to maximise bandwidth and efficiency (Figure 11.5). However, other positions may also work if bandwidth is not a constraint. For antenna boosters (electric), the corner of the ground plane is preferred for maximising bandwidth and efficiency. However, for magnetic boosters (Chapter 3), the central parts of the edges of the ground plane are desired.

The matching network may comprise lumped components such as inductors or capacitors as well as transmission lines. For frequencies below 10 GHz, inductors and capacitors are typically used. However, for much higher frequencies, losses and parasitic behaviour of the packing degrade the performance, and thus, transmission lines are preferred.

The matching network is an essential part of the architecture, providing the flexibility to match the antenna booster from single-band to multiband operation. Once the antenna booster system is matched, the performance is evaluated in terms of S_{11}, total efficiency, radiation patterns, and even TRP and TIS, among the main ones (Figure 11.6).

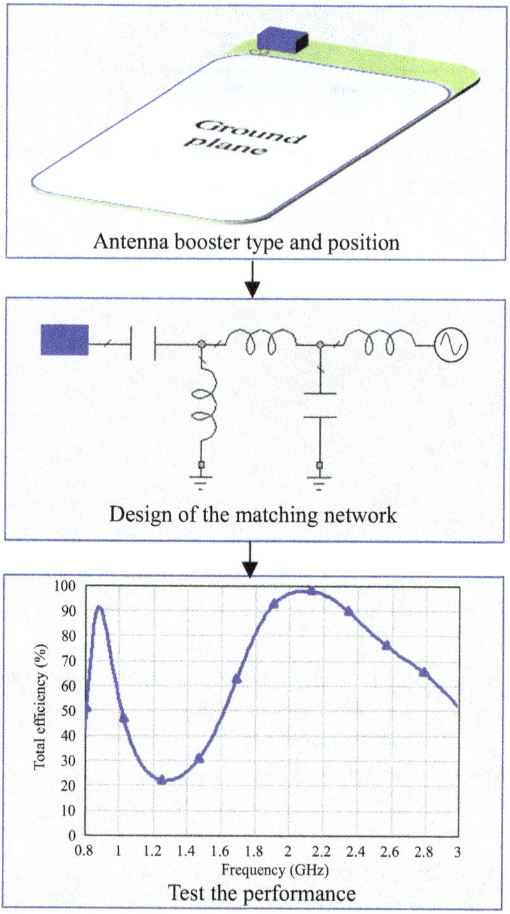

Figure 11.6 Basic steps to design a wireless device embedding an antenna booster: choose the antenna booster and locate it (typically at the corner of the ground plane); design the matching network; simulate/ measure the electromagnetic performance (e.g., S_{11} and total efficiency)

11.4 Designing wireless devices with antenna boosters

The classic design of wireless devices embedding antennas consists of designing or using an antenna based on shaping the geometry of the antenna (Figure 11.7). As explained in Chapter 10, the design process is cumbersome and requires many iterations, either in the simulation or physical domains, to adjust the geometry of the antenna to achieve the required specs. This route is based on self-resonant antennas, where the geometry of the antenna mainly determines the frequency bands of operation. For example, a single-band antenna may comprise a single bending of a monopole. However, the geometry becomes more sophisticated when more frequency bands are needed, as in a multiband case. This requires antenna engineering expertise on how to shape the different conductive portions of an antenna.

Besides the geometry, matching networks are used for fine-tuning the impedance. However, a different, simpler, and faster route is based on antenna boosters, where the frequencies of operation are mainly determined by the design of the matching network. This is a quicker approach than designing an antenna from scratch. This design flow is very flexible since if any mechanical change on the device affects the antenna booster performance, the antenna booster does not need to be redesigned, but only the matching network must be tuned.

Another advantage of this approach is that the design of the matching networks is totally addressed by software-based synthesis algorithms [8–10] (Figure 11.8). For example, a multiband matching network may comprise seven components. At first sight, the number of possible ladder-type topologies assuming inductors and capacitors is 2^7. Moreover, since each inductor and capacitor can vary from a minimum value to a maximum value, the possibilities increase dramatically. However, thanks to matching network synthesis algorithms, the process is addressed by computer force. Therefore, the design does not require one to be an antenna expert, as is the case for self-resonant antennas, but to be familiar with RD design. Therefore, designing wireless devices that embed antenna boosters makes the technology accessible to a broader audience.

Thanks to the unique properties of antenna boosters, they can be embedded in PCBs like any other microwave device, such as capacitors, amplifiers, filters, etc. The input impedance of an antenna booster on a ground plane without any matching network can be included as a library [11]. Therefore, a wireless designer can design a single/multiband matching network with a matching network synthesiser available in a circuit simulator [8,12,13]. If this process is repeated for a broad set of ground planes and antenna boosters, the design does not need any 3D electromagnetic simulator, just a matching network design. Hence, a designer can quickly test at the beginning of the design journey if the performance of the wireless device is suitable or needs to be changed, such as by enlarging the ground plane or using a reconfigurable architecture.

From a practical point of view, two critical parameters for the matching network elements should be considered: the quality factor (Q) and the tolerance. The Q is related to the losses of the component. In this regard, the larger the Q, the better. As a practical reference, a $Q > 100$ is considered acceptable for minimising

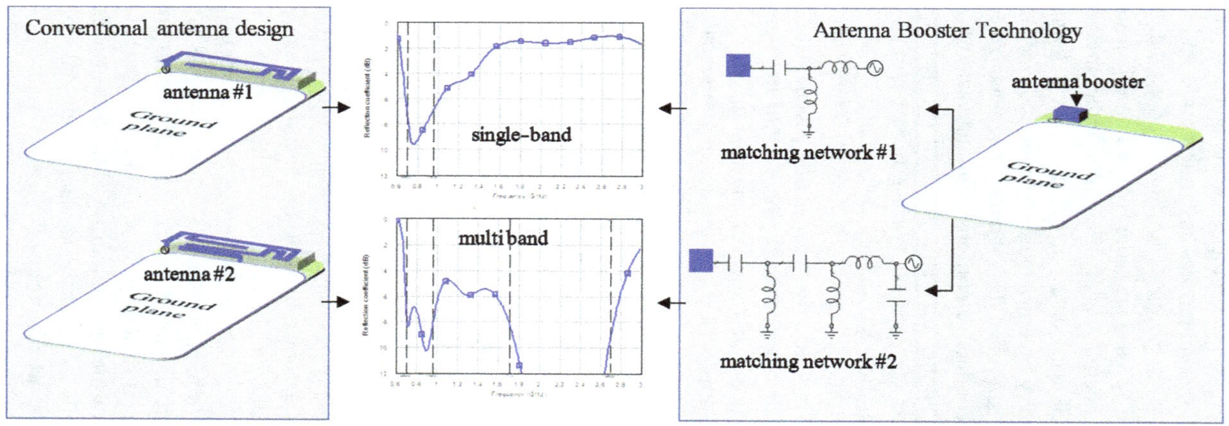

Figure 11.7 *Design with the classic antenna approach: the geometry of the antenna is designed according to the frequency band or bands of operation. In contrast, antenna booster technology relies on an antenna booster, and the matching network is designed to obtain the frequency bands of operation. This process is simpler, faster, and more cost-effective, and it can be driven by software thanks to matching network synthesisers.*

the losses introduced by a matching network. Details on how it impacts efficiency should be reviewed by reading Chapters 3 and 4.

The tolerances of the matching network components should be as tight as possible. Typically, tolerances should be at most 2% to guarantee stable performance when mounting matching networks in mass production.

Since the ground plane of a wireless device is usually not large in terms of the wavelength (size $<\lambda$), radiation patterns are relatively isotropic, that is, able to transmit or receive an electromagnetic signal in any direction. This is a convenient feature since the direction of the incoming signal is unknown, or the wireless device orientation is arbitrary, and thus, having a kind of isotropic radiation pattern guarantees reception at any direction (Figure 11.9).

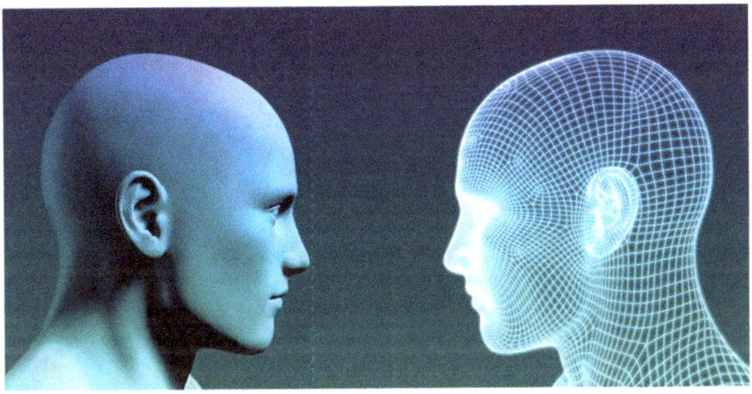

Figure 11.8 The design of wireless devices embedding antenna boosters can be totally addressed by computer algorithms

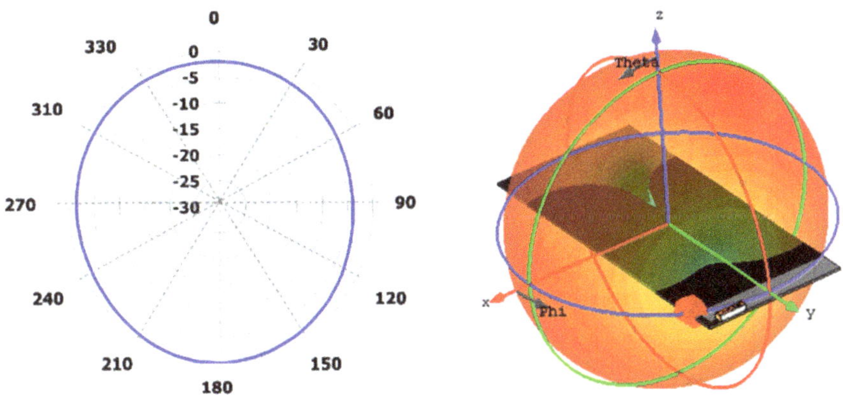

Figure 11.9 The quasi-isotropic radiation pattern ensures similar transmission/reception independently at any direction, making it convenient when the location of the incoming signal is not known or the orientation of the wireless device embedding the antenna booster is unknown

11.5 The role of the ground plane

An antenna booster controls exciting currents on a ground plane. The correct excitation of those currents depends on the position of the antenna booster (typically, the corner of the ground plane is a preferred location – Figure 11.5) and the size of the ground plane. As investigated in Chapter 3, when the length of a rectangular ground plane is $\sim 0.4\lambda$, bandwidth is maximised (Figure 11.10). However, when the maximum dimension is less than $\sim 0.4\lambda$, currents are less excited. The shorter the length, the less the excitation. As a result, bandwidth and efficiency decrease (Figures 11.10 and 11.11).

As far as bandwidth is concerned, a 12 mm × 3 mm × 2.4 mm (height) antenna booster is placed on a ground plane $L \times W$ (Figure 11.3). The potential bandwidth is computed at $S_{11} = -6$ dB at the central frequency of 900 MHz. The potential bandwidth is obtained as the bandwidth obtained with a matching network consisting of one or two lumped components. It can also be approximated with Equations (3.2) and (3.4) – Chapter 3. The size of such a ground plane varies from

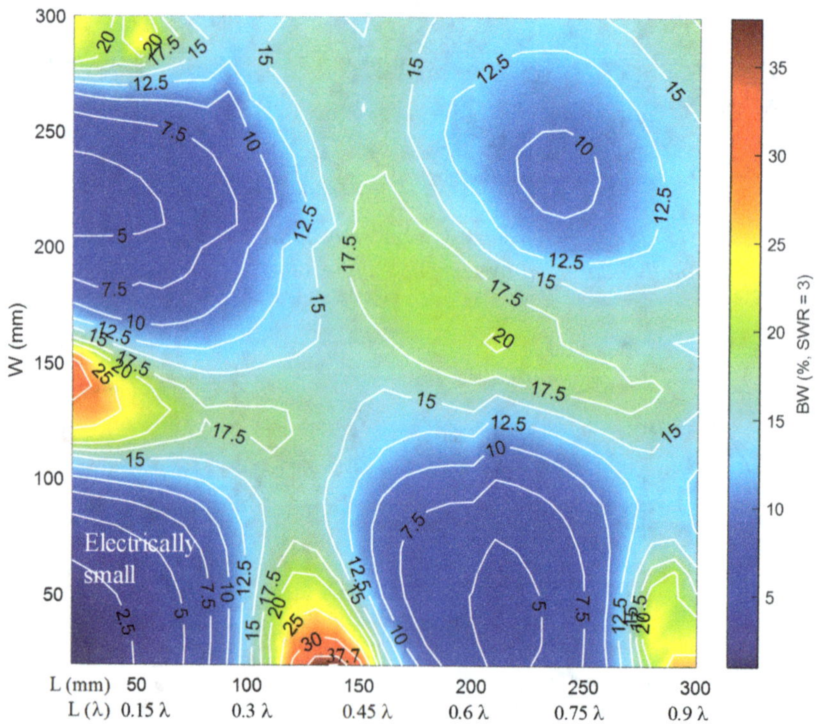

Figure 11.10 The size of the ground plane plays a determinant role in bandwidth. Electrical size is computed at 900 MHz. The antenna booster system is shown in Figure 11.3.

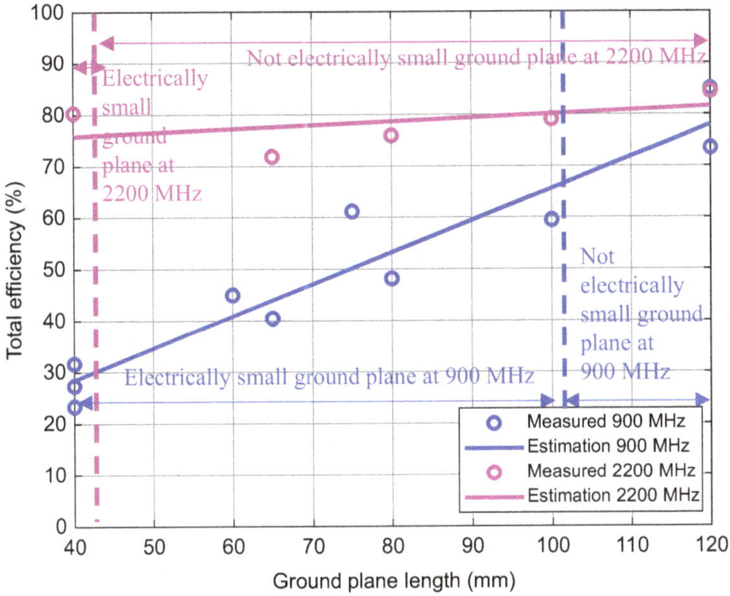

Figure 11.11 Estimation of measured total efficiency for a 12 mm × 3 mm × 2.4 mm antenna booster on different ground plane sizes. Dots represent the measured total efficiency of the antenna booster on a bare PCB.

the smallest one (L, W) = (20 mm, 20 mm) to the largest one (L, W) = (300 mm, 300 mm) – (Figure 11.10).

Several essential conclusions for this setup can be drawn:

- First, there is an optimum ground plane for (L, W) = (140 mm, 20 mm), where the bandwidth is BW = 37.7%. Note that 140 mm corresponds to 0.42λ at 900 MHz.
- When the size of the ground plane is smaller, the bandwidth is reduced. For example, for an (L, W) = (40 mm, 40 mm) bandwidth reduces to BW = 2.2%
- For the area outside of the small zone (maximum dimension $>\lambda/\pi$), bandwidth exceeds 5%.
- It is worth mentioning that what matters is the length of the ground plane and not the area. For example, a ground plane with (L, W) = (140 mm, 20 mm) has an area of 28 cm^2 with a BW = 37.7%. However, a ground plane with (L, W) = (80 mm, 80 mm) has a bigger area of 64 cm^2 but a BW = 8.8%. Therefore, the most critical parameter is not area but length, being $L = 0.42\lambda$, an optimum length for a rectangular ground plane for maximising bandwidth (Figure 11.10). Note that the optimal length of $L = 0.42\lambda$ corresponds to a narrow ground plane of 0.06λ, meaning that width also plays a relevant role. For example, a 140 mm × 20 mm reaches a BW = 37.7%, whereas a

140 mm × 60 mm decreases to 16.5%. From these numbers, one may conclude that 20 mm width is better than 60 mm. Although this is true for single-band operation at 900 MHz, a different result is obtained if multiband operation is needed. For example, if operation at 2 GHz is required, the 60 mm width enables excitation of the fundamental mode along the width since 60 mm is 0.4 λ at 2 GHz. The reader is encouraged to review the bandwidth trade-off for multiband operation for several form factors in Chapter 4.

The size of the ground plane also has a direct impact on efficiency (Figure 11.11). Considering the same 12 mm × 3 mm × 2.4 mm (height) antenna booster on different ground planes on a bare PCB. The results are obtained by averaging the total efficiency in a first frequency region at 900 MHz operating from 824 MHz to 960 MHz and at 2200 MHz ranging from 1710 MHz to 2690 MHz.

As observed, the total efficiency increases with the ground plane length at the low-frequency region (~900 MHz). Note that 40 mm at 900 MHz corresponds to a small electrical size of only 0.12λ. Owing to the scalability principle, 40 mm is electrically small at 900 MHz but not at 2.2 GHz, where 40 mm corresponds to 0.29λ. As a reference and as per the teachings of Chapter 1, a ground plane is electrically small if the radius of the smallest sphere enclosing the ground plane is a $<\lambda/2\pi$, that is, a $<300/(2\pi \cdot f(\text{MHz}))$. Making a simplification, if we consider $W \ll L$ and ignore the antenna booster size, then $L = 2a$, and thus a ground plane can be regarded as electrically small if its length L satisfies:

$$L_{\text{small}} < \frac{\lambda}{\pi} \quad L_{\text{small}} < \frac{300}{\pi \cdot f(\text{MHz})} \tag{11.1}$$

For example, at 900 MHz, $L_{\text{small}} = 106$ mm. Hence, 40 mm is a small ground plane. Now, if we repeat the numbers at 2200 MHz, $L_{\text{small}} = 43$ mm. Thus, 40 mm is electrically small at 2200 MHz but not as small as at 900 MHz. This is translated not only in bandwidth, as discussed above, but also in terms of total efficiency.

Practical tip on efficiency for a given ground plane size
As a rule of thumb, for a ground plane not small (Equation (11.1)) in terms of the wavelength, total efficiencies can reach values >60% (Figure 11.11).

Ground planes typically feature a rectangular shape. However, as detailed in Chapters 3 and 4, the shape of the ground plane can be designed to improve the bandwidth compared to a rectangular shape (Figure 11.12). In those situations where those changes are allowed, shaping the ground plane adds another degree of freedom to further improve the performance in terms of bandwidth and efficiency. The purpose of shaping the ground plane is appropriate for those ground planes featuring a length $L \ll \lambda/\pi$, that is, electrically small ground planes since bandwidth is a challenge. For example, at 900 MHz, a 60 mm × 60 mm rectangular ground plane is electrically small, owing to Equation (11.1). The potential bandwidth is

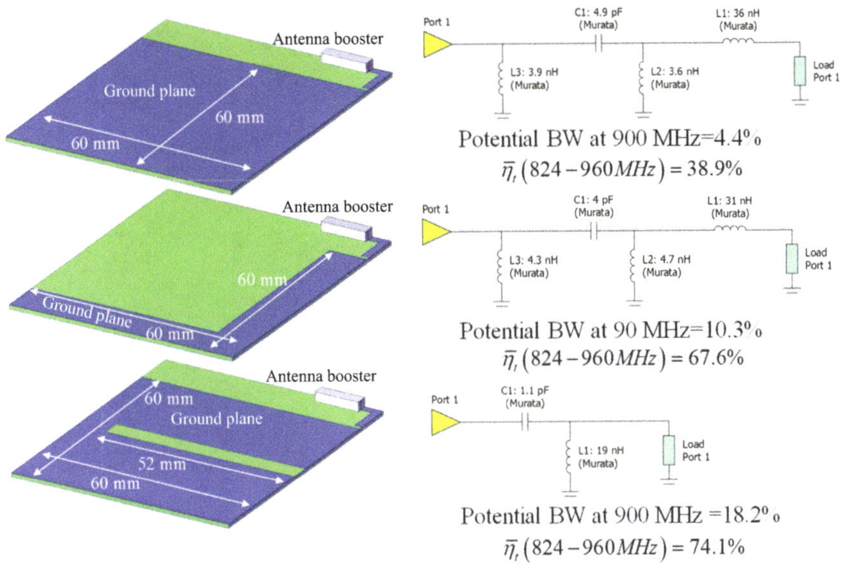

Figure 11.12 The shape of the ground plane can be optimised to improve bandwidth and efficiency

4.4% (Figure 11.12). However, when the rectangular ground plane is shaped as an L having the same size of 60 mm × 60 mm, the potential bandwidth increases to 10.3% and to 18.2% when a slot is included. The main reason is that the current path is enlarged. Thus, the virtual length is larger. To compare the three cases, a matching network using a matching network synthesiser is used to match from 824 MHz to 960 MHz with real lumped components. For the square and L-shaped ground planes, a matching network is needed. However, only two components are required for the slotted ground plane due to its larger bandwidth potential. As a result, the 38.9% average total efficiency for the square ground plane is increased to 67.6% and 74.1% for the L-shaped and slotted ground plane. This last case gives 2.8 dB more efficiency than the square ground plane.

Shaping the ground plane

Shaping the ground plane can increase bandwidth (~5 times) and efficiency (~3 dB), especially when the ground plane is electrically small $L < 300/(\pi \cdot f$ (MHz)).

A ground plane typically takes the form of conductive materials such as copper printed on a thin layer of substrate, forming a PCB. Such a PCB hosts the electronics of a wireless device that can comprise more than one ground plane layer. In practice, a wireless device features a top and a bottom ground plane layer and even intermediate layers.

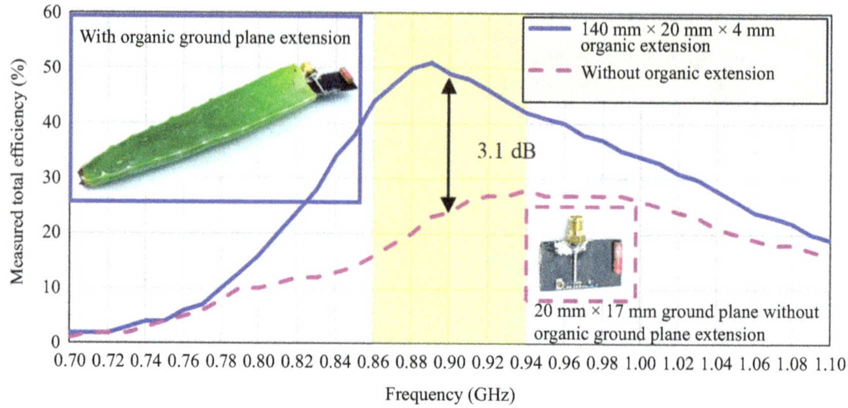

Figure 11.13 Efficiency improvement due to an organic ground plane enlargement

As deeply analysed throughout the book, the ground plane is a crucial player in determining performance (bandwidth, efficiency, and radiation patterns). As previously discussed (Figures 11.10 and 11.11), if the ground plane size decreases, the bandwidth and efficiency decrease. Therefore, in those situations where the size of the device should be compact, the ground plane may be modified to increase bandwidth and efficiency (Figure 11.12).

If we now dream for a moment, we can imagine that instead of having wireless devices spread all around, we could hide those wireless devices in plants. And the question is: why do we not take profit from the plant to improve the efficiency of the wireless device? In this regard, we may think of enlarging the size of a compact wireless device, including a small ground plane, thanks to such an organic implant [14,15]. The composition of leaves is mainly water, and being water, a dielectric with a high relative permittivity in the microwave domain, the ground plane may be enlarged with such a dielectric extension. A proof of concept has been carried out for a 20 mm × 17 mm ground plane with a 12 mm × 3 mm × 2.4 mm antenna booster at 900 MHz. For such a small ground plane, the measured total efficiency, including a matching network to provide impedance matching, is below 30% (Figure 11.13). However, when the ground plane is enlarged with an aloe vera leaf of 140 mm × 40 mm, the measured total efficiency increases by about 3 dB. This result opens the window to a new approach where plants can act as cybernetic elements providing connectivity and, at the same time, feeding the wireless device due to the nutrients present in the soil, which can be enough to power a small sensor and, thus, the device becomes battery-less.

11.6 From single to multiband operation

Once the antenna booster is embedded into a wireless device, the matching network provides the flexibility to obtain the frequency bands of operation by a simple and

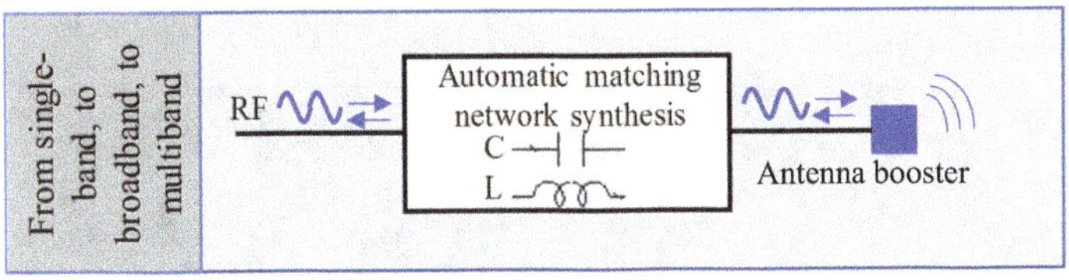

Figure 11.14 Besides the different techniques studied in the book for single or multiband operation, the simplest and more practical approach relies on embedding a single antenna booster and a matching network synthesised through software

fast systematic procedure based on matching network synthesis. Among the techniques explained in Chapters 4 and 5, the simplest and more generic solution comprises one antenna booster and one matching network (Figure 11.14).

Typically, a matching network of two or three lumped components is enough for single-band operation. Suppose the bandwidth of such a frequency band needs to be expanded. Then, a broadband matching network comprising four lumped elements increases the bandwidth compared to a single matching network with only two components. Finally, more than five and less than eight components are typically required for a multiband response. Based on the Bode limit (Chapter 4), a matching network cannot infinitely enhance the bandwidth. From a practical point of view, a broadband matching network can increase the potential bandwidth by a factor of 2.4 for SWR = 3 for a single-band case. For example, if the bandwidth of an antenna booster system with a matching network with two lumped components is 4%, a broadband matching network, as studied in Chapter 4, can increase such bandwidth to 9.6% (2.4 times 4 for SWR = 3). This fundamental limit decreases for a multiband matching network since it has to compromise to match two different frequency regions. In this regard, when a passive matching network cannot squeeze more bandwidth than required, reconfigurable matching networks are one alternative, as explained next. Another alternative relies on shaping the ground plane (Figure 11.12).

Either single-band, broadband, or multiband, the design process is fast, easy, and software-driven thanks to matching network synthesis tools available in most microwave simulators [12,13]. This is an advantage since the synthesis algorithm tests millions of candidates to obtain the best solution or solutions. Therefore, it is easy to know the bandwidth limits with antenna booster technology. However, for a self-resonant approach, once an antenna is designed, the question is: is there any other antenna geometry that can improve the bandwidth? The answer is not easy at all. It can be addressed by synthesising the geometry of the antenna with genetic algorithms, for example. However, it is far more time-consuming than the computation time to synthesise a matching network. For example, a multiband antenna booster system takes one 3D simulation considering the antenna booster and the ground plane to obtain the S_{11}. Then, it takes minutes to synthesise the matching network given the S_{11}. As observed, there is only one 3D simulation involved.

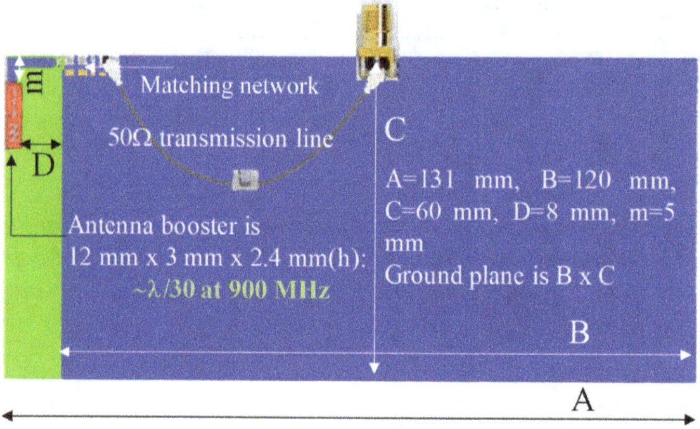

Figure 11.15 A 12 mm × 3 mm × 2.4 mm (height) antenna booster on a 120 mm × 60 mm ground plane. A 50 Ω microcoaxial line ended with a connector allows testing passive parameters such as S_{11}, total efficiency, and radiation patterns.

However, synthesising a self-resonating system for multiband operation takes hours and even days. For example, in [16], it took 50 h for the genetic algorithm to synthesise the multiband antenna shape. Other sources report similar computation times [17]. The issue with this route is that it needs 3D simulation for each antenna candidate, which is not efficient compared to the antenna booster design path.

Besides being fast and easy, the same antenna booster can be reused for any frequency band (Figures 11.15–11.18). For the scope of the book, any frequency band means 400 MHz up to 10.6 GHz. However, applying the frequency scaling concept presented in Chapter 1, antenna booster technology is usable at any frequency as long as the architecture is preserved: the antenna booster, the matching network, and a ground plane. In this regard, an example from single band to multiband is illustrated next.

Let us consider a 12 mm × 3 mm × 2.4 mm (height) antenna booster placed at the corner of a 120 mm × 60 mm ground plane. Close to said corner, pads for a matching network are located. After the matching network, a 50 Ω microcoaxial transmission line is used to test the S_{11} and efficiency (Figure 11.15). Note that the antenna booster is very small in terms of the operating wavelength λ. For example, considering 900 MHz as a reference, the antenna booster is only $\sim\lambda/30$.

In the first case (Figure 11.16), a simple matching with a series and shunt inductor is included for operation at 863–928 MHz. Since the ground plane is 0.36λ in length, the obtained bandwidth at SWR = 3 is greater than the target bandwidth. To be more precise, considering the current 120 mm × 60 mm ground plane, the potential bandwidth (the bandwidth obtained with a simple matching network of one or two components, Chapter 4) is 20.0% (Figure 11.10). If the ground plane is reduced to 60 mm × 60 mm, the potential bandwidth reduces to 3.7%. Although it is smaller, it is still enough for many applications.

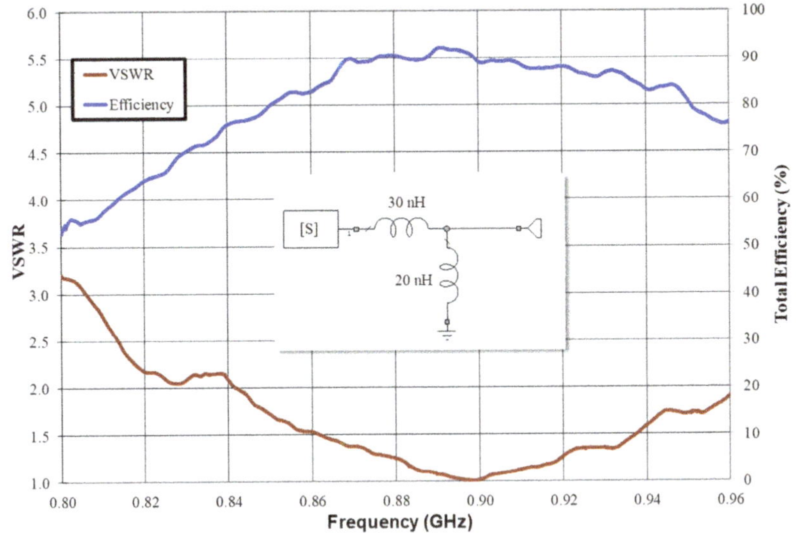

Figure 11.16 *Measured VSWR and total efficiency of the antenna booster system shown in Figure 11.15, adding the series-shunt matching network. [S] represents the impedance of the antenna booster without the matching network.*

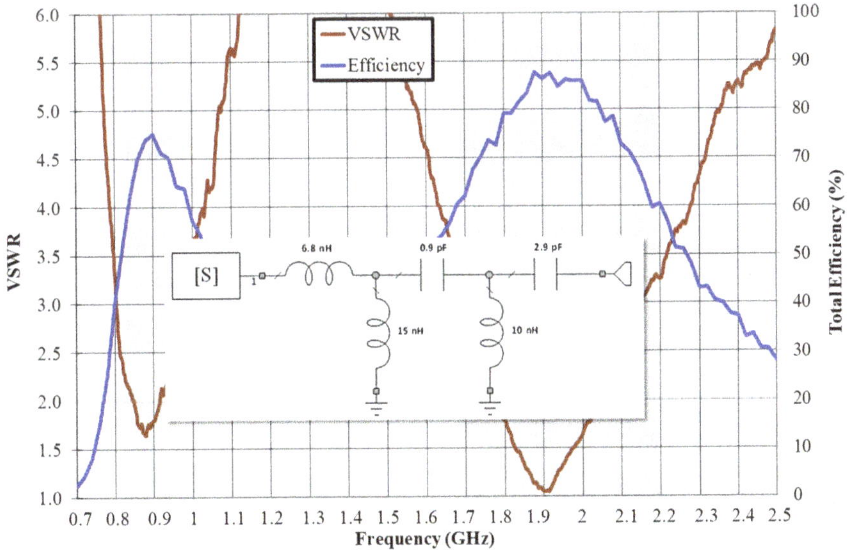

Figure 11.17 *Measured VSWR and total efficiency of the antenna booster system shown in Figure 11.15, adding the matching network with five lumped components. [S] represents the impedance of the antenna booster without the matching network.*

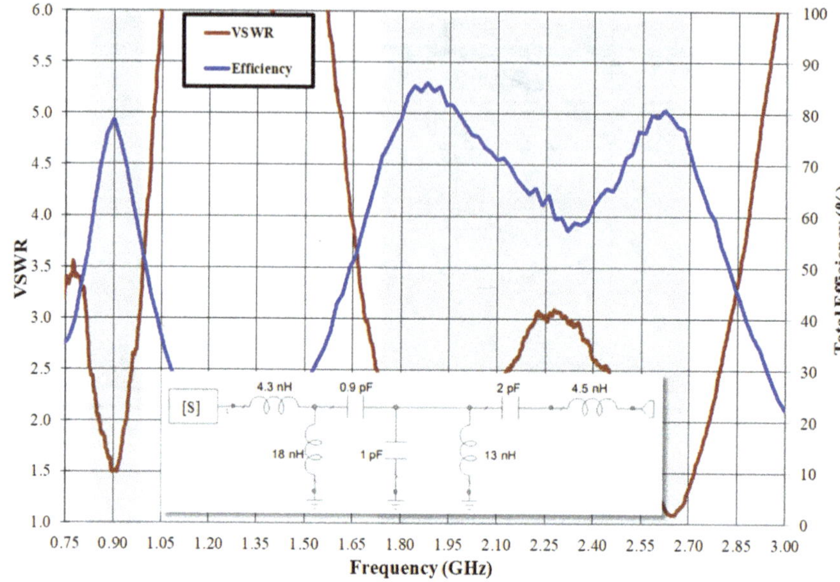

Figure 11.18 Measured VSWR and total efficiency of the antenna booster system shown in Figure 11.15, adding the matching network with seven lumped components. [S] represents the impedance of the antenna booster without the matching network.

Now, let us consider a multiband case where the frequencies of operation are not included in a single frequency region but in more than one [18]. The design is again the same: keep the antenna booster in the corner location and then design a new matching network (Figure 11.17). In this example, the target is to achieve operation from 824 MHz to 960 MHz and 1710 MHz to 2170 MHz. As observed, five lumped components are needed. It is up to the matching network synthesis algorithm to decide the best topology and number of components, avoiding any human intervention to maximise efficiency. In fact, a computer algorithm can test about ten million matching networks per second. This is much more efficient than using human power since it is difficult to know if the solution obtained by a human being is the best one or if there is still one better. This is not the case with the software process since the synthesis algorithm optimises the performance in terms of S_{11} and total efficiency. This process ensures that the best solution is reached, or at least that the likelihood of obtaining the best solution is much higher with the machine-engineered route than with the human way. This is due to the fast computation time that allows for testing an incomparable large number of different matching network topologies.

Finally, another multiband example illustrates bandwidth enhancement at the high-frequency region (Figure 11.18) from 1,710 MHz to 2,170 MHz and from 1710 MHz to 2690 MHz. As explained in Chapter 4, a matching network is not only helpful for matching purposes but also to increase the bandwidth obtained by a simple matching network of one or two components. In this example, seven lumped components are needed. However, since those components are high-Q (and also tight tolerance), the total efficiency reaches 70% and 80% in the low- and high-frequency regions. Owing to the Fano limit detailed in Chapter 4, bandwidth cannot be increased indefinitely by adding more components. Therefore, it is easy to know the maximum achievable bandwidth (Chapters 4 and 5). As a basic rule, the maximum bandwidth achieved by a simple broadband matching network of two components at $S_{11} = -6$ dB is 2.4 times the potential bandwidth, which can be easily computed following the teachings in Chapters 4 and 5.

Software-engineered design of matching networks for antenna boosters
The design of single or multiband matching networks for antenna boosters can be efficiently addressed by software algorithms. This approach ensures that the design process is streamlined, allowing for easy and rapid implementation.

11.7 Multi-radio solutions

With the increasing demand for several radio protocols within a wireless device, there is a need to compact the antenna solution to save space and simplify the integration of antennas. In this regard, a multi-radio antenna booster solution can provide operation at different radios in a compact manner instead of using several antenna boosters spaced apart (typically spaced more and $d > 0.1\lambda$). Multi-radio techniques can be classified as follows (Figure 11.19) – (Chapter 6) [19–22]:

1. One antenna booster using a self-diplexed architecture and a single-band matching for each radio
2. One antenna booster employing a multiband matching network and a multiplexer
3. As many antenna boosters as radios

A triple-radio for 0.9 GHz, 1.57 GHz, and 2.4 GHz is illustrated (Figure 11.20) based on a combination of techniques 1 and 3 detailed above. In this case, a self-diplexed structure at 1.57 GHz and 2.4 GHz with lumped inductors and capacitors is included to split the input port into two highly isolated ports, each one including the respective matching network for 1.57 GHz and 2.4 GHz.

The first port operates at 0.9 GHz, and a second port is split with a self-diplexed stage to provide operation at 1.57 GHz and 2.4 GHz with two independent ports. Isolation at each port is guaranteed thanks to the filtering effect of the matching network plus the reject filters. The reject filter effect can be included in the matching network, and thus, a compact solution is obtained. For example, let us imagine a dual-radio system for 1.57 GHz and 2.4 GHz. The matching network at

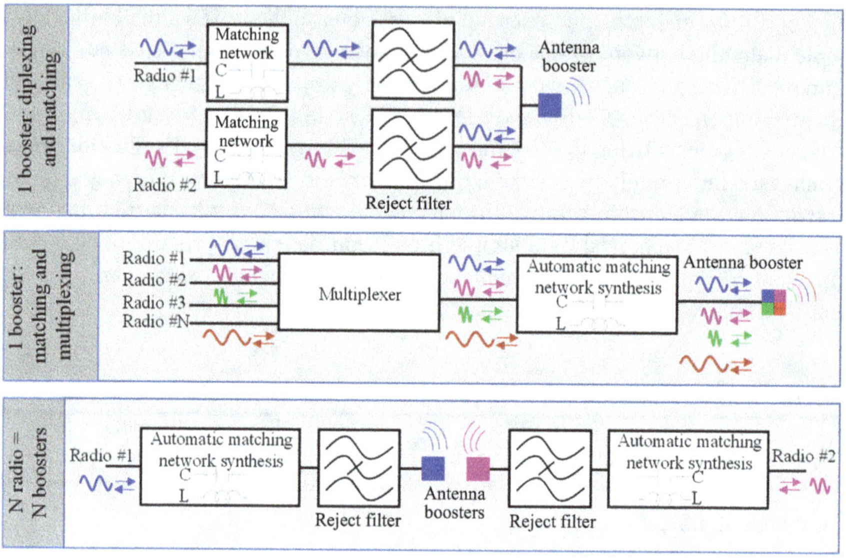

Figure 11.19 Summary of multi-radio techniques: (top) an antenna booster with a self-diplexing stage and single-band matching networks; (middle) a single antenna booster with a multiband matching network and a multiplexer, and (bottom) as many antenna boosters as radios

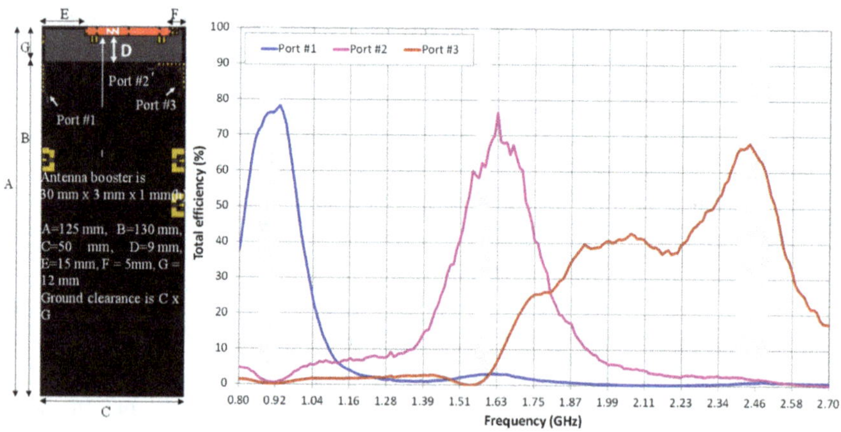

Figure 11.20 Triple radio solution for 0.9 GHz, 1.575 GHz, and 2.450 GHz with a single antenna booster element [21]. The worst-case isolation is 10 dB.

each port can be designed in such a way that it obtains a good match at the corresponding frequency band and, at the same time, is highly mismatched at the other frequency band. Although, at first glance, the N-booster comprises several antenna boosters, a single package embedding different antenna boosters can be adopted, resulting in a compact solution [21].

11.8 Reconfigurable solutions

The bandwidth of an antenna booster system depends mainly on the size of the ground plane. For example, at 900 MHz, a 140 mm × 20 mm ground plane with a 12 mm × 3 mm × 2.4 mm (height) antenna booster can provide a bandwidth of BW = 37.6%. However, if the size of the ground plane is 60 mm × 20 mm, the bandwidth drops to BW = 3.5% (Figure 11.10, Chapters 3 and 4). Therefore, if the ground plane size is small or many frequency bands are needed, or when the external environment of the device may change (device on a conductive asset, human body, etc.) and thus modify the electromagnetic performance, a reconfigurable architecture is preferred. As a reference, if the required bandwidth is larger (e.g., two times) than the one shown in Figure 11.10 for a particular ground plane size, then a reconfigurable architecture may be the solution to obtain competitive bandwidth and efficiency.

As analysed in Chapter 7, antenna boosters with reconfigurable matching networks comprise lumped components (capacitors and inductors) and tuneable components such as DTC (Digitally Tuneable Capacitor) or SPNT switches (Figure 11.21). Those components are controlled by the transceiver, which sets up a

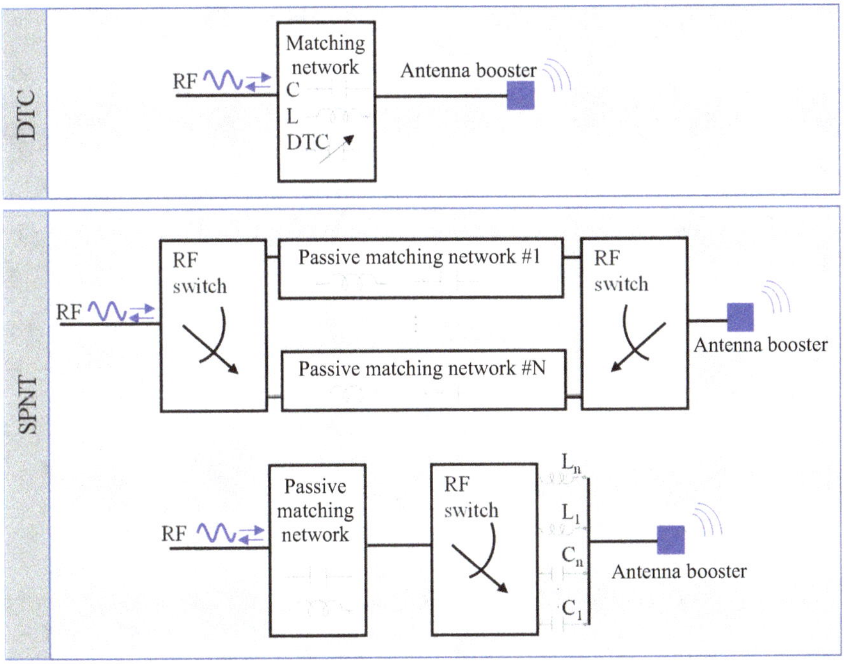

Figure 11.21 Reconfigurable architectures based on DTC and SPNT

particular state of the DTC and SPNT to provide a good match for the requested frequency bands [23–26]. Once a different frequency band is needed, the transceiver sets another state. In addition, a look-up table is stored in the transceiver, mapping a state with a particular frequency band of operation.

Although the reconfigurable solution consumes DC power compared to the passive matching network, the efficiency improvements are considerable (from ~3 dB to ~15 dB, Chapter 7) and ~ make them suitable solutions for improving performance compared to that achieved by a passive matching network.

A reconfigurable antenna booster system is illustrated, comprising a shunt DTC (Figure 11.22). The DTC presents 32 possible states between a C_{min} = 0.4 pF and C_{max} = 1 pF, which allows tuning the total frequency response beyond that of a passive matching network. For this case, the reconfigurable enables a new frequency region of operation from 698 MHz to 824 MHz compared to the case with the passive matching network – 824–960 MHz and 1710–2690 MHz.

The previous architecture provided smooth tuning by changing the state of the capacitor. If coarse frequency tuning is needed, other architectures, such as the one explored with two SPNTs, can be utilised (Figure 11.21). In this case, a set of different matching networks is included between the two SPNTs, so various

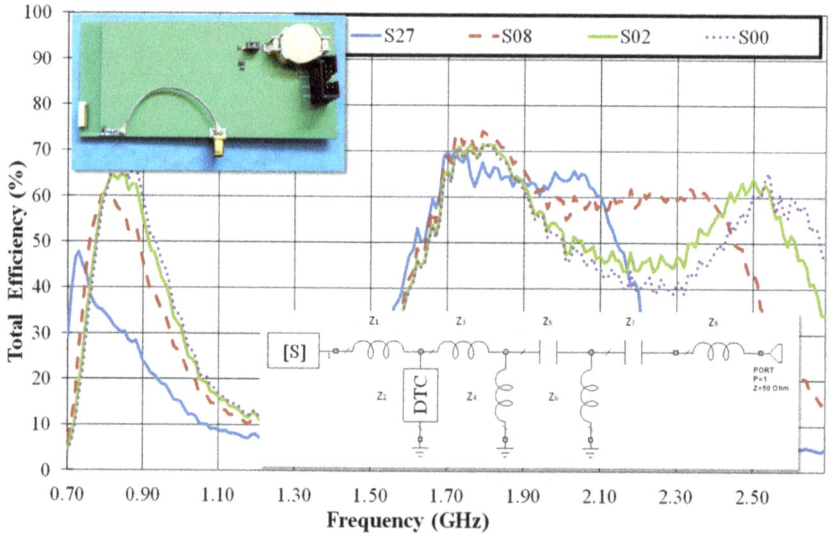

Figure 11.22 Measured total efficiency for a reconfigurable antenna booster system comprising a matching network with a shunt DTC, a 12 mm × 3 mm × 2.4 mm antenna booster, and a 120 mm × 60 mm ground plane. S27, S08, S02, and S00 are the selected four states of the DTC to achieve operation from 698 MHz to 960 MHz and 1710 MHz to 2690 MHz

matching networks can be selected. A simplified architecture with only one SPNT also allows fine-tuning, with the advantage of reducing the bill of materials. As shown in Chapter 7, the proposed architecture provides 65 different matching networks (15 series components and 50 shunt series with only four lumped components connected to the SPNT).

11.9 Practical considerations

The pad layout for allocating the matching network components should be considered when designing the matching network, as far as implementation is concerned. Pad layout can be neglected in an early stage where the objective is to analyse if bandwidth/efficiency can be obtained from the current device.

When embedding antenna boosters into wireless devices, the teachings presented in the book can be applied, such as selecting the appropriate antenna booster size, optimising the antenna booster position, the matching network, the clearance area, the feeding line, the ground plane size, and even its shape. Once the performance is achieved, the pad layout can be considered to obtain the matching network ready to be implemented into a wireless device.

Two techniques have been proposed in Chapter 9 (Figure 11.23). The first one involves the simulation domain [8]. In this scenario, the matching network is synthesised, simulating the wireless device, including the pad layout. In a second route, the matching network is synthesised from an S_{11} measure from the antenna booster in the device [10,27]. In this case, the pad layout is added as a

	Method	Description	Procedure
SIMULATION DOMAIN	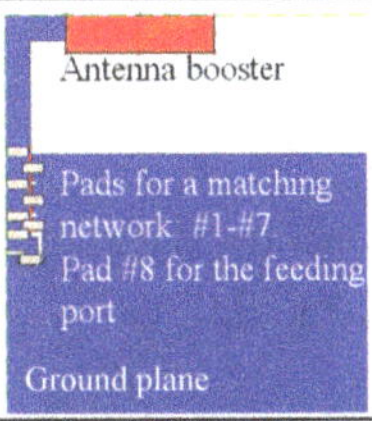	• Parasitic effects of pads and considered in the full-wave electromagnetic simulation • Needs: 1) a full 3D electromagnetic simulator to obtain the [S] data 2) A Network Synthesis Tool 3) Need to know the exact dimensions of the PCB, pads layout, and antenna booster to obtain the [S] data	1) Full-wave simulation to obtain the [S] matrix 2) Import the [S] matrix to a network synthesis tool to obtain the matching network 3) Specify targets 4) Network synthesis
PHYSICAL DOMAIN		• Parasitic effects of pads can be modelled by circuit simulation with short transmission lines • A full 3D simulator is not needed. Only: 1) The physical device 2) Microwave simulator with a Network Synthesis Tool 3) A Vector Network Analyser	1) Measurement of S_{11} of the antenna booster in the real device without any matching network 2) Import the [S] matrix to a network synthesis tool to obtain the matching network. Pads are modelled by short transmission lines 3) Specify targets 4) Network synthesis

Figure 11.23 The pad layout must be taken into account to simplify the physical implementation of the matching networks into the wireless device

block using, for example, transmission lines, which model the effect of the pad layout.

With this procedure, the synthesised matching network is ready to use, minimising human intervention in the design process.

11.10 Boosting the performance of a wireless device embedding an antenna booster

This section presents easy and effective ways to modify the PCB to boost performance in terms of efficiency. Four changes are proposed and analysed here (Figure 11.24):

1. Removing nearby grounded components.
2. Enlarging the clearance area around the antenna booster.
3. Optimising the position of the feeding line to the antenna booster.
4. Extending the size of the PCB.

As illustrated next, applying one or a combination of these changes can significantly improve the radiation performance, with improvements that range from 1 and up to 10 dB in the low-frequency region ($f < 1$ GHz).

These results can be easily replicated using electromagnetic software to model the main parts of the device: the antenna booster, the PCB, including the ground plane, and the substrate (FR4, 1 mm thick with $\varepsilon_r = 4.15$ and $\tan\delta = 0.017$).

The following figure of merit Δ is proposed to quantify the relative improvement of the different design changes. The total efficiency (η_t) averaged at 698–960 MHz and 1710–2690 MHz is calculated for each scenario (Figure 11.24). Then, these averaged values are compared to the original device. The total efficiency η_t is the fraction of power radiated to space over the transmitted power of the radiofrequency source. This efficiency depends on two factors: the system losses on one side η_r, including losses due to plastic covers, components on the PCB, matching network, and losses in the antenna booster itself; and the antenna impedance mismatch (S_{11}) on the other side. Total efficiency is then defined as $\eta_t = \eta_r \cdot (1-|S_{11}|^2)$. For instance, if the power of the front-end module delivered to the antenna is $P_{out} = 23$ dBm and the total efficiency is $\eta_t = 50\%$ (−3 dB), the output radiated power (also known as total radiated power, TRP) is 23 dBm − 3 dB = 20 dBm since TRP(dBm) = P_{out}(dBm)+10log$_{10}(\eta_t)$. The same argument applies to reception, as the antenna is a reciprocal device.

With these considerations, the figure of merit Δ is then defined by Equation (11.2):

$$\Delta(\text{dB}) = 10\log\left(\frac{\bar{\eta}_{t\ \text{new PCB}}}{\bar{\eta}_{t\ \text{original}}}\right) \tag{11.2}$$

Being $\bar{\eta}_{t\,\text{original}}$ and $\bar{\eta}_{t\ \text{new PCB}}$ the average total efficiency in a given frequency range $[f_1, f_2]$ for the original device and the new PCB case, respectively. Two

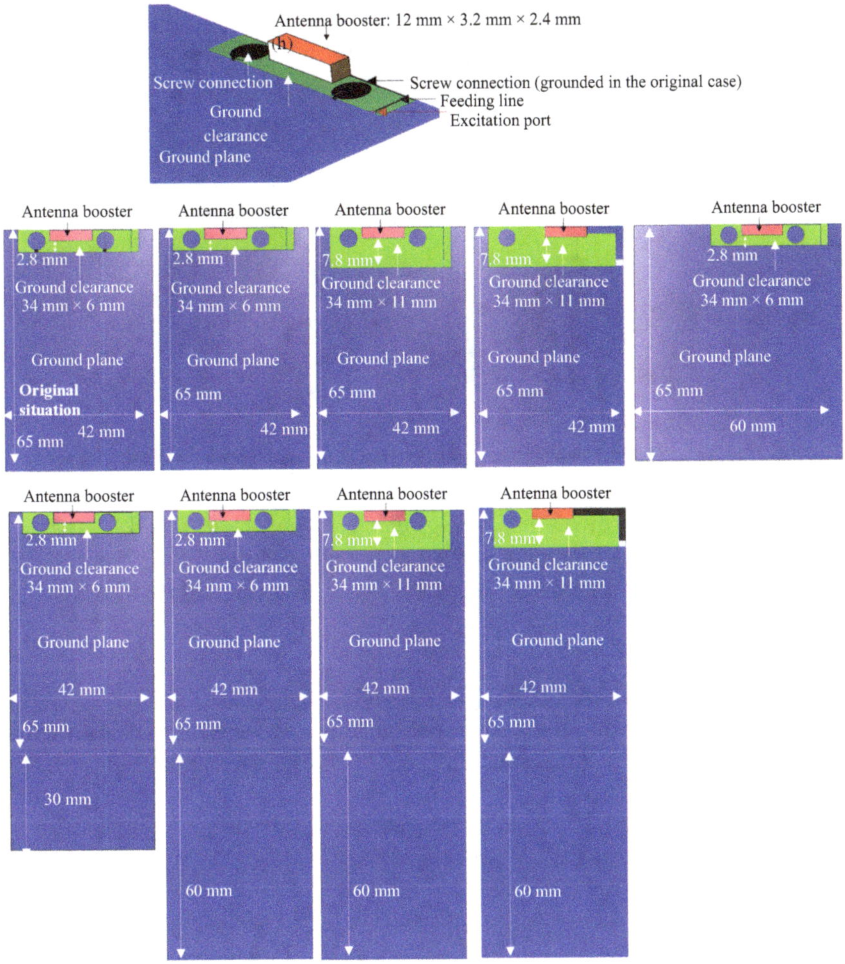

Figure 11.24 Cases considered in the analysis: screw connectors, change ground clearance, change feeding location, and change PCB size. Each scenario has a matching network at the excitation port comprising lumped elements (capacitors and inductors). Only the original case has the screw connectors connected to the ground plane of the PCB; in the other cases, such screw connectors are floating (non-grounded).

frequency ranges are studied: the first one includes $[f_1, f_2]$ = [698–960 MHz], and the second one comprises $[f_3, f_4]$ = [1.71–2.69 GHz].

In this sense, a $\Delta = 3$ dB increment means that total efficiency is increased by 3 dB compared to the original device. This figure of merit Δ helps determine the impact of each scenario when doing a new PCB design to get the best of the antenna performance in the device (Figure 11.25).

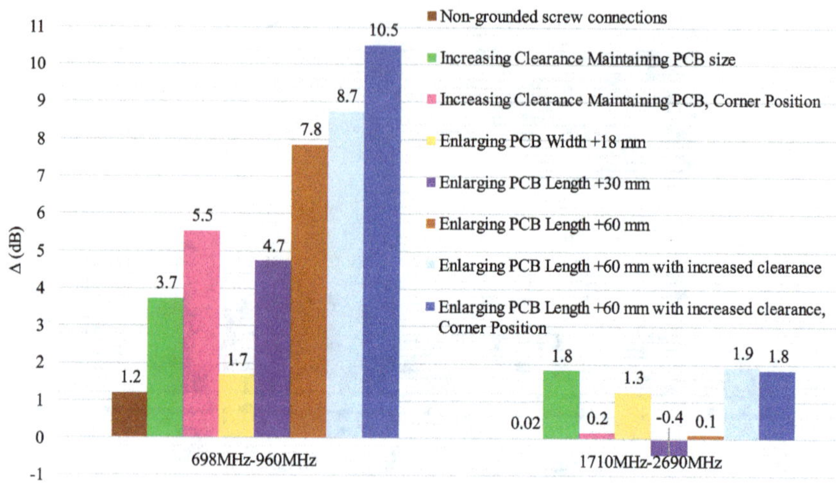

Figure 11.25 The nearby components, ground clearance, and ground plane length impact the low-frequency region (698–960 MHz) and high-frequency region (1710–2690 MHz). The original device PCB is 65 mm × 42 mm.

The matching network at the excitation port for each scenario is optimised to maximise total efficiency. Such matching networks comprise lumped inductors and capacitors, including losses.

Results from Figure 11.25 reveal several interesting design guidelines to be followed when optimising the RF performance:

- Conducting portions close to the antenna should be avoided, but if those cannot be avoided, those portions should be non-grounded. In this situation, when the pads for the screw connectors are non-grounded, the performance at the low-frequency region is improved by $\Delta = 1.2$ dB compared to the original situation where the screw connectors are grounded. From now on, all the remaining cases will have non-grounded screw connectors.
- Increasing ground clearance while maintaining the PCB size (see legend in Figure 11.25) can increase performance $\Delta = 3.7$ dB both at the low and $\Delta = 1.8$ dB at the high-frequency region. This ground clearance increment is due to a larger gap between the antenna booster and the ground plane. For this case, the gap between the antenna booster is increased from 2.8 mm to 7.8 mm. Review Chapter 3, Section 3.13, for more details. Note that the clearance area increment is substantial (3.7 dB) when the ground plane is small ($L \sim 0.15\lambda$ at 830 MHz). For the case of a larger ground plane ($L \sim 0.3\lambda$ at 830 MHz), the clearance increment is not as notorious: from 7.8 dB to 8.7 dB, that is, 0.9 dB increment at the lowest frequency region and 1.8 dB at the highest frequency region.

- Increasing ground clearance while maintaining the PBC size and moving the feeding line to the corner provides further improvement at the low-frequency range of $\Delta = 5.5$ dB.
- PCB width also has a positive effect, although slightly for the case under test. This increment, however, is not as significant as the ones obtained by increasing the ground clearance.
- Increasing PCB length positively impacts the 698–960 MHz region, whereas the 1710–2690 MHz is almost the same. When increasing the ground plane for the original PCB size by 30 mm, the improvement at the low-frequency region is 4.7 dB and 7.8 dB when enlarging 60 mm. Note that ground plane length when enlarging 60 mm is $L = 114$ mm, which corresponds to 0.31λ at 830 MHz (the centre frequency of the 698–960 MHz region).
- Finally, when combining the increase in ground clearance, PCB length, and antenna booster feeding line at the corner of the PCB, the positive effects are added; that is, performance is improved at both frequency regions, in particular, $\Delta = 10.5$ dB at the low-frequency region and $\Delta = 1.8$ dB at the high-frequency region.

These practical tips help optimise the performance of a wireless device at the conception phase when the designer still has the freedom to modify the PCB size. Therefore, these design rules should be made at the beginning of a design project. This will minimise and even avoid cumbersome PCB modifications while the project is in the development phase when all electronics, chipsets, and mechanics are already fixed. Therefore, it is always recommended to think first about these design lines when designing wireless devices embedding antenna boosters to get the most out of it.

Once the PCB size is fixed, embedding an antenna booster in a wireless device follows a simple design flow consisting of three simple steps: (1) placing the antenna booster in the PCB; (2) designing the matching network; (3) testing and fine-tuning the actual device, making the whole embedded antenna design process much faster, cost-effective, easier, and more accessible than the traditional choice of developing or using a customised antenna.

Ground plane length
The length of the ground plane is the essential parameter to maximise bandwidth/efficiency. As a reference, for a rectangular ground plane, if L (mm) $\sim 120/f$ (GHz), bandwidth is optimal.

11.11 Examples of antenna boosters and wireless devices embedding antenna boosters

Since the invention of the Antenna Booster Technology in 2008 [4,5], research has pushed the technology to the commercial field, catalysing the foundation of the company Ignion [7]. Ignion was founded in 2015, with headquarters in Barcelona,

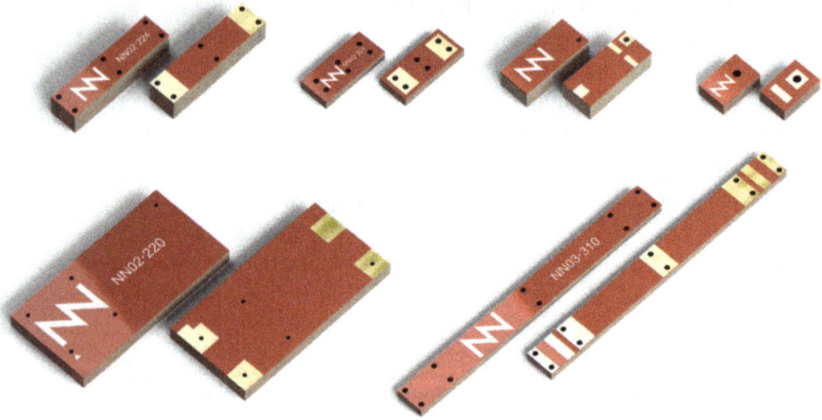

Figure 11.26 Antenna Booster Technology family products from Ignion [7]. Thanks to Ignion.

Spain. The mission of Ignion is to accelerate the IoT (Internet of Things) and empower the ecosystem with antenna booster solutions. Since then, many big players in the IoT field have been using antenna booster technology in their devices to provide connectivity reliably and efficiently. From a commercial standpoint, Antenna Booster Technology is known as Virtual Antenna®. Virtual Antenna® keeps the essence of the antenna booster concept: an antenna booster is not an antenna but a boosting element that excites currents on the ground plane of a wireless device.

Ignion has developed off-the-shelf antenna boosters (Figure 11.26). As observed, the portfolio is simple. In the end, an antenna booster can operate at any frequency (400–10.6 GHz) for any form factor. This simplifies the selection of the chip antenna component by wireless designers. One of these products, the RUN mXTEND™ antenna booster (12 mm × 3 mm × 2.4 mm (height)), has been extensively used throughout the book.

Among the different antenna boosters, two feature multi-radio operation: the DUO mXTEND™ antenna booster features dual radio for 1.57 GHz and 2.5 GHz bands suitable for satellite and short-range communications. The other one, the TRIO mXTEND™ chip antenna component, is usable as a dual radio for cellular communication and is either satellite or short-range. It is also usable as a triple radio, having one port for either cellular or other wide area network protocol, plus satellite and short range (Figure 11.20).

Antenna booster goes first

When designing a wireless device, the placement and clearance area of the antenna booster are crucial (see Figure 11.24). Therefore, to enhance the performance of a wireless device, it is essential to take the antenna booster into account from the outset of the design process.

It is worth mentioning that wireless devices for IoT feature different form factors compared to other wireless devices, such as mobile phones. This is due to the fact that IoT devices serve a diverse range of applications, from tracking, metering, agriculture, and healthcare to industrial monitoring and smart homes. Each application may have unique requirements for the form factor of the devices. For instance, a soil moisture sensor in agriculture needs to be small and unobtrusive, while a smart thermostat in a home may have a different form factor to be user-friendly and aesthetically pleasing. Besides, some companies offer development kits where the size of the PCB can be customised according to the final application. An IoT development kit is a collection of hardware and software tools that are designed to help developers create and prototype IoT devices and applications. These kits typically include a variety of components and resources to simplify the process of building IoT solutions, such as microcontrollers, sensors, transceivers, interface components, power management, and cloud integration. IoT development kits are designed to accelerate the development process by providing a convenient and cohesive set of tools. They are commonly used by professional developers to build a wide range of IoT applications, including smart home devices, industrial monitoring systems, wearable technology, and more. These kits help developers save time and effort by offering pre-integrated components and software stacks, allowing them to focus on the unique aspects of their IoT project.

The following devices embed Antenna Booster Technology for IoT applications, including different radio protocols (Figures 11.27–11.41).

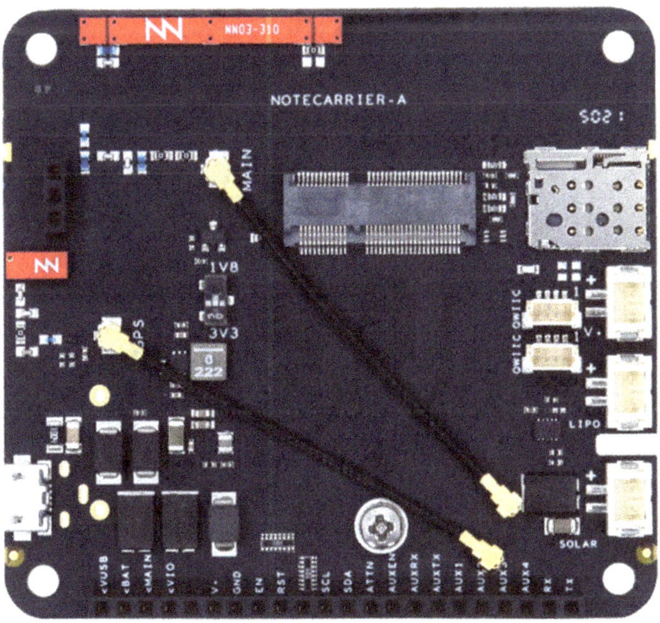

Figure 11.27 Example including cellular 30 mm × 3 mm × 1 mm (height) antenna booster element (top) and 7 mm × 3 mm × 2 mm GPS antenna booster (left) [28]. Thanks to Blues.

Figure 11.28 Tracking device with cellular and satellite connectivity with 30 mm × 3 mm × 1 mm and 7 mm × 3 mm × 2 mm antenna boosters [29]. Thanks to Cactus.

Figure 11.29 2 × 2 MIMO for cellular with two 30 mm × 3 mm × 1 mm (height) antenna booster components and geolocalisation with a 7 mm × 3 mm × 2 mm (height) antenna booster [30]. Thanks to Cavli Wireless.

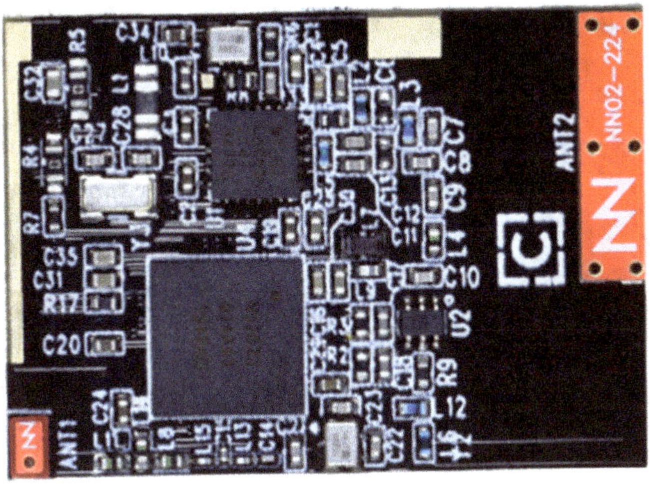

Figure 11.30 Device with LoRaWAN® and Bluetooth: antenna boosters are (right) 12 mm × 3 mm × 2.4 mm (λ/29 at 863 MHz) and (left) 3 mm × 2 mm × 0.8 mm (λ/41 at 2450 MHz) [31]. Thanks to CWD Limited.

Figure 11.31 Device for NB-IoT 700–960 MHz + 1710–2200 MHz and GPS [32]. Thanks to Devalirian Engineering.

Figure 11.32 Application of antenna booster technology for connected electric charge stations, including Wi-Fi and cellular, including 30 mm × 3 mm × 1 mm and 7 mm × 3 mm × 2 mm antenna boosters [33]. Thanks to Floox.

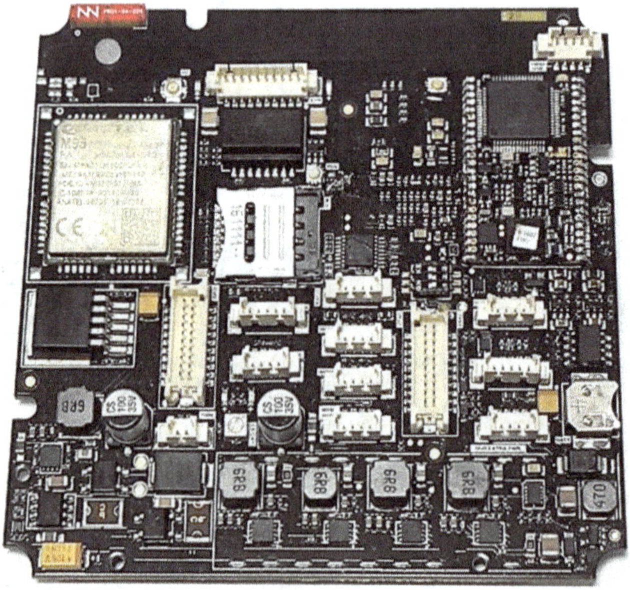

Figure 11.33 12 mm × 3 mm × 2.4 mm (height) antenna booster for cellular connectivity [34]. Thanks to HardwareMe Electronics.

Figure 11.34 *Device embedding a 30 mm × 3 mm × 1 mm antenna booster element with a reconfigurable matching network: cellular at 698–960 MHz and 1710–2170 MHz and geolocalisation at 1575 MHz [35]. Thanks to Nordic Semiconductor.*

Figure 11.35 *Data logger with LoRa connectivity using a 30 mm × 3 mm × 1 mm (height) antenna booster component [36]. Thanks to Nuvathings.*

Figure 11.36 Example of how the same 12 mm × 3 mm × 2.4 mm (height) antenna booster can be reused for any frequency band: cellular, satellite, and short-range wireless [37]. Thanks to RioSH.

Figure 11.37 Tracking device with cellular and satellite connectivity with a dual-port 30 mm × 3 mm × 1 mm (height) antenna booster component [38]. Thanks to Samea.

For the sake of simplicity, only a few are included since the list of devices integrating antenna boosters will need hundreds of pages. Devices are not to scale and are shown in alphabetical order based on the name of the company. The reader is invited to address the references for each device for further details and applications. Some of these devices are customised for a particular application, whereas others are development kits.

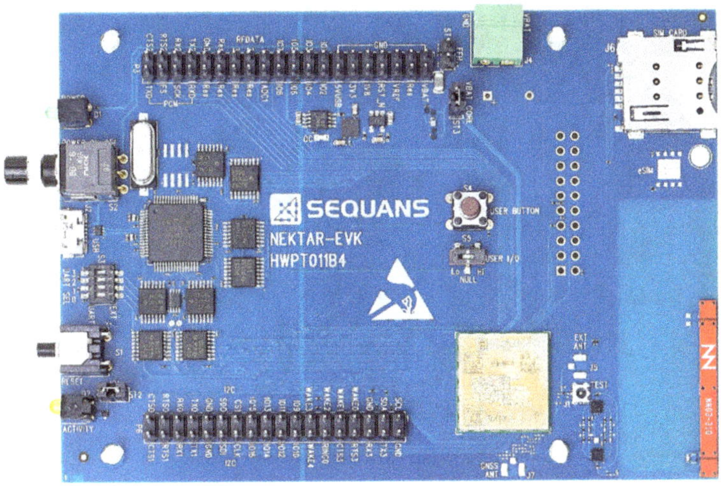

Figure 11.38 Device embedding a 30 mm × 3 mm × 1 mm antenna booster element with a reconfigurable matching network for cellular connectivity at 617–960 MHz and 1695–2200 MHz [39]. Thanks to Sequans.

Figure 11.39 Tracking solar device, including the PCB and the entire enclosure embedding an antenna booster-based element (the red component on the PCB). Operation 699–746 MHz, 880–960 MHz, and 1710–1880 MHz. Thanks to Sodaq [40].

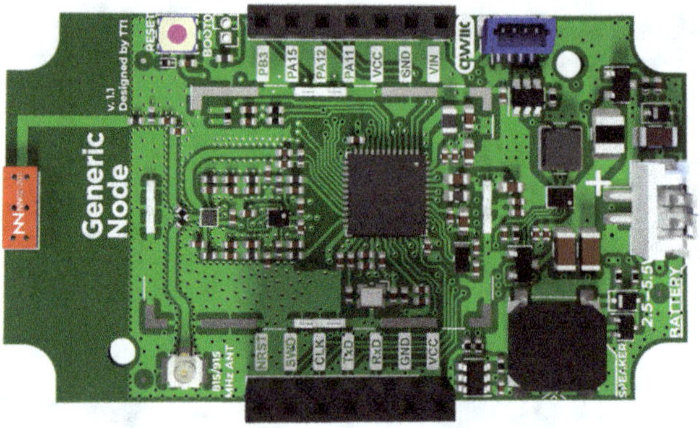

Figure 11.40 Device with single-band antenna booster for LoRaWAN®: antenna booster is 7 mm × 3 mm × 1 mm (λ/49 at 863 MHz) [41]. Thanks to The Things Industries.

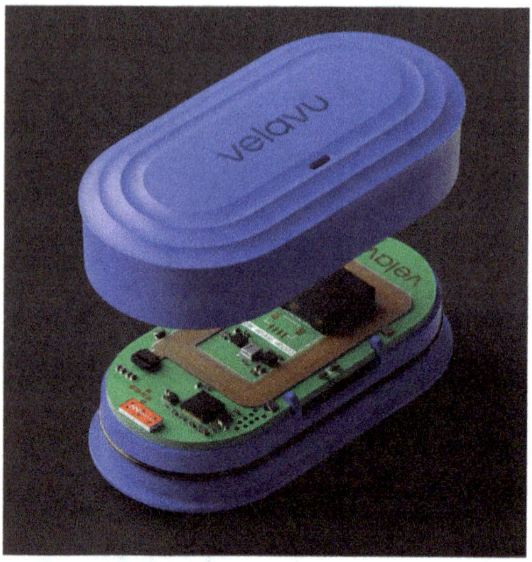

Figure 11.41 Mesh anchor to enable the precise positioning of people or assets within a mesh network. The red component is a 7 mm × 3 mm × 1 mm antenna booster [42]. Thanks to Velavu.

11.12 Antenna booster technology features in one page

11.13 Main takeaways

Besides the main features of Antenna Booster Technology explained in Figure 11.42, this section summarises at a high level the main takeaways of each technique explained throughout the book (Table 11.1 and Figure 11.43).

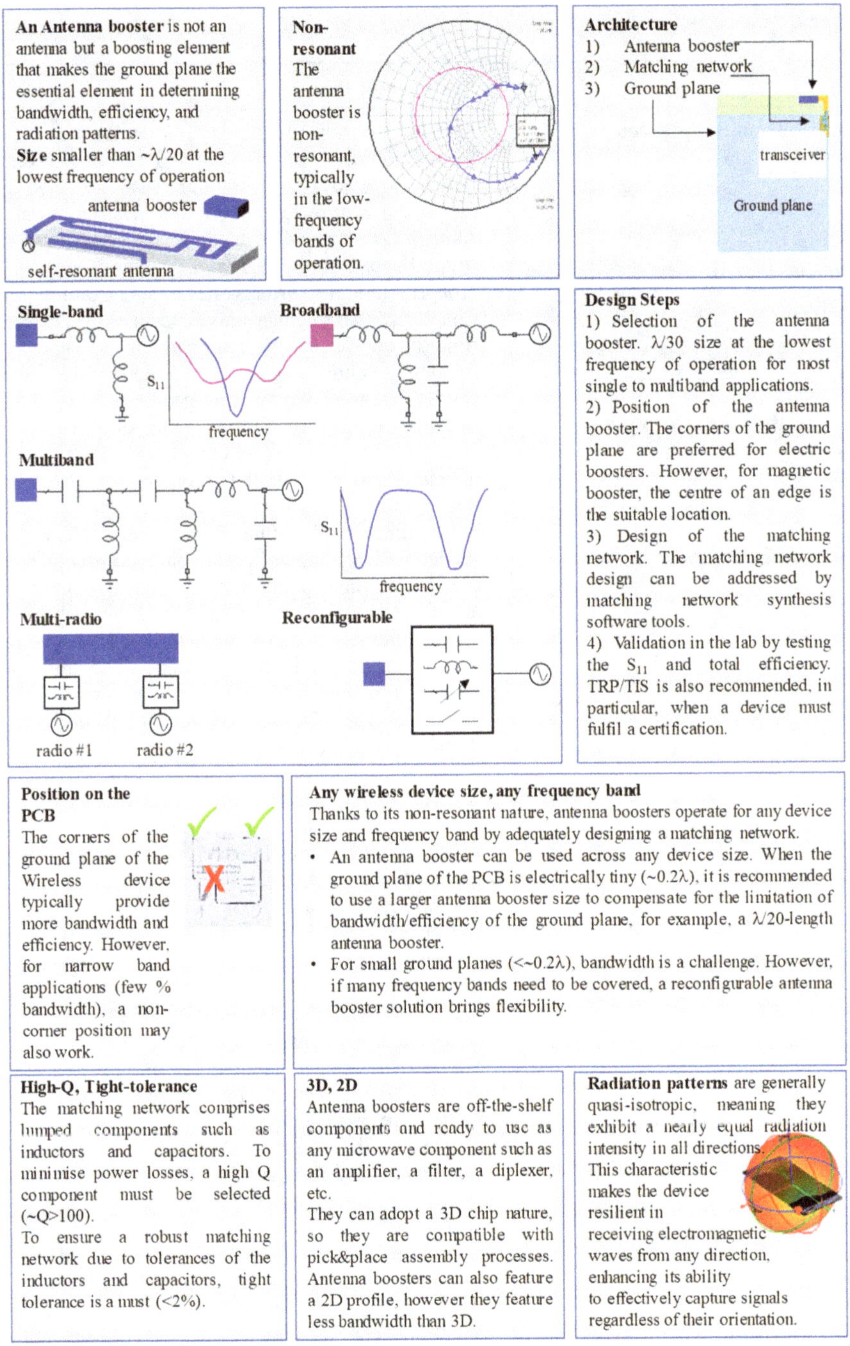

Figure 11.42 The ten commandments of Antenna Booster Technology

Table 11.1 Key aspects to take away

Topic	Review and take away
The selection of the frequency bands of operation	Antenna Booster Technology allows for the accessible design of wireless devices because the frequency bands of operation are easily selected by the design of a matching network. Therefore, an antenna booster can be used for any band, from single band to multiband, and for any size of the ground plane of a wireless device. As examined in Chapter 2, the antenna booster technology differs from the approach of designing antennas, where the frequency bands of operation are determined by a skilled design of the antenna geometry.
Architecture: antenna booster, matching network, and ground plane	Antenna Booster Technology relies on three essential elements: the antenna booster, the matching network, and the ground plane. Thanks to this combination, antenna booster components can be ten times smaller than self-resonant antennas.
The antenna booster	As detailed in Chapter 3, an antenna booster excites currents on the ground plane, as does any other antenna element. However, the advantage of the antenna booster is that there is no need for a large antenna size and complex geometry. The drum analogy illustrates this fact: to make a drum beat, there is no need for a big drumstick but to hit the drum in the correct position. In this sense, the membrane is to a ground plane what the drumstick is to an antenna booster. In this regard, the antenna booster location is relevant to determining bandwidth and efficiency. Typically, the corners of the ground plane work best. The non-resonant nature of an antenna booster is a distinctive feature compared to standard antenna designs based on self-resonant antennas. Such a feature enables an antenna booster to operate at any frequency band by simply adjusting a matching network since the radiation efficiency is flat compared to that of a self-resonant antenna, which typically has efficiency drops between resonances. As a result, one antenna booster can be reused across many frequencies, whereas several resonant antennas are needed for specific frequency ranges.
The matching network	As Chapters 3, 4, and 5 explain, the matching network allows for designing from single-band, broadband, and multiband. Among the several techniques, one of the simplest is using a single antenna booster (Figure 11.43, Single/Multiband) with a matching network designed thanks to synthesis tools. The synthesis tools are software algorithms that allow us to obtain the topology of

(Continues)

Table 11.1 (*Continued*)

Topic	Review and take away
	the matching network as well as real components available in the market. The synthesis tools work quickly, getting the topology that best matches the target specifications in minutes. A typical target is the total efficiency for the frequency bands of interest.
The ground plane	The size of the ground plane is key to determining the performance of the wireless device in terms of bandwidth, efficiency, and radiation patterns. As a simple number, for a rectangular ground plane, a length of $\sim 0.4\lambda$ maximises bandwidth. When the ground plane becomes smaller, it impacts bandwidth, and thus, it is challenging to obtain bandwidth and efficiency.
Multi-radio	Wireless devices operate on more than a single radio protocol, meaning that an antenna booster is needed for each radio, which is also the case for self-resonant antennas. To minimise the coupling between antenna boosters, the basic rule of increasing distance may generally work. In Chapter 6, a way of minimising coupling for closely spaced antenna boosters is described. Basically, it consists of adding a reject filter at each antenna booster port to block the frequency of the other radio protocol. In other situations, a single antenna booster with a multiplexing architecture converts a single antenna booster to a multi-port antenna booster element able to provide multi-radio at each port with the addition of a matching network at each one (Figure 11.43, Multi-radio).
Reconfigurable architectures	Matching networks studied in Chapters 4 and 5 rely on passive components such as lumped inductors and capacitors. However, as theoretically discussed using Fano's limit, a matching network cannot enhance the bandwidth limitlessly. Therefore, in those cases where the frequency bands to cover are large or the ground plane is electrically small (length $<\lambda/\pi$) and thus challenging to obtain bandwidth, a reconfigurable matching network is attractive. Even reconfigurable matching networks are suitable when the device needs to deal with different external factors, such as the device on a conductive body or soil. Since each one impacts impedance, a reconfigurable matching network compensates for the detuning and, thus, keeps the device operating at its maximum level (Figure 11.43, reconfigurable).

(*Continues*)

Table 11.1 (*Continued*)

Topic	Review and take away
	Reconfigurable matching networks for antenna boosters with DTC and RF switches are proposed in Chapter 7. DTC allows for fine tunability whereas RF-switches can also provide coarse tuning of the frequency bands.
Adding intelligence in the ground plane	The third relevant element of Antenna Booster Technology is the ground plane. Not only does the size matter to determine bandwidth, efficiency, and radiation patterns but also its shape. For example, it is shown in Chapters 3 and 11 that other shapes than rectangular ones may provide better bandwidth. From a practical perspective, a rectangular ground plane is preferred due to simplicity when routing traces that need to connect the different components of a wireless device. However, shaping the ground plane should not be underrated due to its potential to enhance bandwidth, as the example explained in Figure 11.12, where the bandwidth is improved four times compared to the square ground plane.
Towards a greener world	Wireless technology should not only facilitate connectivity among people and things but also contribute to a greener ecosystem. In this regard, early steps show how plant leaves can be used as ground planes. This scheme may be attractive to decrease the use of substrates that impact the carbon dioxide footprint and substitute for organic ground planes (Figure 11.43, Organic ground planes). For example, the production of one square metre of standard FR-4 PCB is equivalent to 17.7 kgCO_2e.
Robust to the environment	Wireless devices encounter some scenarios, such as operating close to a conductive object, for example, a smart fridge or washing machine. When the ground plane of a wireless device is electrically close to a conductive object, impedance and efficiency deteriorate. Chapter 8 investigates a technique that adds connections between the ground plane of the wireless device and the conductive body underneath. This technique recovers bandwidth and efficiency as if it were not a conductive body below the wireless device (Figure 11.43, Robust to Environment).
Non-Foster matching networks	The matching networks studied throughout the book are mainly comprised of inductors and capacitors obeying the Foster principle. In simple terms, L and C are positive values. However, since the nature of an antenna booster can be modelled as

(Continues)

Table 11.1 (*Continued*)

Topic	Review and take away
	a series resistor and capacitor, it would be attractive to match with a negative capacitor. In this regard, Chapter 4 briefly introduced non-Foster matching networks to enhance the bandwidth for single-band operation (Figure 11.43, non-Foster).
Layout effects	Since the components of the matching network are located on the PCB, including pads, their effect should be taken into consideration when implanting a matching network in a real wireless device (Chapter 9).
Antenna Booster Technology compared to other antenna technologies	Antenna Booster Technology has also been compared to other antenna techniques, such as self-resonant antennas, FPC antennas, and microstrip patches. In brief, the following conclusions can be drawn, taking into consideration the scenarios considered in the chapter: • Antenna boosters are about ten times smaller than self-resonant antennas. Therefore, antenna boosters are easier to embed in wireless devices and are more eco-compatible since less material is needed to fabricate them. • Antenna boosters are future-proof, meaning that any future protocol does not impact the antenna booster component, only the design of a matching network. However, for a self-resonant approach, a new antenna must be designed. • Antenna boosters show more reusability for a wide set of wireless device sizes. • Antenna boosters are more robust to variations in the permittivity of the substrate compared to small printed antennas. • Compared to FPC antennas, antenna boosters are much smaller, provide better efficiency when mounted on the PCB, and have predictable performance. • For random positions and orientations of a wireless device, antenna boosters show a faster acquisition of the satellites and, thus, reduce battery consumption, which is critical for battery-powered devices compared to microstrip patch antennas.

Table 11.1 answers the objective given at the beginning of the chapter. It also points to relevant figures and equations to consolidate the concepts learnt. It is also advisable to refresh all essentials included in the grey boxes distributed along the chapter.

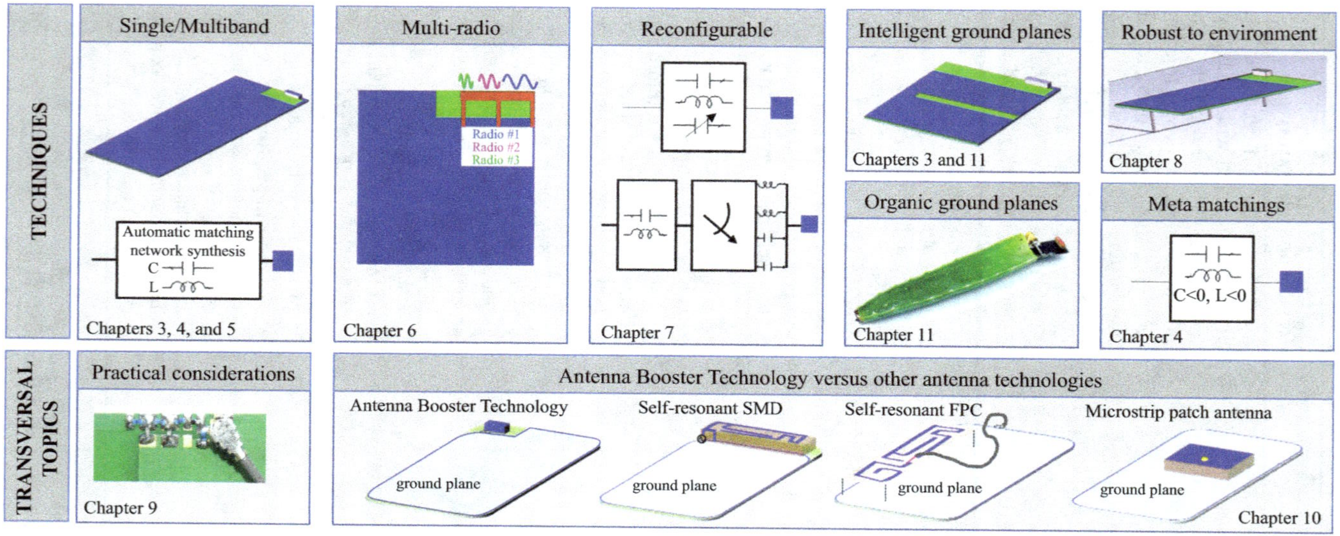

Figure 11.43 Overview of key topics explored in the book

References

[1] J. Anguera, A. Andújar, M. C. Huynh, C. Orlenius, C. Picher, and C. Puente, "Advances in Antenna Technology for Wireless Handheld Devices," *International Journal on Antennas and Propagation,* vol. 2013, Article ID 838364, 2013.

[2] K. L. Wong, *Planar Antennas for Wireless Communications*, Wiley 2003, ISBN 9780471266112.

[3] Z. Zhang, *Antenna Design for Mobile Devices*, John Wiley & Sons, 2011, ISBN 978-0-470-82446-7

[4] J. Anguera, A. Andújar, C. Puente, J. Mumbrú, "Antennaless Wireless Device," *Patent US* 8,203,492 B2, June 2012.

[5] J. Anguera, A. Andújar, C. Puente, and J. Mumbrú, "Antennaless Wireless Device capable of Operation in Multiple Frequency regions," *Patent US. 8*,237,615 B2. August 2012.

[6] J. Anguera, A. Andújar, C. Puente, "Wireless Handheld Devices, Radiation Systems and Manufacturing Methods," *Patent US* 9,331,389 B2, May 2016.

[7] Ignion. www.ignion.io

[8] J. Anguera, A. Andújar, G. Mestre, J. Rahola, and J. Juntunen, "Design of Multiband Antenna Systems for Wireless Devices Using Antenna Boosters," *IEEE Microwave Magazine*, vol. 20, no. 12, pp. 102–114, 2019.

[9] J. Anguera, "Expand the Connectivity of Your LoRa Generic Node Sensor with Virtual Antenna^{TM}," *Webinar at the Things Conference*, 2022. Available at https://www.youtube.com/watch?v=e4YZWIYQ_8M

[10] J. Anguera, A. Andújar, D. Vye, and M. Wien, "Easy Design of IoT Devices with Antenna Boosters," Webinar at EDICON, Beijing, China, 2020. Available https://www.youtube.com/watch?v=W3WpMD1MUw0&t=815s

[11] J. Anguera, C. Puente, A. Andújar, R. M. Mateos, D. Vye and M. Lien, "Antenna Library for IoT Devices with Antenna Boosters," *2020 50th European Microwave Conference (EuMC)*, 2021, pp. 1147–1150, doi:10.23919/EuMC48046.2021.9338246.

[12] D. Vye, "Network Synthesis Wizard Automates Interactive Matching-Circuit Design," *Microwave Journal*, vol. 61, no. 11, pp. 96–102, 2018.

[13] J. Rahola, "Optimization of Matching Circuits for Antennas," *European Conference on Antennas and Propagation*, EuCAP 2011, Rome, 2011.

[14] L. A. M. Geubbels, G. Federico, V. Vidojkovic, J. Anguera, L. A. Bronckers, "Plantenna: Using Plant Leaves to Increase Antenna Performance," *European Conference on Antennas and Propagation*, EUCAP 2022, Madrid, Spain, 2022.

[15] G. Massana "Analysis and Optimisation of a 900 MHz Antenna Booster Using Plan Structures as Alternative Ground Plane," *Engineering Degree Thesis* at La Salle Engineering, Universitat Ramon Llull. Directed by: J. Anguera and A. Andújar. 2023.

[16] J. M. J. W. Jayasinghe, J. Anguera, and D.N. Uduwawala, "A Simple Design of Multi-band Microstrip Patch Antennas Robust to Fabrication Tolerances for GSM, UMTS, LTE, and Bluetooth Applications by Using Genetic Algorithm Optimization," *Progress in Electromagnetics Research M*, vol. 27, pp. 255–269, 2012.

[17] L. A. Griffiths, C. Furse and Y. C. Chung, "Broadband and Multiband Antenna Design Using the Genetic Algorithm to Create Amorphous Shapes Using Ellipses," *IEEE Transactions on Antennas and Propagation*, vol. 54, no. 10, pp. 2776–2782, 2006, doi:10.1109/TAP.2006.882154.

[18] J. Anguera, N. Toporcer, and A. Andújar, "Slim Bar Booster for Electronics Devices," *Patent US* 9,960,478 B2, May 2018.

[19] J. Anguera, C. Picher, A. Andújar, C. Puente, "Concentrated Antennaless Wireless Device Providing Operability in Multiple Frequency Regions," *Patent US* 9,379,443 B2, June 2016.

[20] C. Picher, J. Anguera, A. Bujalance, and A. Andújar, "Ground Plane Booster Antenna Technology Using a Self-Diplexed Matching Network for Multi-band Operation," *Microwave and Optical Technology Letters*, vol. 58, no. 2, pp. 453–461, 2016.

[21] J. Anguera, A. Andújar, C. Puente, and R. Mateos, "Modular Multi-stage Antenna System and Component for Wireless Communications," *Patent US* 11,482,772 B2, October 2022.

[22] J. Anguera and A. Andújar, C. Puente, "Compact Antenna Technology for Wireless Communications," Patent App. WO 2020 0120589A1, June 2020.

[23] J. Anguera, A. Andujar, J. L. Leiva, O. Massó, J. Tonnesen, E. Rindalsholt, R. Brandsegg, R. Gaddi, "Reconfigurable Multiband Operation for Wireless Devices Embedding Antenna Boosters," *Electronics*, vol. 10, no. 7, p. 808, 2021.

[24] J. Anguera, A. Andújar, J. L. Leiva, C. Schepens, R. Gaddi and S. Kahng, "Multiband Antenna Operation with a Non-Resonant Element Using a Reconfigurable Matching Network," *12th European Conference on Antennas and Propagation (EuCAP 2018)*, 2018, pp. 1–4, doi:10.1049/cp.2018.1199.

[25] J. Anguera and A. Andújar, "Wireless Device," *Patent US* 10,122,403 B2, Nov. 2018.

[26] A. Andújar, J. L. Leiva, J. Anguera, C. Schepens, and R. Gaddi, "Chip Antenna–Antenna Tuner Combo Cover LTE Bands," *Microwave Journal*, vol. 62, no. 1, pp. 70–78, 2019.

[27] A. Fernández, J. Rahola, J. Juntunen, A. Andújar, J. L. Pijoan, J. Anguera, "Embedding Antenna Booster in Smart-metering Platforms," *European Conference on Antennas and Propagation,* EUCAP 2021, Madrid, Spain, 2022.

[28] Blues Wireless https://blues.io/

[29] Cactus. https://cactusiot.com/

[30] Cavli Wireless. https://www.cavliwireless.com/

[31] CWD Limited. https://www.cwdin.com/

[32] Devalirian Engineering. https://devalirian.com/en/home/

[33] Floox. https://www.flooxpower.com/
[34] HardwareMe Electronics. https://www.hwmelectronics.com/
[35] Nordic Semiconductor. https://www.nordicsemi.com/
[36] Nuvathings. https://www.nuvathings.com/
[37] Riosh. https://www.rioshtech.com/
[38] Samea. https://www.samea-innovation.com/
[39] Sequans. https://sequans.com/
[40] Sodaq. https://sodaq.com/
[41] The Things Industries. https://www.genericnode.com/
[42] Velavu. https://www.velavu.com/

Index

www.ingramcontent.com/pod-product-compliance
Lightning Source LLC
LaVergne TN
LVHW011545200926
848484LV00001B/290
9781839533006